THE OXFORD ENGINEERING SCIENCE SERIES

The Physics of Fluid Turbulence

W. D. McCOMB

Lecturer, Department of Physics
University of Edinburgh

CLARENDON PRESS · OXFORD

Oxford University Press, Walton Street, Oxford OX2 6DP

Oxford New York Toronto
Delhi Bombay Calcutta Madras Karachi
Kuala Lumpur Singapore Hong Kong Tokyo
Nairobi Dar es Salaam Cape Town
Melbourne Auckland Madrid

and associated companies in
Berlin Ibadan

Oxford is a trade mark of Oxford University Press

Published in the United States
by Oxford University Press Inc., New York

© W. D. McComb 1990
First published 1990
First published in paperback
(with corrections) 1991
Reprinted 1992

British Library Cataloguing in Publication Data
McComb, W. D.
The physics of fluid turbulence.
1. Fluids. Turbulence
I. Title
532'.0527

ISBN 0-19-856256-X (Pbk)

Library of Congress Cataloging in Publication Data
McComb, W. D.
The physics of fluid turbulence / W. D. McComb.
p. cm. — (Oxford engineering science series; 25)

ISBN 0-19-856256-X (Pbk)
1. Turbulence. 2. Fluid mechanics. 3. Perturbation (Physics)
4. Renormalization (Physics) I. Title. II. Series.
QA913.M43 1990
532'.0527—dc20   89-28440

Printed in Great Britain by
Bookcraft (Bath) Ltd.
Midsomer Norton, Avon

FOR MY WIFE

PREFACE

Over the last three decades the various topics making up the subject of engineering science have tended to become ever more scientific and sophisticated. In effect the emphasis has been increasingly on the 'science' aspect of the subject, and the study of turbulence has been no exception.

Up until about the end of the 1950s, most work on turbulence involved the application of *ad hoc* methods to difficult practical problems. That is to say, turbulence was essentially a branch of hydraulics, and hence was treated by the methods of 'handbook engineering'. There was only a relatively small amount of fundamental research and this was described in a rather complete way by the well-known monographs due to Batchelor, Hinze, and Townsend (for references, see Chapter 1), all of which were first published in the period 1953–1959.

However, since 1960 the study of turbulence has seen a great increase in activity of a more scientific kind. The invention of laser anemometry, the development of powerful computers, new methods of digital data processing and signal analysis, and the introduction of exotic (at least, in engineering terms) theoretical methods from quantum physics have all played a part.

It is perhaps not altogether surprising that over this time there have been some important developments in the subject. As examples, one may note the spectacular phenomenon of drag reduction by additives, the recognition of the importance of coherent structures, the direct numerical simulation of turbulence on the computer, and the derivation of closures of the Navier–Stokes hierarchy using renormalized perturbation theory.

As engineering has tended to become more scientific, there has been (pleasingly to one's sense of symmetry!) a comparable trend in physics in the reverse direction. The stunning success of modern physics (i.e. Einsteinian relativity and, more particularly, quantum mechanics) in solving many fundamental problems has led, by way of the search for new worlds to conquer, to a strong applied physics discipline. In turn, mutual interest has resulted in the formation of many interfaces between engineering science and applied physics, especially in the areas of electronics, physical optics, and materials science.

It is at this point that the study of turbulence does prove to be an exception: the applied physics involvement is almost completely absent. In view of the extraordinary practical importance of turbulence (not to mention its intense study by other disciplines in the context of applications), this is at first sight quite astonishing. Yet the reason for such apparent neglect is easily found. Quite simply the fundamental problems of turbulence are still unsolved. That is, the 'turbulence problem' is still a matter of pure physics. Indeed, turbulence is often referred to as the 'unsolved problem of classical physics'.

Considered as a problem in pure physics, turbulence has not been altogether neglected. But, when compared with topics like condensed matter physics or high energy physics, it has been very much a minority cult. It is therefore particularly encouraging to note that over the past decade or so there have been convincing signs of an exponential growth in the study of turbulence as a branch of physics. This is, of course, a predictable trend. As fewer and fewer linear problems are left to solve, there is inevitably a growing pursuit of non-linear problems,[†] and turbulence is the archetypal non-linear non-equilibrium problem of statistical physics.

In this book our aim is to deal with certain topics which form a subset of both engineering science and physics. In doing so, we hope to assist two broad classes of reader. First there are those who are new to the subject (although not necessarily new to research), and secondly there are those who are already familiar with one or more of the traditional branches of the study of turbulence, but whose background and experience does not prepare them for the usages of quantum physics. (In passing, one notes that theoretical physicists who write papers on turbulence often appear to make little or no effort to present their work in a way which would be accessible to the more theoretically inclined members of the turbulence community.)

In order to cater for, in particular, the first of these groups, we have aimed at a reasonable degree of completeness. The first two chapters set out to give a concise summary of the theory and practice of turbulence up to about 1960. This material is based upon lectures, which I have given to final-year undergraduates, originally in mechanical engineering but now, in more recent years, in physics. Chapter 3 then serves two purposes. First, it tries to give the reader a broad picture of what the rest of the book is about. Thus certain topics in this category can be regarded as overviews of their more detailed treatment later in the book. The second purpose of this chapter is to cover topics which would not easily fit into the main part of the book, yet are important. Section 3.1 on anemometry and data processing is an example of such a topic.

Chapters 4–10 constitute the main part of the book and deal with modern (post-1960) turbulence theory. Chapter 4 begins by presenting some background material on the statistical mechanics of the classical N-body system. This introduces useful concepts and terminology, and provides a context for the subsequent rigorous formulation of the turbulence problem as an example of a non-equilibrium statistical system with strong coupling. Chapter 5 aims to 'de-mystify' the application of renormalized perturbation theory (RPT) to turbulence. RPT is introduced for some simpler problems by considering (a) the virial cluster expansion in dilute N-body systems and (b) the Debye-Hückel screened potential for the classical plasma as an example of long-range

[†] An interesting and authoritative discussion of this situation was given by R. Kubo in his opening address to the Oji Seminar on Non-linear, Non-equilibrium Statistical Mechanics (1978: *Suppl. Progr. Theor. Phys.* **64**, 1).

interactions. Then, a general treatment of the perturbation expansion of the Navier–Stokes equation (based on a modified version of Wyld's analysis) follows, and the chapter closes with a consideration of Kraichnan's direct-interaction approximation as an example of a second-order truncation of the renormalized expansion. It should be noted that at this point (and at various others) we do not always keep to the strict chronological order in which the different pieces of work were carried out.

Chapter 6 deals with those RPTs which do not yield the Kolmogorov spectrum as a solution and Chapter 7 deals with those that do.

Chapter 8 attempts to provide a critical assessment of RPTs. The main emphasis is on the comparison of numerical solutions of the spectral and response equations with the results of laboratory and computer experiments. The chapter closes with an appraisal of RPTs in the light of various published critiques, and their relationship with other fields of physics.

In Chapter 9 we introduce the newer method of renormalizing the transport coefficients in turbulence: the renormalization group (RG). Both this and RPT methods crop up again in Chapter 10, where we discuss the numerical simulation of turbulence. This is entirely appropriate, for the general area of large-eddy simulation (which resembles other hybrid areas of physics, such as lattice gauge calculations) provides for the first time an arena where engineers and physicists can meet on equal terms and find much of mutual interest.

The final part of the book, consisting of Chapters 11–14, can be seen as offering some sort of counterpoise to the exclusively theoretical (and often esoteric) nature of most of the preceding chapters. My intention here is to deal with some of the more practical aspects of the subject, while at the same time illustrating the amazing richness of physical phenomena which has emerged from experimental studies of turbulence over the last few years. What is now needed is much more interest from physicists, especially on the experimental side which has for the most part been left to the engineers. If the material contained in Chapters 11–14 does not inspire the requisite interest, then I believe that nothing will!

I took my own first steps in turbulence theory under the guidance of Sam Edwards and David Leslie. In their different ways they taught me much. It is a pleasure to acknowledge my debt to them here. Also, in the course of writing this book, I have received a great deal of help from friends and colleagues. I would like to thank Francis Barnes, Phil Hutchinson, V. Shanmugasundaram, and Alex Watt for reading the manuscript (in various drafts!) and pointing out errors and making suggestions for improvement. Finally, I would like to thank Denis Jones of Dantec Electronics for his help and advice concerning the section on anemometry.

Edinburgh   W. D. McC.
June 1989

ACKNOWLEDGEMENTS

It is a pleasure to thank the various organizations and individuals, who gave permission to reproduce figures, as follows.

American Institute of Chemical Engineers

Figures 3.15 and 14.2 are reprinted by permission from *AIChE J.* **28**, 547–65. Copyright 1982.

American Institute of Physics

Figure 14.1 is reprinted by permission from *Phys. Fluids* **20**, 873–9. Copyright 1977.

Cambridge University Press

Figure 1.7 is reprinted by permission from *J. Fluid Mech.* **48**, 477–505. Copyright 1971.

Figures 3.17 and 3.18 are reprinted by permission from *J. Fluid Mech.* **43**, 689–710. Copyright 1970.

Figures 8.6–8.14 are reprinted by permission from *J. Fluid Mech.* **143**, 95–123. Copyright 1984.

Plate II is reprinted by permission from *J. Fluid Mech.* **110**, 73–95. Copyright 1981.

Dantec Electronics Ltd

Figures 3.5 and 3.6 were reproduced from negatives kindly supplied by Dantec.

Institute of Petroleum

Figure 3.16 is reprinted by permission from *J. Inst. Petrol.* **47**, 329–25. Copyright 1961.

Macmillan Magazines Limited

Figure 14.3 is reprinted by permission from *Nature (Lond.)* **292**, 520–2. Copyright 1981 Macmillan Magazines Ltd.

Pergamon Press

Figure 13.1 is reprinted by permission from *Chem. Eng. Sci.* **26**, 419–39. Copyright 1971.

Figure 13.2 is reprinted by permission from *J. Aerosol Sci.* **6**, 227–47. Copyright 1975.

Figure 13.3 is reprinted by permission from *J. Aerosol Sci.* **9**, 229–313. Copyright 1978

Springer-Verlag New York Inc

Figures 8.4 and 8.5 are reprinted by permission from *Statistical models and turbulence* (eds M. Rosenblatt and C. Van Atta), *Lecture Notes in Physics Series*, Vol. 12, p. 148. Copyright 1972 (article by J. R. Herring and R. H. Kraichnan).

Thanks are also due to Professor F. H. Abernathy (Figs 3.17 and 3.18), Professor Mohamed Gad-el-Hak (Fig. 11.1/Plate II), Dr Jack Herring (Figs 8.4 and 8.5), Professor Phil Hutchinson (Fig. 13.1), and Dr C. J. Lawn (Fig. 1.7).

My thanks are also due to Denise McCluskey who set up the laser-Doppler anemometer shown in Plate I and to Peter Tuffy who took the photograph. In addition, I thank Peter Tuffy for his technical assistance in the preparation of Plate II and Figs 3.5 and 3.6. Lastly, I wish to express my gratitude to V. Shanmugasundaram for a great deal of help in the preparation of many of the figures, and to Alex Watt for his assistance with the preparation of Appendix G.

The author and publishers would like to thank Dantec Electronics Ltd for their generosity in sponsoring the reproduction of the colour plates in this volume.

CONTENTS

Plates I and II fall between pp. 92 and 93

NOTATION

1 General remarks

Vectors, matrices, and tensors are represented by the standard Cartesian notation, i.e. A_α, $B_{\alpha\beta}$, $C_{\alpha\beta\gamma}$, and so on. In all cases we use Greek letters α, β, γ, δ, ... for the tensor subscripts; each subscript can take the value 1, 2, or 3. In certain cases, vectors are shown in bold face type. This is usually the case where the vector concerned is the independent variable, i.e. $U_\alpha(\mathbf{x}, t)$ or even $F[\mathbf{U}(\mathbf{x}, t)]$. In the latter example, the square brackets are used to indicate that F is a functional or 'function of a function'.

We shall usually, but not always, employ the summation convention, in which repeated tensor indices are taken to be summed without the summation symbol being needed. For example, the scalar product of two vectors $\mathbf{X}$ and $\mathbf{Y}$ can be written as

$$\mathbf{X} \cdot \mathbf{Y} = \sum_{\alpha=1}^{3} X_\alpha Y_\alpha$$

or as

$$\mathbf{X} \cdot \mathbf{Y} = X_\alpha Y_\alpha.$$

All averages are denoted by Dirac brackets $\langle\ \rangle$, although an overscore is sometimes used for single variables. Thus the mean velocity can be written as $\bar{U} = \langle U \rangle$. A local (or partial) average is introduced in Chapter 9, and this is denoted by $\langle\ \rangle_0$.

Position vectors are written as $\mathbf{x}, \mathbf{x}', \mathbf{x}'', \dots$ for different positions, while times of different events are usually written as $t, t', t'', \dots$.

Inconsistently, distinct wavevectors are normally written as $\mathbf{k}, \mathbf{j}, \mathbf{l}, \dots$, or $\mathbf{k}_1, \mathbf{k}_2, \mathbf{k}_3, \dots$.

2 Italic symbols

a	semi-width of a plane channel, radius of spherical particles
A_p	cross-sectional area of added particles
c_f	friction (or drag) coefficient for external flow round a solid body
C_μ, C_μ'	empirical constants in 'two equation' turbulence models
C_D, C_1, C_2	
C_1–C_4	constants specifying the trial spectra, as used in Chapter 8
d	diameter of pipe, diameter of diffusing particle, width of a plane jet
D_{LL}, D_{LLL}	structure functions
D	diffusivity due to molecular motion
D_T	eddy diffusivity for fluid points

D_p	eddy diffusivity for added particles
D_c	diffusivity associated with a macroscopic random walk
E_T	total kinetic energy of fluid motion
E	kinetic energy of fluctuating motion per unit mass of fluid
$E(k, t)$	energy spectrum
F	flatness factor of a probability distribution
f	friction factor for pipe or channel flow
$k_\alpha, \mathbf{k}$	wavevector (or position coordinate in k-space)
k_d	Kolmogorov dissipation wavenumber
k	thermal conductivity, as used in Chapter 13
L	integral length scale
$N(\mathbf{x}, t)$	scalar concentration field
$\bar{N}(\mathbf{x}, t)$	mean scalar concentration field
$n(\mathbf{x}, t)$	fluctuating scalar concentration field
N_0	reference level for scalar concentration field
$P(\mathbf{x}, t)$	instantaneous pressure field
$\bar{P}(\mathbf{x}, t)$	mean value of the pressure field
$p(\mathbf{x}, t)$	fluctuating pressure field
$Q(k, t)$	spectral density
$q(k)$	stationary spectral density
$\mathbf{r}$	relative coordinate, $\mathbf{r} = (\mathbf{x} - \mathbf{x}')$
R	Reynolds number for pipe flow, centroid coordinate $\mathbf{R} = (\mathbf{x} + \mathbf{x}')/2$
Rx_1	Reynolds number for boundary layer flow
R_λ	microscale-based Reynolds number (or Taylor–Reynolds number)
$R_{\alpha\beta}$	correlation coefficient
$R_E(\tau)$	Eulerian time-correlation
$R_L(\tau)$	Lagrangian time-correlation
$s_{\alpha\beta}$	deviatoric stress tensor
S	skewness factor of a probability distribution
T_E	Eulerian integral time-scale
T_L	Lagrangian integral time-scale
T_p	Lagrangian integral time-scale for diffusing particles
$T(k, t)$	inertial transfer spectrum
U	bulk mean velocity in pipe or channel flow
U_∞	free stream velocity in external flows
U_c	centre-line velocity in pipe or channel flow
$U_\alpha(\mathbf{x}, t)$	instantaneous velocity field
$\bar{U}_\alpha(\mathbf{x}, t)$	mean value of the velocity field
$u_\alpha(\mathbf{x}, t)$	fluctuating velocity field
$U_\alpha(\mathbf{k}, t)$	Fourier transform of the velocity field
$u_\alpha(\mathbf{k}, t)$	Fourier transform of the fluctuating velocity field

$u_\alpha^<(\mathbf{k}, t)$ }   Fourier components of the velocity field
$u_\alpha^-(\mathbf{k}, t)$ }   restricted to a band of wavenumbers $0 < k < k_c$(say)
$u_\alpha^>(\mathbf{k}, t)$ }   Fourier components of the velocity field
$u_\alpha^+(\mathbf{k}, t)$ }   restricted to a band of wavenumbers $k_c < k < \infty$ (say)

u_α', $\mathbf{u}'$	root mean square value of the fluctuating velocity
u	scalar component of the velocity field in isotropic turbulence
u_τ	friction velocity
v	velocity scale associated with the Kolmogorov length scale and time-scale
$\mathbf{V}(t)$	Lagrangian velocity of a fluid particle
$V_p(t)$	Lagrangian velocity of an added particle
$W(k)$	rate of doing work by Gaussian stirring forces, with delta function autocorrelations in time, per unit mass of fluid and per unit volume of k-space
$w(k; t - t')$	autocorrelation of external stirring forces
x_α, $\mathbf{x}$	position coordinate in configuration space
$\mathbf{X}(t)$	Lagrangian position coordinate for fluid particles
$\mathbf{X}_p(t)$	Lagrangian position coordinate for added particles

3 Greek symbols

α	constant of proportionality in the Kolmogorov spectrum
β	ratio of Lagrangian to Eulerian time-scales, Obukhov–Corrsin constant
γ	intermittency factor
$\delta_{\alpha\beta}$	Kronecker delta
$\delta(x)$	Dirac delta function
δ	boundary-layer thickness
ε	dissipation rate
ϵ	expansion parameter (Chapters 9 and 13)
η	Kolmogorov dissipation length scale
κ	thermal diffusivity of the fluid
λ	Taylor microscale, bookkeeping parameter in the perturbation expansion of the Navier–Stokes equation, Debye–Hückel length
μ	dynamic viscosity of the fluid
v_0, v	alternative notations for the kinematic molecular viscosity of the fluid
v_T, $v(k)$	effective (turbulent) viscosity
ρ	density of the fluid
ρ_p	material density of added particles
σ_E, σ_ϵ	empirical constants in two-equation turbulence models
$\tau_{\alpha\beta}$	total stress tensor in a turbulent fluid
τ_w	shear stress at the wall in duct flows

τ_E Eulerian microscale

τ momentum relaxation time

χ variance of scalar concentration field

ψ stream function

4 Important formulae

$R = Ud/v$: Reynolds number for pipe flow

$\text{Pr} = v/\kappa$: Prandtl number

$\text{Sc} = v/D$: Schmidt number

$E = \frac{1}{2}\sum_\alpha \langle u_\alpha^2 \rangle$: kinetic energy of fluctuations per unit mass of fluid

$u_\tau = (\tau_w/\rho)$: friction velocity

$f = 2\tau_w/\rho U^2$: friction factor

$Q_{\alpha\beta}(\mathbf{x}, \mathbf{x}'; t, t') = \langle u_\alpha(\mathbf{x}, t)u_\beta(\mathbf{x}', t') \rangle$: two-point, two-time correlation of two velocities

$Q_{\alpha\beta\gamma}\ldots(\mathbf{x}, \mathbf{x}', \mathbf{x}'', \ldots; t, t', t'', \ldots) = \langle u_\alpha(\mathbf{x}, t)u_\beta(\mathbf{x}', t')u_\gamma(\mathbf{x}'', t'') \ldots \rangle$

$M_{\alpha\beta\gamma}(\mathbf{k}) = (2i)^{-1}\{k_\beta D_{\alpha\gamma}(\mathbf{k}) + k_\gamma D_{\alpha\beta}(\mathbf{k})\}$: inertial transfer operator occurring in the solenoidal Navier–Stokes equation

$D_{\alpha\beta}(\mathbf{k}) = \delta_{\alpha\beta} - k_\alpha k_\beta/|\mathbf{k}|^2$: projection operator

$\langle u_\alpha(\mathbf{k}, t)u_\beta(\mathbf{k}', t') \rangle = (2\pi/L)^3 \delta_{\mathbf{k}+\mathbf{k}', 0} Q_{\alpha\beta}(\mathbf{k}; t, t')$: defines the spectral density tensor for finite system volume

$\langle u_\alpha(\mathbf{k}, t)u_\beta(\mathbf{k}', t') \rangle = \delta(\mathbf{k} + \mathbf{k}')Q_{\alpha\beta}(\mathbf{k}; t, t')$: defines the spectral density tensor for infinite system volume

$Q_{\alpha\beta}(\mathbf{k}; t, t') = D_{\alpha\beta}(\mathbf{k})Q(k; t, t')$: isotropic spectrum tensor

$Q(k; t, t) = Q(k, t)$: single-time spectral density function

$Q(k; 0) = q(k)$: stationary single-time spectral density

$E(k, t) = 4\pi k^2 Q(k, t)$: the energy spectrum

$E(k) = 4\pi k^2 q(k)$: the stationary energy spectrum

$\langle f_\alpha(\mathbf{k}, t)f_\beta(-\mathbf{k}, t') \rangle = D_{\alpha\beta}(\mathbf{k})w(k; t, t')$: defines autocorrelation of the stirring forces in isotropic turbulence

$w(k; t - t') = W(k)\delta(t - t')$: defines the correlation function $W(k)$ for stirring forces with delta function autocorrelations

$k_d = (\varepsilon/v^3)^{1/4}$: Kolmogorov dissipation wavenumber

$v = (\varepsilon v)^{1/4}$: Kolmogorov dissipation-range velocity scale

$k_B = (\varepsilon/vD^2)^{1/4}$: Batchelor wavenumber

$k_C = (\varepsilon/D^3)^{1/4}$: diffusion cut-off wavenumber

1

THE SEMI-EMPIRICAL PICTURE OF TURBULENT SHEAR FLOW

Turbulence, a phenomenon which is complicated in itself, can be encountered in a wide variety of more or less complicated situations, both in the natural environment and in many industrial processes. Thus, when engineers attempt to predict the behaviour of turbulent systems, they have more to reckon with than the intrinsic difficulty of the turbulence problem. In order to carry out design calculations, they must also tackle the problems inherent in describing fluid flow through complicated physical systems such as turbine rotors, chemical reaction vessels, or tube bundles in heat exchangers.

Nevertheless, as computational methods improve and computers grow in power, there is continuous progress in treating the merely complicated aspects of problems. Therefore, in many engineering applications nowadays, the major problem faced is the irreducible one of the turbulence itself.

It hardly comes as a surprise, therefore, that a survey of the literature of turbulence reveals the classical symptoms of a subject that is of immense practical importance, yet is at the same time poorly understood. That is to say, there is much activity but relatively little consensus. Also, there is a strong tendency to form separate subject schools, each related to a particular industry or type of phenomenon, but with relatively little communication between them. Above all, there is a great variety of predictive methods, ranging from simple empirical correlations, through rigorous (but limited) deductions from the Navier–Stokes equations, to elaborate statistical models which are based on dubious analogies between the chaotic behaviour of turbulence and the molecular chaos of dilute gases.

In this book we are interested only in the physics of the turbulence itself, and not in engineering complications. For this reason we shall restrict our attention to flows with a rather simple geometrical configuration, such as free jets and wakes, the boundary layer next to a flat plate, and flows through pipes and plane channels. These are the classical flows upon which the general subject of fluid dynamics has very largely been based. Each of them has the virtue of allowing a general formulation of the equations of motion to be greatly simplified. They therefore form an important special case, and we shall refer to them generically as 'two-dimensional mean flows'.

(As an aside, we should perhaps emphasize that only the mean velocity is two dimensional in such flows; the turbulent fluctuating velocities are fully three dimensional.)

An additional simplification can come about if the flow configuration is narrow (in comparison with its length in the flow direction), and the cross-stream component of velocity is small compared with the streamwise component. This allows a further approximate reduction of the equations of motion, in what is known as the 'thin shear layer' or 'boundary-layer' approximation. The first of these terms is undoubtedly the more generally correct, but the second is much more commonly employed. We shall follow custom in this matter.

Our aim in this chapter is to present a concise summary of the formulation of the turbulence problem from the point of view of engineering applications. We begin with the equations of fluid motion and take averages in order to form an equation for the mean velocity. At this stage correlations are restricted to values taken at a single point in space and time, as opposed to the multipoint many-time correlations of the fundamental approach in Chapter 2. We immediately encounter the fundamental closure problem of turbulence, in that the mean velocity can only be calculated if the two-velocity correlation is also known, and consider some of the traditional engineering models used to overcome the problem in the context of the special case of two-dimensional mean flows. Although the statistical equations presented are not soluble, we show how they can be used to analyse and interpret experimental results. Not least, we show that simple empirical correlations can be used, not just as the only really reliable predictive methods, but also in order to indicate the presence of universal behaviour.

1.1   The equations of fluid motion

Throughout this book we shall only consider fluid motion which can be regarded as incompressible. For a general discussion of the conditions under which this will be true, reference can be made to Batchelor (1967). For our present purposes, it amounts to a requirement that the fluid density ρ always remains constant, and the equation of continuity (which expresses conservation of mass) takes the form

$$\frac{\partial U_\beta(\mathbf{x}, t)}{\partial x_\beta} = 0 \tag{1.1}$$

where $U_\beta(\mathbf{x}, t)$ is the fluid velocity at position $\mathbf{x}$ and time t. Note that we shall almost invariably use Cartesian tensor notation, and that Greek indices such as α, β, or γ take values of 1, 2, or 3. We shall also employ the summation convention in which, as in eqn (1.1), repeated indices are summed.

For an incompressible fluid, the equation expressing conservation of momentum is

$$\frac{\partial U_\alpha}{\partial t} + U_\beta \frac{\partial U_\alpha}{\partial x_\beta} = -\frac{1}{\rho}\frac{\partial P}{\partial x_\alpha} + \frac{1}{\rho}\frac{\partial s_{\alpha\beta}}{\partial x_\beta} \tag{1.2}$$

where $s_{\alpha\beta}$ is the deviatoric stress tensor. For a Newtonian fluid, $s_{\alpha\beta}$ is given by

$$s_{\alpha\beta} = \rho v \left(\frac{\partial U_\alpha}{\partial x_\beta} + \frac{\partial U_\beta}{\partial x_\alpha} \right) \qquad (1.3)$$

where v is the kinematic viscosity of the fluid.

As a specific example, let us consider steady shearing flow between infinite plates which lie parallel to the $(x_1 x_3)$ plane. The flow is taken to be solely in the direction of x_1, and the velocity field reduces to $U_\alpha(\mathbf{x}, t) = \{U_1(x_2), 0, 0\}$. Under these circumstances, the non-linear term in (1.2) becomes

$$\frac{\partial\{U_1(x_2)U_1(x_2)\}}{\partial x_1} = 0,$$

and the conservation of momentum equation takes the form

$$\frac{\partial P}{\partial x_1} = \frac{\partial s_{12}}{\partial x_2}. \qquad (1.2a)$$

Note that the vanishing of the non-linear term is a general property of unidirectional laminar flows (Batchelor 1967), and that in this particular situation the momentum equation describes a balance between the streamwise pressure gradient and the resistive viscous stresses. For this flow, the viscous stress tensor can be obtained from eqn (1.3) as

$$s_{12} = \rho v \frac{dU_1}{dx_2}, \qquad (1.3a)$$

a result which is usually known as Newton's law. This law provides both a definition for and a method of measuring the dynamic viscosity $\mu = \rho v$. Physically, the effect of viscosity can be interpreted as an irreversible flux of streamwise momentum in the direction of x_2.

With the substitution of $s_{\alpha\beta}$ from (1.3) and the use of (1.1), the resulting equation of motion is known as the Navier–Stokes equation:

$$\frac{\partial U_\alpha}{\partial t} + \frac{\partial U_\alpha}{\partial x_\beta} U_\beta = -\frac{1}{\rho} \frac{\partial P}{\partial x_\alpha} + v\nabla^2 U_\alpha \qquad (1.4)$$

where $\nabla^2 = \partial^2/\partial x_\beta \partial x_\beta$.

These equations can be derived using the methods of continuum mechanics (e.g. Landau and Lifshitz 1959; Batchelor 1967) or, more restrictively, from kinetic theory (e.g. see Reichl 1980). They describe the motion of many common fluids, such as water, alcohol, glycerine, air, and most gases, provided only that the density remains constant. Many other fluids (e.g. particulate suspensions, polymer solutions) require a more complicated constitutive relationship between stress and rate of strain than the linear form given by eqn

(1.3). The subject of non-Newtonian fluids will be considered later in Chapter 14 which deals with turbulent drag reduction by additives

1.2   A brief statement of the problem

The scientific study of turbulence is generally taken as having begun with the work of Osborne Reynolds (1883). The problem which Reynolds studied was the classic one of flow through long straight pipes of constant diameter and circular cross-section. Using his 'method of colour bands', he was the first person to show that, for a given fluid and pipe, the flow would be orderly (laminar) for velocities below a certain critical speed. At the critical speed, the flow abruptly became turbulent at some distance from the pipe entrance. Above the critical speed, turbulence was found to be the normal state of the flow, although laminar flow could be maintained as a metastable state by carefully eliminating all disturbances or perturbations.

Reynolds found that the criterion for the transition from laminar to turbulent flow could be expressed in general (universal) form in terms of the value taken by dimensionless group

$$R = Ud/v \tag{1.5}$$

where R is what we now call the Reynolds number. Here d and U are representative length and velocity scales, in this case the diameter and the (bulk) mean velocity. The latter quantity is obtained by measuring the volumetric flow rate and dividing by the cross-sectional area of the pipe. With this definition, the experiments of Reynolds indicated that the minimum value of R for which turbulence could occur in a tube was about 2000. However, laminar flow in a pipe can be metastable at much larger values of the Reynolds number than this. A fuller, although still concise, discussion of transition to turbulence will be found in Goldstein (1938, pp. 69–74).

If we choose pipe flow as a specific problem on which to focus our attention, then we shall find it helpful to subdivide the basic theoretical problem into two parts:

(1) we would like to solve the Navier–Stokes equation for the critical value of the Reynolds number at which the transition from laminar to turbulent flow occurs;
(2) we would also like to solve the Navier–Stokes equation for the mean values of quantities like the velocity and the pressure, for Reynolds numbers larger than the critical value.

This subdivision may be seen as quite pragmatic. At this point it would be premature to try to offer any more fundamental justification. However, it does reflect the way in which the subject is organized. The two problems are normally treated in the literature as quite separate topics.

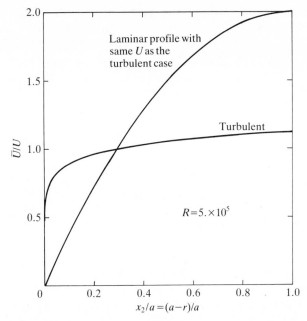

Fig. 1.1. Comparison of laminar and turbulent mean velocity distributions in pipe flow at the same Reynolds number ($R = 5 \times 10^5$).

Of the two, the second is overwhelmingly the more important in practical terms. Again, this shows up in the literature, although the subject of transition has received much more attention in recent years, mainly because of its relationship to other critical phenomena and its relevance to deterministic chaos. We make some brief comments on these developments in Chapter 11, Section 11.5.3. However, that section apart, we shall concentrate throughout on the second problem: the need to obtain a statistical description of fully developed turbulence.

Let us now consider flow through a pipe when the Reynolds number is well above the critical value. In Fig. 1.1 we show the distribution of the mean velocity with radial position in the pipe. The mean flow is taken to be in the x_1 direction and x_2 is the transverse or radial coordinate. For purposes of comparison, we have also plotted the equivalent laminar velocity profile for the same Reynolds number. This is, of course, the well-known parabolic profile which can be calculated directly from the Navier–Stokes equation. A not unreasonable goal for a turbulence theory would be to perform the analogous task for the mean turbulent velocity profile.

For the present we shall take mean values to be given by time averages and, for the particular case of the mean velocity, we shall show this by an overbar. In general, the operation of taking averages (however defined) will be represented by angular brackets $\langle \ \rangle$, and these brackets will be used to indicate higher-order moments. Thus, for the mean velocity, we have

$$\bar{U}_1(x_2, t) = \langle U_1(x_2, t) \rangle = \frac{1}{2T} \int_{-T}^{T} U_1(x_2, t + t') \, dt', \qquad (1.6)$$

where $2T$ is the averaging time which must be long enough to smooth out the turbulent fluctuations but short enough not to average out any imposed time dependence. However, for the most part, we shall only consider flows in which the mean velocity is constant in time.

We can now take a preliminary look at the task of solving the Navier–Stokes equation for the mean velocity profile for steady (mean) flow through a pipe. As we have already seen for the laminar case, eqn (1.2) reduces to the particularly simple form of (1.2a) for steady flow. Similar considerations apply to the equation for the mean velocity in turbulent flow; however, as the fluctuations in velocity are not unidirectional, the mean value of the non-linear term is not zero (the appropriate generalization of eqn (1.2a) to turbulent flow will be found in Section 1.4.5). This average contribution from the turbulent fluctuations gives rise to an additional resistance to flow. It also raises the central problem of the statistical theory of turbulence, as encountered by Reynolds (1895).

We can give a general (if rather simplified) explanation of this as follows. Let us rearrange eqn (1.4) to obtain

$$\left(\frac{\partial}{\partial t} - \nu \nabla^2 \right) U_\alpha = -\frac{\partial U_\alpha}{\partial x_\beta} U_\beta - \frac{1}{\rho} \frac{\partial P}{\partial x_\alpha},$$

which can then be written in the following highly symbolic fashion:

$$L_0 U = L_1 U U + L_2 P$$

where L_0, L_1, and L_2 stand for the respective differential operators, and we have temporarily suppressed the tensor indices for simplicity.

Now average each term to obtain an equation for the mean velocity:

$$L_0 \langle U \rangle = L_1 \langle UU \rangle + L_2 \langle P \rangle.$$

At this stage we should note that P is related to U through the continuity equation: we shall discuss the implications of this in Chapter 2. For the present it implies that a solution for $\langle U \rangle$ depends (in principle) only on the second-order moment $\langle UU \rangle$.

An equation for $\langle UU \rangle$ is readily obtained by multiplying each term of (1.4) by U and then averaging:

$$L_0 \langle UU \rangle = L_1 \langle UUU \rangle + L_2 \langle UP \rangle.$$

Multiplying in turn by UU, UUU, ..., before averaging then generates the hierarchy of moment equations

$$L_0\langle UUU\rangle = L_1\langle UUUU\rangle + L_2\langle UUP\rangle$$

$$L_0\langle UUUU\rangle = L_1\langle UUUUU\rangle + L_2\langle UUUP\rangle$$

and so on.

That is, we now have an open set of n equations for $n + 1$ moments. The problem of closing the moment hierarchy is usually referred to as the 'closure problem' and is the underlying problem of turbulence theory. We shall meet it at many points—and in various guises—throughout the rest of this book.

1.3 The statistical formulation

In this section we follow the procedure devised by Reynolds (1895) and write the instantaneous velocity as the sum of the mean $\bar{U}_\alpha$ and the fluctuation U_α from the mean:

$$U_\alpha(\mathbf{x}, t) = \bar{U}_\alpha(\mathbf{x}, t) + u_\alpha(\mathbf{x}, t). \tag{1.7}$$

Similarly, the instantaneous pressure can be written as

$$P(\mathbf{x}, t) = \bar{P}(\mathbf{x}, t) + p(\mathbf{x}, t) \tag{1.8}$$

where $\bar{P}(\mathbf{x}, t)$ is the mean and $p(\mathbf{x}, t)$ is the fluctuating pressure.

It follows from these definitions that the fluctuations have zero mean, that is

$$\langle u_\alpha(\mathbf{x}, t)\rangle = 0 \qquad \langle p(\mathbf{x}, t)\rangle = 0. \tag{1.9}$$

Physically, this result has the simple interpretation that, on average, the fluctuations are as often positive as they are negative. Correspondingly, the average of the square of a fluctuation will not vanish and we can introduce the root mean square (r.m.s) value u'_α as follows:

$$u'_\alpha(\mathbf{x}, t) = \langle u_\alpha^2(x, t)\rangle^{1/2}. \tag{1.10}$$

In practice u'_α is often used as a convenient measure of fluctuation intensity.

We should remind ourselves that the time dependence exhibited by the above mean quantities refers to slow external variations and is only retained here (and at other points) for generality. When we actually come to specific examples of real flows, we shall restrict our attention to the stationary case.

1.3.1 *Equations for the mean velocity*

Let us first consider the continuity equation, as given by (1.1). If we substitute (1.7) for the mean velocity, and average according to (1.6) and (1.9), we obtain

$$\frac{\partial \bar{U}_\beta(\mathbf{x}, t)}{\partial x_\beta} = 0. \tag{1.11}$$

Subtracting this result from (1.1) yields a similar equation for **u**, and so the mean and fluctuating velocities separately satisfy the continuity equation. This is, of course, a trivial consequence of the linearity of eqn (1.1).

In order to treat the equation of motion the same way, we substitute (1.7) and (1.8) into (1.4) and average term by term. It is easily shown that the operation of time averaging commutes with the operation of differentiating with respect to time.[1]

This time we have the non-linear term to deal with, and so, noting that $\langle \bar{U}u \rangle = \bar{U}\langle u \rangle = 0$, we find that

$$\frac{\partial \bar{U}_\alpha}{\partial t} + \frac{\partial \bar{U}_\alpha}{\partial x_\beta}\bar{U}_\beta + \frac{\partial}{\partial x_\beta}\langle u_\alpha u_\beta \rangle = -\frac{1}{\rho}\frac{\partial \bar{P}}{\partial x_\alpha} + \nu \nabla^2 \bar{U}_\alpha. \tag{1.12}$$

Comparison with (1.4) shows that the equation for the mean velocity is just the Navier–Stokes equation written in terms of the mean variables, but with the addition of the term involving $\langle u_\alpha u_\beta \rangle$. Thus, the equations of mean motion—in this case (1.11) and (1.12)—involve three independent unknowns $\bar{U}_\alpha$, $\bar{P}$, and $\langle u_\alpha u_\beta \rangle$. This is perhaps the best-known version of the closure problem referred to in the previous section.

Equation (1.12) is the Reynolds equation and the term $\langle u_\alpha u_\beta \rangle$ is the (kinematic) Reynolds stress. This term represents the transport of momentum due to turbulent fluctuations. It was noted by Reynolds that it effectively augmented the viscous stresses due to random molecular motions. The hypothesis that an analogy can be drawn between these two processes, with the implication that $\langle u_\alpha u_\beta \rangle$ could be expressed as a linear relationship in terms of the mean rate of strain and an effective (or apparent) coefficient of viscosity, has been one of the dominant themes of research in turbulence.

A detailed derivation of the stress tensor of apparent turbulent friction will be found in Schlichting (1968, p. 527). Here, we shall find it convenient to introduce a total shear stress tensor by means of the relation

$$\tau_{\alpha\beta} = s_{\alpha\beta} - \rho \langle u_\alpha u_\beta \rangle \tag{1.13}$$

where the viscous stress tensor is given by (1.3).

The equation of motion for the fluctuating velocity is obtained by subtracting (1.12) from (1.4):

$$\frac{\partial u_\alpha}{\partial t} + \frac{\partial \bar{U}_\alpha}{\partial x_\beta}u_\beta + \frac{\partial \bar{U}_\beta}{\partial x_\beta}u_\alpha + \frac{\partial}{\partial x_\beta}\{u_\alpha u_\beta - \langle u_\alpha u_\beta \rangle\}$$

$$= \frac{-1}{\rho}\frac{\partial p}{\partial x_\alpha} + \nu \nabla^2 u_\alpha. \tag{1.14}$$

Clearly, each term in this equation vanishes when averaged. But if we multiply through by $u_\alpha(\mathbf{x}, t)$ and then average, we have the basis for studying the single-point single-time moment hierarchy as used in engineering approaches.

Or, if we multiply through by $u_\gamma(\mathbf{x}', t')$ before averaging, we generate the two-point two-time moment hierarchy which underlies the fundamental approach. We shall defer the latter case to Chapter 2 and concentrate on the single-point form here. In particular, we shall consider the energy balance for the fluctuations.

1.3.2 *The energy balance equation for fluctuations*

We can learn quite a lot about the physics of turbulence by simply considering the ways in which energy is transported from one place to another, or from one range of eddy sizes to another. It should be noted that by 'energy' we shall invariably mean the kinetic energy of macroscopic fluid motions.

For the general case of a fluid occupying a volume V bounded by a surface S, the total kinetic energy of fluid motion E_T can be expressed in terms of the instantaneous velocity field $U_\alpha(\mathbf{x}, t)$ as

$$2E_T = \sum_\alpha \int_V \rho U_\alpha^2 \, \mathrm{d}V. \tag{1.15}$$

It should be noted that, as here, we shall sometimes use a summation sign rather than write $U_\alpha^2 = U_\alpha U_\alpha$.

In Appendix A we show that an equation for E_T can be derived from the Navier–Stokes equation in the form

$$\frac{\mathrm{d}E_T}{\mathrm{d}t} = \int_V \rho U_\alpha f_\alpha \, \mathrm{d}V - \int_V \rho \varepsilon \, \mathrm{d}V \tag{1.16}$$

where $f_\alpha(\mathbf{x}, t)$ is an externally applied force (per unit mass of fluid) and ε is the energy dissipation per unit time and per unit mass of fluid. It is given (see Appendix A) by

$$2\varepsilon = \nu \sum_\alpha \sum_\beta \left\{ \frac{\partial U_\alpha}{\partial x_\beta} + \frac{\partial U_\beta}{\partial x_\alpha} \right\}^2. \tag{1.17}$$

Equation (1.16) tells us that the rate of change of energy is equal to the rate at which the external forces do work on the fluid less the rate at which viscous effects convert kinetic energy into heat. Referring to the derivation of (1.16) as given in Appendix A, we should note two points.

First, the non-linear terms in the Navier–Stokes equation do not contribute to (1.16). That is, these terms do no net work on the system. (This remark also applies to the pressure; as we shall see later, the pressure is in effect a non-linear term.) Mathematically, this is because any term which appears as a divergence in the local energy equation vanishes when we integrate over the system volume in order to obtain the global equation (see Appendix A).

Second, the viscous term can be divided into two parts, of which one is diffusive in character and vanishes when integrated, and the other is dissipative and is given by eqn (1.17).

In the turbulent case eqn (1.15) for the total kinetic energy is readily generalized to the form

$$2E_T = \sum_\alpha \int_V \rho \bar{U}_\alpha^2 \, dV + \sum_\alpha \int_V \rho \langle u_\alpha^2 \rangle \, dV \qquad (1.18)$$

where, again, we have used the relationship $\langle \bar{U}u \rangle = 0$.

Normally we shall only be interested in the energy associated with the velocity fluctuations $u_\alpha(\mathbf{x}, t)$. We can derive the appropriate balance equation, via an equation for the Reynolds stress tensor, as follows.

Consider the equation for the fluctuating component $u_\alpha(\mathbf{x}, t)$ as given by (1.14). Rewrite this as an equation for $u_\gamma(\mathbf{x}, t)$, keeping all other indices and variables the same, and give it the notional equation number (1.14a). Now multiply (1.14) through by $u_\gamma(\mathbf{x}, t)$ and average, then multiply (1.14a) through by $u_\alpha(\mathbf{x}, t)$ and average, and, making repeated use of the rule for differentiating a product, add the resulting equations together to obtain

$$\frac{\partial}{\partial t} \langle u_\alpha u_\gamma \rangle + \bar{U}_\beta \frac{\partial}{\partial x_\beta} \langle u_\alpha u_\gamma \rangle + \langle u_\alpha u_\beta \rangle \frac{\partial \bar{U}_\gamma}{\partial x_\beta} +$$

$$+ \langle u_\gamma u_\beta \rangle \frac{\partial \bar{U}_\alpha}{\partial x_\beta} + \frac{\partial}{\partial x_\beta} \langle u_\alpha u_\beta u_\gamma \rangle$$

$$= \frac{-1}{\rho} \left\{ \left\langle u_\alpha \frac{\partial p}{\partial x_\gamma} \right\rangle + \left\langle u_\gamma \frac{\partial p}{\partial x_\alpha} \right\rangle \right\} + \nu \{ \langle u_\alpha \nabla^2 u_\gamma \rangle + \langle u_\gamma \nabla^2 u_\alpha \rangle \}. \qquad (1.19)$$

The viscous terms can be simplified (see the comparable procedure in Appendix A) and the rest of the equation rearranged a little to give

$$\frac{\partial}{\partial t} \langle u_\alpha u_\gamma \rangle + \bar{U}_\beta \frac{\partial}{\partial x_\beta} \langle u_\alpha u_\gamma \rangle$$

$$= - \langle u_\alpha u_\beta \rangle \frac{\partial \bar{U}_\gamma}{\partial x_\beta} - \langle u_\gamma u_\beta \rangle \frac{\partial \bar{U}_\alpha}{\partial x_\beta} - \frac{\partial}{\partial x_\beta} \langle u_\alpha u_\beta u_\gamma \rangle -$$

$$- \frac{1}{\rho} \left\{ \left\langle u_\alpha \frac{\partial p}{\partial x_\gamma} \right\rangle + \left\langle u_\gamma \frac{\partial p}{\partial x_\alpha} \right\rangle \right\} + \nu \frac{\partial}{\partial x_\beta} \left(\frac{\partial}{\partial x_\beta} \langle u_\alpha u_\gamma \rangle \right) - 2\nu \left\langle \frac{\partial u_\alpha}{\partial x_\beta} \frac{\partial u_\gamma}{\partial x_\beta} \right\rangle. \qquad (1.20)$$

The left-hand side gives the total rate of change with time (i.e. local plus convective time derivatives) of the kinematic Reynolds stress tensor $\langle u_\alpha u_\gamma \rangle$. On the right-hand side the presence of the term involving $\langle u_\alpha u_\beta u_\gamma \rangle$ serves to remind us that the moment closure problem is still with us.

If we set $\alpha = \gamma$ in the above equation then we obtain the equation for the mean square excitation for each component of the fluctuating velocity as follows:

$$A$$

$$\frac{\partial}{\partial t}\langle u_\alpha^2 \rangle + \bar{U}_\beta \frac{\partial}{\partial x_\beta}\langle u_\alpha^2 \rangle = -\frac{\partial}{\partial x_\beta}\left\{\langle u_\alpha^2 u_\beta \rangle + \frac{2}{\rho}\langle u_\beta p \rangle\right\} -$$

$$B \qquad\qquad C \qquad\qquad D$$

$$-2\langle u_\alpha u_\beta \rangle \frac{\partial}{\partial x_\beta}\bar{U}_\alpha + v\frac{\partial^2}{\partial x_\beta^2}\langle u_\alpha^2 \rangle - 2v\left\langle\left(\frac{\partial u_\alpha}{\partial x_\beta}\right)^2\right\rangle.$$

$$(1.21)$$

Furthermore, if we sum over α, then (1.21) gives us the balance equation for the turbulent kinetic energy E per unit fluid mass, which is defined by

$$2E = \sum_\alpha \langle u_\alpha^2 \rangle. \qquad\qquad (1.22)$$

Clearly, the only manipulation required is to divide across by 2 and replace $\langle u_\alpha^2 \rangle$ on the left-hand side (l.h.s.) by E.

In words then, eqn (1.21) tells us that the total rate of change of turbulent energy with time is given by the net effect of the terms on the right-hand side (r.h.s.). To interpret these we note that the first and third terms (A and C) can each be written as a divergence and accordingly do not contribute to the global energy balance (see the analogous step in Appendix A on going from eqn (A.10) to (A.12)). Thus their physical effect is the diffusion of turbulent energy through space by non-linear and viscous actions respectively.

The difficulty with the second term B is that apparently it cannot be written as a divergence. However, the reader may find it instructive to form an energy-balance equation for the mean velocity, where he will find a corresponding term. The two terms, taken together, can be integrated and shown to conserve energy jointly. Thus the term B on the r.h.s. of (1.21) can be interpreted as a flow of energy from the mean field to the fluctuating velocity field. It is often referred to as the 'production term' (see Hinze (1975) for a fuller discussion).

Obviously the last term D on the r.h.s. represents the irreversible dissipation of kinetic energy into heat. A more detailed discussion of this, and the other terms described above, will be found in Appendix A.

Finally, we should note that (1.21) cannot be solved: the moment closure problem rules that out. However, the individual terms can be measured experimentally and the energy balance studied in this way. We shall return to this subject in Section 1.6.2.

1.4 Two-dimensional mean flow as a special case

Our general problem is to solve the equations of mean motion for turbulent flows. However, if we consider very simple flows, we can reduce the number of variables involved. We can do this by means of symmetries (a rigorous method) and by 'order-of-magnitude' arguments (an approximate method). In this way the overall size of the problem can be reduced, although its intrinsic difficulty is unaffected.

As mentioned earlier, we shall consider two-dimensional mean flows, and historically the most important example of these is the boundary layer. This concept was introduced by Prandtl in 1904 (e.g. see Schlichting 1968), and takes its simplest form when a fluid flows at zero incidence over a flat plate.

To be specific, we shall take x_1 to be the direction of flow (this will be our convention throughout this book) and the plate to lie in the $(x_1 x_3)$ plane. The plate is also supposed to be of infinite extent in the x_3 direction and the incident stream has velocity U_∞, which is constant in time and uniform in space. With these restrictions the free stream is said to be an irrotational flow. That is, its vorticity (curl $\mathbf{U}$) is everywhere zero.

However, when the fluid encounters the plate, the 'no-slip' boundary condition ensures that vorticity is generated where the fluid is in contact with the surface. As the fluid moves downstream, this vorticity diffuses outwards (at right angles to the plate) with its rate of diffusion controlled by the viscosity of the fluid. In this way we obtain a physical interpretation of the idea of a boundary layer; that is, the distance travelled by vorticity away from the plate in the x_2 direction is a measure of the boundary-layer thickness.

If we take the leading edge of the plate to lie along the x_3 axis, then x_1 is distance measured along the plate and x_2 is distance measured perpendicular to it. We restrict our attention to one side of the plate ($x_2 > 0$) and consider the evolution of the boundary layer for $x_1 > 0$. It is known from experiment that the boundary layer is (in the absence of disturbances) laminar for small values of x_1 but, as downstream distance increases, it passes through a short transition length and thereafter is fully turbulent. It is conventional to define the Reynolds number (based on distance along the plate) as $Rx_1 = U_\infty x_1 / \nu$. With this form of Reynolds number, transition to turbulence normally occurs when $Rx_1 = 3.2 \times 10^5$. (As an aside, it may be of interest to note that, if the Reynolds number for the boundary layer is based on the thickness of the layer, then the value at transition is much closer to that for pipe flow based on diameter.)

The great simplification introduced by this concept is that the Navier–Stokes equations need only be solved inside the boundary layer where approximations can be made. In the laminar case, this means that we only have to solve for U_1 and U_2, as functions of x_1 and x_2, on the interval $0 < x_2 < \delta(x_1)$ where the boundary-layer thickness is arbitrarily defined by the relationship

$$x_2 = \delta(x_1) \quad \text{when} \quad U_1(x_1, x_2) = 0.99 U_\infty. \tag{1.23}$$

From symmetry considerations, it follows that $U_3 = 0$ and that variations of mean quantities with x_3 can be neglected. The problem is then further reduced by order-of-magnitude arguments based on the smallness of $\delta(x_1)$ in comparison with the length of the plate in the streamwise direction. For instance $U_2 \ll U_1$, and variations with x_2 are much larger than those with x_1. Finally, the existence of similarity solutions of the form

$$U_1(x_1, x_2) = U_\infty f(x_2/\delta), \tag{1.24}$$

where f is a universal function, reduces the dimensionality still further.

These ideas can be extended to the turbulent part of the boundary layer by applying the same arguments to the mean velocity $\bar{U}_\alpha(\mathbf{x}) = (\bar{U}_1, \bar{U}_2, 0)$. Of course the fluctuating field $u_\alpha(\mathbf{x}, t)$ is fully three-dimensional, but even here derivatives of mean quantities with respect to x_3 vanish, as do off-diagonal elements of the Reynolds stress tensor $\langle u_\alpha u_\beta \rangle$ in which $\alpha = 3$ or $\beta = 3$. The latter point also follows from symmetry (see Appendix C).

1.4.1   The boundary-layer equations of motion

In this section we shall briefly summarize a few central aspects of boundary-layer theory. We shall restrict our attention to steady flow over a flat plate and invoke the various simplifying assumptions made in the discussion above. It is then easily shown that eqn (1.11) for continuity reduces to

$$\frac{\partial \bar{U}_1}{\partial x_1} + \frac{\partial \bar{U}_2}{\partial x_2} = 0. \tag{1.25}$$

Similarly, eqn (1.12) becomes

$$\bar{U}_1 \frac{\partial \bar{U}_1}{\partial x_1} + \bar{U}_2 \frac{\partial \bar{U}_1}{\partial x_2} = -\frac{1}{\rho}\frac{\partial \bar{P}}{\partial x_1} + v\frac{\partial^2 \bar{U}_1}{\partial x_2^2} - \frac{\partial}{\partial x_2}\langle u_1 u_2 \rangle - \frac{\partial}{\partial x_1}\langle u_1^2 \rangle \tag{1.26}$$

for conservation of mean momentum in the x_1 direction.

The latter equation is approximate, but a further approximation can be made. Experimental results suggest that the streamwise variation of $\langle u_1^2 \rangle$ can be neglected, provided that the pressure gradient is not too large.

The momentum equation for the x_2 direction reduces to the condition that the pressure variation between the boundary and the free stream is of the second order of small quantities. Hence the streamwise pressure gradient in the boundary layer can be found approximately from the Bernoulli equation applied to the irrotational flow outside the boundary layer:

$$\bar{P} + \rho U_\infty^2/2 = \text{a constant along a streamline,}$$

or

$$\frac{d\bar{P}}{dx_1} = -\rho U_\infty \frac{dU_\infty}{dx_1}.$$

It follows that the pressure gradient in (1.26) can be put equal to zero for the case where the free stream is parallel to the plate, U_∞ is constant, and there is no externally applied pressure gradient.

1.4.2   *The turbulent boundary layer: length and velocity scales*

In Fig. 1.2 we show a schematic view of a turbulent boundary layer in a plane normal to the flow. The first point to be noticed is the irregularity of the outer edge. This means that we must work with the average value of the boundary-layer thickness, and so this is what we shall mean when we refer to $\delta(x_1)$, or indeed to any of its various subdivisions, which we shall now consider.

First of all, it is usual to divide the turbulent boundary layer into an inner layer for (approximately) the range $0 < x_2 < 0.2\delta$, and an outer layer bounded by $0.2\delta < x_2 < \delta$. This is based on the experimental observation that the total shear stress τ_{12} is nearly constant over the inner layer and is approximately equal to τ_w, its value at the surface (or wall).

The inner (or constant stress) layer can be further subdivided according to the relative magnitude of the viscous and turbulent parts of the total shear stress. In the present case, eqn (1.13) takes the simple form

$$\tau_{12} = \rho\nu\frac{d\bar{U}_1}{dx_2} - \rho\langle u_1 u_2\rangle \tag{1.27}$$

where the viscous part is given by Newton's law applied to the mean rate of

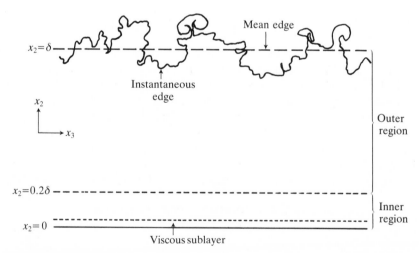

Fig. 1.2. Schematic view of a turbulent boundary layer (not to scale; flow direction normal to the page).

strain and the turbulent contribution is just the appropriate component of the Reynolds stress tensor.

Near the wall, the boundary conditions on the velocities, $u_1, u_2 \to 0$ as $x_2 \to 0$, imply that the product $u_1 u_2$ vanishes rapidly as the wall is approached. Hence the stress at the wall is entirely due to viscous shear and can be expressed as

$$\tau_w = \rho v \left[\frac{d\bar{U}_1}{dx_2} \right]_{x_2=0}. \tag{1.28}$$

We can now define the viscous sublayer as the region next to the wall where the first term on the r.h.s. of (1.27) is dominant. For larger values of x_2, the second term on the r.h.s. of (1.27) will ultimately become dominant, and this region is usually referred to as the fully turbulent constant-stress layer. Evidently there will be an intermediate region where the two stresses will be of equal magnitude, and this is called the transition sublayer (or, often, the buffer layer).

The physical extent of each of these sublayers is most conveniently expressed in terms of the so-called 'inner layer variables', which can introduced as follows.

Dimensional analysis (confirmed by experiment) indicates that the relevant velocity scale for the inner region is given by

$$u_\tau = (\tau_w/\rho)^{1/2}, \tag{1.29}$$

which then allows the relevant length scale to be written as

$$\text{'inner layer' length scale} = v/u_\tau \tag{1.30}$$

where u_τ is called the 'friction velocity'. The rather odd name can make this a puzzling concept for, at first sight, it seems to have little to do with any physical velocity. However, it is hallowed by tradition and, as we shall see, it is of the same order of magnitude as the r.m.s. value of the velocity fluctuations.

With (1.29) and (1.30), it is usual to define scaled variables for the inner region as

$$x_2^+ = \frac{x_2 u_\tau}{v} \qquad U_1^+ = \frac{\bar{U}_1}{u_\tau}. \tag{1.31}$$

This amounts to measuring distance from the wall in units of v/u_τ and, for small v (i.e. large Reynolds number), it has the effect of expanding the wall region.

The experimental results (to be discussed in more detail in Section 1.6) then suggest the following classification:

$$\text{inner layer } 0 < x_2 < 0.2\delta$$

$$\text{outer layer } 0.2\delta < x_2 < 1.0\delta.$$

The inner layer is subdivided as follows:

$$\text{viscous sublayer } 0 < x_2^+ < 5$$

$$\text{transition sublayer } 5 < x_2^+ < 30$$

$$\text{turbulent constant stress layer } 30 < x_2^+.$$

It should be noted that the values chosen above for x_2 or x_2^+ to classify the various layers will vary from one source to another in the literature. This, of course, reflects the difficulty in establishing precise criteria for the boundary between any two sublayers.

1.4.3   *Universal mean velocity distribution near a solid surface*

Phenomenological theories for the mean velocity have been much helped by experimental observations of scaling behaviour in the sort of simple situation that we are considering here. For instance, in the inner region of the boundary layer, measurements of mean velocities can be reduced to the universal form

$$U_1^+ = f(x_2^+), \tag{1.32}$$

which is known as the 'law of the wall'.

(We are working on the assumption that the wall has a smooth surface. If the height (however defined) of any roughness elements on the wall is less than the thickness of the viscous sublayer, the surface is said to be 'hydraulically smooth', and eqn (1.32) holds. If the roughness height is greater than the sublayer thickness, then it is the roughness height which determines the inner-region length scale.)

However, for the outer region, experimental results take the self-preserving form

$$U_\infty^+ - U_1^+ = g(x_2/\delta), \tag{1.33}$$

which is known as the 'velocity defect law'.

The functions f and g can be determined (at least for much of the boundary layer) by requiring there to be a region where the two forms (or, rather, their first derivatives) are continuous. A detailed treatment is given by Hinze (1975). The result is that f and g must be logarithms, and that eqn (1.32) becomes

$$U_1^+ = A \ln(x_2^+) + B, \tag{1.34}$$

where A and B are constants which have to be determined from a comparison of (1.34) with experimental results.

This logarithmic mean velocity distribution is strongly supported by experiment, to such an extent that it is virtually accorded the status of a law of nature by fluid dynamicists. Yet, as we shall see, it breaks down both near the wall and in the outer parts of the boundary layer. (It is of course evident that eqn (1.34) cannot satisfy the boundary condition $\bar{U}_1 = 0$ at $x_2 = 0$.)

However, we can establish the limiting form of the mean velocity at the wall by considering eqn (1.27) for the total shear stress when restricted to the viscous sublayer. Recalling that the total stress is constant in this region ($\tau_{12} = \tau_w$) and that the Reynolds stress tends to zero, we obtain the result

$$\frac{\tau_w}{\rho} = v\frac{d\bar{U}_1}{dx_2} \qquad \text{for } x_2^+ < 5. \tag{1.35}$$

Then, integrating with respect to x_2, and using (1.29), we find

$$\bar{U}_1 = \frac{u_\tau^2 x_2}{v} \tag{1.36}$$

or, in scaled variables,

$$U_1^+ = x_2^+ \tag{1.37}$$

where the constant of integration has been set equal to zero in order to satisfy the boundary condition. This linear law only applies to the viscous sublayer and, as we shall see in Section 1.6, has received ample experimental confirmation.

1.4.4 *A simple way of calculating the friction drag due to a boundary layer*

The simplest possible practical calculation method is to follow the approach taken with laminar boundary layers and use the equation of motion to derive an integral relation based on a similarity solution of the type

$$\bar{U}_1(x_1, x_2) = U_\infty h(x_2/\delta) \tag{1.38}$$

where h is chosen to satisfy the boundary conditions.

Setting the gradients of $\bar{P}$ and $\langle u_1^2 \rangle$ to zero, and integrating each term of eqn (1.26) with respect to x_2, then yields (e.g. see Goldstein 1938) the von Karman momentum-integral equation

$$\tau_w = \rho U_\infty^2 \frac{d}{dx_1}\left\{\int_0^\delta (1 - h)h\,dx_2\right\}. \tag{1.39}$$

In the analogous laminar case, we would fit h by a low-order polynomial, solve (1.39) simultaneously with (1.28) for the viscous shear stress to obtain δ, and hence obtain the wall shear stress as a function of x_1.

This is not possible in the turbulent case, mainly because the shape of the mean velocity profile is not easily represented by a polynomial. For this reason, we are forced to take two rather arbitrary steps.

First, we introduce an empirical formula for the shear stress at the wall. This is due to Blasius (Goldstein 1938, p. 340) and takes the form

$$\tau_w = 0.023\rho U_\infty^2 \left(\frac{\nu}{U_\infty \delta}\right)^{1/4}. \tag{1.40}$$

The r.h.s. of this expression can be equated to the r.h.s. of (1.39) in order to solve for the boundary-layer thickness. The second step is to use the empirical mean velocity profile given by

$$h = (x_2/\delta)^{1/n}. \tag{1.41}$$

The value of n depends on the Reynolds number and, for example, for moderate Reynolds numbers $(Rx_1 \approx 10^7)$ we can take $n = 7$.

If it is assumed, for simplicity, that the boundary layer is fully turbulent from $x_1 = 0$, the above procedure yields the result

$$\delta(x_1) = 0.38 \left(\frac{U_\infty}{\nu}\right)^{1/5} x_1^{4/5}, \tag{1.42}$$

and the drag force can be obtained by integrating the r.h.s. of either (1.39) or (1.40) over the surface of the plate. Alternatively, the result of the calculation is often expressed in terms of the friction (or drag) coefficient, as defined by the relationship

$$C_f = \frac{2\tau_w}{\rho U_\infty^2}. \tag{1.43}$$

Thus even the simplest practical calculation of a turbulent flow depends on some prior input from experiments. It has to be said that, at the time of writing, this is true of all practical methods used in engineering calculations.

1.4.5   Empirical relationships for resistance to flow through ducts

In the preceding sections we have been concerned with developing flow, where mean quantities vary (however weakly) with the streamwise coordinate x_1. We shall devote this section to a brief consideration of well-developed flows, in which mean quantities are independent of x_1. In particular, we shall discuss the two-dimensional (mean) channel flow obtained by considering a second plate in the $(x_1 x_3)$ plane which is placed at $x_2 = 2a$, above our original plate at $x_2 = 0$. Then, sufficiently far downstream (i.e. for large values of x_1), where the upper and lower boundary layers have merged, the turbulent flow will be well developed.

The approximations used in deriving eqn (1.26) remain valid, but we can now simplify this equation further by putting the gradients of $\bar{U}_1$ and $\langle u_1^2 \rangle$ with respect to x_1 equal to zero. Also, from (1.25) we deduce that $\bar{U}_2$ is constant and hence, from the boundary conditions, equal to zero. Thus, with some rearrangement, eqn (1.26) reduces to

$$\nu \frac{d^2 \bar{U}_1}{dx_2^2} - \frac{d\langle u_1 u_2 \rangle}{dx_2} = \frac{1}{\rho} \frac{d\bar{P}}{dx_1}, \tag{1.44}$$

which is a well-known form of the Reynolds equation. Integration of each term with respect to x_2 then yields

$$\rho\nu \frac{d\bar{U}_1}{dx_2} - \rho\langle u_1 u_2 \rangle = \frac{(x_2 - a)\,d\bar{P}}{dx_1} = \tau_{12}, \tag{1.45}$$

where the last step follows from (1.27) and we have used the condition $d\bar{U}_1/dx_2 = 0$, $\langle u_1 u_2 \rangle = 0$ at $x_2 = a$ (the channel centre).

For x_2 not too near the wall, we can neglect the viscous stress and write the Reynolds stress as

$$\rho\langle u_1 u_2 \rangle = -(x_2 - a)\frac{d\bar{P}}{dx_1}. \tag{1.46}$$

However, at the wall (e.g. for $x_2 = 2a$) we have the important result

$$\tau_w = a\frac{d\bar{P}}{dx_1}, \tag{1.47}$$

which offers a simple method of determining the shear stress at the wall in terms of two quantities which are readily measured.

Also, we can adapt the definition of the friction coefficient (see eqn (1.43)) to duct flows if we replace the free-stream velocity by the bulk mean velocity U. In this case the friction coefficient is renamed the friction factor f and eqn (1.43) becomes

$$f = 2\tau_w/\rho U^2. \tag{1.48}$$

In fact, we should mention that many of the deductions in preceding sections about the boundary layer at zero incidence—especially the mean velocity distributions—apply equally to flow in ducts. We can extend these results to pipe and channel flows by replacing the outer-region variables U_∞ and δ by the bulk mean velocity U and by the radius (circular pipes) or the semi-width (rectangular channels).

Lastly, for completeness, we should note that, for flow through ducts, there is an alternative to the Blasius formula as given by eqn (1.40). This is the logarithmic (or Prandtl–Karman) law for the resistance to flow through ducts. It takes the form (Goldstein 1938, p. 338)

$$1/f^{1/2} = 4.0 \log_{10}(Rf^{1/2}) - 0.40 \tag{1.49}$$

where f is the friction factor as given by eqn (1.48).

1.5   Semi-empirical theoretical methods

Since the early days of research into turbulence, there have been many attempts to reconcile the basic ideas that underlie the kinetic theory of gases
with the continuum concepts (especially vorticity and vortex motion in general) encountered in macroscopic fluid motion. The result has been many
theories of turbulence based on analogies between the chaotic motion of
eddies and the random motion of molecules in dilute gases. The mixing-length
model of Prandtl (e.g. see Schlichting 1968; Hinze 1975) is perhaps the best
known, and provides us with a representative example to discuss here. We
begin by considering the associated concept of an effective turbulent viscosity.

1.5.1   *The eddy-viscosity hypothesis*

The notion that the collective interaction of eddies can be represented by an
increased coefficient of viscosity is an attractive one. Traditionally it is introduced by analogy with the result from kinetic theory (e.g. as in eqn (1.3a)):

$$\text{mean viscous shear stress} = \rho v \frac{\mathrm{d}\bar{U}_1}{\mathrm{d}x_2}.$$

It is then tempting to guess that the turbulent shear stress can be written in
the analogous form

$$-\rho \langle u_1 u_2 \rangle = \rho v_T(x_2) \frac{\mathrm{d}\bar{U}_1}{\mathrm{d}x_2} \tag{1.50}$$

where $v_T(x_2)$ is the kinematic eddy viscosity.

Despite the fact that the apparent unsoundness of such analogies was clear
even to the earliest workers in turbulence, this hypothesis still receives much
critical attention. At a later stage we shall be considering the support given
to the idea of eddy viscosity by recent developments in the theory of renormalization, albeit with some restriction. At this stage, we merely note the
pragmatic criticism that, in flows where $\mathrm{d}\bar{U}_1/\mathrm{d}x_2$ and $\langle u_1 u_2 \rangle$ do not have their
zeros at the same point, the eddy viscosity (as defined by (1.50)) may be either
zero or infinite at certain points in the flow (e.g. Bradshaw 1972). Evidently—if
we were to pursue an analogy with the subject of continuum mechanics, rather
than kinetic theory—the 'constitutive relationship' for turbulence must in
general be more complicated than the purely 'Newtonian' one implied by eqn
(1.50).

1.5.2   *The Prandtl mixing-length model for flow near a solid boundary*

The mixing-length model is a more ambitious attempt to build on analogies
with kinetic theory. We start from the recognition that the Reynolds shear

stress $\rho\langle u_1 u_2 \rangle$ represents a flux of x_1-momentum in the direction of x_2. Prandtl assumed that this momentum was transported by discrete 'lumps' of fluid, which moved in the x_2 direction over a distance l without interaction (i.e. momentum is assumed to be conserved over distance l) and then mixed with existing fluid at the new location. Clearly l, which is called the mixing length, is supposed to play the part of a mean free path in this process.

The essentials of the analysis are then as follows. A fluid element dV is carried from x_2 to $x_2 + l$ by a velocity fluctuation u_2. It carries net x_1-momentum to its new location, owing to the difference between $\bar{U}_1(x_2)$ and $\bar{U}_1(x_2 + l)$. The result is a fluctuation in x_1-momentum and hence in x_1-velocity u_1.

This can be expressed as

$$\rho u_1 \, dV = \rho[\bar{U}_1(x_2) - \bar{U}_1(x_2 + l)]\, dV$$

$$= -\rho\left[l\frac{d\bar{U}_1}{dx_2} \right] dV \tag{1.51}$$

to first order in l, and hence

$$u_1 = -l\frac{d\bar{U}_1}{dx_2}. \tag{1.52}$$

We note that $\bar{U}_1$ is an increasing function of x_2, and so a fluid 'lump' moving in the direction of positive x_2 (i.e. one which corresponds to a positive fluctuation u_2) causes a negative fluctuation in u_1. Thus the Reynolds stress will be negative, and the correlation can be written in terms of the r.m.s. velocity components as

$$\langle u_1 u_2 \rangle = -R_{12}u_1' u_2'$$

$$= -C\langle u_1^2 \rangle \tag{1.53}$$

where R_{12} is the correlation coefficient. The second step follows from the experimental observation that u_2' is of the same order of magnitude as u_1' in the constant-stress layer. Also, R_{12} has been absorbed into the constant of proportionality C. Hence, from (1.52),

$$\rho\langle u_1 u_2 \rangle = -\rho l^2 \left(\frac{d\bar{U}_1}{dx_2}\right)^2 \tag{1.54}$$

where now the constant C has been absorbed into l.

At this stage we need three further assumptions:

(a) in the constant stress layer we can put $\tau_{12} = \tau_w$;
(b) for $x_2^+ > 5$, we can neglect the viscous term in the shear stress;
(c) $l = kx_2$, where k is known as the von Karman constant.

Then, using (1.27) and (1.29), eqn (1.53) becomes

$$\frac{\tau_w}{\rho} = -\langle u_1 u_2 \rangle = l^2 \left(\frac{d\bar{U}_1}{dx_2}\right)^2,$$

(1.55)

and hence

$$k^2 x_2^2 \left(\frac{d\bar{U}_1}{dx_2}\right)^2 = u_\tau^2.$$

(1.56)

Finally, taking the square root of both sides and integrating with respect to x_2, we obtain the logarithmic profile in the form

$$\bar{U}_1 = \frac{u_\tau}{k} \ln x_2 + D$$

(1.57)

where D is a constant of integration. This result can be matched to the linear profile (see eqn (1.37)) by making an appropriate choice of the constant D. The result is that the logarithmic profile given by eqn (1.57) can be seen to satisfy the 'law of the wall' form, as shown in eqn (1.34). In Section 1.6.1 we shall see that there is substantial experimental support for the logarithmic mean velocity distribution.

1.5.3   *The mixing-length model applied to a free jet*

Jets, wakes, and mixing layers are other flows in which the criteria for a thin shear layer are often satisfied. Thus we can employ the boundary-layer approximation, despite the fact that the absence of rigid boundaries may make the name seem something of a misnomer. Also, we shall use the mixing-length model again, as it will be instructive to apply this theory to a free shear flow.

The boundary-layer equations for the mean velocity, as given by (1.25) and (1.26), can again be applied:

$$\frac{d\bar{U}_1}{dx_1} + \frac{d\bar{U}_2}{dx_2} = 0$$

(1.25)

while (1.26) reduces to

$$\bar{U}_1 \frac{d\bar{U}_1}{dx_1} + \bar{U}_2 \frac{d\bar{U}_1}{dx_2} = -\frac{\partial \langle u_1 u_2 \rangle}{\partial x_2},$$

(1.58)

where we have followed the usual practice (established on experimental grounds) of neglecting the streamwise variation of the pressure and the Reynolds stress. For the same reasons, we have also neglected the viscous term on the r.h.s. in comparison with the term involving the Reynolds stress.

If we wished to study wake flows, then it would be convenient to express (1.58) in terms of the velocity defect, that is, the amount by which the velocity in the wake is reduced below that of the free stream. However, here we shall take the case of the plane turbulent jet as an example, and (1.58) can be used as it stands.

Let us consider a jet emerging from a thin slot into a still fluid. As always, we take the mean velocity to be in the x_1 direction and the jet spreads out in the x_2 direction. We assume similarity in different sections sufficiently far downstream (an assumption borne out in practice) and look for a general solution of the form

$$\bar{U}_1 = \frac{\partial \psi}{\partial x_2} \qquad \bar{U}_2 = -\frac{\partial \psi}{\partial x_1}. \qquad (1.59)$$

and

$$\psi = x_1^n f\left(\frac{x_2}{x_1^m}\right) = x_1^n f(\eta), \qquad (1.60)$$

where the introduction of the stream function ψ ensures that the continuity equation is automatically satisfied (see Batchelor (1967) for a general discussion of the use of stream functions) and the similarity variable η is given by

$$\eta = x_2/x_1^m. \qquad (1.61)$$

We need a 'closure' for eqn (1.58), and we shall take the mixing-length form as given by (1.54). Then (1.58) becomes

$$\bar{U}_1 \frac{d\bar{U}_1}{dx_1} + \bar{U}_2 \frac{d\bar{U}_1}{dx_2} = -\frac{\partial\{l^2(\partial\bar{U}_1/\partial x_2)^2\}}{\partial x_2}. \qquad (1.62)$$

The next step is to fix values for the exponents n and m in (1.60). The first condition for this is that eqn (1.62) should be self-consistent when we substitute (1.59) and (1.60) for $\bar{U}_1$ and $\bar{U}_2$. This requirement gives $m = 1$. Our second condition is obtained by integrating (1.62) with respect to x_2 from $-\infty$ to $+\infty$, which yields the condition of constant momentum flux at different streamwise positions along the jet:

$$\frac{d\{\int_{-\infty}^{\infty} \bar{U}_1 \, dx_2\}}{dx_1} = 0. \qquad (1.63)$$

Satisfying this condition then gives us $n = 1/2$.

In the previous section, it was assumed that the mixing length was linearly proportional to the distance from the wall. In free-shear flows it is usual to assume that l is proportional to the width of the layer. In the present case, with $m = 1$ and $n = 1/2$, we have

$$l = ax_2 \qquad \eta = x_2/x_1 \qquad \psi = x_1^{1/2} f(\eta), \qquad (1.64)$$

and eqn (1.62) can be shown to reduce to

$$f' + ff'' = 4af''f''' \qquad (1.65)$$

where the primes denote differentiation with respect to η and (1.65) has to be solved subject to the conditions of symmetry and of zero mean velocity at the edge of the jet.

A first integral of (1.65) is easily obtained, but thereafter it has to be solved numerically. Tolmien (1926; see Goldstein 1938, p. 593) found that conditions for the edge of the jet must be satisfied at

$$\eta = x_2/x_1 = 3.04a^{2/3}, \tag{1.66}$$

and comparison with experiment gave a value for the mixing-length constant of $a = 0.0165$. In Fig. 1.8 (Section 1.6.3) we make a comparison of Tolmien's calculation with some recent results for the mean axial velocity distribution in a plane jet, and it can be seen that the agreement is quite good. The same analysis can also be carried out for the round free jet, in which case $n = 1$ and $m = 1$.

Full discussions of the various mixing-length types of theory will be found in Goldstein (1938), along with good discussions of their application to various flows. From our present point of view, an interesting feature of the application discussed above is that the mixing-length model is used quite explicitly as a closure of the statistical equations of motion. This was less apparent in the previous section, where we relied quite heavily on the idea of the constant-stress layer.

1.6   Some experimental results for shear flows

There is an enormous amount of data on turbulent flows, much of it amassed a long time ago. Indeed, the interested reader can obtain quite a good historical impression of the subject by consulting the book *Modern developments in fluid dynamics* (Goldstein 1938, two volumes).

In the hope of presenting an uncluttered picture, we shall discuss only a very few investigations. We shall also mainly restrict our attention to duct flows, although we shall also give some results for the free jet.

In order to give a representative picture of flow through ducts, we have drawn on the investigations of Nikuradse (1932; see Goldstein 1938), Laufer (1954), and Lawn (1971), all of which deal with flow through a pipe of circular cross-section. Results for the other main duct configuration—plane channel flow—are not very different from those for pipe flow. But, for completeness, we should mention the investigations into channel flow by Laufer (1951), Hussain and Reynolds (1975), and Kreplin and Eckelmann (1979).

There is also no shortage of results for free jets, but we would mention in particular the papers by Wygnanski and Fiedler (1969) on the axisymmetric jet, and by Bradbury (1965) and Ramaprian and Chandrasekhara (1985) on the plane jet. The latter reference is especially interesting, as it involves the use of a frequency-shifted laser-Doppler anemometer. The importance of this measuring device is that it determines the direction, as well as the magnitude, of the fluid velocity and hence (unlike the traditional pitot tube and hot-wire devices) can take proper account of the effect of the flow reversals which occur at the edge of a jet where it entrains ambient fluid. A discussion of anemometry will be found in Chapter 3.

Lastly, before turning to a discussion of the actual experimental results, we should define our coordinate systems for the different geometries of flow. For duct flows, x_1 is the axial (streamwise) coordinate, x_2 is the distance from the wall in the radial direction, and x_3 is the circumferential (spanwise) coordinate. In jets, x_1 is again the flow direction, but now x_2 is the distance from the centre line in the radial or transverse direction.

1.6.1 *Mean and root mean square velocity distributions in duct flow*

In Fig. 1.3 we show mean velocity distributions in pipe flow for three widely separated Reynolds numbers. The results have been selected from those published by Nikuradse (1932; see Goldstein 1938) and are quite characteristic of turbulence, with a steep rise near the wall and a flatter profile in the core region. Clearly, this behaviour becomes more marked as the Reynolds number increases.

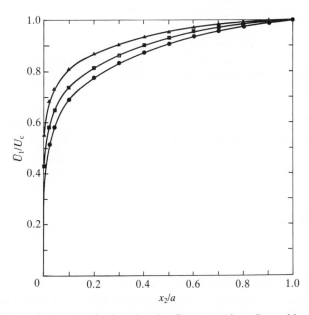

Fig. 1.3. Mean velocity distributions in pipe flow at various Reynolds numbers: ●, $R = 4000$; ■, $R = 110\,000$; ▲, $R = 3\,200\,000$ (Nikuradse 1932; see Goldstein 1938).

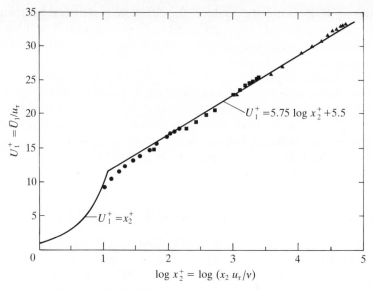

Fig. 1.4. Logarithmic mean velocity distributions in pipe flow: universal form in law of the wall coordinates (symbols for experimental data are the same as those in Fig. 1.3).

These results of Nikuradse can be replotted (along with those of other workers (Goldstein 1938)) to verify the 'velocity defect law' as given by eqn (1.33), but we shall not pursue that here.

The reduction to the universal 'law of the wall' form is demonstrated in Fig. 1.4, using the same three sets of results. As the abscissa is the log (i.e. logarithm to base 10) of x_2^+, the straight line indicates a satisfactory logarithmic dependence for most of the data.

This result has been confirmed by many investigators (e.g. see Goldstein 1938; Hinze 1975), which means that the velocity distribution given by eqn (1.34) is in good agreement with experimental data, except close to the wall. However, there is less than complete agreement in the literature on the values of the constants A and B. In Fig. 1.4, the straight line represents the equation

$$U = 5.75 \log_{10}(x_2^+) + 5.5,$$

and, converting to natural logarithms, this implies that $A = 2.50$ and hence the von Karman constant k is 0.40. But, even with the restricted data presented here, it is clear that experimental scatter would permit some variation in the values of A and B.

A specific indication of the extent of disagreement on this point can be obtained from the discussion in Hinze (1975), where it is stated that taking $A = 2.44$ and $B = 4.9$ represents 'a good average' of various investigations. Yet Hinze (1975, p. 626) also suggests that many workers in the field are now inclined to prefer the values $A = 2.5$ and $5.2 < B < 5.5$, with a slight bias in favour of $B = 5.5$. In other words, these values are very much in line with the above equation, which represents the original conclusion based on Nikuradse's results.

Given the central position of the 'log law' in the phenomenology of turbulence, it is really rather surprising that there should be so little unanimity on the experimental value of the universal constant A. Indeed, as recently as the early 1970s it was suggested that, for low Reynolds numbers, A actually depended on the Reynolds number, although this seems to have been something of a false alarm (Huffman and Bradshaw 1972; Purtell, Klebanoff, and Buckley 1981).

In contrast, the linear part of the mean velocity profile in the viscous sublayer is a rigorous result. In addition, although we do not show any experimental points in this region of Fig. 1.4, the linear law of the wall has also been verified experimentally (for a fairly recent reference, see Bakewell and Lumley (1967)).

For turbulence quantities the classic measurements were those of Laufer (1954), who used a hot-wire anemometer to obtain values of the three components of fluctuating velocity in air flow through a pipe. In Fig. 1.5, we show a sketch of the r.m.s. velocities u_1', u_2', and u_3', each divided by the friction velocity u_τ, plotted against x_2/a. Evidently each r.m.s. component is roughly of the same order as the friction velocity.

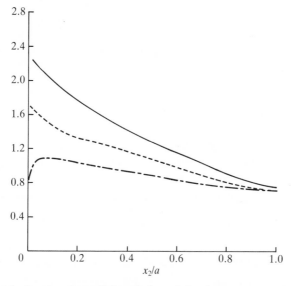

Fig. 1.5. Sketch showing the radial variation of the three components of the r.m.s. velocity in turbulent flow through a round pipe of diameter $2a$ (note that $x_2 = 0$ corresponds to the position of the wall): ———, u_1'/u_τ; ———, u_2'/u_τ; ————, u_3'/u_τ).

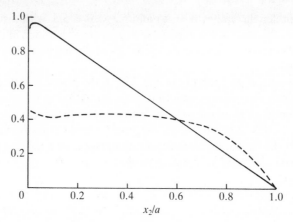

Fig. 1.6. Sketch showing the radial variation of the Reynolds shear stress and its associated correlation coefficient in turbulent flow through a pipe: ———, $\langle u_1 u_2 \rangle / u_\tau^2$; ————, $\langle u_1 u_2 \rangle / u_1' u_2'$.

Other points we might note include the disparity between the three components, their increasing trend towards the wall, and their tendency to the same value at the centre of the pipe. A discussion of these aspects will be deferred until we consider the production and transport of turbulence in the next section.

Figure 1.6 shows the distribution across the pipe of $\langle u_1 u_2 \rangle / u_\tau^2$ and $\langle u_1 u_2 \rangle / u_1' u_2'$. The first of these is the ratio of Reynolds stress to wall shear stress, and bears out the linear relationship predicted in eqn (1.46). The second quantity is the correlation coefficient, and we note that this takes a value of about 0.4 almost irrespective of position. The near constancy of this correlation is noteworthy in itself, given the distinct non-uniformity of most mean quantities in shear flow.

1.6.2   The turbulent energy balance

In Section 1.3.2 we derived the energy balance for fluctuating velocities (eqn 1.21). We have called the third term on the r.h.s. of this equation the production term because it represents the transfer of kinetic energy from the mean field to the fluctuating field and is therefore interpreted as the rate of generation of turbulence. Let us now consider this term for the particular case of steady well-developed duct flow. Recalling that these restrictions imply that derivatives with respect to t, x_1, and x_3 vanish, as do off-diagonal correlations involving u_3, we can reduce the general form of the production term in (1.21) to

$$\text{production term} = -2\langle u_1 u_2 \rangle \frac{d\bar{U}_1}{dx_2}. \tag{1.67}$$

We are now in a position to understand some of the qualitative features of the results for the r.m.s. velocities in the preceding section. From a consideration of eqns (1.21) and (1.58) we can make the following points.

(a) Only the streamwise turbulent excitation $\langle u_1^2 \rangle$ is produced by direct conversion of kinetic energy of mean motion, and so it is not really surprising that u_1' is larger than either u_2' or u_3'.

(b) The radial and circumferential excitations $\langle u_2^2 \rangle$ and $\langle u_3^2 \rangle$ are generated by inertial transfer from $\langle u_1^2 \rangle$ through the triple correlation and, more specifically, through the term involving the pressure fluctuations.

(c) The rate of generation of $\langle u_1^2 \rangle$ will be very much peaked near the wall, where both the Reynolds stress and the mean velocity gradient are large, and will fall off rapidly away from it. Thus the role of the triple correlations is to transport energy in the radial direction (tendency to homogeneity), and to transfer energy from $\langle u_1^2 \rangle$ to $\langle u_2^2 \rangle$ and $\langle u_3^2 \rangle$ (tendency to isotropy). This is borne out by Fig. 1.5, which indicates that the three r.m.s. components are approximately equal and relatively uniform near the centre of the pipe where the production term is zero.

We can consider the energy balance by summing each term in eqn (1.21) over α and making the appropriate simplifications for well-developed duct flow to obtain

$$
-\langle u_1 u_2 \rangle \frac{d\bar{U}_1}{dx_2} - \frac{d}{dx_2} \left\{ \frac{1}{2} \sum_\alpha \langle u_\alpha^2 u_2 \rangle + \frac{1}{\rho} \langle u_2 p \rangle \right\}
$$

$$
= \nu \sum_{\alpha,\beta} \left\langle \left(\frac{\partial u_\alpha}{\partial x_\beta} \right)^2 \right\rangle = \varepsilon. \tag{1.68}
$$

Here we have neglected the viscous diffusion on the experimental grounds that it is small, and only of appreciable importance very close to the wall.

Lawn has measured the individual terms in this equation (see Lawn (1971), eqn (2.1) for the same equation in cylindrical coordinates) and his results are reproduced in Fig. 1.7. Measurement of the production is quite straightforward, but the dissipation rate presents some formidable difficulties, involving as it does the measurement of the nine separate components of the fluctuating rate-of-strain tensor. In Fig. 1.7 the dissipation curve has been obtained by taking the difference between the production and the inertial transfer (or diffusion). Unfortunately, this procedure suffers from the snag that the contribution from the fluctuating pressure to the diffusion cannot be measured directly and has to be estimated, with consequent loss of precision.

Lawn also obtained values for the dissipation rate by two other methods, both relying on an assumption of local isotropy in the small scales which are largely responsible for the dissipation. We shall not pursue these methods here, as they involve topics which we will deal with in the next chapter. However,

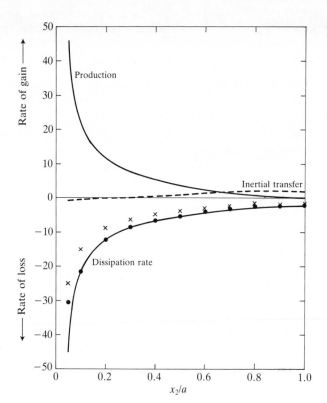

Fig. 1.7. Turbulent energy balance in the core region of pipe flow (after Lawn 1971).

it is clear from Fig. 1.7 that the three methods of obtaining the rate of dissipation agree well enough for us to find this figure a convincing picture of the turbulent energy balance in pipe flow.

The production term in (1.68) is always positive; this follows from the arguments of Section 1.5.2, where we deduced that the Reynolds stress is negative when the mean velocity gradient is positive. As a closing topic, we briefly consider the case where the mean velocity gradient $d\bar{U}_1/dx_1$ in the streamwise direction is not zero (e.g. flow in a converging or diverging duct).

Under these circumstances, the production term in eqn (1.21) becomes

$$\text{production} = -\langle u_1 u_2 \rangle \frac{d\bar{U}_1}{dx_2} - \langle u_1^2 \rangle \frac{d\bar{U}_1}{dx_1}. \tag{1.69}$$

If it is assumed that $d\bar{U}_1/dx_1$ is not large enough to produce inflections in $d\bar{U}_1/dx_2$, the first term remains positive. But, as $\langle u_1^2 \rangle$ is always positive, the sign of the second term depends on the sign of $d\bar{U}_1/dx_1$. Thus, if $d\bar{U}_1/dx_1 > 0$, production due to this term is negative and the turbulence decreases. In turn, this can lead to relaminarization.

Striking examples of this effect have been demonstrated, with the help of flow visualization, in boundary-layer flows with strong pressure gradients (e.g. Van Dyke 1982, Plate 164).

1.6.3 *Mean and root mean square velocity distributions in a plane jet*

We have already referred to Fig. 1.8, where we compare the mixing-length calculation of Tolmien with the experimental results of Ramaprian and Chandrasekhara (1985). These authors found that their mean axial velocity profiles exhibited self-similarity for downstream distances (as measured from the jet exit) greater than $10D$, where D is the width of the jet at its exit. These self-similar profiles took a Gaussian form:

$$\frac{\bar{U}_1(\eta)}{\bar{U}_1(0)} = \exp\{-C\eta^2\}$$

where η is the similarity variable, as given by eqn (1.64), and the experimental value of the constant is $C = \ln(0.5)$. The measured velocities differed significantly from the Gaussian profile at the jet edges, where the authors noted that the mean velocities were slightly negative. As we pointed out earlier, traditional measuring techniques would be insensitive to the sign of the velocity.

As the jet spreads out, there must also be a lateral component of the mean velocity, and we sketch a typical result for the profile of this quantity in Fig.

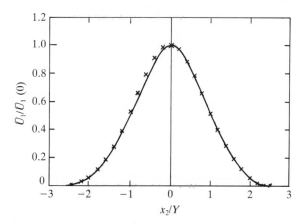

Fig. 1.8. Distribution of axial mean velocity in a plane jet. Comparison of the mixing-length theory with some representative results: ———, Mixing-length theory, as calculated by Tollmien (1926; see Goldstein 1938, p. 593); X, representative experimental value of Ramaprian and Chandrasekhara (1985). Note that the mean velocity is divided by its own value at the centre line and plotted against transverse distance divided by Y. Here Y is the value of x_2 where $\bar{U}_1(x_2) = 0.5\bar{U}_1(0)$.

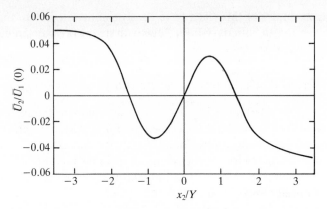

Fig. 1.9. Sketch showing the distribution of the transverse mean velocity in a free jet.

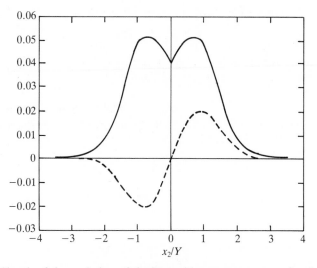

Fig. 1.10. Sketch of the variation of the Reynolds stresses across a free jet: ————, normal stress $\langle u_1^2 \rangle / \bar{U}_1^2(0)$; –––, shear stress $\langle u_1 u_2 \rangle / \bar{U}_1^2(0)$.

1.9. This result can actually be calculated from the continuity equation (1.25) and the experimental result for the axial mean velocity, as presented in Fig. 1.8.

In Fig. 1.10 we show a sketch of typical results for the turbulent normal and shear stresses in a jet. Comparison with the corresponding results for pipe flow in Figs. 1.5 and 1.6 reveals a general qualitative similarity. However, the quantitative differences are worth noting as they underline the essential distinctions which must be drawn between confined and free turbulent flows. In particular, it should be noted that the peaks of the Reynolds stress profiles in duct flow are rather sharp and close to the wall. This reflects the intense localized nature of the turbulent production near the wall. In contrast, the peaks of the turbulent stress profiles in the jet are much more spread out and much closer to the symmetry axis of the flow. This can be seen as reflecting the more diffuse nature of the production process in a free shear flow which involves the entrainment of the surrounding fluid into the jet.

Finally, in view of the reliance on the conservation of momentum flux in Section 1.5.3, we should note the growing speculation that momentum flux is not conserved in jets (e.g. Kotsovinos 1978). Ramaprian and Chandrasekhara (1985) found that the axial momentum flux actually increased with down-stream distance, reaching an asymptotic value at $x_1/D = 30$ of about 50 per cent greater than the value at the jet exit. However, this violation of the conservation law for axial momentum flux did not apparently affect the general form of the classical jet scaling law as given in eqn (1.64).

1.7 Further reading

The purpose of this chapter has been to cover just enough of the traditional semi-empirical approach to turbulence to provide a background for the more fundamental work which will be the subject of most of the rest of this book. We believe that we have at least touched on the most important central points, but otherwise we make no claims to completeness. In a subject which is (in some ways) as much art as science, each new variant on the standard problem (e.g. fluid compressibility, wall curvature, complicated geometries, wall rough-ness, two or more phases, heat and mass transfer, and many more) brings its own particular remedies. We do not pretend to have even scratched the surface of this vast field, nor would it be appropriate for us to try.

To sum up, our main theme has been to formulate the statistical problem in a general way, using correlations which are worked out at a single point in space and time. Then we have shown how the size of the problem can be reduced by specializing these general equations to various two-dimensional mean flows.

This approach should be distinguished from that to be adopted in Chapter 2. There we repeat the statistical formulation, but with greater emphasis on generality. In particular, we are concerned with the hierarchy of many-point and many-time moments. Then we reduce the size of the fundamental problem by specializing our general moment equations to the idealized case of homo-geneous isotropic turbulence.

The reader who wishes to learn more about boundary layers, or the aero-dynamic aspects, will find the books by Schlichting (1968) and Thwaites (1960) helpful, while those interested in the general phenomenology of turbulence could refer to the standard works by Hinze (1975) and Townsend (1976). The first of these is very comprehensive; the second, although shorter, is inclined to concentrate more on the underlying structure of turbulence. Those readers who have become tired of relentless experimental detail may find some relief in the readable discussions of the physics of the subject by Bradshaw (1971).

Lastly, any newcomer to the subject is strongly recommended to glance through the many beautiful photographs in *An album of fluid motion* (Van Dyke 1982). Turbulence can be a difficult subject, and it may be useful to remind oneself that the phenomenon is not without its aesthetic side.

Note

1. We require to show that the operations of differentiation and taking an average commute, that is

$$\left\langle \frac{\partial A}{\partial t} \right\rangle = \frac{\partial}{\partial t} \langle A \rangle.$$

We begin with the l.h.s.:

$$\left\langle \frac{\partial A}{\partial t} \right\rangle = \frac{1}{2T} \int_{t-T}^{t+T} \left(\frac{\partial A}{\partial t} \right) dt$$

$$= \frac{1}{2T} \{ A(t + T) - A(t - T) \}.$$

We now need the formula for an integral differentiated with respect to its upper limit:

$$\frac{\partial}{\partial y} \int_0^{ay} g(x)\, dx = a g(ay).$$

It follows from this that we can write the r.h.s. of our first expression as

$$\frac{\partial}{\partial t} \langle A \rangle = \frac{\partial}{\partial t} \left(\frac{1}{2T} \int_{t-T}^{t+T} A\, dt \right)$$

$$= \frac{\partial}{\partial t} \left(\frac{1}{2T} \int_0^{t+T} A\, dt - \frac{1}{2T} \int_0^{t-T} A\, dt \right)$$

$$= \frac{1}{2T} \{ A(t + T) - A(t - T) \},$$

which is equal to the final expression above for the l.h.s.

References

BAKEWELL, H. P. and LUMLEY, J. L. (1967). *Phys. Fluids* **10**, 1880.

BATCHELOR, G. K. (1967). *An introduction to fluid dynamics*. Cambridge University Press, Cambridge.

BRADBURY, L. J. S. (1965). *J fluid Mech.* **23**, 31.

BRADSHAW, P. (1971). *An introduction to turbulence and its measurement*. Pergamon Press, Oxford.

—— (1972). *Aeronaut. J.* **76**, 403.

GOLDSTEIN, S. (1938). *Modern developments in fluid dynamics*. Clarendon Press, Oxford. (Reprinted by Dover Publications, New York, 1965).

HINZE, J. O. (1975). *Turbulence* (2nd edn). McGraw-Hill, New York.

HUFFMAN, D. G. and BRADSHAW, P. (1972). *J. fluid Mech.* **53**, 45.

HUSSAIN, A. K. M. F. and REYNOLDS, W. C. (1975). *J. fluids Eng.* **97**, 568.

KOTSOVINOS, N. E. (1978). *J. fluid Mech.* **87**, 55.

KREPLIN, H.-P. and ECKELMANN, H. (1979) *Phys. Fluids* **22**, 1233.

LANDAU, L. D. and LIFSHITZ, E. M. (1959). *Fluid mechanics*. Pergamon Press, London.

LAUFER, J. (1951). *NACA Rep. 1053.*

—— (1954). *NACA Tech. Rep. 1174.*

LAWN, C. J. (1971). *J. fluid Mech.* **48**, 477.

PURTELL, L. P., KLEBANOFF, P. S., and BUCKLEY, F. T. (1981). *Phys. Fluids* **24**, 802.

REICHL, L. E. (1980). *A modern course in statistical physics*. Edward Arnold, London.

RAMAPRIAN, B. R. and CHANDRASEKHARA, M. S. (1985). *J. fluids Eng.* **107**, 264.

REYNOLDS, O. (1883). *Phil. Trans. R. Soc. A* **175**, 935.

—— (1895). *Phil. Trans. R. Soc. A* **186**, 123.

SCHLICHTING, H. (1968). *Boundary layer theory* (6th edn). Pergamon Press, London.

THWAITES, B. (1960). *Incompressible aerodynamics*. Clarendon Press, Oxford.

TOWNSEND, A. A. (1976). *The structure of turbulent shear flow* (2nd edn). Cambridge University Press, Cambridge.

VAN DYKE, M. (1982) *An album of fluid motion*. Parabolic Press, Stanford, Ca.

WYGNANSKI, I. and FIEDLER, H. (1969). *J. fluid Mech.* **38**, 577.

2

THE FUNDAMENTAL APPROACH

Superficially the background material presented in this chapter may seem very similar to the topics covered in Chapter 1. That is, we begin with the Navier–Stokes equations and use the Reynolds averaging procedure to formulate the statistical approach. Then (just as in Chapter 1) we are concerned with the mean square excitation of fluctuations, with energy balances, and with the moment closure problem. We shall also consider the use of both dimensional methods and approximate models, and conclude by considering some representative experimental results.

However, there will be two important differences of emphasis. First of all, our main interest will be in the underlying structure of the turbulence. That is, we shall consider the correlation of velocities at two or more points (and times). The foundations of this approach were laid by Taylor (1935) in a paper which also introduced the concepts of statistical homogeneity and isotropy: a step which took turbulence theory into the realm of physics, rather than engineering. In a subsequent paper (Taylor 1938a) the introduction of the energy spectrum in wavenumber (i.e. the Fourier transform of the two-point correlation in space) virtually completed this process, and, as we shall see, the calculation of this spectrum provides a major goal for fundamental turbulence theory.

Secondly, we shall also give some attention to the establishment of certain general results and procedures which will provide the basis for the symbolic manipulations needed in the later theoretical work. As an example, we begin by eliminating the pressure from the equations of motion. The procedure to be followed is based on the work of Chou (1945), and the resulting divergence-less form of the Navier–Stokes equation provides the starting point for all modern theories of turbulence.

2.1 The Navier–Stokes equation in solenoidal form

Consider an incompressible fluid of density ρ and Newtonian viscosity v occupying a volume V which is bounded by a surface S. Parts of the boundary surface may be at infinity (e.g. flow through long pipes or a jet emerging into a large reservoir), and we take the boundary condition on the velocity field to be

$$U_\alpha(\mathbf{x}, t) = 0 \qquad (\mathbf{x} \text{ on } S). \qquad (2.1)$$

As we saw in the previous chapter, the velocity U_α and the pressure P are determined by eqns (1.1) and (1.4). The pressure term in (1.4) can be eliminated by using the continuity equation as follows.

We begin by rearranging (1.4) with the linear terms on the l.h.s.:

$$\frac{\partial U_\alpha}{\partial t} - \nu \nabla^2 U_\alpha = -\frac{1}{\rho}\frac{\partial P}{\partial x_\alpha} - \frac{\partial}{\partial x_\beta} U_\alpha U_\beta. \tag{2.2}$$

Now take the divergence of each term in (2.2). From (1.1) we see that the linear terms vanish and we are left with an equation for the pressure:

$$\frac{1}{\rho}\nabla^2 P = -\frac{\partial^2}{\partial x_\alpha \partial x_\beta} U_\alpha U_\beta, \tag{2.3}$$

which is a form of Poisson's equation (just as encountered in electrostatics for example).

The boundary condition on the pressure can be obtained by taking (2.2) on S and using (2.1) to obtain

$$\frac{1}{\rho}\frac{\partial P}{\partial x_\alpha} = \nu \frac{\partial^2}{\partial x_\beta \partial x_\beta} U_\alpha \qquad (\mathbf{x} \text{ on } S). \tag{2.4}$$

Alternatively, we can express this in terms of the normal derivatives

$$\frac{\partial}{\partial n} = n_\beta \frac{\partial}{\partial x_\beta} \quad \text{and} \quad \frac{\partial^2}{\partial n^2} = n_\beta n_\gamma \frac{\partial^2}{\partial x_\beta \partial x_\gamma},$$

where $n_\alpha(\mathbf{x})$ is the unit inward normal at $\mathbf{x}$ on S, to obtain

$$\frac{1}{\rho}\frac{\partial P}{\partial n} = \nu n_\beta \frac{\partial^2}{\partial n^2} U_\beta. \tag{2.5}$$

Equation (2.5) has the advantage of being in standard (Neumann) form. Also, it relies only on the normal component of $U_\alpha(\mathbf{x}, t)$ vanishing on S. This allows us to extend the formalism to include the case where parts of the boundary surface may be in motion, which would be the situation in (for instance) Couette flow.

We can solve (2.3) for the pressure, subject to the boundary condition (2.5), in terms of Green's function $G(\mathbf{x}, \mathbf{x}')$ which satisfies Laplace's equation in the form

$$\nabla^2 G(\mathbf{x}, \mathbf{x}') = \delta(\mathbf{x} - \mathbf{x}') \tag{2.6}$$

subject to the condition

$$\frac{\partial G(\mathbf{x}, \mathbf{x}')}{\partial n} = 0 \qquad (\mathbf{x} \text{ on } S). \tag{2.7}$$

Discussions of Green's function methods can be found in the book by Roach (1970). The formal solution of (2.3) can be written down immediately as

$$P(\mathbf{x}, t) = -\rho \int_V d^3x' G(\mathbf{x}, \mathbf{x}') \frac{\partial^2 \{U_\beta(\mathbf{x}', t) U_\alpha(\mathbf{x}', t)\}}{\partial x'_\beta \partial x'_\alpha} +$$

$$+ \rho v \int_S d^2x' G(\mathbf{x}, \mathbf{x}') n_\beta \frac{\partial^2 U_\beta(\mathbf{x}', t)}{\partial n^2}. \tag{2.8}$$

The first term on the r.h.s can be modified in two ways which will later prove helpful. First, we note that the index α is now repeated and hence is to be summed. Accordingly we rename this dummy index γ in order to avoid confusion with the labelling index α in eqn (2.2). Second, we carry out a partial integration (twice), and use the boundary conditions and the symmetry of $G(\mathbf{x}, \mathbf{x}')$ under interchange of $\mathbf{x}$ and $\mathbf{x}'$ to take the differentials outside the integration. Our final expression for the pressure then takes the form

$$P(\mathbf{x}, t) = -\rho \frac{\partial^2}{\partial x_\beta \partial x_\gamma} \int_V d^3x' G(\mathbf{x}, \mathbf{x}') U_\beta(\mathbf{x}', t) U_\gamma(\mathbf{x}', t) +$$

$$+ \rho v \int_S d^2x' G(\mathbf{x}, \mathbf{x}') n_\beta \frac{\partial^2 U_\beta(\mathbf{x}', t)}{\partial n^2}. \tag{2.9}$$

Clearly, our next step will be to use (2.9) to express the pressure gradient in (2.2) in terms of the velocity field. This will take the form of a non-linear tensor expression involving tensor indices α, β, and γ. We can partly anticipate this process by rewriting (2.2) as

$$\frac{\partial U_\alpha}{\partial t} - v\nabla^2 U_\alpha = -\delta_{\alpha\gamma} \frac{\partial}{\partial x_\beta} U_\beta U_\gamma - \frac{1}{\rho} \frac{\partial P}{\partial x_\alpha}, \tag{2.2a}$$

where the Kronecker delta $\delta_{\alpha\gamma}$ acts as a substitutional symbol. Then, with (2.9) for the pressure, (2.2a) becomes

$$\frac{\partial U_\alpha}{\partial t} - v\nabla^2 U_\alpha = -\frac{\partial}{\partial x_\beta} D_{\alpha\gamma}(\mathbf{V})[U_\beta(\mathbf{x}, t) U_\gamma(\mathbf{x}, t)] - L_{\alpha\beta}(\mathbf{V})[U_\beta(\mathbf{x}, t)], \tag{2.10}$$

where the operators $D_{\alpha\gamma}(\mathbf{V})$ and $L_{\alpha\beta}(\mathbf{V})$ can be defined in terms of their effect on an arbitrary function $f(\mathbf{x})$:

$$D_{\alpha\gamma}(\mathbf{V})[f(\mathbf{x})] = \delta_{\alpha\gamma} f(\mathbf{x}) - \frac{\partial^2}{\partial x_\alpha \partial x_\gamma} \int_V d^3x' G(\mathbf{x}, \mathbf{x}') f(\mathbf{x}') \tag{2.11}$$

and

$$L_{\alpha\beta}(\mathbf{V})[f(\mathbf{x})] = v \frac{\partial}{\partial x_\alpha} \int_S d^2x' G(\mathbf{x}, \mathbf{x}') n_\beta(\mathbf{x}') \frac{\partial^2 f(\mathbf{x}')}{\partial n^2}. \tag{2.12}$$

Two final steps are now needed. First, we can put the non-linear term on the r.h.s. of (2.10) in a more symmetric form. Noting that the interchange of the dummy indices β and γ must leave this term unchanged, we introduce the symmetric operator $M_{\alpha\beta\gamma}(\mathbf{V})$ such that

$$M_{\alpha\beta\gamma}(\mathbf{V}) = -\frac{1}{2}\left\{\frac{\partial}{\partial x_\beta}D_{\alpha\gamma}(\mathbf{V}) + \frac{\partial}{\partial x_\gamma}D_{\alpha\beta}(\mathbf{V})\right\}. \qquad (2.13)$$

Second, we can extend this formulation to include flows driven by a constant externally imposed pressure gradient. In this case the external pressure P_{ext} would be such that

$$\nabla P_{\text{ext}} = \text{constant}$$

and hence would satisfy Laplace's equation

$$\nabla^2 P_{\text{ext}} = 0. \qquad (2.14)$$

Thus P_{ext} can be added on to $P(\mathbf{x}, t)$, as given by (2.9), without affecting the solution of Poisson's equation.

With these points in mind, we can now write (2.10) as

$$\frac{\partial}{\partial t}U_\alpha - \nu\nabla^2 U_\alpha = M_{\alpha\beta\gamma}(\mathbf{V})[U_\beta(\mathbf{x}, t)U_\gamma(\mathbf{x}, t)] -$$

$$- L_{\alpha\beta}(\mathbf{V})[U_\beta(\mathbf{x}, t)] - \frac{1}{\rho}\frac{\partial P_{\text{ext}}}{\partial x_\alpha}, \qquad (2.15)$$

which is the Navier–Stokes equation in divergenceless form.

This may seem to be a rather cumbersome procedure, and, although we have eliminated the pressure from the problem, we do seem to have paid the price of additional complication in the equation of motion. Fortunately, this disadvantage is more apparent than real. In most of the fundamental work to be considered in this book, we shall be able to neglect the surface integral in (2.15). Also, as we shall see, for most purposes the non-linear term in (2.15) is really no more complicated than the original non-linearity in eqn (2.2).

2.2   The general statistical formulation

In eqn (1.6) we have previously introduced the time average as the basis of a statistical approach. This is the method of averging normally used by experimenters, but from now on we shall base all our theoretical work on the ensemble average. It is usual to assume that the two methods of taking averages are equivalent: this property is known as ergodicity.

We shall defer a discussion of ergodic theory, along with a consideration of what constitutes a turbulent ensemble, until Chapter 4. For the moment let us consider how we would obtain an ensemble average for some specific experiment. We can imagine carrying out N simultaneous independent identical versions of our experiment and then wishing to know the mean value of some quantity $f(x, t)$ for particular values of x and t. If we associate a superscript n with each version (or realization) of our experiment, then we can write the ensemble average as

$$\langle f(x,t) \rangle = \lim_{N \to \infty} \frac{1}{N} \sum_{n=0}^{N} f^{(n)}(x,t) \qquad (2.16)$$

for any given values of x and t. Evidently this definition satisfies a necessary condition for an average:

$$\langle\langle f \rangle\rangle = \langle f \rangle. \qquad (2.17)$$

The essential feature of (2.16) is its linearity, and various important results follow from this. For example, it is now obvious that the operations of differentiation and taking averages commute; that is,

$$\left\langle \frac{\partial f}{\partial t} \right\rangle = \frac{\partial \langle f \rangle}{\partial t} \qquad \left\langle \frac{\partial f}{\partial x} \right\rangle = \frac{\partial \langle f \rangle}{\partial x}, \qquad (2.18)$$

and of course this result can readily be extended to any independent variable. Also, if A is a constant, it is easy to show that

$$\langle Af \rangle = A \langle f \rangle \qquad (2.19)$$

and hence

$$\langle\langle f \rangle g \rangle = \langle f \rangle \langle g \rangle, \qquad (2.20)$$

where $g(x,t)$ is another random variable in the domain. Finally, the linearity of (2.16) implies that

$$\langle f + g \rangle = \langle f \rangle + \langle g \rangle \qquad (2.21)$$

even when f and g are not statistically independent.

Of course, in our study of turbulence we shall normally be concerned with variables which are not statistically independent or which may only become independent of each other under certain restricted circumstances. In order to quantify the degree of statistical dependence of two or more variables, we make use of the correlation, of which $\langle fg \rangle$ is an example. This concept will be developed further in the next section. Also, a brief general introduction to some relevant aspects of probability and statistics will be found in Appendix B.

2.2.1   *Statistical equations and the closure problem*

We now follow the Reynolds averaging procedure again. But this time we use eqn (1.7) to set up the formal many-point many-time moment hierarchy based on the solenoidal form of the Navier–Stokes equation.

Recalling that the velocity can be decomposed into mean and fluctuating parts (eqn (1.7))

$$U_\alpha(\mathbf{x},t) = \bar{U}_\alpha(\mathbf{x},t) + u_\alpha(\mathbf{x},t),$$

where

$$\bar{U}_\alpha(\mathbf{x}, t) = \langle U_\alpha(\mathbf{x}, t) \rangle \qquad (2.22)$$

and

$$\langle u_\alpha(\mathbf{x}, t) \rangle = 0, \qquad (2.23)$$

we can set up the infinite sequence of velocity field moments as follows:

$$Q_{\alpha\beta}(\mathbf{x}, \mathbf{x}'; t, t') = \langle u_\alpha(\mathbf{x}, t) u_\beta(\mathbf{x}', t') \rangle \qquad (2.24)$$

$$Q_{\alpha\beta\gamma}(\mathbf{x}, \mathbf{x}', \mathbf{x}''; t, t', t'') = \langle u_\alpha(\mathbf{x}, t) u_\beta(\mathbf{x}', t') u_\gamma(\mathbf{x}'', t'') \rangle \qquad (2.25)$$

and, to any order,

$$Q_{\alpha\beta\gamma\delta\cdots}(\mathbf{x}, \mathbf{x}', \mathbf{x}'', \mathbf{x}''', \ldots; t, t', t'', t''', \ldots)$$

$$= \langle u_\alpha(\mathbf{x}, t) u_\beta(\mathbf{x}', t') u_\gamma(\mathbf{x}'', t'') u_\delta(\mathbf{x}''', t''') \ldots \rangle.$$

Two quite general points can be made about each of these moments. First, their value cannot be affected by the order in which we take the measuring points or times. Second, the constituent velocities each satisfy the continuity condition so that the moments must also satisfy eqn (1.1). Taking the second-order moment as an example, we can express these two conditions in turn as

$$Q_{\alpha\beta}(\mathbf{x}, \mathbf{x}'; t, t') = Q_{\beta\alpha}(\mathbf{x}', \mathbf{x}; t', t) \qquad (2.26)$$

$$\frac{\partial}{\partial x_\alpha} Q_{\alpha\beta}(\mathbf{x}, \mathbf{x}'; t, t') = \frac{\partial}{\partial x'_\beta} Q_{\alpha\beta}(\mathbf{x}, \mathbf{x}'; t, t') = 0. \qquad (2.27)$$

We now wish to derive a more general form of the Reynolds equations which we obtained in Chapter 1. Essentially we follow the same procedure and substitute (1.7) for the velocity field into the Navier–Stokes equation, although this time we use the solenoidal form as given by (2.15). Averaging then yields the equation for the mean velocity:

$$\left(\frac{\partial}{\partial t} - \nu\nabla^2 \right) \bar{U}_\alpha(\mathbf{x}, t) = -\frac{1}{\rho} \frac{\partial P_{\text{ext}}}{\partial x_\alpha} + M_{\alpha\beta\gamma}(\mathbf{V})[\bar{U}_\beta(\mathbf{x}, t)\bar{U}_\gamma(\mathbf{x}, t) +$$

$$+ Q_{\beta\gamma}(\mathbf{x}, \mathbf{x}; t, t)] - L_{\alpha\beta}(\mathbf{V})\bar{U}_\beta(\mathbf{x}, t). \qquad (2.28)$$

The equation for the fluctuating velocity $u_\alpha(\mathbf{x}, t)$ is obtained (as in the preceding chapter) by subtracting (2.28) for the mean velocity from (2.15) for the total velocity. It is easy to show that this takes the form

$$\left(\frac{\partial}{\partial t} - \nu\nabla^2 \right) u_\alpha(\mathbf{x}, t) = 2M_{\alpha\beta\gamma}(\mathbf{V})[\bar{U}_\beta(\mathbf{x}, t)u_\gamma(\mathbf{x}, t)] +$$

$$+ M_{\alpha\beta\gamma}(\mathbf{V})[u_\beta(\mathbf{x}, t)u_\gamma(\mathbf{x}, t) -$$

$$- Q_{\beta\gamma}(\mathbf{x}, \mathbf{x}; t, t)] - L_{\alpha\beta}(\mathbf{V})[u_\beta(\mathbf{x}, t)] \qquad (2.29)$$

where we have used the property $M_{\alpha\beta\gamma} = M_{\alpha\gamma\beta}$.

Equation (2.29) can be used to obtain an equation for a moment of any order. We shall illustrate this process by forming the governing equation for the vitally important second-order moment.

Multiply each term in (2.29) by $u_\sigma(\mathbf{x}', t')$ and average throughout. The result is

$$\left(\frac{\partial}{\partial t} - \nu\nabla^2\right) Q_{\alpha\sigma}(\mathbf{x}, \mathbf{x}'; t, t')$$

$$= 2M_{\alpha\beta\gamma}(\mathbf{V})[\bar{U}_\beta(\mathbf{x}, t) Q_{\gamma\sigma}(\mathbf{x}, \mathbf{x}'; t, t')] +$$

$$+ M_{\alpha\beta\gamma}(\mathbf{V})[Q_{\beta\gamma\sigma}(\mathbf{x}, \mathbf{x}, \mathbf{x}'; t, t, t')] -$$

$$- L_{\alpha\beta}(\mathbf{V})[Q_{\beta\sigma}(\mathbf{x}, \mathbf{x}'; t, t')]. \tag{2.30}$$

Of course, in order to solve this for the second-order moment we first have to solve a similar equation for the third-order moment which in turn requires the fourth-order moment, and so on. We have previously referred to this as the closure problem, and this is its most general form.

2.2.2   Two-point correlations

The second-order two-point correlations (or moments) play a leading part in turbulence theory. Indeed, much of the rest of this book will be concerned in one way or another with these correlations or (more usually) their Fourier transforms. Accordingly we shall spend a little time (in this section and in Section 2.3) in considering the relationship between the two-point (fundamental) and the single-point (engineering) approaches to the turbulence problem. In particular, as the multipoint theory is the more general, we should be interested in knowing under what circumstances it can be expected to reduce to the single-point form.

Let us begin by considering how we would actually measure the correlation of velocities at two points in, say, a pipe flow. We can assume that we have an anemometer which will measure all three scalar components of the fluctuating velocity at a single point $\mathbf{x}$. (The subject of anemometry will be discussed in the next chapter.) If we have a second such anemometer at the point $\mathbf{x}'$, then in principle all we have to do is multiply each pair of signals together and average the resultant product in order to obtain the nine scalar components of $Q_{\alpha\beta}$.

Each component of the correlation tensor is itself a function of eight scalar variables. Therefore before we start generating a mountain of data, we should carefully consider how we can approach this task in the most systematic fashion.

In general terms, the velocity correlations can be expected to depend on two things. First, as we move the measuring points apart, we would expect

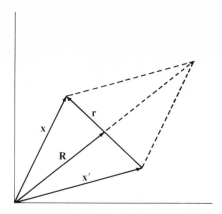

Fig. 2.1. Centroid (**R**) and relative (**r**) coordinate system for two-point correlations.

the correlation to die away. Thus correlations will depend on the distance between the measuring points. Second, the magnitude of the correlations must (at fixed separation) depend on the absolute values of the constituent velocities. If we refer to the experimental results given in the previous chapter (e.g. Fig. 1.5), then clearly the value of the correlation at a given separation can be expected to depend on absolute radial position in the pipe.

In practice, quite a good way of tackling this task would be to mount the two anemometers rigidly on a traversing system which would allow the distance between them to be varied in a controlled way. Then this system could itself be attached to a second traversing mechanism which would move it bodily to various different locations in the pipe, with the effect of variable separation being investigated independently at each such location.

Formally this procedure corresponds to a change of variables, as illustrated schematically in Fig. 2.1. That is, we express the two-point correlations in terms of

$$\mathbf{R} = (\mathbf{x} + \mathbf{x}')/2 \tag{2.31}$$

$$\mathbf{r} = (\mathbf{x} - \mathbf{x}'), \tag{2.32}$$

where **R** is the centroid (or absolute) coordinate and **r** is the difference (or relative) coordinate.

A similar transformation can be carried out for the basic time variables t and t', and in general we have

$$Q_{\alpha\beta}(\mathbf{x}, \mathbf{x}'; t, t') = Q_{\alpha\beta}(\mathbf{r}, \mathbf{R}; \tau, T) \tag{2.33}$$

where

$$T = (t + t')/2 \tag{2.34}$$

and

$$\tau = t - t'. \tag{2.35}$$

Clearly if we put $\mathbf{x} = \mathbf{x}'$ in (2.33) we recover the type of single-point correlation which was of interest to us in the previous chapter, where we met it as the Reynolds stress. In this case its dependence on absolute position was the important feature.

However, if we wish to achieve a fundamental understanding of turbulence, then we should concentrate on how the correlations depend on the relative coordinate. Let us try to see why this is so.

For example, consider two variables A and B, each a random function of time and each with zero mean. In simple terms, the mean of A vanishes because A is as often negative as it is positive and so its integral over all time is zero. The same is true of B. However, the mean squares of A and B do not vanish because the integrands are always positive. What then is the implication for $\langle AB \rangle$?

The answer to this question obviously depends on the extent to which there is some connection between the random variables A and B. If the two variables are closely related in some way, then at one extreme we have A and B always going negative together, so that their product is always positive. (Or, of course, the two variables might be out of phase, in which case their product would always be negative. But $\langle AB \rangle$ would still exist.) At the other extreme, there might be no connection at all between A and B, in which case the product AB would be as often positive as negative, and hence $\langle AB \rangle = 0$.

In turbulence the answer is normally somewhere in between the above extremes. We have already seen in the previous chapter that the single-point correlation coefficient R_{12} is about 0.4 in pipe flow (see Fig. 1.6). If we now consider the case where $A = B = u_1$, say, then it is obvious that the associated single-point correlation coefficient will be unity. What we would like to know in this case is: over what distance (or time) will the correlation fall to zero as we move the measuring points apart in space (or time)? In order to provide a quantitative answer we introduce the integral length scale and time-scale. Essentially, we wish to establish where turbulence fits in the range between the purely deterministic process (e.g. laminar flow) and the completely random process (e.g. tossing a coin, throwing a die).

Formally, we can introduce the general correlation coefficient $R_{\alpha\beta}$ through the relationship

$$Q_{\alpha\beta}(\mathbf{x}, \mathbf{x}'; t, t') = R_{\alpha\beta}(\mathbf{x}, \mathbf{x}'; t, t') u'_\alpha(\mathbf{x}, t) u'_\beta(\mathbf{x}', t'), \tag{2.36}$$

where the primes on the velocities again denote r.m.s. values and the repeated indices are not summed. From the above discussion, it is clear that $R_{\alpha\beta}$ must satisfy the condition

$$\max|R_{\alpha\beta}(\mathbf{x}, \mathbf{x}; t, t)| = 1, \tag{2.37}$$

and intuitively we would expect that the correlation coefficient would also satisfy

$$R_{\alpha\beta} \to 0 \qquad \text{for } |\mathbf{x} - \mathbf{x}'| \to \infty$$

$$\text{and/or } |t - t'| \to \infty. \tag{2.38}$$

If we set $\mathbf{x} = \mathbf{x}'$ and $t = t' = 0$, then with $\alpha = 1$ and $\beta = 2$ the above definition of $R_{\alpha\beta}$ reduces to the form given by eqn (1.39).

The work of this section can be completed by giving general definitions of the integral scales. Let R be any component of the correlation tensor, and for convenience choose $\mathbf{x} = \mathbf{x}' = 0$ and $T = (t + t')/2 = 0$; then the integral time scale T_E is defined by

$$T_E = \int_0^\infty R(\tau)\, d\tau. \tag{2.39}$$

The subscript E stands for 'Eulerian', meaning that the correlation has been measured in the Eulerian (or laboratory) frame of reference. Later in the book we shall we interested in the diffusion of fluid particles, which will require a Langrangian (or material coordinate) frame of reference, and so it will be necessary to draw a distinction between the two.

In a similar way we can define the integral length scales, now choosing $t = t' = 0$ and $\mathbf{R} = (\mathbf{x} + \mathbf{x}')/2 = 0$ for convenience, to write

$$L_{\alpha\beta}(\hat{\mathbf{r}}) = \int_{0 < |\mathbf{r}| < \infty} R_{\alpha\beta}(\mathbf{r})\, d\mathbf{r} \tag{2.40}$$

where $\hat{\mathbf{r}}$ is the unit vector in the direction of $\mathbf{r}$.

For a specific example, we might wish to know the integral correlation for the streamwise velocity u_1 in channel flow, when the measuring points are separated in the cross-stream (x_2) direction. Then from (2.40) we would use the definition

$$L_{11}(\hat{r}_2) = \int_0^\infty R_{11}(r_2)\, dr_2. \tag{2.41}$$

Roughly speaking, a value for L_{11} would tell us how far we would have to go in the cross-stream direction for the streamwise velocity to become uncorrelated with itself.

2.3 Reduction of the statistical equations to the form for channel flow as an example

In Section 1.4.5 we considered the case of pressure-driven two-dimensional mean flow between infinite parallel planes. In this section we shall make a temporary digression from the fundamental approach to show how our general formulation can be applied specifically to that problem. In particular, we shall consider eqn (2.28) for the mean velocity, and show that it reduces to the Reynolds equation as given by (1.44).

Referring to Section 1.4.5 for the details of the flow, we begin by noting that for steady mean flow in the x_1 direction, the mean velocity becomes

$$\bar{U}_\alpha(\mathbf{x}, t) = \bar{U}_1(x_2) \tag{2.42}$$

and hence the l.h.s. of (2.28) reduces to

$$\text{l.h.s. of (2.28)} = -v\frac{d^2\bar{U}_1}{dx_2^2}. \tag{2.43}$$

Now let us consider the r.h.s. of (2.28) with $\alpha = 1$ throughout. Both terms involving the mean velocity are zero. Take the quadratic term first. From (2.42) we have $\beta = \gamma = 1$, and hence $M_{111}(\mathbf{V})$, containing only differentials with respect to x_1, acts on a function of x_2 only to give zero. The linear term can be eliminated by means of similar arguments.

The term containing the correlation $Q_{\alpha\beta}(\mathbf{x}, \mathbf{x}; t, t)$ requires a more general treatment based on the symmetries of the problem, and this provides us with a first look at the type of argument that will be an essential feature of the fundamental approach in subsequent sections. For this reason we shall discuss it as a special case of the general two-point two-time form as given by (2.33).

First we deal with the time variables by noting that the act of putting $t = t'$ eliminates dependence on τ and the condition that the flow is steady eliminates dependence on T. Thus there is no time dependence at all, and incidentally this conclusion applies to moments of all orders.

The spatial variables are a little more complicated. We begin by noting that the flow domain extends to $\pm\infty$ in the x_1 and x_3 directions. As a result, mean quantities do not depend on absolute position in these directions, and hence the correlation is a function of $r_1 = x_1 - x_1'$ and $r_3 = x_3 - x_3'$. In the x_2 direction all moments are a function of absolute position (e.g. see the experimental results for channel flow in Section 1.5) because of the no-slip boundary conditions. Thus the correlation tensor also depends on $R_2 = (x_2 + x_2')/2$, as well as on $r_2 = x_2 - x_2'$, and the general two-point form for steady channel flow becomes

$$\langle u_\alpha(\mathbf{x}, t)u_\beta(\mathbf{x}', t')\rangle = Q_{\alpha\beta}(\mathbf{r}, R_2) \tag{2.44}$$

Now we use the fact that the mean properties of the flow must be unaffected by reflections of the velocity vectors in any one of the three coordinate planes. Details can be found in Appendix C, where it is shown that

$$Q_{13} = -Q_{31} = \text{constant} = 0$$

$$Q_{23} = -Q_{32} = \text{constant} = 0 \tag{2.45}$$

but

$$Q_{12} = -Q_{21} = \text{an antisymmetric function of } x_2. \qquad (2.46)$$

Thus, with (2.44), (2.45), (2.46), $\alpha = 1$, and $\mathbf{x} = \mathbf{x}'$, the r.h.s. of (2.28) reduces to

$$\text{r.h.s of (2.28)} = -\frac{1}{\rho}\frac{dP_{\text{ext}}}{dx_1} - \frac{dQ_{12}}{dx_2}. \qquad (2.46)$$

It is readily seen that (2.43) and (2.46) taken together are equivalent to eqn (1.52) with

$$Q_{12}(x_2) = \langle u_1 u_2 \rangle.$$

We conclude this section by noting that the above results and arguments can be used to reduce eqn (2.30)—the general equation of motion for the two-point correlation tensor—to the single-point energy balance equation for channel flow. However, it should also be noted that this form will not be identical with that previously given in eqn (1.68), as we have now eliminated the pressure as a dependent variable.

2.4 Homogeneous isotropic turbulence

In the previous section we saw how the statistical equations obtained by averaging the Navier–Stokes equation could be very much simplified by specializing them to the case of channel flow. In order to study the physics of turbulence we need to concentrate on the simplest non-trivial problems, and it has long been widely agreed that homogeneous isotropic turbulence presents the very simplest such problem. We shall begin by briefly discussing the two concepts in turn. It should be noted that we shall not need the time variables in this section, and accordingly we can make the algebra more compact by not showing them explicitly.

Homogeneity is really a contraction of 'spatial homogeneity' and indicates that mean properties do not vary with absolute position in a particular direction. Thus the channel flow considered previously was assumed to be homogeneous in the x_1 and x_3 directions. However, the unqualified term 'homogeneous' is normally applied to fields which are translationally invariant in all three mutually perpendicular coordinate directions, and this is the case which will concern us here.

The most important implication of this restriction is that (2.24) for the correlation can be written as a function of relative position only:

$$\langle u_\alpha(\mathbf{x})u_\beta(\mathbf{x}')\rangle = Q_{\alpha\beta}(\mathbf{x} - \mathbf{x}') = Q_{\alpha\beta}(\mathbf{r}), \qquad (2.47)$$

with analogous results for higher-order moments. Also, the correlation must be unaffected by the interchange of $\mathbf{x}$ and $\mathbf{x}'$, and hence will be a symmetric function of $\mathbf{r}$ with

$$Q_{\alpha\beta}(\mathbf{r}) = Q_{\alpha\beta}(-\mathbf{r}).\tag{2.48}$$

The additional restriction to isotropy implies independence of direction as well as independence of position in the fluid. Formally, this means that all velocity moments are invariant under rotations of the coordinate frame and under reflections in coordinate planes. The principal implication is the additional symmetry requirement

$$Q_{\alpha\beta}(\mathbf{r}) = Q_{\beta\alpha}(\mathbf{r}).\tag{2.49}$$

We can begin to explore the further implications of isotropy by first considering single-point forms. It can be shown (see Appendix C) that all off-diagonal elements of the single-point correlation tensor vanish, that is,

$$\langle u_1 u_2 \rangle = \langle u_2 u_3 \rangle = \langle u_3 u_1 \rangle = 0,\tag{2.50}$$

and that the diagonal elements are all equal, that is,

$$\langle u_1^2 \rangle = \langle u_2^2 \rangle = \langle u_3^2 \rangle = \langle u^2 \rangle,\tag{2.51}$$

where $\langle u^2 \rangle$ is the mean square of the fluctuating velocity in any arbitrary direction. It follows from eqn (1.22) that the single-point isotropic correlation tensor can be written as

$$Q_{\alpha\beta}(0) = \frac{2E}{3}\delta_{\alpha\beta},\tag{2.52}$$

where E is the kinetic energy of turbulent fluctuations per unit mass of fluid and $\delta_{\alpha\beta}$ is the Kronecker delta symbol.

Thus the single-point correlation tensor has been expressed in terms of only one scalar constant. As we shall see shortly, the two-point correlation tensor can also be reduced to a single scalar, which is in this case not a constant but a function of the distance between the points $\mathbf{x}$ and $\mathbf{x}'$.

Now let us consider the question of how realizable is isotropic turbulence? Referring back to the results for pipe flow as a typical example, we see from Fig. 1.5 that the three r.m.s. velocity components are quite different from each other, and hence (2.51) is not satisfied. Likewise, Fig. 1.6 shows that $\langle u_1 u_2 \rangle$ is not zero, except at the axis of symmetry, and hence (2.50) is not satisfied either. Clearly, pipe flow is in fact highly anisotropic, and this will generally be true of most flows as the presence of rigid boundaries and externally imposed pressure gradients inevitably implies a preferred direction.

All of this would tend to suggest that we should look for isotropic turbulence in flows that are of large physical extent where appreciable regions can be found that are remote from the boundaries. Obviously examples are the large-scale geophysical flows such as occur in the atmosphere or the ocean.

Alternatively, in laboratory-scale shear flows, we might take the opposite approach and concentrate on a range of very small eddy sizes in the hope that they would be relatively unaffected by the rigid boundaries. This could be achieved by keeping the distances between measuring points small compared with length scales over which there was appreciable inhomogeneity; this concept is known as 'local isotropy'.

As we shall see later, when we consider some experimental results, both approaches can lead to quite good approximations to isotropic turbulence. However, the development of the subject has been greatly helped by the contrivance of a rather artificial kind of turbulence. This is grid-generated turbulence. For brevity, we normally just refer to it as 'grid turbulence'. It can be produced in the laboratory in the following way.

Suppose that the air flowing through a wind tunnel is made to pass through a mesh or grid. The physical situation we should envisage is one where the boundary layers on the tunnel walls are thin, so that much of the flow constitutes a potential core (in other words, the flow in a wind tunnel corresponds to the entrance region in a pipe flow). Under these circumstances, a vortex street is generated by each bar (or rod) which makes up the grid and, provided that all the grid parameters have been sensibly chosen, the many vortex streets coalesce downstream to form a turbulent field. Experiments have shown that such fields are approximately isotropic (see Goldstein 1938, pp. 228–9).

Unfortunately grid turbulence cannot be fully homogeneous as it decays in the direction of flow. However, by changing to a coordinate system which is moving with the free-stream velocity U_∞, we can make it mathematically equivalent to the problem of homogeneous isotropic turbulence which is freely decaying with time. If the flow is taken to be in the x_1 direction, then the correspondence is made through the relation

$$t = x_1/U_\infty \qquad (2.53)$$

where t is the decay time.

In practice it is found that the early stages of the decay can depend quite strongly on the design of the grid which is producing the turbulence. This is not very surprising, but we would expect that, sufficiently far downstream of the grid, the turbulence would become independent of the way in which it was formed and assume a universal character controlled only by the equation of motion. As we shall see, experimental results suggest that this is indeed the case.

2.4.1 Isotropic form of the two-point correlation tensor

Returning to the two-point correlation, we have already said that we can reduce the nine scalar functions, which (in principle) are needed to describe this, to only one. The general method of doing this was invented by Robertson (1940) and is based on the idea that an isotropic tensor can be expressed in terms of the invariants of the rotation group. A full discussion can be found in the original paper by Robertson (1940) or in the books by Batchelor (1971) and Hinze (1975). Here we shall just give some of the more interesting points.

Robertson showed that the isotropic two-point correlation could be written as

$$Q_{\alpha\beta}(\mathbf{r}) = A(r)r_\alpha r_\beta + B(r)\delta_{\alpha\beta} \tag{2.54}$$

where A and B are even functions of $r = |\mathbf{r}|$. We should note that this result reduces to the appropriate single-point form, as given by (2.52), when we set $r = 0$.

Although (2.45) satisfies all the requisite symmetries, we can further invoke the continuity condition in order to relate A to B. First, however, it is usual to express the coefficients A and B in terms of the transverse and longitudinal correlation coefficients. These can be defined by reference to Fig. 2.2. Once we introduce two measuring points, we also introduce a preferred direction, that is, the direction of the vector joining the two points. Thus although the three possible correlation coefficients are all the same when referred to the Cartesian coordinate system, two different types are possible when the coordinate system is based on the configuration of the measuring points. That is, the correlation can be based on the longitudinal velocity component u_L parallel to r or on the transverse component u_T which is normal to r, with two (identical) correlations possible in the latter case.

The longitudinal and transverse correlation coefficients, f and g respectively, can be introduced through the relationships

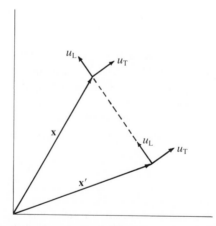

Fig. 2.2. Definition of the longitudinal (L) and transverse (T) components of the fluctuating velocity in an isotropic turbulent field.

$$\langle u^2 \rangle f(r) = \langle u_L(\mathbf{x}) u_L(\mathbf{x}') \rangle \tag{2.55}$$

and

$$\langle u^2 \rangle g(r) = \langle u_T(\mathbf{x}) u_T(\mathbf{x}') \rangle, \tag{2.56}$$

where f and g are even differentiable functions of r and satisfy $f(0) = g(0) = 1$ as well as $f'(0) = g'(0) = 0$. We should also recall that $\mathbf{x} = \mathbf{x}' + \mathbf{r}$.

We can relate f and g to the coefficients A and B in (2.54) by considering the special case where we take $\mathbf{r}$ to lie along the x_1 axis. That is, we put $\mathbf{x} = 0$ and $\mathbf{r} = (r_1, 0, 0)$. Then, putting $\alpha = \beta = 1$ in (2.54), we obtain

$$Q_{11}(r) = Ar^2 + B = \langle u^2 \rangle f \tag{2.57}$$

where the last step follows from (2.55). Similarly, we can put $\alpha = \beta = 2$ in (2.54) to obtain

$$Q_{22}(r) = B = \langle u^2 \rangle g. \tag{2.58}$$

Evidently (2.58) gives us B in terms of g, and if we substitute this result into (2.57) we readily find A in terms of f and g. Then eqn (2.54) can be written as

$$Q_{\alpha\beta}(\mathbf{r}) = \langle u^2 \rangle (f - g) \frac{r_\alpha r_\beta}{r^2} + \langle u^2 \rangle g \delta_{\alpha\beta}. \tag{2.59}$$

Our last step is to eliminate g in terms of f. We do that by differentiating (2.59) with respect to r_α and invoking the continuity relation in the form of eqn (2.27). With some straightforward algebra and a little rearrangement we readily find that the two-point isotropic correlation can be expressed in terms of a single scalar function f as

$$Q_{\alpha\beta}(\mathbf{r}) = \langle u^2 \rangle f \delta_{\alpha\beta} + \frac{1}{2} \langle u^2 \rangle r f' \left(\delta_{\alpha\beta} - \frac{r_\alpha r_\beta}{r^2} \right) \tag{2.60}$$

where the prime denotes differentiation with respect to r. For the particular case $r = 0$, we note that

$$\sum_\alpha Q_{\alpha\alpha}(0) = \langle u^2 \rangle f(0) \mathrm{tr}\, \delta_{\alpha\beta} = 3 \langle u^2 \rangle$$

$$= 2E, \tag{2.61}$$

where tr stands for 'trace', that is, the sum of the diagonal elements of a matrix.

2.4.2 *Length scales for isotropic turbulence*

Two important length scales can be formed from the longitudinal correlation coefficient $f(r)$. The first of these is the Taylor microscale λ. This is a differential length scale and is obtained in the following way. Suppose that we expand $f(r)$ about $r = 0$. Then, recalling that f must be a symmetric function of r

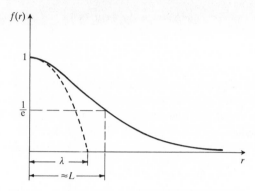

Fig. 2.3. Sketch of the longitudinal correlation coefficient (note that the integral scale is based on the erroneous assumption that the correlation coefficient can be approximated by an exponential function, and is intended to be illustrative only).

and hence $f'(0) = 0$, we can write

$$f(r) = 1 - \frac{r^2}{2\lambda^2} + O(r^4) \tag{2.62}$$

where

$$\frac{1}{\lambda^2} = -f''(0). \tag{2.63}$$

That is, we define the microscale by fitting a parabola to the correlation function for small values of r. This is shown schematically in Fig. 2.3.

The physical importance of the Taylor microscale is indicated by the relationship connecting it to the rate of dissipation of energy in isotropic turbulence, which is (e.g. Batchelor 1971)

$$\varepsilon = \frac{15\nu\langle u^2 \rangle}{\lambda^2}. \tag{2.64}$$

We shall discuss this further in Section 2.8.

The longitudinal integral scale L is just another particular case of the general integral length scale as defined by eqn (2.40). It is given by

$$L = \int_0^\infty f(r)\,dr. \tag{2.65}$$

We can illustrate the physical meaning of L (although not altogether rigorously) by fitting an exponential form to the correlation function. (In practice this can be quite a good approximation, although clearly it is invalid at $r = 0$, where we require $f' = 0$, from symmetry.)

Purely for this purpose, therefore, let us take $f = \exp(-br)$, where b is a parameter with dimensions of inverse length. Substituting this form into (2.65), we readily find that $L = 1/b$. Or, in other words, if the correlation were exponential in form then the integral length scale would be the distance over which the correlation coefficient declined from $f(0) = 1$ to $f = 1/e$. Again, we show this schematically in Fig. 2.3.

2.5 Stationary turbulence

Turbulence is intrinsically a time-dependent phenomenon. Nevertheless we can distinguish between those situations where mean velocities are dependent on time and those where they are not. For instance, an everyday example is when we use a hose pipe connected to a water tap. Now suppose that we turn the tap up or down. While we are doing this, the mean velocity of the water through the pipe will vary with time. But if we allow external factors (such as tap setting, nozzle adjustment, or ambient conditions) to remain constant, then normally the mean velocity of the water will also remain constant, a condition we referred to as 'steady mean flow' in Chapter 1.

The extension of this idea from single-time means to many-time moments brings us to the concept of stationarity. Formally, we say that $u_\alpha(\mathbf{x}, t)$ is a stationary random variable if the associated many-time probability distribution depends only on the differences between measuring times and not on their absolute values.

As an example let us consider the two-time correlation. Temporarily suppressing space variables and tensor indices as irrelevant to the work of this section, we have from eqns (2.24) and (2.33)

$$\langle u(t)u(t')\rangle = Q(\tau, T) \tag{2.66}$$

where τ and T are the relative (or difference) and absolute times respectively. Then if $u(t)$ is a stationary random function, (2.66) becomes

$$\langle u(t)u(t')\rangle = Q(\tau) = Q(t - t') \tag{2.67}$$

such that

$$Q(t - t') = Q(t' - t). \tag{2.68}$$

Thus stationarity is, in effect, homogeneity in time.

We have already defined an integal time-scale in eqn (2.39). For stationary fields we can also define a microscale in the same way as the Taylor microscale was introduced for homogeneous turbulence in the previous section. If $R_E(\tau)$ is any Eulerian correlation function (e.g. $f(r, t)$ in isotropic turbulence, R_{11} in channel flow, etc), then for small values of $\tau = t - t'$, we can write

$$R_E(\tau) = 1 - \frac{\tau^2}{2\tau_E^2} + O(\tau^4) \tag{2.69}$$

where the microscale τ_E is given by

$$\frac{1}{\tau_E^2} = -R_E''(0) \tag{2.70}$$

and the primes denote differentiation with respect to τ.

In order to carry out time averages, we require the turbulence to be stationary, or at least approximately so, on time-scales which are long compared with, say, the integral time-scale. It is often argued that small eddies may evolve so rapidly, compared with large eddies, that they can be regarded as being in a quasi-stationary state, rather like the idea of local isotropy which we discussed previously. In this context reference is often made to an 'equilibrium range' of eddy sizes. However, it seems to the present writer that this usage invites confusion with the idea of thermal equilibrium. Therefore we shall only use the word equilibrium in the latter sense and not as a synonym for 'steady state'.

2.6 Fourier analysis of the turbulent velocity field

The introduction of Fourier analysis leads to three main benefits. It converts differential operators into multipliers, it gives us a relatively simple picture of the physics of turbulence, and it allows us to define the degrees of freedom of the turbulent system. This last aspect will turn out to be particularly important both for discussion of turbulence from the point of view of statistical mechanics and for our consideration of the problems of the numerical solution of the equations of motion.

We begin by considering the turbulent fluid to be occupying a cubic box of side L. The velocity field (or any of the other dynamical variables) can be expanded in a Fourier series as follows:

$$u_\alpha(\mathbf{x}, t) = \sum_{\mathbf{k}} u_\alpha(\mathbf{k}, t) \exp\{i\mathbf{k} \cdot \mathbf{x}\} \tag{2.71}$$

where the wavevector $\mathbf{k}$ is given by

$$\mathbf{k} = \frac{2\pi}{L} \{n_1, n_2, n_3\} \tag{2.72}$$

and n_1, n_2, and n_3 are integers, each of which is summed over the range from $-\infty$ to $+\infty$.

It should be noted that we use the same notation for fields, irrespective of whether we are referring to the actual velocity or its Fourier component. This, although contrary to the normal usage to be found in mathematical textbooks, is quite usual in this sort of work. In practice there should be no confusion, and indeed we shall find it helpful to think of $u_\alpha(\mathbf{k}, t)$ as the velocity field in wavenumber space.

For completeness, we present a brief summary of Fourier methods in Appendix D, with emphasis being given to those results which are most relevant to the theory of turbulence. However, we shall not consider the more profound questions concerning the validity of the Fourier representation of stationary random functions, and the interested reader who wishes to pursue such questions should refer to the books by Batchelor (1971) or Monin and Yaglom (1975).

Nevertheless, one particular point should perhaps be mentioned here. In order to have rigorous isotropy, we must allow the box size to go to infinity. This means that Fourier sums become Fourier integrals in the usual way. However, the random nature of the velocity field causes difficulties for the implementation of the limiting process inherent in the transition from a summation to Riemannian integration.

In our view, the simplest way of dealing with this is to defer taking the infinite system limit until after we have taken averages. Then the problem simply does not arise. This also seems quite an appropriate procedure for the additional reason that homogeneity (a necessary precondition for isotropy) is purely a statistical property of the velocity field.

2.6.1 *The solenoidal Navier–Stokes equations*

We begin by transforming the continuity equation, as this provides us with a relatively easy start. As we are restricting our attention to isotropic fields with zero mean, eqn (1.7) now becomes

$$U_\alpha(\mathbf{x}, t) = u_\alpha(\mathbf{x}, t).$$

Then substituting (2.71) into (1.1) and carrying out the differentiation, we obtain

$$\sum_{\mathbf{k}} (ik_\beta)u_\beta(\mathbf{k}, t)\exp\{i\mathbf{k} \cdot \mathbf{x}\} = 0 \tag{2.73}$$

which must hold for arbitrary $\exp\{i\mathbf{k} \cdot \mathbf{x}\}$. Thus the continuity relation becomes

$$k_\beta u_\beta(\mathbf{k}, t) = 0. \tag{2.74}$$

Or, $\mathbf{u}(\mathbf{k})$ and $\mathbf{k}$ are mutually orthogonal.

The general solenoidal equation of motion is given by (2.15). We wish to apply this to a system where the boundary is at infinity and there are no mean velocities or externally applied pressure gradients. If we then invoke the boundary condition that $u_\alpha(\mathbf{x}, t)$ goes to zero as $\mathbf{x}$ goes to infinity, eqn (2.15) reduces to

$$\left(\frac{\partial}{\partial t} - v\nabla^2\right)u_\alpha(\mathbf{x}, t) = M_{\alpha\beta\gamma}(\nabla)[u_\beta(\mathbf{x}, t)u_\gamma(\mathbf{x}, t)]. \tag{2.75}$$

The technicalities of Fourier transforming this equation are discussed in Appendix D. However we can do it straightaway if we bear in mind just two points:

(a) as we saw in the process of deriving (2.74), when we substitute (2.71) into (2.75) the net effect will be to replace each differential operator by its analogous wavenumber expression;

(b) the non-linear term involves the product of two velocities in **x**-space and hence, by the Convolution Theorem, will give rise to the convolution of their respective Fourier transformer in **k**-space.

Thus, by inspection, we can write down the result obtained by substituting (2.71) into (2.75) and carrying out the appropriate differentiations:

$$\left(\frac{\partial}{\partial t} + vk^2\right)u_\alpha(\mathbf{k}, t) = M_{\alpha\beta\gamma}(\mathbf{k})\sum_{\mathbf{j}} u_\beta(\mathbf{j}, t)u_\gamma(\mathbf{k} - \mathbf{j}, t), \qquad (2.76)$$

where

$$M_{\alpha\beta\gamma}(\mathbf{k}) = (2i)^{-1}\{k_\beta D_{\alpha\gamma}(\mathbf{k}) + k_\gamma D_{\alpha\beta}(\mathbf{k})\} \qquad (2.77)$$

and

$$D_{\alpha\beta}(\mathbf{k}) = \delta_{\alpha\beta} - \frac{k_\alpha k_\beta}{|\mathbf{k}|^2}. \qquad (2.78)$$

Clearly, $M_{\alpha\beta\gamma}(\mathbf{k})$ and $D_{\alpha\beta}(\mathbf{k})$ are just the Fourier transforms of $M_{\alpha\beta\gamma}(\mathbf{V})$ and $D_{\alpha\beta}(\mathbf{V})$, as defined by eqns (2.13) and (2.11) respectively.

It is easy to verify that solutions of (2.76) also satisfy the continuity condition as given by (2.74). If we multiply each side by k_α, the vanishing of the l.h.s. follows trivially from (2.74). The r.h.s. of (2.76) also vanishes, and this follows from an important property of $D_{\alpha\beta}(\mathbf{k})$, namely that

$$k_\alpha D_{\alpha\beta}(\mathbf{k}) = 0. \qquad (2.79)$$

2.6.2 *Homogeneity and the velocity-field moments*

In order to develop a formalism based on eqn (2.76), we need to know something about the general properties of the moment hierarchy in wavenumber space. We begin by considering the implications of homogeneity, meaning that moments are translationally invariant in configuration (**x**) space.

Temporarily dropping the time arguments, we can write the expression for the Fourier coefficients in (2.71) as

$$u_\alpha(\mathbf{k}) = \left(\frac{1}{L}\right)^3 \int d^3x\, u_\alpha(\mathbf{x})\exp\{-i\mathbf{k}\cdot\mathbf{x}\}, \qquad (2.80)$$

from which it follows that the two-point correlation in **k**-space can be related to the corresponding form in x-space by

$$\langle u_\alpha(\mathbf{k})u_\beta(\mathbf{k}')\rangle = \left(\frac{1}{L}\right)^6 \int\int d^3x \, d^3x' \langle u_\alpha(\mathbf{x})u_\beta(\mathbf{x}')\rangle \times$$

$$\times \exp\{-i\mathbf{k}\cdot\mathbf{x} - i\mathbf{k}'\cdot\mathbf{x}'\}$$

$$= \left(\frac{1}{L}\right)^6 \int\int d^3x \, d^3r \langle u_\alpha(\mathbf{x})u_\beta(\mathbf{x}-\mathbf{r})\rangle \times$$

$$\times \exp\{-i(\mathbf{k}+\mathbf{k}')\cdot\mathbf{x}\}\exp\{i\mathbf{k}'\cdot\mathbf{r}\}. \qquad (2.81)$$

We now invoke the property of invariance under translation. That is,

$$\langle u_\alpha(\mathbf{x})u_\rho(\mathbf{x}-\mathbf{r})\rangle = \langle u_\alpha(0)u_\beta(\mathbf{r})\rangle,$$

and (2.81) becomes

$$\langle u_\alpha(\mathbf{k})u_\beta(\mathbf{k}')\rangle = \left(\frac{1}{L}\right)^6 \int\int d^3x \, d^3r \langle u_\alpha(0)u_\beta(\mathbf{r})\rangle \times$$

$$\times \exp\{-i(\mathbf{k}+\mathbf{k}')\cdot\mathbf{x}\}\exp\{i\mathbf{k}'\cdot\mathbf{r}\}. \qquad (2.82)$$

At this stage we can perform the integration with respect to x, noting (e.g. see Appendix D) that

$$\left(\frac{1}{L}\right)^3 \int d^3x \exp\{-i(\mathbf{k}+\mathbf{k}')\cdot\mathbf{x}\} = \delta_{\mathbf{k}+\mathbf{k}',0},$$

so that the correlation in wavenumber space takes the form

$$\langle u_\alpha(\mathbf{k})u_\beta(\mathbf{k}')\rangle = \delta_{\mathbf{k}+\mathbf{k}',0}\left(\frac{1}{L}\right)^3 \int d^3r \, Q_{\alpha\beta}(\mathbf{r})\exp\{i\mathbf{k}\cdot\mathbf{r}\}. \qquad (2.83)$$

Thus, in general terms, if we correlate velocity fields corresponding to two different modes $\mathbf{k}$ and $\mathbf{k}'$, we only obtain a non-vanishing contribution when $\mathbf{k}+\mathbf{k}'=0$.

Similarly, it can be shown for the third-order moment that

$$\langle u_\alpha(\mathbf{k})u_\beta(\mathbf{j})u_\alpha(\mathbf{l})\rangle = 0, \quad \text{unless } \mathbf{k}+\mathbf{j}+\mathbf{l}=0, \qquad (2.84)$$

and, in general, the homogeneity property for a moment of any order can be written as

$$\langle u_\alpha(\mathbf{k})u_\beta(\mathbf{j})u_\gamma(\mathbf{l})\ldots u_\sigma(\mathbf{p})\rangle = 0, \quad \text{unless } \mathbf{k}+\mathbf{j}+\mathbf{l}+\cdots+\mathbf{p}=0. \qquad (2.85)$$

That is, the n-order homogeneous moment, involving the set of n distinct wavevectors $\{\mathbf{k},\mathbf{j},\mathbf{l},\ldots,\mathbf{p}\}$, is zero unless all n wavevectors sum to zero.

It should be noted that all these results apply to denumerable $\mathbf{k}, \mathbf{j}, \mathbf{l}, \ldots, \mathbf{p}$. In the next section, we shall consider the transition to continuous wavevector variables.

2.6.3   Limit of infinite system volume

The Fourier decomposition of the velocity field into harmonic components is given by eqn (2.71), with (2.80) as the inverse relationship which specifies the Fourier coefficients. It should be noted that the integral in the latter equation is over the volume L^3. What we now wish to obtain is the corresponding pair of relationships connecting the two-point correlations, as a preliminary to taking the limit as $L \to \infty$.

In the previous section we found that the fact that the two-point correlation $\langle u_\alpha(\mathbf{x})u_\beta(\mathbf{x}') \rangle$ depended on only one space variable was enough to ensure that $\langle u_\alpha(\mathbf{k})u_\beta(\mathbf{k}') \rangle$ in turn depended on only one wavenumber variable. Thus, by direct substitution from (2.71) and then averaging, it is readily shown that the analogue of (2.71) for the correlation is

$$Q_{\alpha\beta}(\mathbf{r}) = \sum_{\mathbf{k}} \langle u_\alpha(\mathbf{k})u_\beta(-\mathbf{k}) \rangle \exp\{i\mathbf{k}\cdot\mathbf{r}\}, \qquad (2.86)$$

where we should bear in mind that $\mathbf{r} = \mathbf{x} - \mathbf{x}'$. The equivalent result for the Fourier coefficients can be obtained from (2.83) by setting $\mathbf{k} = -\mathbf{k}'$:

$$\langle u_\alpha(\mathbf{k})u_\beta(-\mathbf{k}) \rangle = \left(\frac{1}{L}\right)^3 \int d^3r\, Q_{\alpha\beta}(\mathbf{r})\exp\{-i\mathbf{k}\cdot\mathbf{r}\}. \qquad (2.87)$$

It should be noted that (2.87) is analogous to eqn (2.80) for the Fourier coefficients in the expansion of the velocity field.

Formally, we now make the transition from (2.86) and (2.87) to the Fourier integral representation in the usual way (e.g. see Appendix D) by defining the (continuous) spectral density tensor $Q_{\alpha\beta}(\mathbf{k})$ as follows:

$$Q_{\alpha\beta}(\mathbf{k}) = \left(\frac{L}{2\pi}\right)^3 \langle u_\alpha(\mathbf{k})u_\beta(-\mathbf{k}) \rangle. \qquad (2.88)$$

Then, taking the limit as $L \to \infty$, we can introduce the Fourier transform pair $Q_{\alpha\beta}(\mathbf{r})$ and $Q_{\alpha\beta}(\mathbf{k})$ through the relationships

$$Q_{\alpha\beta}(\mathbf{r}) = \int d^3k\, Q_{\alpha\beta}(\mathbf{k})\exp\{i\mathbf{k}\cdot\mathbf{r}\} \qquad (2.89)$$

and

$$Q_{\alpha\beta}(\mathbf{k}) = \left(\frac{1}{2\pi}\right)^3 \int d^3r\, Q_{\alpha\beta}(\mathbf{r})\exp\{-i\mathbf{k}\cdot\mathbf{r}\}. \qquad (2.90)$$

This procedure can be extended to all higher-order moments. Choosing the constituent wavevectors to be such that they satisfy the general homogeneity property (2.85), we can write down the higher-order analogues of (2.88) as

$$Q_{\alpha\beta\gamma}(\mathbf{k},\mathbf{j}) = \left(\frac{L}{2\pi}\right)^6 \langle u_\alpha(\mathbf{k})u_\beta(\mathbf{j})u_\gamma(-\mathbf{k}-\mathbf{j}) \rangle \qquad (2.91)$$

and

$$Q_{\alpha\beta\gamma\delta}(\mathbf{k},\mathbf{j},\mathbf{l}) = \left(\frac{L}{2\pi}\right)^9 \langle u_\alpha(\mathbf{k})u_\beta(\mathbf{j})u_\gamma(\mathbf{l})u_\delta(-\mathbf{k}-\mathbf{j}-\mathbf{l})\rangle. \tag{2.92}$$

Clearly, the moment of any order can then be found inductively.

Now that we have made the transition to the infinite system, sums over wavevectors can be replaced by integrals according to

$$\lim_{L\to\infty} \left(\frac{2\pi}{L}\right)^3 \sum_{\mathbf{k}} = \int d^3k.$$

Lastly, a generalization of eqn (2.88) to the case of distinct wavevectors $\mathbf{k}$ and $\mathbf{k}'$ can be obtained by combining eqns (2.83) and (2.90), with the results

$$\langle u_\alpha(\mathbf{k},t)u_\beta(\mathbf{k}',t')\rangle = \left(\frac{2\pi}{L}\right)^3 \delta_{\mathbf{k}+\mathbf{k}',0} Q_{\alpha\beta}(\mathbf{k};t,t') \tag{2.93a}$$

where, for complete generality, we have restored the time arguments. Then, if we take the limit $L \to \infty$, it follows that

$$\langle u_\alpha(\mathbf{k},t)u_\beta(\mathbf{k}',t')\rangle = \delta(\mathbf{k}+\mathbf{k}')Q_{\alpha\beta}(\mathbf{k};t,t'), \tag{2.93b}$$

which is a form that will prove useful later on.

2.6.4   *The isotropic spectrum tensor*

In Section 2.4.1 we gave a short account of the application of Robertson's (1940) theory of invariants to the isotropic correlation tensor in configuration space. The same methods can be applied to the isotropic spectrum tensor (e.g. see Batchelor 1971), and indeed are much easier to use in wavenumber space.

Employing the same arguments as before, we obtain the analogous result to equation (2.54):

$$Q_{\alpha\beta}(\mathbf{k}) = B(k)\delta_{\alpha\beta} + A(k)k_\alpha k_\beta, \tag{2.94}$$

which satisfies all the appropriate symmetry requirements. The scalars A and B are even functions of $\mathbf{k}$, but otherwise arbitrary.

Again, we invoke the continuity condition in order to eliminate one of the two scalars. Multiplying both sides of (2.94) by k_β and summing repeated indices as usual, it follows from (2.74) that

$$k_\beta Q_{\alpha\beta}(\mathbf{k}) = 0 = B(k)k_\alpha + A(k)k^2 k_\alpha.$$

As this must hold for arbitrary k_α, the required relationship is

$$B(k) = -k^2 A(k). \tag{2.95}$$

In order to conform with later notation, we shall rename B as q, that is,

$$q(k) = B(k) = -k^2 A(k), \tag{2.96}$$

and, substituting from (2.96) for A and B, we obtain (2.94) in the form

$$Q_{\alpha\beta}(\mathbf{k}) = q(k)\delta_{\alpha\beta} - q(k)k_\alpha\frac{k_\beta}{k^2} = D_{\alpha\beta}(\mathbf{k})q(k) \qquad (2.97a)$$

where $D_{\alpha\beta}(\mathbf{k})$ is given by eqn (2.78). Or, restoring the time arguments, we can make the generalization

$$Q_{\alpha\beta}(\mathbf{k}; t, t') = D_{\alpha\beta}(\mathbf{k})Q(\mathbf{k}; t, t'). \qquad (2.97b)$$

Then we make connection with the time-independent forms by writing

$$Q(\mathbf{k}; t, t) = Q(k; t)$$

$$Q(k; 0) = q(k).$$

It follows from (2.79) that the above form of the isotropic spectrum tensor satisfies the continuity condition $k_\beta Q_{\alpha\beta}(\mathbf{k}) = 0$ for arbitrary $q(k)$. This procedure can be extended to higher moments, but, as our interest will be in closures in terms of the second-order moment, we shall not pursue this here. Purely for completeness we should mention that Orszag (1969) has considered the general problem of representing the isotropic moment of any order by scalar functions.

Now let us consider the physical interpretation of $q(k)$, and hence justify our reference to $Q_{\alpha\beta}(\mathbf{k})$ as the spectral tensor.

First we take the trace of $Q_{\alpha\beta}(\mathbf{k})$, as given by (2.97). That is,

$$\operatorname{tr} Q_{\alpha\beta}(\mathbf{k}) = \operatorname{tr} D_{\alpha\beta}(\mathbf{k})q(k) = 2q(k). \qquad (2.98)$$

Now we can also relate $\operatorname{tr} Q_{\alpha\beta}(\mathbf{k})$ to the energy E per unit mass of fluid, as follows. From (2.61) we have

$$2E = 3\langle u^2 \rangle = \operatorname{tr} Q_{\alpha\beta}(\mathbf{r})|_{r=0} = \operatorname{tr} \int d^3k \, Q_{\alpha\beta}(\mathbf{k})$$

$$= \operatorname{tr} \int_0^\infty q(k)k^2 \, dk \int D_{\alpha\beta}(\mathbf{k}) \, d\Omega_\mathbf{k}, \qquad (2.99)$$

where the second step follows from (2.89) with $r = 0$, the third step relies on eqn (2.97), and $d\Omega_\mathbf{k}$ stands for the elementary solid angle in wavenumber space.

The angular integration in (2.99) is readily carried out (e.g. see Leslie 1973, p. 20), with the result that

$$E = \frac{4\pi}{3} \operatorname{tr} \delta_{\alpha\beta} \int_0^\infty q(k)k^2 \, dk$$

$$= \int_0^\infty 4\pi k^2 q(k) \, dk$$

$$= \int_0^\infty E(k) \, dk, \qquad (2.100)$$

where $E(k)$ is the contribution to the total energy from harmonic components with wavevectors lying within the spherical shell between k and $k + dk$, and is given by

$$E(k) = 4\pi k^2 q(k). \tag{2.101}$$

Usually $E(k)$ is referred to as the 'wavenumber spectrum'. More formally, it is the distribution of energy with wavenumber (or angular spatial frequency) and, in view of eqn (2.101), we should interpret $q(k)$ as the density of contributions in wavenumber space to the total energy. Therefore we shall refer to $q(k)$ as the spectral density.

2.6.5 The Taylor hypothesis and the one-dimensional spectrum

We have been discussing the three-dimensional energy spectrum, which is one of the key concepts in turbulence theory. However, in practice experimentalists find it most convenient to measure the spectrum of the single fluctuating velocity in the streamwise direction. The reasons for this are discussed in Chapter 3, along with the subject of anemometry. In the meantime, we have to face the fact that the vast majority of measured spectra are of this type, and so we must consider the problem of relating the measured spectrum to the theoretical one.

As usual we take the flow to be in the x_1 direction and imagine an anemometer to be positioned at a fixed point and measuring the fluctuating velocity u_1. If the anemometer signal is passed through a spectrum analyser, then the velocity fluctuations can be decomposed into their harmonic components with respect to the (angular) frequency ω. Then, if the output of the spectrum analyser is squared and averaged, the resulting frequency spectrum $E_{11}(\omega)$ must have the property

$$\langle u_1^2 \rangle = \int_0^\infty E_{11}(\omega) \, d\omega. \tag{2.102}$$

The physical significance of this spectrum can be seen more clearly as follows. Let us assume that the signal representing u_1 is being fed to a filter of centre frequency ω and bandwidth $\Delta\omega$. After squaring and averaging, the filter output can be represented (in a compressed notation) by $\langle \Delta u_1^2(\omega) \rangle$. Then, if we vary ω over the entire range of frequencies, we can construct a histogram showing how $\langle \Delta u_1^2(\omega) \rangle$ is distributed among the various frequency bands.

A continuous spectrum can be obtained by taking the limit where the filter bandwidth shrinks to zero:

$$E_{11}(\omega) = \lim_{\Delta\omega \to 0} \frac{\langle \Delta u_1^2(\omega) \rangle}{\Delta\omega}. \tag{2.103}$$

Clearly, this form satisfies the condition specified by (2.102).

The possibility of a relationship between frequency and wavenumber spectra rests entirely on the validity of the hypothesis of 'frozen convection' (Taylor 1938a). Taylor argued that the changes in u_1 with time at the fixed measuring point could be assumed to be due to the passage of a frozen pattern of turbulent motion past the point, provided that the mean (or free-stream) velocity carrying the turbulent eddies was much larger than the turbulent velocity fluctuations. That is to say, the velocity field at different instants could be related by the transformation

$$u(\mathbf{x}, t) = u(\mathbf{x} - \bar{U}_1 t, 0),$$

and hence the local time derivative at a point could be replaced by the convective derivative

$$\frac{\partial}{\partial t} = -\bar{U}_1 \frac{\partial}{\partial x_1} \qquad \text{if } u' \ll \bar{U}_1. \tag{2.104}$$

The Taylor hypothesis is widely regarded as having been validated in practice, and is extensively used in experimental work on turbulence. However, at best it is still an approximation and hence, as experimental techniques improve, it is still subject to critical scrutiny; see for example the recent papers by Zaman and Hussain (1981) and Brown, Antonia, and Rajagopalan (1983).

In the context of our present interest, eqn (2.104) is equivalent to

$$k_1 = \omega / \bar{U}_1, \tag{2.105}$$

or for grid turbulence we could use U_∞ instead of $\bar{U}_1$. Correspondingly, we can define

$$E_{11}(k_1) = \bar{U}_1 E_{11}(\omega), \tag{2.106}$$

and it follows from (2.105), (2.106), and (2.102) that

$$\int_0^\infty E_{11}(\omega)\, d\omega = \int_0^\infty \frac{E_{11}(k_1)\, d\omega}{\bar{U}_1} = \int_0^\infty E_{11}(k_1)\, dk_1 = \langle u_1^2 \rangle. \tag{2.107}$$

Thus the frequency spectrum can be related to one component of the spectrum tensor in wavenumber space, albeit as a function of one scalar variable k_1 only. The extension to three dimensions is straightforward: we simply integrate out the dependence of $Q_{11}(\mathbf{k})$ on k_2 and k_3 to find

$$E_{11}(k_1) = \int_{-\infty}^\infty \int_{-\infty}^\infty Q_{11}(k_1, k_2, k_3)\, dk_2\, dk_3$$

$$= \frac{1}{2\pi} \int_{-\infty}^\infty \langle u_1^2 \rangle f(r_1) \cos(k_1 r_1)\, dr_1, \tag{2.108}$$

where the second line follows from (2.90) and (2.57), and the cosine Fourier transform is employed because f is an even function of r_1.

From (2.108) it can be shown further that $E_{11}(k_1)$ is related to the three-dimensional spectrum for isotropic turbulence by

$$E_{11}(k_1) = \int_{k_1}^{\infty} \left\{ \left(1 - \frac{k_1^2}{k^2}\right) \frac{E(k)}{k} \right\} dk, \tag{2.109}$$

with the inverse relation

$$E(k) = \frac{k^3 \, d[k^{-1} \, dE_{11}(k)/dk]}{dk}. \tag{2.110}$$

Derivations of these results can be found in the book by Batchelor (1971) or, in more algebraic detail, in that by Hinze (1975, pp. 208–9).

2.6.6 Statistical equations in wavenumber space: many-time moments

We shall complete the work of this section by using eqn (2.76) as a basis for the formulation of the closure problem for isotropic turbulence in wave-number space. We begin by considering an equation for the mean velocity.

If we average both sides of (2.76), the result is

$$\left(\frac{\partial}{\partial t} + vk^2 \right) \langle u_\alpha(\mathbf{k}, t) \rangle = M_{\alpha\beta\gamma}(\mathbf{k}) \sum_{\mathbf{j}} \langle u_\beta(\mathbf{j}, t) u_\gamma(\mathbf{k} - \mathbf{j}, t) \rangle.$$

Now, from (2.83) we see that

$$\langle u_\beta(\mathbf{j}, t) u_\gamma(\mathbf{k} - \mathbf{j}, t) \rangle = 0 \qquad \text{unless } \mathbf{j} + \mathbf{k} - \mathbf{j} = 0,$$

and hence the r.h.s. of the equation for $\langle u_\alpha(\mathbf{k}, t) \rangle$ is zero as $M_{\alpha\beta\gamma}(0) = 0$. This result is, of course, consistent with our earlier claim that we could only have homogeneous isotropic turbulence if the mean velocity was zero or, at worst, constant over space.

The equation for the two-time correlation can be formed by multiplying each term of (2.76) by $u_\sigma(-\mathbf{k}, t')$ and averaging:

$$\left(\frac{\partial}{\partial t} + vk^2 \right) \langle u_\alpha(\mathbf{k}, t) u_\sigma(-\mathbf{k}, t') \rangle$$

$$= M_{\alpha\beta\gamma}(\mathbf{k}) \sum_{\mathbf{j}} \langle u_\beta(\mathbf{j}, t) u_\gamma(\mathbf{k} - \mathbf{j}, t) u_\sigma(-\mathbf{k}, t') \rangle. \tag{2.111}$$

Essentially this is just eqn (2.30), specialized to homogeneous turbulence and then Fourier transformed. Further, we can express this in terms of the correlation tensors. Invoking (2.88) and (2.91), and taking the infinite system limit, we obtain

$$\left(\frac{\partial}{\partial t} + vk^2 \right) Q_{\alpha\sigma}(\mathbf{k}; t, t') = M_{\alpha\beta\gamma}(\mathbf{k}) \int d^3j \, Q_{\beta\gamma\sigma}(\mathbf{j}, \mathbf{k} - \mathbf{j}; t, t'). \tag{2.112}$$

Our overall goal is to obtain a closed form of eqn (2.112) solely in terms of $Q_{\alpha\sigma}(\mathbf{k}; t, t')$, and this is something which we shall consider in great detail later on. Therefore, although it is an easy matter to derive the next and higher-order equations in the hierarchy, we shall not pursue this here. However, we can complete this section by writing down the form of (2.112) which is appropriate to isotropic turbulence. Invoking eqn (2.97b) for the isotropic correlation tensor, setting $\sigma = \alpha$, and summing over α leads to

$$\left(\frac{\partial}{\partial t} + vk^2\right)Q(k; t, t') = P(k; t, t') \qquad (2.113a)$$

$$P(k; t, t') = \tfrac{1}{2}M_{\alpha\beta\gamma}(\mathbf{k})\int d^3j\, Q_{\beta\gamma\alpha}(\mathbf{j}, \mathbf{k} - \mathbf{j}; t, t'), \qquad (2.113b)$$

where we have used the property $\operatorname{tr} D_{\alpha\beta}(\mathbf{k}) = 2$.

2.6.7 Statistical equations in wavenumber space: single-time moments

When we consider single-time moments, the situation is rather different. To begin with, the derivation is slightly less straightforward than in the two-time case. Also, at several later stages in the book we shall need higher numbers of the single-point hierarchy in explicit form. Accordingly, at this stage we shall derive the equivalent of (2.112) and the next-highest equation in the sequence.

Once again our starting point is eqn (2.76) for $u_\alpha(\mathbf{k}, t)$. We multiply each term in this by $u_\sigma(-\mathbf{k}, t)$, but, before we average, we form a second equation from (2.76) for $u_\sigma(-\mathbf{k}, t)$, multiply through by $u_\alpha(\mathbf{k}, t)$, and add the two resulting equations together. The important step is that we can now write

$$\left\langle u_\alpha(\mathbf{k}, t)\frac{\partial}{\partial t}u_\sigma(-\mathbf{k}, t)\right\rangle + \left\langle u_\sigma(-\mathbf{k}, t)\frac{\partial}{\partial t}u_\alpha(\mathbf{k}, t)\right\rangle = \frac{\partial}{\partial t}\langle u_\alpha(\mathbf{k}, t)u_\sigma(-\mathbf{k}, t)\rangle,$$

and the equation for the single-time correlation is readily found as

$$\left(\frac{\partial}{\partial t} + 2vk^2\right)\langle u_\alpha(\mathbf{k}, t)u_\sigma(-\mathbf{k}, t)\rangle$$

$$= M_{\alpha\beta\gamma}(\mathbf{k})\sum_{\mathbf{j}} \langle u_\beta(\mathbf{j}, t)u_\gamma(\mathbf{k} - \mathbf{j}, t)u_\sigma(-\mathbf{k}, t)\rangle +$$

$$+ M_{\sigma\beta\gamma}(-\mathbf{k})\sum_{\mathbf{j}} \langle u_\beta(\mathbf{j}, t)u_\gamma(-\mathbf{k} - \mathbf{j}, t)u_\alpha(\mathbf{k}, t)\rangle \qquad (2.114)$$

or, taking the infinite system limit,

$$\left(\frac{\partial}{\partial t} + 2vk^2\right)Q_{\alpha\sigma}(\mathbf{k}, t)$$

$$= M_{\alpha\beta\gamma}(\mathbf{k})\int d^3j\, Q_{\beta\gamma\sigma}(\mathbf{j}, \mathbf{k} - \mathbf{j}; t) + \{\mathbf{k} \leftrightarrow -\mathbf{k}; \alpha \leftrightarrow \sigma\} \qquad (2.115)$$

where the second term on the right is generated from the first by interchanging $\mathbf{k}$ with $-\mathbf{k}$, and α with σ.

Similarly we can obtain the equation for the triple moment by first generating three equations from (2.76) for $u_\alpha(\mathbf{k}, t)$, $u_\rho(\mathbf{l}, t)$, and $u_\sigma(\mathbf{p}, t)$. Then each of these equations is multiplied through by the other pair of velocities. When all three equations are added together and averaged, the result is easily found to be

$$\left(\frac{\partial}{\partial t} + vk^2 + vl^2 + vp^2\right)\langle u_\alpha(\mathbf{k})u_\rho(\mathbf{l})u_\sigma(\mathbf{p})\rangle$$

$$= M_{\alpha\beta\gamma}(\mathbf{k}) \sum_\mathbf{j} \langle u_\beta(\mathbf{j})u_\gamma(\mathbf{k} - \mathbf{j})u_\rho(\mathbf{l})u_\sigma(\mathbf{p})\rangle +$$

$$+ \{\mathbf{k} \leftrightarrow \mathbf{l}; \alpha \leftrightarrow \rho\} + \{\mathbf{k} \leftrightarrow \mathbf{p}; \alpha \leftrightarrow \sigma\}, \qquad (2.116)$$

where the time arguments have been dropped in the interests of conciseness.

Further equations in this sequence will be found in the paper by Orszag and Kruskal (1968).

2.7 The energy cascade in isotropic turbulence

In this section we shall again (as in Section 1.3.2) be concerned with the energy equation for turbulent fluctuations. However, it is probably as well to remind ourselves that we are now dealing with a situation where the turbulent energy is constant throughout space. Thus when we consider the transport of turbulent energy, this will be in wavenumber, rather than configuration, space. Alternatively, we can envisage a transfer from one range of eddy sizes to another; this process is known as the energy cascade.

2.7.1 The energy balance equation

Our starting point is eqn (2.115) for the single-time correlation. We specialize the l.h.s. to the isotropic case by using (2.97b) to reduce the spectral tensor to its isotropic form

$$\left(\frac{\mathrm{d}}{\mathrm{d}t} + 2vk^2\right)D_{\alpha\sigma}(\mathbf{k})Q(k, t)$$

$$= M_{\alpha\beta\gamma}(\mathbf{k}) \int \mathrm{d}^3j\, Q_{\beta\gamma\sigma}(\mathbf{j}, \mathbf{k} - \mathbf{j}, -\mathbf{k}; t) -$$

$$- M_{\sigma\beta\gamma}(\mathbf{k}) \int \mathrm{d}^3j\, Q_{\beta\gamma\sigma}(\mathbf{j}, -\mathbf{k} - \mathbf{j}, \mathbf{k}; t), \qquad (2.117)$$

where we have also used the property $M_{\alpha\beta\gamma}(-\mathbf{k}) = -M_{\alpha\beta\gamma}(\mathbf{k})$. It should also be noted that we have added the third vector argument to the triple correlation. Normally we only give the two arguments, as the third one can be deduced from the homogeneity requirement that all the vector arguments add up to zero. However, in this section it will be helpful to be reminded which vector corresponds to which tensor suffix.

Now we can form the equation for the energy spectrum $E(k, t)$, as defined by (2.101), but now generalized to the time-dependent case. If we set $\sigma = \alpha$, sum over α and (remembering that tr $D_{\alpha\beta}(\mathbf{k}) = 2$) multiply each term of (2.117) by $2\pi k^2$, we obtain

$$\left(\frac{\mathrm{d}}{\mathrm{d}t} + 2\nu k^2\right) E(k, t) = T(k, t) \tag{2.118}$$

where the non-linear term $T(k, t)$ is given by

$$T(k, t) = 2\pi k^2 M_{\alpha\beta\gamma}(\mathbf{k}) \int \mathrm{d}^3 j \{Q_{\beta\gamma\alpha}(\mathbf{j}, \mathbf{k} - \mathbf{j}, -\mathbf{k}; t) -$$

$$- Q_{\beta\gamma\alpha}(\mathbf{j}, -\mathbf{k} - \mathbf{j}, \mathbf{k}; t)\}. \tag{2.119}$$

When we discussed the turbulent energy balance in Section 1.3.2, we saw that the non-linear terms conserved energy: a global result proved in Appendix A. The implication of this is that $T(k, t)$ can only redistribute energy in wavenumber space, and so if we integrate each term in (2.119) over $\mathbf{k}$, it follows from (2.100) that we must have

$$\frac{\mathrm{d}E}{\mathrm{d}t} + \int_0^\infty 2\nu k^2 E(k, t)\,\mathrm{d}k = 0. \tag{2.120}$$

The rate of decay of the total kinetic energy of fluctuations (per unit mass of fluid) is just the dissipation rate (see Appendix A). Hence (2.120) provides us with a simple result for the dissipation rate ε for isotropic turbulence:

$$\frac{\mathrm{d}E}{\mathrm{d}t} = -\varepsilon = -\int_0^\infty 2\nu k^2 E(k, t)\,\mathrm{d}k. \tag{2.121}$$

It is quite easily checked that (2.121) can also be obtained by Fourier transforming the result given at the end of Appendix A.

It is of some interest to verify that the integral of $T(k, t)$ does indeed vanish when integrated over all wavenumbers. We can do this in the following way. From the definition of $D_{\alpha\beta}(\mathbf{k})$, as given by (2.78), it can be shown that

$$D_{\alpha\beta}(\mathbf{k})u_\beta(\mathbf{k}) = u_\alpha(k), \tag{2.122}$$

and thus

$$D_{\alpha\beta}(\mathbf{k})Q_{\beta\gamma\alpha}(\mathbf{j}, \mathbf{l}, -\mathbf{k}) = Q_{\beta\gamma\beta}(\mathbf{j}, \mathbf{l}, -\mathbf{k}), \tag{2.123}$$

where we have temporarily introduced the new dummy variable $\mathbf{l} = \mathbf{k} - \mathbf{j}$.

Then, from (2.119), (2.123), and (2.77), we can put the integral of $T(k,t)$ in the form

$$\int_0^\infty 2T(k,t)\,\mathrm{d}k = \int \mathrm{d}^3k \int \mathrm{d}^3j(2\mathrm{i})^{-1}\{k_\gamma Q_{\beta\gamma\beta}(\mathbf{j},\mathbf{l},-\mathbf{k}) - $$

$$- k_\gamma Q_{\beta\gamma\beta}(\mathbf{j},\mathbf{l}-2\mathbf{k},\mathbf{k})\}, \tag{2.124}$$

where dummy variables have been interchanged and dummy tensor indices renamed as appropriate. At this point we again invoke the continuity condition in the form

$$l_\gamma u_\gamma(\mathbf{l}) = 0,$$

and hence

$$l_\gamma Q_{\beta\gamma\beta}(\mathbf{j},\mathbf{l},-\mathbf{k}) = 0.$$

Clearly this allows us to replace k_γ in the first term on the right of (2.124) by $k_\gamma - l_\gamma = j_\gamma$, and so find

$$\int_0^\infty T(k,t)\,\mathrm{d}k = \int \mathrm{d}^3k \int \mathrm{d}^3j(2\mathrm{i})^{-1}\{j_\gamma Q_{\beta\gamma\beta}(\mathbf{j},\mathbf{l},-\mathbf{k}) - $$

$$- k_\gamma Q_{\beta\gamma\beta}(\mathbf{j},\mathbf{l}-2\mathbf{k},\mathbf{k})\}. \tag{2.125}$$

As each triple moment is symmetric under interchange of $\mathbf{k}$ and $\mathbf{j}$, it follows that the above integrand is antisymmetric under interchange of $\mathbf{k}$ and $\mathbf{j}$ and therefore vanishes when integrated over all space with respect to these variables, from which we conclude that

$$\int_0^\infty T(k,t)\,\mathrm{d}k = 0. \tag{2.126}$$

It will be seen later that this an important result, and also the form of the proof will be found to be very helpful in later sections of the book.

2.7.2 Spectral picture: the Kolmogorov hypotheses

The usual interpretation of eqn (2.118) is that the energy in the system at small k (large scales) is transferred by the non-linear term $T(k,t)$ to large k (small scales), where it is dissipated (i.e. turned into heat) by the viscous term. Evidently the non-linear term, which represents the collective action of all the other modes on one particular mode $\mathbf{k}$, is intrinsically complicated. However, despite this, its overall effect is the energy cascade and this makes an appealingly simple picture.

In this section we shall find it helpful to rewrite eqn (2.118) as

$$\frac{\mathrm{d}E(k,t)}{\mathrm{d}t} = \int_0^\infty S(k,j,|\mathbf{k}-\mathbf{j}|)\,\mathrm{d}j - 2\nu k^2 E(k,t) \tag{2.127}$$

where the definition of S can be inferred from (2.119). It follows from (2.125) and (2.126) that S has the property

$$\int dk \int dj\, S(k,j,|\mathbf{k} - \mathbf{j}|) = 0 \qquad (2.128)$$

for $k_1 < k, j < k_2$, where k_1 and k_2 are arbitrary.

It is instructive to begin by considering (2.127) with the non-linearity set equal to zero. Then the solution of (2.127) takes the form

$$E(k,t) = E(k,t_0)\exp\{-2\nu k^2(t - t_0)\} \qquad (2.129)$$

for an arbitrary starting time t_0. Thus each mode $\mathbf{k}$ (or, more strictly, the amount of energy in mode $\mathbf{k}$) decays individually with inverse time $2\nu k^2$.

It is clear that the higher the value of k the faster the decay. This is equivalent to the situation in shear flows, which we discussed in Chapter 1. There we argued that, as the molecular viscosity is a small parameter, viscous effects would only become important when the velocity gradients were appreciable. Thus the effect of the non-linearity could be deduced as transferring energy from where there was net production (e.g. near a solid boundary) to where there was net dissipation (at the centre). Similarly, when we consider the full form of (2.127), we can expect the non-linearity to transfer energy from where it enters the system (typically at small values of k, where $2\nu k^2$ will be small) up to large values of k, where the viscous damping will be very rapid. Hence the effect of the non-linear inertial transfer should be to concentrate the dissipation process in those regions of wavenumber space where it will be most efficient.

Before discussing the process of inertial transfer, we shall first consider the range of wavenumbers involved. In the case of the lower bound this is quite straightforward. The largest possible eddy will be bounded by the size of the system, and so the smallest possible wavenumber $k_{\min}$ will be given by

$$k_{\min} = 2\pi/L, \qquad (2.130)$$

where L is the largest relevant linear dimension of the turbulent system.

We can expect the upper cut-off in wavenumber to be determined by the viscous dissipation. The only relevant physical parameters available to us are the kinematic viscosity ν and the dissipation rate ε; therefore on dimensional grounds we introduce a characteristic length scale

$$\eta = (\nu^3/\varepsilon) \qquad (2.131)$$

and (for later convenience) an associated velocity scale

$$\nu = (\nu\varepsilon)^{1/4}. \qquad (2.132)$$

The inverse of (2.131) is normally taken as an approximate measure of the maximum possible wavenumber. We shall call this k_d and accordingly we have

$$k_{\rm d} = 1/\eta = (\varepsilon/v^3)^{1/4}. \tag{2.133}$$

It is interesting to note that if we define a local (in wavenumber) Reynolds number based on the above scales, then we obtain

$$R(k_{\rm d}) = v\eta/v = 1, \tag{2.134}$$

thus indicating that viscous (rather than inertial) processes are dominant for $k \approx k_{\rm d}$.

The important point to note at this stage is that the smallest wavenumbers are determined by the nature and size of the particular turbulent system under consideration, whereas the largest ones are determined by the general properties v and ε. Thus the ratio of $k_{\rm d}$ to $k_{\rm min}$ can be made as large as we please and, in the limit (of infinite Reynolds number), infinitely large.

Now, it has long been known from experiment (Taylor 1938a) that the energy is determined (i.e. through eqn (2.100)) by the lowest wavenumbers, while the dissipation rate is determined (through (2.121)) by the highest wavenumbers, and that the two ranges do not overlap even at quite modest values of the Reynolds number. Thus it follows that the inertial term (which is the link between the two ranges) can be made to dominate over as large a range of wavenumbers as we like simply by increasing the Reynolds number and hence (through (2.133)) the dissipation wavenumber. The consequences of this fact are crucial for the physics of turbulence, for it follows us to consider the inertial transfer of energy without worrying about the details of input (at low $\mathbf{k}$) or output (at high $\mathbf{k}$).

As a first step in our discussion of the non-linear inertial transfer term we return briefly to eqn (2.76), which is the Navier–Stokes equation for the Fourier components of the velocity field. Here the non-linear term is readily interpreted in the following way. Two velocity coefficients with different frequencies $\mathbf{j}$ and $\mathbf{l} = \mathbf{k} - \mathbf{j}$ are coupled together to make a contribution to the Fourier coefficient with frequency $\mathbf{k}$. The total contribution from the non-linearity is then given by the sum over all such interactions. This coupling of wavevectors in triads is usually referred to as the 'triangle condition', and is an example of a phenomenon familiar in other fields (e.g. electronics) as non-linear mixing.

The collective nature of the non-linearity is inescapable. In principle each Fourier mode of the velocity field is coupled to every other mode and therefore we are faced with a very difficult physical problem.

In such circumstances it is natural to look for some simplifying assumption, and clearly it would be helpful if the sum over modes were limited by an inherent localness or 'peakiness' in the non-linearity. That is, we would like to argue that distant wavevectors are only weakly coupled, and that any particular mode $\mathbf{k}$ would only effectively interact with modes $\mathbf{j}$ and $\mathbf{l}$ such that $\mathbf{k}$, $\mathbf{j}$, and $\mathbf{l}$ were of the same order of magnitude. In other words, to take an analogy with (say) spins on a lattice, we would like to be able to restrict the sum over modes to nearest neighbours.

The physical basis for such an assumption can be seen by considering the effect of the non-linearity in configuration space. The interaction of two eddies can be decomposed into (a) the convection of one by the other and (b) the shearing of one by the other. As we shall see later, the first of these effects results only in a phase change of the associated Fourier coefficient and is not dynamically significant. The second effect results in the internal distortion of the eddies with the transfer of energy to a smaller size of disturbance. If the interacting eddies differ greatly in size, then it is physically plausible to argue that the dynamically irrelevant phase change is the main effect. From there it is only a step to argue that we can assume that the non-linear coupling of modes is to some degree local in wavenumber space.

If we now return to the energy equation, we can interpret the non-linear term in terms of energy flows between the modes, with each such mode coupling being the average effect of many eddy–eddy interactions of the kind discussed above. The combination of some degree of localness in the basic interaction with the effect of averaging leads to the important idea that, after a number of steps, the energy cascade may become independent of the way in which the turbulence was created. Therefore the energy spectrum at high wavenumbers may take a universal form.

These ideas were first formalized by Kolmogorov in two famous hypotheses (Kolmogorov 1941a, b). The Kolmogorov hypotheses are essentially similarity principles for the energy spectrum and can be expressed in the following way. First we argue that, at sufficiently high wavenumbers, the spectrum can only depend on the fluid viscosity, the dissipation rate, and the wavenumber itself. Then, on dimensional grounds, the energy spectrum can be written

$$E(k) = v^2 \eta f(k\eta) = v^{5/4} \varepsilon^{1/4} f(k\eta),$$ (2.135)

where the second line follows from (2.131) and (2.133), f is an unknown function of universal form, and the dissipation length scale η is given by eqn (2.131).

The second similarity principle is that $E(k)$ should become independent of the viscosity as the Reynolds number tends to infinity. It is easily seen that this amounts to a requirement that the unknown function in (2.135) must take the form

$$f(k\eta) = \alpha(k\eta)^{-5/3} = \alpha v^{-5/4} \varepsilon^{5/12} k^{-5/3}$$ (2.136)

where α is a constant. Hence, with the substitution of (2.136) for $f(k\eta)$, eqn (2.135) becomes

$$E(k) = \alpha \varepsilon^{2/3} k^{-5/3}$$ (2.137)

in the limit of infinite Reynolds number.

In theoretical approaches we often work with the spectral density function $q(k)$, rather than $E(k)$. Thus for later convenience we use (2.101) to write down the equivalent of (2.137) for $q(k)$, which is

$$q(k) = \frac{\alpha}{4\pi} \varepsilon^{2/3} k^{-11/3}. \tag{2.138}$$

For large, but finite, Reynolds numbers we can adapt the above arguments as follows. If the Reynolds number is sufficiently large, we can postulate the existence of an inertial subrange of wavenumbers such that

$$2\pi/L \ll k \ll k_d,$$

for which the spectrum, as given by (2.135), is independent of the viscosity. Then we can modify (2.136) to take the form

$$f(k\eta) = \alpha(k\eta)^{-5/3} F\left(\frac{k}{k_d}\right), \tag{2.139}$$

where F is another universal function which satisfies the condition

$$F(0) = 1, \tag{2.140}$$

and, as a result, eqn (2.135) becomes

$$E(k) = \alpha \varepsilon^{2/3} k^{-5/3} F\left(\frac{k}{k_d}\right), \tag{2.141}$$

for $k \gg 2\pi/L$, and tends asymptotically to the form given by eqn (2.137) in the inertial subrange of wavenumbers.

There is no agreed theoretical form for F, but, as we shall see later, experimental results suggest that some from of exponential law might be appropriate.

Discussions of the Kolmogorov hypotheses often refer to the universal range of wavenumbers (in which (2.135) is supposed to be valid) as the 'equilibrium range of wavenumbers'. This terminology is based on the argument that the small eddies will evolve much more rapidly than the large eddies which contain most of the energy. Thus eddies in the universal range can adjust so quickly to changes in external conditions that they can be assumed to be always in a state of 'local equilibrium'. In fact this seems to be a rather confusing way of referring to a process which is, in the thermodynamic sense, very far from equilibrium. Possibly a better term would be 'quasi-stationary'.

Of course in practice many real flows are stationary, and this is therefore an important class of flows. We can consider the stationary state even within the concept of isotropic turbulence by means of an artifice. Let us introduce an input term $W(k)$ to the energy equation. This is an arbitrary step which we shall discuss in more detail in Chapter 4. For the moment we merely specify the input by

$$\int_0^\infty W(k)\,dk = \varepsilon. \tag{2.142}$$

Clearly, once a stationary state has been reached under the influence of our arbitrary stirring forces, the rate of doing work will be equal to the rate of dissipation.

Under these circumstances, eqn (2.127) can be written as

$$\int_0^\infty dj\,S(k,j,|\mathbf{k}-\mathbf{j}|) + W(k) - 2vk^2 E(k) = 0. \tag{2.143}$$

Now, in order to have a well-posed problem with well-separated input and dissipation ranges of wavenumbers, we choose a wavenumber k' such that

$$\int_0^{k'} W(k)\,dk \simeq \varepsilon \simeq -\int_{k'}^\infty 2vk^2 E(k)\,dk. \tag{2.144}$$

This means that we require the input term to be peaked near $k = 0$, and that the Reynolds number should not be too low.

With all this in mind, we can obtain two energy-balance equations by first integrating each term of (2.143) from zero up to k', and then integrating each term from infinity down to k'. The result can be written as

$$\int_0^{k'} dk \int_{k'}^\infty dk\,S(k,j,|\mathbf{k}-\mathbf{j}|) + \int_0^{k'} W(k)\,dk = 0 \tag{2.145}$$

$$\int_{k'}^\infty dk \int_0^{k'} dj\,S(k,j,|\mathbf{k}-\mathbf{j}|) - \int_{k'}^\infty 2vk^2 E(k)\,dk = 0, \tag{2.146}$$

where we have used (2.128) to eliminate contributions from the double integral for $0 \leqslant k, j \leqslant k'$ in the first case, and for $k' \leqslant k, j \leqslant \infty$ in the second.

The first of these equations tells us that energy supplied directly by the production term to modes $k < k'$ is transferred by the non-linearity to modes $j > k'$. Thus, in this range of wavenumbers, $T(k)$ behaves like a dissipation and absorbs energy.

Similarly, from eqn (2.146), we see that the non-linearity transfers energy from modes $j < k'$ to modes $k > k'$. Therefore in this range of wavenumbers $T(k)$ behaves like a source of energy, and this input is dissipated by the viscous term.

A consideration of these energy-balance equations, taken in conjunction with (2.144), allows us to put another interpretation on ε as the rate at which the inertial forces transfer energy from low to high wavenumbers, assuming that the turbulence is stationary.

Lastly, we should note that the various arguments and hypotheses presented in this section are not limited to the mathematical idealization of isotropic turbulence. Indeed it is usual to argue that, as the cascade proceeds to ever smaller eddy sizes, the conversion of velocity fluctuations to scalar pressure fluctuations (and vice versa) will lead to the ironing out of directional preferences. Thus, provided that the eddies concerned are small compared with any spatial inhomogeneity of the flow, we can think of a state of 'local isotropy'. Under these circumstances (which, again, amount to the require-ment that the Reynolds number be large enough), the Kolmogorov spectrum can be expected to apply to shear flows as well.

2.7.3   Interpretation in terms of vortex stretching

High levels of fluctuating vorticity are characteristic of all turbulent flows. This fact (which is obvious from a glance at any of the well-known photo-graphs showing turbulent flow visualization) taken in conjunction with the historical development of fluid dynamics as a subject very much concerned (through classical hydrodynamics) with vortex motions, made it natural to interpret turbulence in terms of vortex stretching (e.g. Taylor 1938b).

Nowadays the vortex-stretching interpretation of turbulence may seem a rather underdeveloped branch of the subject compared with, for instance, the statistical theory based on the velocity field. Yet it is a topic which is rapidly growing in importance, particularly in the study of the so-called coherent structures. We shall be returning to this, and other applications of vortex models, in later chapters. Here we will consider only some of the basic ideas.

Let us return to configuration space. Reminding ourselves that the vorticity ω is just the curl of the velocity field, we take the curl of each term of (2.2) to obtain (Batchelor 1967)

$$\frac{D\omega_\alpha}{Dt} = \omega_\beta \frac{\partial u_\alpha}{\partial x_\beta} + \nu \nabla^2 \omega_\alpha, \tag{2.147}$$

where D/Dt stands for the total time derivative and contains the convective derivative $u_\beta \partial/\partial x_\beta$ which acts on ω_α. This is the equation of motion for the vorticity. It tells us that the rate of change of vorticity is controlled by the interaction of the vorticity with the velocity gradients and by direct viscous dissipation.

It will help us to understand the effect of the interaction term on the r.h.s. of (2.147) if we consider a vortex tube in the direction of x_1. We can envisage this structure as a long thin cylinder of fluid, rotating about its own axis, which lies parallel to x_1. This means that we take $\beta = 1$ in (2.147) and examine what happens physically when $\alpha = 1, 2,$ or 3.

In what follows, viscous effects will be neglected on the grounds that inertial processes will be much faster, provided that the scales involved are appreciably larger than those of the dissipation range. This allows us to invoke Kelvin's theorem (Batchelor 1967), which states that the circulation (in effect, the vorticity) moves with the (inviscid) fluid. Thus we can argue that the action of the velocity gradients on the vorticity is (to a good approximation) much the same as that on the fluid.

Now consider the cases $\alpha = 2$ or $\alpha = 3$. In either case the gradient is shearing, and will tend to rotate a fluid element. Clearly, the effect on our hypothetical vortex tube will be to tilt its axis towards the x_2 or x_3 axes respectively. Thus the effect of this kind of interaction is to exchange vorticity between the three scalar components of ω.

The remaining case corresponds to $\alpha = 1$, and here the gradient is extensional. If we assume that $\partial u_1/\partial x_1$ is positive then it will cause the vortex tube to be streched out in the x_1 direction, with a consequent decrease in its cross-sectional area. Conservation of angular momentum (per unit fluid mass) can then be expressed in the form

$$\omega_1^2 r = \text{constant}$$

where r is the radius of the vortex tube. Therefore, as r decreases under the influence of the extensional gradient, the angular velocity (proportional to ω_1) will increase, and accordingly the energy $(\omega_1 r)^2$ associated with scale r will also increase. Hence energy is transferred to the small scales.

Of course, in this picture of turbulence the velocity gradients in the above discussion should be regarded as belonging to some other vortex tube. Thus we can interpret the interaction term in (2.147) as really being a vorticity–vorticity interaction. Bradshaw (1971) has presented an ingenious 'family tree' to illustrate how a series of such interactions on decreasing scales can lead to an energy cascade.

An important question hanging over the above discussion is: what is the overall effect of negative values of $\partial u_1/\partial x_1$? Clearly, any individual interaction involving negative strain rate would have the opposite effect to those discussed above, with the vortex tube being compressed rather than stretched.

Discussions of this aspect often resort to the argument that it is a general property of random walks that (on average) the distance between any two marked points will increase with time. Or, alternatively, we could rely on the proof that infinitesimal line elements are on average stretched in isotropic turbulence (Batchelor 1952; Cocke 1969; Orszag 1970a; Corrsin 1972; Dhar 1976), although the proof cannot be rigorously extended to vorticity or to finite line lengths. However, as only a statistical answer has any meaning for us, it seems more logical to pose the question in a more formal way by considering the equation for the mean square vorticity.

We can do this by multiplying eqn (2.147) through by $2\omega_\alpha$ and averaging to obtain

$$\frac{D\langle \omega_\alpha^2 \rangle}{Dt} = 2\left\langle \omega_\alpha \omega_\beta \frac{\partial u_\alpha}{\partial x_\beta} \right\rangle - 2v\left\langle \frac{\partial \omega_\alpha}{\partial x_\beta} \frac{\partial \omega_\alpha}{\partial x_\beta} \right\rangle, \qquad (2.148)$$

where the first term on the r.h.s. represents the mean rate of production of vorticity from the fluctuating velocity field. It was shown by Taylor (1938b) that this production term was always positive, thus leading to an increase in mean square vorticity limited only by the viscous effects represented by the second term on the r.h.s. of (2.148).

It was also shown by Taylor (1938b) that the above vorticity budget would hold for the more general case of turbulent shear flows, provided that the Reynolds number was large enough. That is, terms involving the mean velocity can be neglected and the vorticity is predominantly generated by the fluctuating velocity field. As the only possible mechanism for the energy cascade is some form of vortex stretching, this result reinforces the concept of universal behaviour in the small scales for all flows.

Although the assumption that the turbulent cascade consists of a tangle of vortex tubes, interacting through induced velocity fields, seems very natural, other geometric forms have also been proposed. For example, Townsend (1951) has used both vortex sheets and vortex lines to model energy transfer processes in the far dissipation region. The basic hypothesis is that of a balance between straining processes tending to concentrate vorticity and the tendency of viscous effects to diffuse it. The result is a prediction for the energy spectrum at very large wavenumbers, and certainly a plot of $k_1^2 E_{11}(k_1)$ against wavenumber agrees rather well with experimental results, with the model based on vortex sheets performing better than the one based on vortex lines.

In fact we have no firm reason for choosing one geometrical form rather than another. However, it is interesting to note that Kuo and Corrsin (1972) analysed experimental data from grid-generated turbulence in terms of various hypotheses about vortex shape (i.e. 'blobs', 'rods', and 'slabs'). They concluded that fine-structure regions in the decaying turbulence were more 'rod like' than like either of the other two shapes.

2.8 Closure approximations

There have been many attempts to derive a closed form of the equation for the energy spectrum. The situation is in fact very much like that described in Chapter 1, in connection with single-point equations, with a similar variety of *ad hoc* methods being employed. As in that case, the concept of a turbulent or effective viscosity has proved helpful. We shall discuss the effective viscosity method of Heisenberg (1948a, b) as a representative example.

In this section we shall also consider the quasi-normality hypothesis. Unlike the Heisenberg model, this was not even a particularly successful theory. But it is of historical importance as the first truly general analytical theory of turbulence. In other words, it operates by an appeal to a general principle rather than by making very specific assumptions. The distinction that we are making should become clearer as we discuss the two theories. Also, an important—if pragmatic—reason for treating the quasi-normality hypothesis in detail at the present stage is that in the process we derive the main algebraic results which are then needed for nearly all the turbulence theories to be discussed in the main part of the book.

2.8.1   The Heisenberg effective viscosity theory

We begin by writing eqn (2.118) for the energy spectrum in the form

$$\frac{dE(k,t)}{dt} = T(k,t) - 2vk^2 E(k,t)$$

$$= \int_0^\infty S(k,j,|\mathbf{k} - \mathbf{j}|)\,dj - 2vk^2 E(k,t),$$

where the second line is the same as the r.h.s. of eqn (2.127). The detailed structure of $S(k,j,|\mathbf{k} - \mathbf{j}|)$ can be inferred from (2.119), although at this point we shall only need to know that it is antisymmetric under interchange of $\mathbf{k}$ and $\mathbf{j}$.

The simplest way of closing the above equation is just to assume that T is proportional to E. We can find some physical justification for such a step as follows. Integrate each term of the spectrum equation up to some arbitrary wavenumber k':

$$\int_0^{k'} \left\{ \frac{dE}{dt} \right\} dk = \int_0^{k'} dk \int_{k'}^\infty dj\, S(k,j,|\mathbf{k} - \mathbf{j}|) - \int_0^{k'} 2vk^2 E(k,t)\,dk \quad (2.149)$$

where the integral of S over the range $0 \leqslant k, j \leqslant k'$ vanishes (e.g. see eqn (2.128)).

Now, as we have already seen in connection with eqn (2.145) for stationary turbulence, the effect of the non-linear term in the above equation can be interpreted as a drain of energy from modes $k < k'$ due to inertial transfer to wavenumbers $k < k'$. Accordingly, if we wish to model this energy drain as analogous to viscous dissipation, we can introduce an effective turbulent viscosity through the hypothesis that the inertial transfer term can be written as

$$\int_0^{k'} T(k,t)\,dk = -2v(k',t) \int_0^{k'} E(k,t)k^2\,dk, \quad (2.150)$$

where $v(k',t)$ is the kinematic eddy viscosity and represents the effect of an integral over wavenumbers from k' up to infinity.

This suggests that we look for an expression for the eddy viscosity which involves such an integral. Hence, if we write

$$v(k',t) = \int_{k'}^{\infty} f[j, E(j,t)]\,dj, \tag{2.151}$$

then the unknown function $f[j, E(j,t)]$ can be determined on dimensional grounds:

$$f[j, E(j,t)] = Aj^{-3/2}[E(j,t)]^{1/2} \tag{2.152}$$

where A is a constant.

Collecting together (2.150), (2.151), and (2.152), we can now express eqn (2.149) for the energy spectrum as

$$\int_{0}^{k'} \left\{\frac{dE}{dt}\right\} dk = -2\{v + v(k',t)\} \int_{0}^{k'} E(k,t)k^2\,dk, \tag{2.153}$$

and the Heisenberg-type eddy viscosity is given by

$$v(k',t) = A \int_{k'}^{\infty} j^{-3/2}[E(j,t)]^{1/2}\,dj. \tag{2.154}$$

These equations have been solved for the stationary case (e.g. see Batchelor 1971). We shall only note here that the resulting spectrum reduces to the Kolmogorov $-5/3$ law in the inertial range of wavenumbers, but behaves like k^{-7} in the dissipation range. This latter form of behaviour is known to be incorrect, as experimental results show that the energy spectrum at large wavenumbers falls off exponentially (i.e. faster than any power).

2.8.2 The quasi-normality hypothesis

Once again we take eqn (2.118) as our starting point, that is,

$$\left\{\frac{d}{dt} + 2vk^2\right\} E(k,t) = T(k,t),$$

but this time we are interested in the detailed structure of $T(k,t)$ and indeed in the moment hierarchy that (in principle) determines $T(k,t)$.

Noting that the r.h.s. of (2.119) must be homogeneous and hence unaffected by changing the sign of $\mathbf{k}$, we can rewrite the equation for $T(k,t)$ as

$$T(k,t) = 4\pi k^2 M_{\alpha\beta\gamma}(\mathbf{k}) \int d^3j\, Q_{\alpha\beta\gamma}(-\mathbf{k}, \mathbf{j}, \mathbf{k} - \mathbf{j}; t), \tag{2.155}$$

where $Q_{\alpha\beta\gamma}$ can be obtained from (2.116).

We can do this explicitly, renaming dummy variables to avoid confusion and invoking (2.91) in order to introduce the triple correlation function:

$$\left\{\frac{d}{dt} + v(k^2 + j^2 + |\mathbf{k} - \mathbf{j}|^2)\right\} \left(\frac{2\pi}{L}\right)^6 Q_{\alpha\beta\gamma}(-\mathbf{k}, \mathbf{j}, \mathbf{k} - \mathbf{j}; t)$$

$$= M_{\alpha\rho\delta}(-\mathbf{k}) \sum_{\mathbf{m}} \langle u_\rho(\mathbf{m})u_\delta(-\mathbf{k} - \mathbf{m})u_\beta(\mathbf{j})u_\gamma(\mathbf{k} - \mathbf{j})\rangle +$$

$$+ M_{\beta\rho\delta}(\mathbf{j}) \sum_{\mathbf{m}} \langle u_\rho(\mathbf{m})u_\delta(\mathbf{j} - \mathbf{m})u_\alpha(-\mathbf{k})u_\gamma(\mathbf{k} - \mathbf{j})\rangle +$$

$$+ M_{\gamma\rho\delta}(\mathbf{k} - \mathbf{j}) \sum_{\mathbf{m}} \langle u_\rho(\mathbf{m})u_\delta(\mathbf{k} - \mathbf{j} - \mathbf{m})u_\beta(\mathbf{j})u_\alpha(-\mathbf{k})\rangle, \quad (2.156)$$

As before, we suppress time arguments in the interests of conciseness. Then, making the helpful abbreviation

$$\text{r.h.s. of (2.156)} = \left(\frac{2\pi}{L}\right)^6 H_{\alpha\beta\gamma}(-\mathbf{k}, \mathbf{j}, \mathbf{k} - \mathbf{j}; t), \quad (2.157)$$

we can formally write down the solution of (2.156) as

$$Q_{\alpha\beta\gamma}(-\mathbf{k}, \mathbf{j}, \mathbf{k} - \mathbf{j}; t) = \int_0^t ds \exp\{-v(k^2 + j^2 + |\mathbf{k} - \mathbf{j}|^2)(t - s)\} \times$$

$$\times H_{\alpha\beta\gamma}(-\mathbf{k}, \mathbf{j}, \mathbf{k} - \mathbf{j}; s). \quad (2.158)$$

At this stage we introduce the quasi-normality hypothesis (Proudman and Reid 1954; Tatsumi 1957), which is essentially to the effect that all even-order moments are assumed to be related as if for a normal distribution. It should be noted that this is a much weaker step than assuming that the distribution of turbulent velocities is normal. That would be unphysical, if only because it would be inconsistent with the existence of the triple correlation, which, as we have seen, is responsible for turbulent energy transfer. In contrast, the limited assumption of quasi-normality is quite well supported by experimental measurements (Frenkiel and Klebanoff 1967; Van Atta and Chen 1969; Van Atta and Yeh 1970; Frenkiel and Klebanoff 1973), with even-order correlations apparently being related in this way, except perhaps at small separations of the measuring points.

The quasi-normality hypothesis is used to close the moment hierarchy by expressing the quadruple moments on the r.h.s. of (2.156) in terms of products of second moments. If we denote the quadruple correlation symbolically by $\langle 1, 2, 3, 4 \rangle$, then for a normal distribution we have (e.g. see Birnbaum 1964)

$$\langle 1, 2, 3, 4 \rangle = \langle 1, 2 \rangle \langle 3, 4 \rangle + \langle 1, 3 \rangle \langle 2, 4 \rangle + \langle 1, 4 \rangle \langle 2, 3 \rangle.$$

Then application of this relationship to each of the three fourth-order moments on the r.h.s. of (2.156) generates nine such products of second-order moments—a degree of algebraic complexity which may seem rather formidable.

However, in practice this will turn out not to be as bad as it seems at first sight. We shall see that three of these terms are identically zero, and, of the six remaining, two terms can be combined into one, as can (ultimately) the other four, leaving only two separate and distinct terms at the end of the calculation. We can begin by taking the first term on the r.h.s. of (2.156) as an example:

$$M_{\alpha\rho\delta}(-\mathbf{k}) \sum_{\mathbf{m}} \{ \langle u_{\rho}(\mathbf{m})u_{\delta}(-\mathbf{k} - \mathbf{m}) \rangle \langle u_{\beta}(\mathbf{j})u_{\gamma}(\mathbf{k} - \mathbf{j}) \rangle +$$

$$+ \langle u_{\rho}(\mathbf{m})u_{\beta}(\mathbf{j}) \rangle \langle u_{\delta}(-\mathbf{k} - \mathbf{m})u_{\gamma}(\mathbf{k} - \mathbf{j}) \rangle +$$

$$+ \langle u_{\rho}(\mathbf{m})u_{\gamma}(\mathbf{k} - \mathbf{j}) \rangle \langle u_{\delta}(-\mathbf{k} - \mathbf{m})u_{\beta}(\mathbf{j}) \rangle \}.$$

We now invoke the homogeneity condition for second-order moments (see eqn (2.83)) in order to deal with each of the above products as follows:

First product: homogeneity implies $\mathbf{m} - \mathbf{k} - \mathbf{m} = 0$ and so $k = 0$, but as $M_{\alpha\beta\gamma}(\mathbf{k}) = 0$ for $k = 0$, this term is zero.
Second product: homogeneity implies $\mathbf{m} + \mathbf{j} = 0$.
Third product: homogeneity implies $\mathbf{m} + \mathbf{k} - \mathbf{j} = 0$.

Hence, summing over $\mathbf{m}$ (i.e. replacing $\mathbf{m}$ by $-\mathbf{j}$ in the second product and by $-\mathbf{k} + \mathbf{j}$ in the third), using (2.88) to introduce the pair-correlation tensors, and exploiting the symmetry

$$M_{\alpha\rho\delta}(-\mathbf{k}) = M_{\alpha\delta\rho}(-\mathbf{k})$$

under interchange of ρ and δ, we reduce the above expression for the first term on the r.h.s. of (2.156) to

$$2\left(\frac{2\pi}{L}\right)^{6} M_{\alpha\rho\delta}(-\mathbf{k})Q_{\rho\beta}(\mathbf{j})Q_{\delta\gamma}(\mathbf{k} - \mathbf{j}).$$

By applying the same methods to the other two quadruple moments, we obtain for $H_{\alpha\beta\gamma}(-\mathbf{k}, \mathbf{j}, \mathbf{k} - \mathbf{j}; t)$ as defined by (2.157)

$$H_{\alpha\beta\gamma}(-\mathbf{k}, \mathbf{j}, \mathbf{k} - \mathbf{j}; t) = 2M_{\alpha\rho\delta}(-\mathbf{k})Q_{\rho\beta}(\mathbf{j})Q_{\delta\gamma}(\mathbf{k} - \mathbf{j}) +$$

$$+ 2M_{\beta\rho\delta}(\mathbf{j})Q_{\rho\alpha}(\mathbf{k})Q_{\delta\gamma}(\mathbf{k} - \mathbf{j}) +$$

$$+ 2M_{\gamma\rho\delta}(\mathbf{k} - \mathbf{j})Q_{\rho\alpha}(\mathbf{k})Q_{\delta\beta}(\mathbf{j}). \qquad (2.159)$$

Nw we can combine (2.159), (2.158), and (2.155) to give the inertial transfer term as

$$T(k,t) = 4\pi k^2 M_{\alpha\beta\gamma}(\mathbf{k}) \times$$

$$\times \int d^3 j \int_0^t ds \exp\{-v(k^2 + j^2 + |\mathbf{k} - \mathbf{j}|^2)(t - s)\} \times$$

$$\times [2M_{\alpha\rho\delta}(-\mathbf{k})Q_{\rho\beta}(\mathbf{j},s)Q_{\delta\gamma}(\mathbf{k} - \mathbf{j},s) +$$

$$+ 2M_{\beta\rho\delta}(\mathbf{j})Q_{\rho\alpha}(\mathbf{k},s)Q_{\delta\gamma}(\mathbf{k} - \mathbf{j},s) +$$

$$+ 2M_{\gamma\rho\delta}(\mathbf{k} - \mathbf{j})Q_{\rho\alpha}(\mathbf{k},s)Q_{\delta\beta}(\mathbf{j},s)]. \tag{2.160}$$

At this stage we take the following two steps. First, we restrict the formulation to the isotropic case. That is, we use eqn (2.97b) to express the spectral tensor $Q_{\alpha\beta}(\mathbf{k},t)$ in terms of the projection operator $D_{\alpha\beta}(\mathbf{k})$ and the scalar spectral density function $Q(k,t)$. Second, we note that inside the integration with respect to $\mathbf{j}$, the variable $\mathbf{k} - \mathbf{j}$ can be treated on an equal footing with $\mathbf{j}$ as a dummy variable. Thus, in the last term of (2.160) we can interchange $\mathbf{k} - \mathbf{j}$ with $\mathbf{j}$.

The algebraic details can be found in Appendix E, where it is shown that (2.160) reduces to the comparatively simple form

$$T(k,t) = 4\pi k^2 \int d^3 j \int_0^t ds \exp\{-v(k^2 + j^2 + |\mathbf{k} - \mathbf{j}|^2)(t - s)\} \times$$

$$\times 2L(\mathbf{k},\mathbf{j})Q(|\mathbf{k} - \mathbf{j}|,s)[Q(j,s) - Q(k,s)], \tag{2.161}$$

where

$$L(\mathbf{k},\mathbf{j}) = -M_{\alpha\beta\gamma}(\mathbf{k})M_{\alpha\rho\delta}(-\mathbf{k})D_{\rho\beta}(\mathbf{j})D_{\delta\gamma}(\mathbf{k} - \mathbf{j})$$

$$= 2M_{\rho\beta\gamma}(\mathbf{k})M_{\beta\rho\delta}(\mathbf{j})D_{\delta\gamma}(\mathbf{k} - \mathbf{j})$$

$$= \frac{[\mu(k^2 + j^2) - kj(1 + 2\mu^2)](1 - \mu^2)kj}{k^2 + j^2 - 2kj\mu} \tag{2.162}$$

and μ is the cosine of the angle between the vectors $\mathbf{k}$ and $\mathbf{j}$.

Formally (2.161) gives us a closed equation for the energy spectrum when we substitute for $T(k,t)$ in (2.118). However, in order to be consistent, we have the choice of either using (2.101) to convert all the spectral density functions in (2.161) to spectra or of expressing the l.h.s. of (2.118) in terms of the spectral density function $Q(k,t)$.

The latter course seems easier, and also fits in better with the sort of theoretical work we shall be considering later on in the book. Accordingly, we substitute (2.161) into (2.118), divide both sides by $4\pi k^2$, and, using (2.101) to introduce the spectral density function on the l.h.s., obtain the closure approximation based on the hypothesis of quasi-normality:

$$\left\{\frac{d}{dt} + 2\nu k^2\right\} Q(k, t)$$

$$= 2 \int d^3j \int_0^t ds \exp\{-\nu(k^2 + j^2 + |k - j|^2)(t - s)\} \times$$

$$\times L(k, j)Q(|k - j|, s)[Q(j, s) - Q(k, s)]. \tag{2.163}$$

If we prescribe $Q(k, t)$ at $t = 0$, then (2.163) can be integrated forward in time for the case of freely decaying turbulence. Unfortunately, despite the apparently reasonable nature of the basic assumption, when (2.163) was solved numerically (O'Brien and Francis 1962; Ogura 1963; for a more modern critical account, see Orszag 1970b) it was found that $Q(k, t)$ became negative for certain values of the wavenumber k. This unphysical result had a profound effect on theorists, which persists to the present day as a slightly excessive concern with the 'physical realizability' of theories. As Orszag points out, there is no reason why an approximate theory should not violate realizability to some small extent. The problem with quasi-normality is that the negative spectra do not constitute a small effect.

Our treatment of quasi-normality is justified by more than its historical interest. As we shall see later, there is much to be learned about the physics of turbulence from a consideration of why quasi-normality failed (and indeed from the methods that can be used to patch it up; currently this is a very active field of research and is discussed further in Section 7.5). But, not least, in deriving eqn (2.163) we have already covered the algebra which many people find quite intimidating when encountered in conjunction with modern closure approximations. Thus the results of this section are worth mastering, as they will be extremely useful later on.

2.9 Some representative experimental results for spectra

At various later points in the book, we shall look at experimental measurements of correlations and spectra in detail. In this section we shall confine our attention to some very general results about spectra. In particular, we shall be interested in the Kolmogorov hypotheses and the question of how universal they are. We shall also take a preliminary look at the experimental picture of the energy transfer processes in wavenumber, and, lastly, a neat method of using the Kolmogorov spectrum to obtain a measurement of the local dissipation rate at a point in pipe flows.

2.9.1 One-dimensional energy spectra

In Section 2.6.5 we introduced the one-dimensional spectrum $E_{11}(k_1)$, which is defined on the interval $-\infty \leqslant k_1 \leqslant \infty$ and which is related to the mean

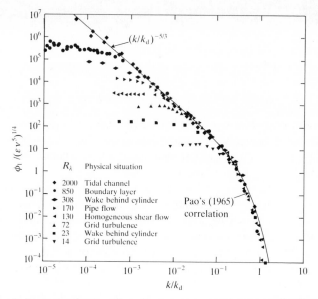

Fig. 2.4. Measured one-dimensional spectra for a wide range of Reynolds numbers and physical situations (note that both spectra and wavenumbers have been scaled by the appropriate Kolmogorov variables): ▼, Stewart and Townsend (1951); ■, Uberoi and Freymuth (1969); ▲, Comte-Bellot and Corrsin (1971); ◄, Champagne, Harris, and Corrsin 1970; ► Laufer (1954); ◆, Uberoi and Freymuth (1969); ●, Coantic and Favre (1974); ◆, Grant *et al.* 1962.

square velocity through (2.107). In practice experimentalists often use $\phi(k_1)$, which is defined on the interval $0 \leqslant k_1 \leqslant \infty$ and satisfies

$$\int_0^\infty \phi(k_1)\,dk_1 = \langle u_1^2 \rangle. \tag{2.164}$$

Clearly $\phi(k_1)$ is just twice our $E_{11}(k_1)$, but for simplicity we shall follow the experimentalists' practice when presenting results.

In Fig. 2.4 we present experimental data for $\phi(k_1)$ obtained from many diverse experimental situations, ranging from measurements in laboratory wind-tunnels to the classic sea-borne investigation of Grant, Stewart, and Moilliet (1962) in a tidal channel. As the physical sources of the data are so diverse (and we are really only interested in small-scale structures), we follow the usual practice and characterize each data set by a Reynolds number R_λ based on the Taylor microscale, that is,

$$R_\lambda = \lambda u / \nu \tag{2.165}$$

where u is the r.m.s. velocity, ν is the kinematic viscosity of the fluid, and the Taylor microscale λ is given by (2.63).

It can be seen that the spectra have been made dimensionless in terms of the Kolmogorov variables, as given by (2.132) and (2.133), and plotted against the dimensionless ratio k/k_d. Clearly, the spectra at the higher wavenumbers collapse to a universal form, thus supporting Kolmogorov's first similarity hypothesis as summarized in eqn (2.135).

As the Reynolds number increases, it is also clear that the spectra show longer ranges (in wavenumber) of universal behaviour, with a tendency to depart asymptotically from the $k^{-5/3}$ law (as predicted by the second Kolmogorov similarity hypothesis; see (2.137)) at low wavenumbers.

Incidentally, it should be noted that the constant asymptote of each spectrum at low wavenumbers is purely an artefact, resulting from the fact that the one-dimensional spectrum is merely a 'slice' through the three-dimensional spectrum. In physical terms this means that the low wavenumber part of $\phi(k_1)$ is strongly affected by 'aliasing' from higher wavenumbers moving at an angle to the x_1 axis (e.g. see Tennekes and Lumley 1972, p. 249). Referring to eqn (2.108), it is easy to show that $E_{11}(k_1)$ is finite at $k = 0$ and depends on the mean square level of the turbulence and on the integral length scale.

2.9.2 The Kolmogorov constant

The constant of proportionality α in the Kolmogorov spectrum has long been a target for theoretical predictions and accordingly its experimental value is a matter of some importance. To begin with, if we substitute (2.137) into (2.109) we find

$$E_{11}(k_1) = \frac{9}{55}\alpha\varepsilon^{2/3}k_1^{-5/3}, \tag{2.166}$$

and (recalling that $\phi(k_1)$ is twice $E_{11}(k_1)$) it follows that an experimental measurement of

$$\phi(k_1) = \alpha_1\varepsilon^{2/3}k_1^{-5/3} \tag{2.167}$$

implies that the spectral constant in (2.137) is given by

$$\alpha = \frac{55}{18}\alpha_1. \tag{2.168}$$

The results of Grant et al. (1962), which would be regarded by many as among the most reliable, suggest that $\alpha_1 = 0.47 \pm 0.02$ and hence, from (2.168), $\alpha = 1.44 \pm 0.06$. Other investigations have given slightly different results, but there appears to be wide agreement among workers in the field that the Kolmogorov constant is about 1.5.

However, this agreement is not entirely unanimous. Kraichnan (1966) has commented that the estimate made of the spectral constant will depend to some extent on where the boundary between the inertial and dissipation ranges is chosen. Normally this is taken to be about $k = 0.1k_d$ but, as reference to Fig. 2.4 will show, it must be difficult to be at all precise about this. Evidently we could do with an analytical form which could be fitted to both ranges of wavenumbers.

Various models and correlations have been proposed for this purpose. Probably the best known is that due to Pao (1965). Essentially, Pao's argument was that the rate at which energy is transferred through wavenumber space has the same dependence on viscosity as the energy spectrum. Thus the ratio of these two quantities is independent of viscosity: this is true in the inertial range (on the Kolmogorov picture) and hence the $-5/3$ spectrum is recovered there. If this hypothesis is extended to the dissipation range as well, the result is that (2.141) becomes

$$E(k) = \alpha \varepsilon^{2/3} k^{-5/3} \exp\left\{-\left(\frac{3\alpha}{2}\right)\left(\frac{k}{k_d}\right)^{4/3}\right\} \tag{2.169}$$

which, as Fig. 2.4 shows, agrees quite well with experiment.

The basic assumption in Pao's work appears to be quite imponderable and it is not presented here as a spectral theory (although attempts have been made to refine it (Pao 1968; Tennekes 1968; Lin 1972) or extend it to lower wavenumbers (Driscoll and Kennedy 1983)). Essentially it seems to be a good way of analysing experimental results. From our point of view, perhaps the most noteworthy feature of this analysis is the way in which it draws attention to the fact that values of α as large as 2.2 (or possibly larger) are not incompatible with the data.

2.9.3   Energy transfer in wavenumber space

The turbulent energy balance for wavenumber k during free decay is given by eqn (2.118). Hence the energy transfer spectrum $T(k)$, which determines the rate at which energy is transferred from large to small eddies, can be obtained experimentally for grid turbulence by measuring the rate of decay dE/dt and the dissipation spectrum $D(k, t) = 2\nu k^2 E(k, t)$, and taking the difference of the two using (2.118). This was first done by Uberoi (1963).

Later Van Atta and Chen (1969) pioneered the use of fast Fourier transforms to calculate the transfer spectrum directly from measured third-order correlation functions. This method has the virtue that it can also be applied to steady shear flows, and this has permitted a comparison of spectral energy budgets between non-isotropic turbulence (in a free jet) and (nearly) isotropic grid turbulence (Helland, Van Atta, and Stegen 1977).

In Fig. 2.5 we show the three terms which make up the spectral energy balance as given by eqn (2.118). Each term has been multiplied by a factor $(k/k_d \nu)^3$ in order to make it dimensionless and also to spread it out in **k**-space. These particular results were obtained by computing a version of the local

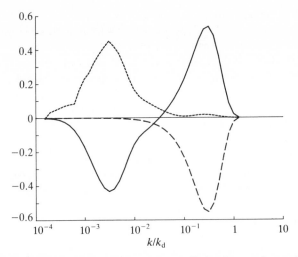

Fig. 2.5. Sketch of the three-dimensional energy, dissipation, and transfer spectra in isotropic turbulence: (——— $T(k,t)$; ----- $-\partial E(k,t)/\partial t$; —— $-2\nu k^2 E(k,t)$; each multiplied by a factor $(k/k_d v^3)$, where k_d and v are the Kolmogorov dissipation wavenumber and velocity scale, respectively).

energy transfer (LET) theory (see Chapter 7). At this stage we are only interested in a qualitative picture, and so it is instructive to draw an analogy with Fig. 1.7 which shows the turbulent energy balance in x-space for pipe flow. In that case, we saw that the role of the non-linear term was to transfer energy from the neighbourhood of the pipe wall (where production is greater than viscous dissipation) to the core region where the reverse is the case. In other words, the inertial transfer is from source to sink. Similarly we can see from Fig. 2.5 that $T(k,t)$ absorbs energy at small wavenumbers (where dE/dt or the production spectrum can be assumed to be large) and re-emits it at higher wavenumbers where it is 'absorbed' by the dissipation spectrum.

However, in drawing this analogy we should bear in mind that the energy cascade is actually from one eddy size to another, and in (for example) pipe flow such a cascade would be going on at every point in the flow.

For our last topic in this chapter we again hark back to our discussion in Section 1.6.2 of the turbulent energy balance in pipe flow. There we discussed the experiments of Lawn (1971) with pipe flows, and noted that he used two methods of checking his dissipation curve, both based on assumptions of isotropy.

The first of these was to measure the Taylor microscales and use a generalization of eqn (2.64) in order to relate the dissipation rate to the mean square turbulence level and the microscales; for further details the interested reader should refer to Lawn's paper. The corresponding results are denoted by crosses in Fig. 1.7.

The second method relied on the fact that measured spectra were found to have the Kolmogorov $-5/3$ dependence for a limited range of wavenumbers. There is, of course, no reason why the Kolmogorov hypotheses should not apply to pipe flows. We merely require local isotropy, and Bradshaw (1967) has shown that the conditions for that to hold in shear flows are less restrictive than had originally been supposed. He also proposed that this fact could be made the basis of a measurement of the dissipation rate.

Thus, by fitting the region of his spectrum which had a $-5/3$ slope to the theoretical one-dimensional spectrum (see eqn (2.167) in the present work), and taking a specific value for the spectral constant ($\alpha_1 = 0.55$), Lawn was able to estimate the dissipation rate. Full circles denote these particular results in Fig. 1.7, and clearly all three methods are in reasonable agreement.

2.10　Further reading

The monograph by Batchelor (1971) is widely regarded as a classic. For the newcomer it provides a lucid introduction to the subject of homogeneous isotropic turbulence, although the Fourier–Stieltjes notation failed to catch on and now looks rather dated. This work would be admirably complemented by the encyclopaedic book by Monin and Yaglom (1975), which also employs the same notation for the Fourier integral.

Other more general books on turbulent shear flows (e.g. Hinze 1975; Townsend 1976) have sections on isotropic turbulence which can be helpful, if only to give another viewpoint. The first of these references is particularly good at giving rather full mathematical details of the various derivations.

Lastly, for those who find the generally abstract treatment of the subject rather daunting, the book by Tennekes and Lumley (1972) may be helpful. These authors make a point of working out many specific examples in detail, especially in order to illustrate new or difficult concepts.

References

BATCHELOR, G. K. (1952). *Proc. R. Soc. A* **213**, 349.
—— (1967). *An introduction to fluid dynamics*. Cambridge University Press, Cambridge.
—— (1971). *The theory of homogeneous turbulence* (2nd edn). Cambridge University Press, Cambridge.
BIRNBAUM, Z. W. (1964). *Introduction to probability and mathematical statistics*. Harper, New York.
BRADSHAW, P. (1967). *NPL Aero Report 1220*.
—— (1971) *An introduction to turbulence and its measurement*. Pergamon Press, Oxford.
BROWN, L. W. B., ANTONIA, R. A., and RAJAGOPALAN, S. (1983). *Phys. Fluids* **26**, 1222.
CHAMPAGNE, F. H., HARRIS, V. G., and CORRSIN, S. (1970). *J. fluid Mech.* **41**, 81.
CHOU, P. Y. (1945) *Q. appl. Math.* **3**, 38.
COANTIC, M. and FAVRE, A. (1974). *Adv. Geophys.* **18A**, 391.

COCKE, W. J. (1969). *Phys. Fluids* **12**, 2488.

COMTE-BELLOT, G. and CORRSIN, S. (1971). *J. fluid Mech.* **48**, 273.

CORRSIN, S. (1972). *Phys. Fluids* **13**, 1370.

DHAR, D. (1976). *Phys. Fluids* **19**, 1059.

DRISCOLL, R. J. and KENNEDY, L. A. (1983). *Phys. Fluids* **26**, 1228.

FRENKIEL, F. N. and KLEBANOFF, P. S. (1967). *Phys. Fluids* **10**, 1737.

—— and —— (1973). *Phys. Fluids* **16**, 725.

GOLDSTEIN, S. (1938). *Modern developments in fluid dynamics.* Clarendon Press, Oxford. (Reprinted by Dover Publications, 1965.)

GRANT, H. L., STEWART, R. W., and MOILLIET, A. (1962). *J. fluid Mech.* **12**, 241.

HEISENBERG, W. (1948a). *Z. Phys.* **124**, 628.

—— (1948b). *Proc. R. Soc. A* **195**, 402.

HELLAND, K. N., VAN ATTA, C. W., and STEGEN, G. R. (1977) *J. fluid Mech.* **79**, 337.

HINZE, J. O. (1975). *Turbulence* (2nd edn). McGraw-Hill, New York.

KOLMOGOROV, A. N. (1941a). *C. R. Acad. Sci. URSS* **30**, 301.

—— (1941b). *C.R. Acad. Sci. URSS* **32**, 16.

KRAICHNAN, R. H. (1966). *Phys. Fluids* **9**, 1728.

KUO, A. Y. and CORRSIN, S. (1972). *J. fluid Mech.* **56**, 447.

LAUFER, J. (1954). *NACA Tech. Rep.* 1174.

LAWN, C. J. (1971). *J. fluid Mech.* **48**, 477.

LESLIE, D. C. (1973). *Developments in the theory of turbulence.* Clarendon Press, Oxford.

LIN, J.-T. (1972). *Phys. Fluids* **15**, 205.

MONIN, A. S. and YAGLOM, A. M. (1975). *Statistical fluid mechanics*, Vol.. 2, *Mechanics of turbulence.* MIT Press, Cambridge, MA.

O'BRIEN, E. E. and FRANCIS, G. C. (1962). *J. fluid Mech.* **13**, 369.

OGURA, Y. (1963). *J. fluid Mech.* **16**, 33.

ORSZAG, S. A. (1969). *Stud. appl. Maths.* **48**, 275.

—— (1970a). *Phys. Fluids* **13**, 2203.

—— (1970b). *J. fluid Mech.* **41**, 363.

—— and KRUSKAL, M. D. (1968). *Phys. Fluids* **11**, 43.

PAO, Y.-H. (1965) *Phys. Fluids* **8**, 1063.

—— (1968) *Phys. Fluids* **11**, 1371.

PROUDMAN, I. and REID, W. H. (1954). *Phil. Trans. R. Soc. Lond. A* **247**, 163.

ROACH, G. F. (1970). *Green's functions: introductory theory with applications.* Van Nostrand Reinhold, London.

ROBERTSON, H. P. (1940). *Proc. Camb. Phil. Soc.* **36**, 209.

STEWART, R. W. and TOWNSEND, A. A. (1951). *Phil. Trans. R. Soc. Lond. A* **243**, 359.

TATSUMI, T. (1957). *Proc. R. Soc. Lond. A* **239**, 16.

TAYLOR, G. I. (1935). *Proc. R. Soc. Lond. A* **151**, 421.

TAYLOR, G. I. (1938a). *Proc. R. Soc. Lond. A* **164**, 476.

TAYLOR, G. I. (1938b). *Proc. R. Soc. Lond. A* **164**, 15.

TENNEKES, H. (1968). *Phys. Fluids* **11**, 246.

TENNEKES, H., and LUMLEY, J. L. (1972). *A first course in turbulence.* MIT Press, Cambridge, MA.

TOWNSEND, A. A. (1951). *Proc. R. Soc. Lond. A* **208**, 534.

—— (1976). *The structure of turbulent shear flow* (2nd edn). Cambridge University Press, Cambridge.

UBEROI, M. S. (1963). *Phys. Fluids* **6**, 1048.

—— and FREYMUTH, P. (1969). *Phys. Fluids* **12**, 1359.

VAN ATTA, C. W. and CHEN, W. Y. (1969). *J. fluid Mech.* **38**, 743.

—— and YEH, T. T. (1970). *J. fluid Mech.* **41**, 169.

ZAMAN, K. B. Q. and HUSSAIN, A. K. M. F. (1981). *J. fluid Mech.* **112**. 379.

3

SOME RECENT DEVELOPMENTS IN THE STUDY OF TURBULENCE

This chapter is somewhat miscellaneous in character and is intended to serve two purposes. The first of these is to cover certain topics which would not easily fit into the main part of the book, yet which are important. The sections on anemometry, data processing, and engineering models are the main topics in this category.

The second purpose is to try to give the reader a broad picture of what the rest of the book is about. That is, certain sections of this chapter should act as both an introduction to and an overview of more detailed and advanced treatments later in the book. For example, the section on renormalized perturbation methods is mainly descriptive, using only rather simple mathematics to explain the general method.

3.1 Measurement techniques and data analysis

The subject of flow measurement is vast, but here we are only concerned with the quantitative measurement of the fluid velocity at a point. Our main objective will be to discuss the new developments of laser anemometry and digital data processing, which together are in the process of revolutionizing the experimental study of turbulence. We begin by considering the older methods of making local measurements of fluid velocity in order to provide a background for the newer material.

3.1.1 *Anemometry*

Historically, methods of continuously monitoring fluid velocities have depended on restricting the flow in some way, and using the Bernoulli equation to relate the flow through the restriction to the pressure difference across it. Discussions of such methods can be found in almost any elementary text on fluid mechanics. From our present point of view, the only one of these methods to qualify as measuring local velocities is the pitot tube. Essentially this is just a thin tube placed with its open end facing the oncoming flow and its other end connected to a manometer in order to measure the total stagnation pressure. The other side of the manometer is then connected to a pressure tapping which is positioned such that it registers the free-stream pressure.

The accuracy of the pitot tube can be established by comparing its value for the bulk mean velocity in (say) pipe flow with the result obtained by measuring the volumetric flow rate and dividing by the cross-section area. Surprisingly, even for turbulent flows, the pitot tube turns out to be quite an accurate measuring device. The only proviso is that we must restrict its use to mean velocities, as the inertia of the system is such that it cannot respond rapidly to turbulent fluctuations.

However, for anything other than the simplest of practical applications, it is precisely the fluctuations that we are interested in. Thus virtually the whole quantitative treatment of turbulence during most of this century has depended on the hot-wire anemometer, which has been developed to the point where it can handle the fastest turbulent fluctuations with ease.

In its simplest form, this consists of a short fine metal wire which is heated by the passage of an electric current through it. If the wire is cooled by a flowing fluid, its electrical resistance is reduced and (all other parameters being constant) this reduction in resistance can be related to the fluid velocity. Typically wires for this purpose are made of platinum or tungsten and have a diameter of a few microns.

The operating principle is readily understood as follows. We assume that the wire is initially at the same temperature T_f as the fluid and has electrical resistance R_f. If we then heat up the wire to some temperature T_w, the resulting resistance R_w will be given by the well-known formula

$$R_w = R_f\{1 + \sigma(T_w - T_f)\} \tag{3.1}$$

where σ is the temperature coefficient of resistance.

If a current I flows in the wire, then heat is generated at a rate given by

$$\text{Joule heating rate} = I^2 R_w.$$

This heat is then transferred from the wire to the fluid at a rate given by

$$\text{rate of cooling} = hS(T_w - T_f)$$

where h is the heat transfer coefficient and S is the surface area of the wire. For thermal equilibrium, these two rates can be equated:

$$I^2 R_w = hS(T_w - T_f) = \frac{hS(R_w - R_f)}{\sigma R_f}, \tag{3.2}$$

where the last step follows upon substitution from (3.1) for the temperature difference.

Evaluation of the heat transfer coefficient h is a rather specialized matter and we shall not go into the details here. However, provided that free convection can be neglected and the cooling process is entirely due to the fluid velocity U_N perpendicular to the wire, a reasonably general result is

$$\frac{I^2 R_w}{R_w - R_f} = A + B\sqrt{U_N} \tag{3.3}$$

where A and B are assumed to be independent of the fluid velocity. This formula is often referred to as 'King's law', although strictly that term should be applied to the case of a cylindrical wire, where (3.3) can be obtained from first principles (King 1914).

In practice it is always best to treat eqn (3.3) only as a guide and to calibrate the hot-wire anemometer against a pitot tube in laminar flow. Then the resulting I^2 can be plotted against $\sqrt{U_N}$ in order to obtain values for A and B. This calibration can be extended to low intensity turbulance, but it is best to use an electronic linearizer such that the output of the anemometer becomes proportional to U_N.

The usual method of measuring the change of resistance in the hot wire is to incorporate the device in one arm of a Wheatstone bridge. Then the use of a power supply with a high internal resistance ensures that changes in bridge resistance do not affect the current and the voltage differences across the bridge can be detected. This is known as the constant-current mode of operation. It is limited to low turbulent intensities by its poor response at high frequencies.

Nowadays the intrinsic sluggishness of thermal devices can be overcome by using electronic feedback to keep the temperature (and hence the resistance) of the wire constant. In this way the upper frequency limit can be increased from about 500 Hz to about 1.2 MHz, which is adequate for almost any turbulent flow.

The operating circuit of the constant temperature anemometer is shown schematically in Fig. 3.1. The mode of operation can be summarized as follows:

(a) The bridge is initially in balance: no error voltage.
(b) Flow changes, probe resistance changes: hence an error voltage is generated.
(c) The error voltage is fed to the servo amplifier.

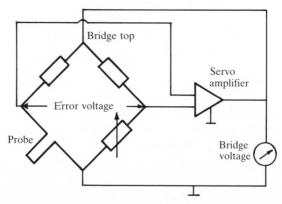

Fig. 3.1. Circuit diagram for a constant-temperature hot-wire anemometer.

(d) The servo amplifier increases or decreases the operating voltage at the bridge top in order to restore the balance.

(e) The velocity change appears as a change in the bridge voltage.

It shared be noted that the higher the gain of the amplifier, and the faster its response, the lower is the error voltage needed to compensate a temperature change in the sensor. This is the basis of the fast frequency response, despite the slow thermal response of the wire.

Hot wires (and hot films) can be used in all sorts of configurations, some of them very complicated, depending on the precise purpose of the anemometer. The interested reader will find much further information in the relevant chapter in Hinze (1975), in the review article by Comte-Bellot (1976), and in the specialist texts by Perry (1982) and Lomas (1986). We shall conclude this section with a discussion of the simplest, and most important, application.

Let us consider a single hot wire in a turbulent flow with mean velocity $\bar{U}_1$. Although the mean flow is unidirectional, we know that the fluctuating velocity has three scalar components, which we denote by (u_1, u_2, u_3). Our problem now is to distinguish between these three components, but this turns out to be a lot simpler than it might appear.

We align the wire with the x_3 axis. The physical situation is illustrated schematically in Fig. 3.2. This means that the u_3 component does not (at least to a good approximation) contribute to the cooling of the wire. That is, we are assuming that the rate of cooling is entirely determined by the velocity normal to the major axis of the wire, which is given by the vector sum of $\bar{U}_1 + u_1$ and u_2, or

$$U_N = \{(\bar{U}_1 + u_1)^2 + u_2^2\}^{1/2}$$

$$= \bar{U}_1 \left(1 + \frac{2u_1}{\bar{U}_1} + \frac{u_1^2 + u_2^2}{\bar{U}_1^2}\right)^{1/2}.$$ (3.4)

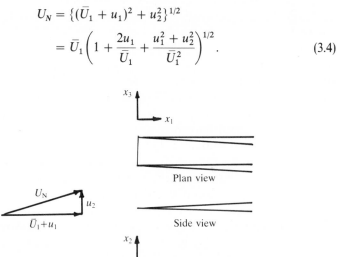

Fig. 3.2. Hot-wire configuration to measure the streamwise turbulent fluctuation in a unidirectional mean flow.

Then, provided that the turbulent intensities (i.e. $u_1/\bar{U}_1$ or $u_2/\bar{U}_1$) are small, we can usefully approximate this result by

$$\bar{U}_N = \bar{U}_1\left\{1 + \frac{u_1}{\bar{U}_1} + o\left(\frac{u_1^2}{\bar{U}_1^2}\right)\right\}. \tag{3.5}$$

Thus, to the first order of small quantities,

$$U_N \approx \bar{U}_1 + u_1 \tag{3.6}$$

or, in other words, the effective cooling velocity is approximately equal to the instantaneous streamwise velocity, provided that the velocity fluctuations are small compared with the mean.

3.1.2   The laser anemometer

The invention of the laser anemometer is usually attributed to Yeh and Cummins (1964), who showed that steady fluid velocities could be measured by observing the Doppler shift in the frequency of laser light scattered by small particles moving with the fluid. Over the following decade the method was extended, refined, analysed, and generally improved to the point where it is now routinely used in research in fluid dynamics and (in particular) in turbulence. (Plate I)

Many of the pioneering workers were content to interpret the basic operating principle in terms of the Doppler shift of scattered light, and this fact has left us with the legacy that the method is normally referred to as 'laser-Doppler anemometry' (LDA) or (sometimes) as 'laser-Doppler velocimetry' (LDV). However it is now widely accepted that the operation of the laser anemometer can be more easily understood in terms of conventional optical concepts like interference and diffraction. Thus, although we are stuck with the above acronyms, in this chapter we shall introduce laser anemometry without relying on the Doppler effect. However, in the fuller discussion of the background to the subject given in Appendix F we shall find it helpful to make use of the general analysis in terms of the Doppler shift of the scattered light.

The laser anemometer can be set up in many different optical configurations. In Fig. 3.3 we show an optical arrangement which is probably the most widely used and is certainly the easiest to understand. The incident beam of laser light is split into two separate beams by a suitable beam splitter. There are, of course, many ways of doing this, and the precise form of device is not critical. However, it is generally advantageous if the resulting beams are of about equal intensity and symmetrically disposed about the incident beam.

A convex lens is used to focus the two beams to a point where they form a pattern of interference fringes. This is the measuring point and, for convenience, we take it as the origin of coordinates. The angle of intersection is 2θ and this is simply related to the beam separation and the focal length of

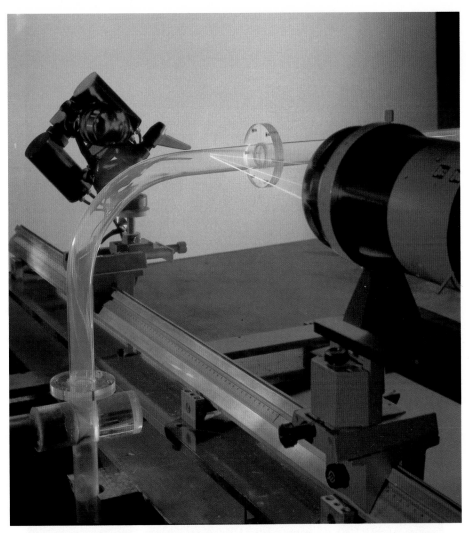

Plate I. A laser-Doppler anemometer being used to measure the velocity of a fluid. This device measures the fluid velocity at the point where the light beams intersect. (Courtesy of Dantec Electronics Ltd)

Flow

Plate II. A turbulent spot in a laminar boundary layer (Gad-el-Hak *et al.* 1981).

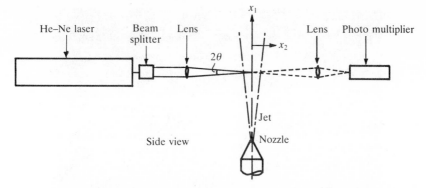

Fig. 3.3. The laser anemometer: a typical optical arrangement.

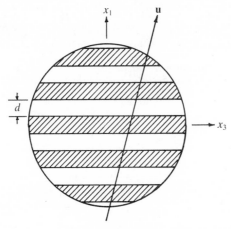

Fig. 3.4. Schematic view of a particle with velocity vector **u** crossing the interference fringes.

the lens. The set-up is completed by a second lens which focuses scattered light onto a photodetector, which in turn outputs a voltage proportional to the incident light intensity.

In order to have a definite example, we show the anemometer set up to measure the axial component of velocity u_1 in a free jet from a nozzle. The incident beams lie in the $x_1 x_2$ plane. In Fig. 3.4 we show a cross-section through the fringe pattern in the $x_1 x_3$ plane (of course, the fringe pattern consists of an ellipsoidal volume of small but finite extent; this is often referred to as the probe volume and it is only on the overall scale of the flow that we can regard it as a point). When a particle in the jet moves through the fringe pattern with velocity **u**, it scatters light from each bright fringe. This light is

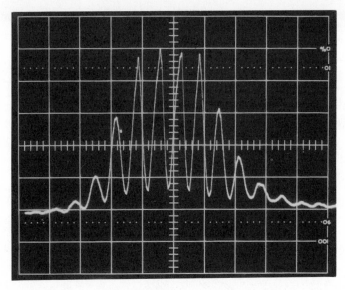

Fig. 3.5. Oscilloscope trace showing the photomultiplier output voltage corresponding to the light scattered by one particle crossing the fringe pattern.

collected and focused onto the photodetector, which produces an output current proportional to the scattered light intensity.

Figure 3.5 shows a typical laser anemometer signal in the form of an oscillograph. This is for one particle crossing the fringes and reflects the sinusoidal intensity variation (modified by the laser beam's own Gaussian profile) in the fringe pattern. We should also note that the alternating signal is on a pedestal or d.c. level. Details of these optical technicalities can be found in Appendix F; here we shall concentrate on the relationship between the static fringe pattern and the time-dependent signal on the oscilloscope.

As the particle crosses the fringes, clearly only the u_1 component of its velocity can contribute to the oscilloscope signal; the u_2 and u_3 components do not cause fringe crossings. If the distance between the fringes is d, then the time between fringes at a speed u_1 is d/u_1. Hence it follows at once that the frequency observed on the oscilloscope is

$$f_D = u_1/d \tag{3.7}$$

where the subscript D stands for Doppler. It is shown in Appendix F that the fringe spacing is given by

$$d = \frac{\lambda}{2\sin\theta}$$

where λ is the wavelength of the laser light (typical value 633 nm for a He–Ne laser). Thus, eqn (3.7) can be written as

$$u_1 = \frac{f_D \lambda}{2 \sin \theta}. \tag{3.8}$$

This is a special case of the general result of an analysis in terms of light scattering and the Doppler effect. In that context, this particular optical arrangement is known as the 'differential Doppler mode'.

Equation (3.8) is the operating relationship for the laser anemometer, and two points stand out at once. First, unlike the relationship for a hot-wire anemometer, it is a linear expression. Second, the constant of proportionality in the relationship between the measured Doppler frequency and the particle velocity depends only on the wavelength of the laser light and the angle of intersection of the beams, both of which can be established in advance. Hence the instrument does not require any calibration and can be regarded as a primary standard.

These points, taken in conjunction with the obvious fact that the laser anemometer does not perturb the flow in any way, make this a formidable instrument for research in turbulence. However, on the debit side, the fluid must be transparent, there may be some uncertainty about how well the scattering particle is representative of the actual fluid motion, and the signal-to-noise ratio is not usually as good as for the best hot-wire anemometers (e.g. see Melinand and Charnay 1978; Lau, Whiffen, Fisher, and Smith 1981).

Normally when a laser anemometer is being used, there will be several particles in the probe volume at any one time, and the effect on the output signal can be seen from Fig. 3.6. It should be noted that the trace in Fig. 3.6(a) is just that of Fig. 3.5 after high-pass filtering to remove the d.c. pedestal level. In Fig. 3.6(b), the effect of scattering from several particles is seen to reduce the signal-to-noise ratio, but, by way of compensation, the signal takes a more or less continuous form which can be helpful when it comes to signal processing. Some ways in which the continuously varying Doppler frequency (due to the continuously varying fluid velocity) output can be turned into a voltage are discussed in Appendix F.

We conclude this section by considering the important technique of frequency shifting. It is helpful to introduce this concept through a consideration of the signal spectrum.

If we denote the output current of the photodetector by $i(t)$, then we can obtain the corresponding spectrum $i(f)$ in the usual way by Fourier transformation. Let us begin by considering the single particle signal as exemplified by Fig. 3.5. If this signal were just a cosine wave and a d.c. pedestal level, then the Fourier transform would consist of a spike at zero frequency and another at the Doppler frequency. In fact there is also a finite-size effect, which in this case shows up as the Gaussian modulation due to the intensity profile of the laser beams. As a Gaussian is self-reciprocal under Fourier transformation,

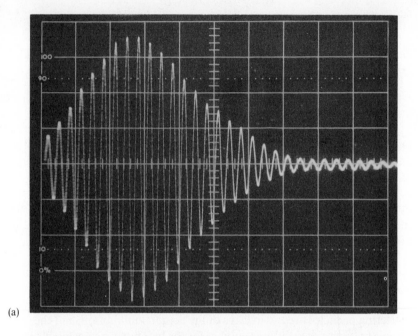

(a)

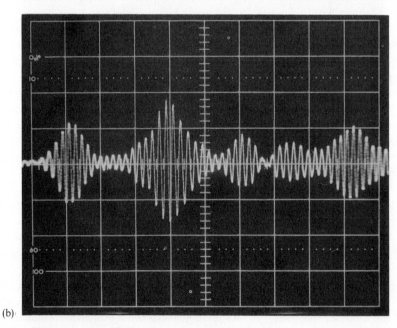

(b)

Fig. 3.6. High-pass-filtered photomultiplier signals for (a) one particle and (b) several particles crossing the fringe pattern.

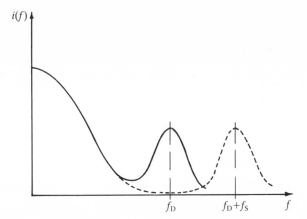

Fig. 3.7. Sketch showing the frequency spectrum of the output photocurrent for a given mean velocity: the broken line shows the effect of a frequency shift f_s.

this means that $i(f)$ for a single-particle signal consists of Gaussian peaks at $f = 0$ and $f = f_D$.

In practice the photocurrent $i(f)$ will be due to a number of particles (as in Fig. 3.6) and this leads to additional sources of spectral broadening such as the random arrival rate of particles, variations in particle speed, velocity gradients across the probe volume, and (not least) turbulent fluctuations. A typical result is sketched in Fig. 3.7, where we also show the effect of a frequency offset f_s. Clearly such an offset improves the signal by making it easier to discriminate between the d.c. and a.c. peaks. However, it has a greater importance than that, as we can see by considering how such a frequency shift is obtained.

Let us return to the basic optical configuration, as in Fig. 3.3. Normally the incident laser beam is split into two parts by some arrangement of mirrors and/or prisms. However, suppose that we were to use a diffraction grating, with all but, say, the two first-order beams masked off. Then the fringe pattern could be interpreted as an image of the lines on the grating. Hence, if the grating were to move, the fringes would also move, and the apparent velocity of a particle would now depend on whether it was travelling in the same or the opposite direction to the fringes. In this way the sign of the particle velocity can be detected, as well as its magnitude.

A practical method of achieving a frequency shift is to use a radial grating which is rotated by a small electric motor. However, the usual way nowadays is to use the method illustrated in Fig. 3.8. A travelling-phase grating is created by passing a sound wave through an acousto-optic (Bragg) cell. The principle is the same, but there is a definite advantage in not having any moving parts.

Apart from the material presented in Appendix F, the subject can be

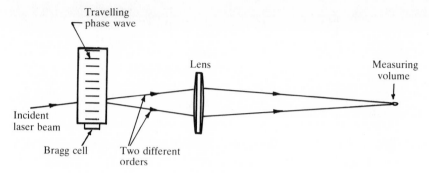

Fig. 3.8. Optical arrangement to impose a frequency shift on the laser anemometer output signal.

explored further in the review by Buchave, George, and Lumley (1979), and in the books by Durst, Melling, and Whitelaw (1976), Durrani and Greated (1977), Watrasiewicz and Rudd (1977), and Drain (1980). A more general look at what can be done with lasers in flow measurement can be found in the review by Lauterborn and Vogel (1984).

3.1.3   Time series analysis and computer data processing

Traditionally the output voltage of an anemometer would be processed by a variety of specialized analogue instruments. That is, the mean velocity would be obtained from a digital voltmeter, the r.m.s. velocity from an r.m.s. voltmeter, correlations from a correlator, energy spectra from a spectrum analyser, and so on. Nowadays there is a growing tend to convert the anemometer output into digital form (indeed in a modern system it is probably already in such a form and is only converted back into an analogue output signal to keep old-fashioned customers happy!), and then all the familiar signal-processing operations can be replaced by their mathematical equivalent on the computer. Given the power (and ubiquity) of the modern microcomputer, it is now possible to have on-line data processing of unprecedented versatility. Accordingly, we complete this section with some brief notes on this subject.

In order to have a specific example, we assume that we are concerned with one scalar component of the turbulent velocity field at a fixed point in the flow. We represent this by $u(t)$, which we take to be a continuous random variable defined everywhere on the interval $-\infty \leqslant t \leqslant \infty$. We obtain a discrete representation u_n of this continuous function by sampling it at regular intervals Δt such that

$$u_n = u(n \, \Delta t). \tag{3.9}$$

Then the sampling theorem (e.g. Otnes and Enochson 1972) essentially states that the infinite sequence of numbers $\{u_n\}$ can be used to reconstruct $u(t)$, provided that the sampling rate $1/\Delta t$ is more than twice as large as the maximum frequency in the energy spectrum of $u(t)$. If this condition is not satisfied, then the reconstructed function $u(t)$ will be subject to aliasing errors.

Now let us restrict our attention to stationary processes such that time average of eqn (1.6) can be written as

$$\bar{u} = \lim_{T \to \infty} \frac{1}{2T} \int_{-T}^{T} u(t)\, \mathrm{d}t. \tag{3.10}$$

Evidently the equivalent result obtained from the digitized data will take the form

$$\bar{u} = \lim_{N \to \infty} \frac{1}{N} \sum_{n=0}^{N-1} u_n. \tag{3.11}$$

Of course, in practice $u(t)$ is normally defined on some finite interval and normally the limit in (3.10) is satisfied for values of T which are not inconveniently large. Clearly it is important to obtain an adequate number of samples in the time series $\{u_n\}$ in order to ensure that $\bar{u}$ as defined by (3.11) is a good estimate of $\bar{u}$ as defined by (3.10).

These results can be extended in obvious ways to obtain estimates for the mean square and the autocorrelation of $u(t)$ (Otnes and Enochson 1972). In practice the minimization of errors is a pragmatic matter of choosing a sufficiently long record length. However, in the case of spectra, the situation is a little more complicated and we shall touch on this briefly here.

There are various ways of calculating spectra from the digitized record $\{u_n\}$. Formerly, the favoured method was to Fourier transform the correlation function (Blackman and Tukey 1958), but nowadays the use of the fast Fourier transform (FFT) algorithm (Cooley and Tukey 1965) seems to be universal. However, irrespective of the method, there are two technical difficulties which have to be taken into account.

The first of these arises from the finite record length, which means that the time series has effectively been pre-filtered. The Fourier transform of this filter is therefore convolved with the estimate of the spectrum and tends to distort it. In practice it is usual to impose an artificial pre-filter in order to minimize the effects on the spectrum. The second problem arises from the statistical variability of the spectrum and boils down to the fact that the error in the spectral estimate cannot be reduced by increasing the record length. Again, this requires some preliminary processing of the digital data, and the interested reader should refer to one of the specialist texts such as the book by Otnes and Enochson (1972).

3.2   Intermittency and the turbulent bursting process

Anyone who has carried out the classical Reynolds pipe-flow experiment (and this is often done as part of an undergraduate course in fluid dyamics) will have encountered a very strange effect. The usual way to do this experiment is to set the pump to give some particular rate of flow, which is then noted, as is the accompanying pressure drop along the pipe. The latter can be measured on a manometer connected to pressure tappings in the wall of the pipe. Then the pump speed is increased and the new flow rate and pressure drop noted, and so on.

Eventually a speed is reached where the manometer levels begin to oscillate wildly. This behaviour continues over a range of speeds, but ultimately a speed is reached where the manometer reading steadies again and thereafter remains steady (however much the speed is increased). This wild oscillation of the manometer is the 'strange effect' referred to above. It can be explained by the fact that the transition from laminar flow to turbulence is not an abrupt phenomenon, occurring at a particular Reynolds number, but takes the form of an alternation between the two states. That is, for a range of values of the Reynolds number (known as the 'transition range'), patches of turbulence are interspersed with patches of laminar flow. Hence, if an anemometer were to be set up in the flow it would register a turbulent signal alternating with a steady signal. This is an example of what we mean by intermittency.

Another example can be found in the boundary layer, as discussed in Chapter 1. Referring to Fig. 1.1, it is apparent that if we were to position an anemometer at the mean edge of the boundary layer then it would only spend part of its time in the turbulent fluid. The rest of the time the anemometer would be in the irrotational free stream. Again, the output signal representing the velocity would have a binary character, alternating between periods of random variation on the one hand and the steady levels indicative of irrotational flow on the other. The relative amount of time spent in each phase would of course itself be a random variable.

In recent years it has been discovered that intermittencies of these kinds are associated with various types of regular or quasi-deterministic behaviour. The generic term is 'coherent structures', and a particularly important one is the bursting phenomenon which appears to be the way in which turbulence is generated in the neighbourhood of a solid surface. We shall discuss the subject of coherent structures in some detail later on: it is nowadays regarded as a crucial aspect of shear flow turbulence. But for the moment we shall turn our attention to what is probably a more fundamental kind of intermittency (in the sense that it is intrinsic to the turbulent cascade), that is, 'fine-structure' intermittency.

This was first reported by Batchelor and Townsend (1949), who used a hot-wire anemometer to study turbulence behind a grid. They obtained oscil-

Look @

11 - 13 14

Recommendations
of Amit !

lograms corresponding to the velocity field and (by differentiating the output signal of the anemometer) its derivatives, which they claimed showed that the energy associated with small wavenumbers was not distributed uniformly throughout space. Instead, there appeared to be regions which were active and those which were (relatively) quiescent, this tendency becoming more marked with increasing order of the derivative. As the viscous dissipation of turbulent kinetic energy is mainly associated with regions where the velocity gradients are large, this fine-structure intermittency also implied that the dissipation may be distributed through the fluid in a rather spotty way.

On a vortex-stretching picture of the turbulent cascade, this behaviour is not too difficult to understand. Irrespective of the precise details of the vortex structures involved, it is clear that there must be some drawing out of (say) vortex tubes such that, not only is energy concentrated in the small scales, but also, concomitantly, in small regions of space. However, the most interesting question (which has aroused a great deal of interest in recent years) is what effect this has on the Kolmogorov predictions about the energy spectrum in the inertial range of wavenumbers. We shall consider this question in the next two subsections, and again at various points later in the book.

3.2.1 *Fine-structure intermittency*

In discussing intermittency it is helpful to work with structure functions, rather than the correlations which we have used in the preceding chapters. In fact structure functions are just correlations of two-point velocity differences, and we shall define them formally after first introducing the concepts of local homogeneity and local isotropy as follows.

Let us take new field variables to be the differences between field values at pairs of points, such as

$$\Delta \mathbf{u} = \mathbf{u}(\mathbf{x}, \mathbf{r}) - \mathbf{u}(\mathbf{x}) \qquad (3.12)$$

for the pair of points $\mathbf{x} + \mathbf{r}$ and $\mathbf{x}$. Then, if the probability distribution of the difference variable $\Delta \mathbf{u}$ is invariant under translations, reflections, and rotations of the pairs of points, it follows that $\Delta \mathbf{u}$ is a homogeneous isotropic field. We then say that $\mathbf{u}$ is locally homogeneous and isotropic.

For convenience let us take u to be the streamwise component of the turbulent fluctuating velocity and r to be the separation of measuring points in the direction of flow. This conforms to the normal experimental practice which allows time and space variations to be related by the Taylor hypothesis of frozen convention, as discussed in Section 2.6.5. If we also let u' be the field value at $x_1 + r$ and u corresponds to the point x_1 (where x_1 is the direction of flow), then the moments of the difference field are the structure functions of the velocity field

$$D_{LL}(r) = \langle (u' - u)^2 \rangle, \tag{3.13a}$$

$$D_{LLL}(r) = \langle (u' - u)^3 \rangle, \tag{3.13b}$$

$$D_{LLLL}(r) = \langle (u' - u)^4 \rangle, \tag{3.13c}$$

and so on. Note that the subscript L stands for longitudinal, as these are longitudinal correlations in the sense of Section 2.4.1. Note also that the connection between these structure functions and the correlations which we have used hitherto is readily established by expanding the products and averaging. For example, we can easily show in this way that the second-order structure function $D_{LL}(r)$ can be expressed in terms of the longitudinal correlation $f(r)$, as defined by eqn (2.55), and of course can be further related to the energy spectrum through eqns (2.59), (2.90), and (2.101).

The real interest in the study of fine-structure intermittency is its implications for the physics of the energy cascade and, in particular, the Kolmogorov theory. In 2.7.2 we gave the Kolmogorov similarity principles in terms of the wavenumber spectrum. However, in his original papers, Kolmogorov (1941a, b) worked with the structure functions and obtained the general result

$$\langle (u' - u)^n \rangle = C_n \varepsilon^{n/3} r^{n/3}, \tag{3.14}$$

where the C_n are constants and n takes integer values.

The constant C_3 can be evaluated using the energy-balance equation (see Hinze 1975) and the first three structure functions written explicitly as

$$D_{LL}(r) = C_2 \varepsilon^{2/3} r^{2/3} \tag{3.15a}$$

$$D_{LLL}(r) = \frac{4}{5} \varepsilon r \tag{3.15b}$$

$$D_{LLLL}(r) = C_4 \varepsilon^{4/3} r^{4/3}. \tag{3.15c}$$

As we shall see presently, the phenomenon of intermittency becomes apparent in the time derivatives of the velocity field, and its effects increase with the order of the derivative. As usual we remind ourselves that the time derivative is equivalent to the spatial derivative in the streamwise direction (in the present context, the longitudinal direction) when we invoke the Taylor hypothesis of frozen convection.

The most sensitive statistical measures of intermittency in a random variable are the skewness (S) and flatness (F) factors of its probability distribution. For the specific case of the longitudinal velocity difference these are given by (Batchelor 1971)

$$S = \frac{\langle (u' - u)^3 \rangle}{\langle (u' - u)^2 \rangle^{3/2}} \tag{3.16}$$

and

$$F = \frac{\langle (u' - u)^4 \rangle}{\langle (u' - u)^2 \rangle^2}. \tag{3.17}$$

Then, substituting from eqns (3.13a)–(3.13c) and (3.15a)–(3.15c), it follows that S and F are constants provided that the separation variable r is in the inertial range of length scales.

Returning to the pioneering paper by Batchelor and Townsend (1949), we can now be more specific about their results. These authors measured the longitudinal velocity and its derivatives up to third order in grid turbulence and a wake flow. In the case of the velocity field u they found flatness factors of about 3, in agreement with the well-known result that the distribution of turbulent velocities at a point is approximately Gaussian or normal. However, the values found for the flatness factors of the velocity derivatives were always significantly greater than 3, and increased with both the order of the derivative and the Reynolds number up to a value of about 7.

The existence of fine-structure intermittency received further confirmation from the investigations of Sandborn (1959) on boundary layers and Kennedy and Corrsin (1961) on free-shear layers. In both cases measurements were made in the fully turbulent part of the layer, and in both cases their results agreed quite well with those of Batchelor and Townsend (1949). However, Wyngaard and Tennekes (1970) found very much larger values of flatness (up to $F \approx 40$) in an atmospheric boundary layer.

Kuo and Corrsin (1971) carried out additonal work on grid turbulence and analysed the results of eight different investigations. By plotting the various results in the form of flatness against Reynolds number, they showed that most of the investigations were in fairly good agreement for values ranging over $10 < R_\lambda < 2 \times 10^3$ with $4 \leqslant F \leqslant 40$. They also remind the reader of the danger of relying on the flatness factor alone as an indication of intermittency, and advocate direct measurement of the intermittency factor γ. This was first introduced by Townsend (1948) in the context of free-surface intermittency, and is defined by

$\gamma =$ factor of time the detection probe sees the variable in its higher amplitude state.

The effect of intermittency on the measured flatness factor can be illustrated using a simple example. Let us consider a field $u(x, t)$ which has a normal probability distribution in its active state, and is intermittent between this and a completely inactive state when u and all its derivatives are zero. If we denote the global average by $\langle \ \rangle$ and the local average over the statistically active state by $[\ ]$, then it is clear that, with the conditions imposed, the moments of u will satisfy

$$\langle u^n \rangle = \gamma [u^n],$$

irrespective of the order of the moment involved. It then follows at once from

(3.17) that the globally averaged flatness factor is given by

$$F = 3/\gamma,$$

and evidently this would result in increased apparent flatness factors in our simplified example.

Now our main interest in this topic lies in its relevance to the physics of the energy cascade. Therefore at this point we shall consider the effect of intermittency on the major feature of the cascade: the energy dissipation rate ε. From eqn (2.121), it is clear that we have been using the concept of a mean dissipation rate and that our treatment of the Kolmogorov picture of the cascade, for example, was based on such a concept. There is nothing particularly unusual in our slurring over the distinction between a global and a mean quantity; after all, no one ever refers to 'mean boundary-layer thickness', although that is of course just what it is.

However, when we return to our basic definition of the dissipation rate and generalize (1.17) for the instantaneous field to a fluctuating field with zero mean

$$2\varepsilon(\mathbf{x}, t) = \nu \sum_{\alpha, \beta} \left(\frac{\partial u_\alpha}{\partial x_\beta} + \frac{\partial u_\beta}{\partial x_\alpha} \right)^2, \tag{3.18}$$

it follows from the nature of $u_\alpha(\mathbf{x}, t)$ that $\varepsilon(\mathbf{x}, t)$ is also a random variable. It also follows that, as ε depends on the velocity gradients, the dissipation rate will itself be intermittent with a rather spotty spatial distribution.

This poses a problem of principle for the derivation of the Kolmogorov spectrum, as given in Section 2.7.2, where the dissipation rate is treated as if it were a constant. Evidently the distribution of the dissipation rate must be taken into account, and there is the possibility that this may depend on the nature of the flow and its external length scales.

Apparently these points were first made by Landau shortly after the Kolmogorov theory was published (e.g. see Kolmogorov (1962), or the footnote on p. 126 of the book by Landau and Liftshitz (1959)), although Grant, Stewart, and Moilliet (1962) also came to this conclusion on the basis of their experimental observation that ε varied widely in oceanic flows. Using a simple model for the distribution of ε, these latter authors concluded that the effects of intermittency on the energy spectrum would be on the intensity level rather than on the power law, and even then would be small. We shall examine these topics further in the next section.

3.2.2   *Intermittency and the energy cascade*

Theories of the effect of fine-scale intermittency on the non-linear transfer of energy to the small scales can be divided into two broad classes: vortex-stretching theories and scale-similarity theories.

In Section 2.7.3 we have already touched on the use of mechanistic models based on assumptions about the shape of the vortex structures making up the turbulent vorticity field. In order to model fine-scale intermittency, Corrsin (1962) assumed that the fine structure was made up of vortex sheets of thickness comparable with the dissipation length scale, and with a mean separation equal to the integral length scale. Tennekes (1968) refined this model by taking the basic structures to be vortex tubes, and Saffman (1968) produced an even more complicated model using an assembly of vortex sheets and tubes. As we mentioned in the previous chapter, this approach is be-devilled by lack of *a priori* knowledge of the geometry of the vortex field (or indeed whether the continuous field really can be represented by such discrete structures), although Kuo and Corrsin (1972) reported that the geometry of the fine structure was more filament like than anything else, a result supported by Kerr's numerical simulation.

The scale-similarity theories (like the vortex models) have their origins in remarks made in the pioneering paper by Batchelor and Townsend (1949). These authors suggested that the simplest possible model of turbulent inter-mittency was to imagine that space is divided into a number of regions, each with its own value of the energy dissipation rate and in each of which the energy spectrum takes the universal Kolmogorov form (2.135). Thus varia-tions in $E(k)$ from one region to another are merely variations of excitation level and are proportional to the local dissipation rate.

More specifically, Batchelor and Townsend (1949) discussed a stepwise process in which the dissipation associated with increasing wavenumber becomes increasingly concentrated in smaller regions of space. Later analy-tical theories have put a simple geometric interpretation on this picture. Consider the fluid to occupy a volume with characteristic length l_0. In order to have a specific example, we shall take this to be a cube. Now divide this cube up into an arbitrary large number of smaller cubes of side $l_1 (l_1 < l_0)$. The simplest picture of intermittency is where the dissipation is not uniformly distributed over all the cubes, but is only found in some of them. Similarly, further subdivision into smaller cubes with side $l_2 < l_1$ would show that the dissipation is only contained in some small number of these smaller cubes, and so on.

The first attempts to modify the Kolmogorov similarity hypotheses to take account of the dissipation rate fluctuations were due to Kolmogorov (1962) and Obukhov (1962), who proposed that the logarithm of the dissipation rate was normally distributed. In particular, if the dissipation rate is averaged over a sphere of radius r to give a mean rate ε_r, then ε_r is assumed to have a log-normal distribution with variance given by

$$\sigma^2_{\ln \varepsilon_r} = A(\mathbf{x}, t) + \mu \ln\left(\frac{L}{r}\right), \tag{3.19}$$

where A may depend on the large-scale motion, μ is a universal constant, and L is the external length scale. Kolmogorov (1962) argued that the log-normality of the locally averaged dissipation rate (along with (3.19)) amounted to a third similarity principle and rederived the expression for the n-order structure function, so that (3.14) was replaced by

$$\langle (u' - u)^n \rangle = \tilde{C}_n(\mathbf{x}, t) \varepsilon^{n/3} r^{n/3} \left(\frac{L}{r} \right)^{\mu n(n-3)/18} , \qquad (3.20)$$

where the $\tilde{C}_n(\mathbf{x}, t)$ are no longer constants and may depend on the large-scale details of the flow. Fourier transformation shows that there is a corresponding effect for spectra (we shall discuss this shortly), while the modified skewness and flatness factors become

$$S(r) = -\frac{4}{5} \tilde{C}_2^{-3/2} \left(\frac{L}{r} \right)^{\mu/6} \qquad (3.21)$$

and

$$F(r) = \tilde{C}_4 \tilde{C}_2^{-2} \left(\frac{L}{r} \right)^{4\mu/9} . \qquad (3.22)$$

Although Gurvich and Yaglom (1967) have presented a derivation of the log-normal distribution of the dissipation, it is not free from assumption, and indeed this is really the overall status of the log-normality hypothesis. Nevertheless it has attracted a good deal of attention over a period of about two decades, particularly from experimentalists with a considerable outpouring of papers on the subject.

We shall not attempt a comprehensive summary: the interested reader can consult Monin and Yaglom (1975). Here we shall just give a representative sample, beginning with the measurements of Wyngaard and Tennekes (1970, mixing layer), Gibson, Stegen, and McConnell (1970, atmospheric boundary layer over the ocean), Van Atta and Chen (1970, atmospheric boundary layer over the ocean), and Sheih, Tennekes, and Lumley (1971, air-borne measurements in the atmosphere), where each experimental environment is given in parentheses with the date of the investigation. In each case these reports supported the log-normality hypothesis. In particular, Van Atta and Chen (1970) noted that second- and third-order structure functions were consistent with the original Kolmogorov theory, especially at large separations. However, the fourth-order structure function gave better agreement with the refined (log-normal) theory.

The case against (so to speak) comes from Gibson and Masiello (1972, atmospheric boundary layer over the ocean), who found departures from log-normality, and Frenkiel and Klebanoff (1975, grid turbulence, boundary on a flat plate), who found that their measurements bore out log-normality provided that the constant μ in Kolmogorov's third hypothesis was not in fact constant. Moreover, from a survey of many experimental investigations, Van Atta and Yeh (1973) remark that, in general, probability distributions have not been found to be log-normal, although they qualify this by pointing out that log-normality is often a good approximation in certain restricted ranges and that it may also depend on the way in which averages are taken. Further, from an analysis of Wyngaard's data, they conclude that there is a lack of support for scale-similarity theories, such as that of Gurvich and Yaglom (1967). However, more recently, Van Atta and Antonia (1980) have made a more refined analysis of earlier data, and have concluded that the predictions of the modified similarity hypotheses fit the experimental results rather well, provided that the constant μ is taken to be about 0.25, rather than about 0.50 as the earlier investigators had suggested.

The refined Kolmogorov–Obukhov theory often appears to be regarded in the literature as if the third (i.e. log-normal) hypothesis had the same physical plausibility as the other two. Yet log-normality is not a necessary ingredient in a scale-similar theory of intermittency (e.g. Novikov and Stewart 1964; Mandelbrot 1972, 1974; Frisch, Sulem, and Nelkin 1978). As an example, we shall briefly discuss the latter theory.

Frisch et al. (1978) begin by adopting the Kolmogorov (1941a, b) picture. In turbulence, energy is injected into eddies of size l_0 and cascaded down through intermediate scales l_n to the dissipation scale l_d. They assume that this process can be represented by the discrete series of length scales

$$l_n = 2^{-n} l_0 \qquad n = 0, 1, 2, \ldots \tag{3.23}$$

with the corresponding discrete wavenumbers $k_n = 1/l_n$. The argument is then conducted in terms of magnitudes, with numerical factors being dropped except when they would accumulate multiplicatively. Thus, if the energy (i.e. kinetic energy of turbulent fluctuations per unit mass of fluid) in scales approximately equal to l_n is E_n, we can define the r.m.s. velocity difference across a distance l_n by v_n, where

$$E_n \approx v_n^2. \tag{3.24}$$

(We should note that, with this definition, the correlations of the v_n will be closely related to the structure functions, as defined by (3.13).) The eddy turnover time can also be defined as

$$t_n \approx l_n/v_n. \tag{3.25}$$

Frisch et al. then make the fundamental assumption that a sizeable fraction of the energy in scales l_n is transferred to scales l_{n+1} in time t. Hence the energy transfer rate (for the n-order eddies) is given by

$$\varepsilon_n \approx E_n/t_n \approx v_n^3/l_n. \tag{3.26}$$

For stationary turbulence, conservation of energy then implies that

$$\varepsilon_n = \langle \varepsilon \rangle \qquad l_0 > l_n > l_d \tag{3.27}$$

where $\langle \varepsilon \rangle$ is the mean dissipation rate.

At this point, Frisch *et al.* remind us that $\langle \varepsilon \rangle$ can be interpreted equally well as a rate of energy injection or as a rate of energy transfer, and that the latter is the dynamically relevant quantity for the inertial range. In this respect, they go some way to meeting Kraichnan's (1974) critique of the refined Kolmogorov (1962) theories in which he argues that a central role for the dissipation is arbitrary since conservation of energy alone provides no link between the local dissipation rate and the local rate of energy transfer.

From (3.26) and (3.27) we have

$$v_n \approx \langle \varepsilon \rangle^{1/3} l_n^{1/3}, \tag{3.28}$$

and from (3.24)

$$E_n \approx \langle \varepsilon \rangle^{2/3} l_n^{2/3}, \tag{3.29}$$

which, after Fourier transformation, yields the familiar Kolmogorov spectrum as in, for instance, eqn (2.137).

The β model is introduced in terms of the rather arbitrary assumption that the average number of offspring of any eddy is N (i.e. an eddy of scale l_n is assumed to give rise to N eddies of scale l_{n+1}, irrespective of the value of n). Then the fractional reduction in volume from one generation to the next is given by

$$\beta = \frac{N l_{n+1}^3}{l_n^3} = \frac{N}{2^3} \leqslant 1, \tag{3.30}$$

where the second step follows from eqn (3.23). If it is further assumed that the largest eddies fill all the space available to them, then in the nth generation only a fraction

$$\beta_n = \beta^n \tag{3.31}$$

of the space will be occupied by eddies of scale l_n.

Now we repeat our previous arguments, but restrict them to the active volume of fluid for each generation n. That is, eqn (3.28) still holds locally for v_n in an active region. But the relationship between the globally averaged energy density and the locally averaged v_n is postulated to be

$$E_n \approx \beta_n v_n^2 \tag{3.32}$$

and, using (3.28), (3.30), and (3.31), this becomes

$$E_n \approx \langle \varepsilon \rangle^{2/3} l_n^{2/3} \left(\frac{N}{2^3} \right)^{n/3}. \tag{3.33}$$

With the further assumption

$$N = 2^D \qquad (3.34)$$

and the use of (3.23) to eliminate n, eqn (3.33) can be written as

$$E_n \approx \langle \varepsilon \rangle^{2/3} l_n^{2/3} \left(\frac{l_n}{l_0} \right)^B, \qquad (3.35)$$

where

$$B = (D - 3)/3 \qquad (3.36)$$

and D is identified as the (non-integer) fractal dimension (Mandelbrot 1974); hence the exponent can also be identified as $B = \mu/3$.

The β model has attracted quite a lot of attention, if only because it is easily followed and was presented in a more than usually readable paper. It has inspired theoretical attempts to find a value of μ, and, for example, Fujisaka and Mori (1979) have used a maximum entropy principle to calculate $\mu = 1/3$. Experimental attempts to distinguish between the β model and log-normality (e.g. Press 1981; Antonia, Satyaprakash, and Hussain 1982) seem to favour the former, but are certainly not conclusive. A particular point of interest about the last reference is that these authors conclude that μ should have a value of about 0.2, and criticize some of the criteria used in early work that led to a generally accepted value of $\mu = 0.5$. More recently, Anselmet, Gagne, Hopfinger, and Antonia (1984) have found $\mu = 0.20 \pm 0.05$ and power-law behaviour of moments which is incompatible with either the β model or log-normality.

3.2.3   *Intermittent generation of turbulence: the bursting process*

In recent years it has become increasingly evident that turbulent flows are not just random or chaotic but can contain more deterministic features, now known as 'coherent structures'. This term is difficult to define in a succinct way, although it is universally employed and (presumably) universally understood. If we were to attempt a general definition, then we would say that, essentially, coherent structures are recognizable by most people when some appropriate method of flow visualization is employed. That is, we are talking about some discernible pattern in the flow, which may have many random features but nevertheless occurs with sufficient regularity, in space or time, to be recognizable as quasi-periodic or near-deterministic.

A specific example should help to make this clearer. Consider the mixing layer formed when two streams of different velocity run parallel to each other, as shown schematically in Fig. 3.9. It was found by Brown and Roshko (1974) that shadowgraphs of such mixing layers revealed the presence of vortex-like structures which, in general appearance, resembled breaking waves. A typical shadowgraph of this kind is sketched in Fig. 3.10.

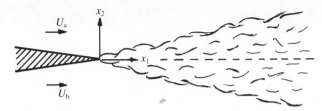

Fig. 3.9. Definition sketch of a plane mixing layer between two parallel streams with different velocities U_a and U_b.

Fig. 3.10. Sketch of a typical shadowgraph showing a regular wave-like structure in the plane mixing layer.

Results like this have been one of the factors in stimulating a renewed interest in vortex interactions. And, indeed, the dynamics of the interface have been analysed in terms of both vortex pairing (Winant and Browand 1974) and vortex tearing (Moore and Saffman 1975).

Although the mixing layer shows the most strikingly regular behaviour, quasi-deterministic structures are found in other flows such as wakes, free jets, boundary layers, and duct flows. We shall discuss these in some detail in a later chapter; here we shall concentrate on the particular coherent structure which is found in turbulent flows bounded by a solid wall. This is the 'bursting process'. In visual terms, this is one of the least spectacular of the coherent structures, yet in many ways it is the most interesting of these phenomena.

Our account of this subject must begin with the paper by Kline, Reynolds, Schraub, and Rundstadler (1967) who studied the turbulent boundary layer (in water) by placing a hydrogen-bubble wire parallel to, and at various distances above, the wall. They found that, even when the wire was as close to the wall as $x_2^+ = 2.7$ (which is well within the viscous sublayer), the bubbles did not move in straight lines along the plate but had a tendency to move sideways to some extent, such that they accumulated in a series of elongated regions known as 'streaks'. Various other measurements showed that the streaks are regions where the streamwise velocity is relatively low, and so they are often referred to as 'low speed streaks'.

The pictures obtained by Kline et al. (1967) may seem rather disappointing compared with, say, the roll vortices of the mixing-layer. Nevertheless, it should be appreciated that there must be some coherent motion in the spanwise (i.e. cross-stream) direction in order to organize the bubbles into streaks.

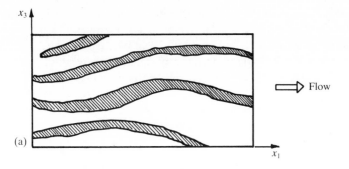

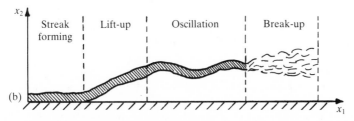

Fig. 3.11. Schematic view of turbulent boundary-layer structure near the wall: (a) top view of low speed streaks; (b) side view of the sequence of events leading to the break-up of a streak.

This is something which we shall discuss in some detail in a later chapter. Here we shall now consider the sequence of events when a streak is viewed from the side.

Kline *et al.* observed that the streaks interacted with the outer parts of the flow through the following sequence of events: lift up, oscillation, and break-up. They called this sequence a 'burst', and we have illustrated it schematically in Fig. 3.11. They also found that a negative pressure gradient (i.e. where pressure falls in the flow direction) tended to reduce the rate of bursting, with the reverse being the case with a positive gradient.

Noting that the first of these cases (i.e. $dp/dx_1 < 0$) corresponds to $dU_1/dx_1 > 0$, and recalling the discussion in Section 1.6.2 of the production term (1.69) in the energy balance, we see that the rate of bursting decreases in circumstances where the rate of production of turbulent kinetic energy also decreases. Hence it was natural for Kline *et al.* to conjecture that the bursting process plays a dominant role in the production of turbulent kinetic energy in the boundary layer. They also speculated that the subsequent ejection of the fluid from the burst into the outer parts of the boundary layer was the main mechanism for transferring energy (and other turbulent quantities) to these outer regions.

It should be emphasized that the bursting process just described is not an artefact of the free-surface intermittency of the turbulent boundary layer. Shortly after the paper by Kline *et al.* (1967), an independent investigation by Corino and Brodkey (1969) reported the existence of an 'ejection-sweep' cycle in pipe flow, which was clearly equivalent to the bursting process of Kline *et al.* Since then many investigations have confirmed the importance of the bursting process for the production of turbulence in the presence of solid boundaries. We shall discuss these in some detail in Chapter 11. For the moment it is perhaps not too fanciful to observe that our traditional picture of a once and for all catastrophic breakdown from laminar flow to turbulence has to be replaced by a picture where at any point in, say, a pipe flow, the laminar–turbulent transition is taking place in a rather regular way several times a second.

3.3 Numerical computation of turbulent flows

The problem of calculating turbulence can be seen as part (albeit, a very large one) of the more general field of computational fluid dynamics (CFD). As a glance at any good general text on the subject (e.g. Roache 1982, Peyret and Taylor 1983) will show, CFD is extensively concerned with the numerical representation and computation of the partial differential equations which govern the motion of real fluids.

The subject tends to divide into two areas: the development of numerical methods, and the creation of algorithms to implement these methods. At the same time, progress in CFD must necessarily depend on developments in computing (in all its aspects—hardware, software, languages, and operating systems). In particular, we should perhaps mention the growing use of concurrent computer architectures, which offer large increases in memory and speed.

There seems to be a consensus among writers in the field, that progress in all these topics has been very rapid over the last decade or two, so that the main obstacle to many practical applications of CFD is the turbulence aspect of the problem. That is to say, if the theoretical approach to any given task is based on the time-averaged Navier–Stokes equations, then the question of how to specify the Reynolds stresses normally turns out to be the crucial problem for CFD.

We shall presently give a brief review of current methods of closing the Navier–Stokes moment hierarchy in the context of engineering applications. But, first, it is interesting to consider whether we can evade this problem altogether, by looking for numerical solutions to the (unaveraged) Navier–Stokes equations. Of course such solutions can be expected to vary randomly with time and position. The question then is: would the extra work of representing random functions, and then forming averages numerically, outweigh the fundamental difficulties implicit in the moment closure problem? We shall consider this question in the next two subsections.

3.3.1   Direct numerical simulation (DNS)

If we were carrying out a numerical simulation of the behaviour of molecules in an ideal gas, for instance, we would first want to decide how large our system was to be and how many molecules it would contain. Obviously we would have to keep the size of our computer in mind when we made these decisions. The question of how similar decisions are made for turbulence was considered by Landau and Lifshitz (1959, p. 123) who estimated the number of degrees of freedom in turbulence as a function of the Reynolds number.

We consider isotropic turbulence and, as before, take it to be occupying a cubical box of side L. We associate with this length scale a velocity scale Δu, which is the variation of the velocity over a distance L. We can argue with some plausibility that over small distances $(<L)$ the velocity differences will also be small $(<\Delta u)$. Thus Δu can be taken as the external velocity scale, just as L is the external length scale. It can also be argued that, although the dissipation is ultimately determined by the action of viscosity on the smallest eddies, in the first place it is determined by the input from the largest eddies. Thus, on dimensional grounds, we can estimate the dissipation rate as

$$\varepsilon \approx \frac{(\Delta u)^3}{L}. \tag{3.37}$$

We can also introduce the Reynolds number R for the system as

$$R = \frac{L\,\Delta u}{v}. \tag{3.38}$$

Landau and Liftshitz (1959) estimated the number of degrees of freedom per unit volume of fluid on dimensional grounds. Denoting this by n, they found

$$n \approx \left(\frac{\varepsilon}{v^3}\right)^{3/4} = \frac{1}{\eta^3}, \tag{3.39}$$

where the last step follows from eqn (2.131) and η is of course the dissipation length scale. Then it follows that the total number of degrees of freedom of the whole system N is given by

$$N \approx \frac{L^3}{\eta^3}. \tag{3.40}$$

Intuitively, this seems quite a natural result. In a numerical simulation, η is the smallest scale that we would have to resolve. Thus in one direction we would have to do this L/η times, and clearly (3.40) follows for all three directions. We now need a relationship between η and L. This can be obtained by combining equation (2.131) for η with eqns (3.37) and (3.38):

$$\eta = R^{-3/4}L, \tag{3.41}$$

and hence, from (3.40),

$$N \approx R^{9/4}. \tag{3.42}$$

It should be appreciated that the number of degrees of freedom becomes zero for $R < R_{\text{crit}}$, where R_{crit} is the value of the Reynolds number for which the flow becomes turbulent. Landau and Liftshitz (1959) suggested that R should be replaced in (3.42) by its ratio to the critical value. However, we shall not pursue that here. Equation (3.42), as it stands, allows us to assess the effect of increasing Reynolds number.

If we now return to the analogy with the ideal gas—N non-interacting particles of mass m in a box—then the N degrees of freedom are the N kinetic energies of translation, which contribute additively to the total energy of the system E:

$$E = \sum_{i=0}^{N} \frac{mv_i^2}{2}. \tag{3.43}$$

Correspondingly, for turbulence, we can expect the N degrees of freedom to be the independently excited Fourier modes which also satisfy an additive law for kinetic energies,

$$E = \sum_{k=0}^{N} \frac{\langle u_k^2 \rangle}{2}, \tag{3.44}$$

where N is given by

$$N \approx \left(\frac{k_d}{k_{\text{min}}}\right)^3 \approx \left(\frac{L}{\eta}\right)^3 \tag{3.45}$$

and hence by eqn (3.42).

As a specific example, we shall now consider a method of simulating isotropic turbulence and obtaining the energy spectrum by numerical means. Our treatment of the subject is loosely based on the pioneering work of Orszag (1971), but is very much simplified in order to let the main ideas stand out. The reader who wishes to learn more about the detailed aspects of the subject should consult the original papers cited. The review papers by Orszag and Israeli (1974) and by Gottlieb, Hussaini, and Orszag (1984) may also prove helpful.

We begin by writing the Navier–Stokes equation, as given by (2.76), in the form

$$\left(\frac{\partial}{\partial t} + vk^2\right)u_\alpha(\mathbf{k}, t) = W_\alpha(\mathbf{k}, t) \tag{3.46}$$

and

$$W_\alpha(\mathbf{k}, t) = \sum_{\mathbf{j}+\mathbf{l}=\mathbf{k}} M_{\alpha\beta\gamma}(\mathbf{k})u_\beta(\mathbf{j}, t)u_\gamma(\mathbf{l}, t), \tag{3.47}$$

where $M_{\alpha\beta\gamma}(\mathbf{k})$ is given by (2.77) and the notation indicates that the right-hand side of (3.47) is summed over all pairs $(\mathbf{j}, \mathbf{l})$ which are such that $\mathbf{j} + \mathbf{l} = \mathbf{k}$.

We now move to a discrete representation in time and wavenumber. The former can be treated by putting $t = 0, 1, 2, \ldots$, in arbitrary units, and we concentrate here on the latter, which must be related to the computational box. First the discrete wavenumbers are introduced through the definition

$$k_\alpha = \frac{2\pi}{L} n_\alpha, \tag{3.48}$$

where

$$n_\alpha = 0, +1, +2, \ldots + (K - 1) \tag{3.49}$$

and K is an integer.

Thus the computation is to be carried out on the cubical mesh specified in this way. For isotropic turbulence we can go further and restrict wavenumbers to the sphere centred on $k = 0$ and bounded by the condition

$$|k_\alpha, j_\alpha, l_\alpha| < \frac{2\pi}{L} K. \tag{3.50}$$

This condition can be generalized to any intermediate wavenumber and hence the isotropic case can be calculated in a series of spherical shells.

A method of taking averages will be needed—even just to set the initial conditions—but we can exploit the isotropy of the turbulence to economize on computer time by averaging over spherical shells in wavenumber space. If we take the general shell to have a mean wavenumber k_n, and thickness $2\Delta k$, then the shell-averaged energy spectrum is defined by

$$E(k_n, t) = \frac{1}{2\Delta k} \sum_{|\mathbf{k}|} \left\{\frac{u_\alpha(\mathbf{k}, t)u_\alpha(-\mathbf{k}, t)}{2}\right\}, \tag{3.51}$$

such that

$$k_n - \Delta k < |\mathbf{k}| < k_n + \Delta k.$$

The general procedure now is to set up some arbitrary initial velocity and integrate eqn (3.46) forward in time. At each time step the shell averages can be worked out for each statistical quantity of interest and the calculation carried on until some appropriate criterion has been satisfied. For instance, the simulation might be terminated when the spectrum had achieved a self-preserving form: we shall discuss such details in Chapter 8 in connection with the evaluation of turbulence theories.

The initial conditions are specified by using a standard random number generator to provide three numbers for each mesh point and hence set up a random vector field $f_\alpha(\mathbf{k})$. Then we can introduce the random velocity field at $t = 0$ by writing

$$u_\alpha(\mathbf{k}, 0) = D_{\alpha\beta}(\mathbf{k})f_\beta(\mathbf{k}), \tag{3.52}$$

where the projection operator $D_{\alpha\beta}(\mathbf{k})$ is given by eqn (2.78). Recalling eqns (2.78) and (2.97), along with the discussion in Section 2.6.4, we see that this manoeuvre ensures that the initial velocity satisfies the continuity equation.

In order to ensure also that the initial velocity field is physically reasonable, we can set values for the mean and variance of the random numbers generated, such that the initial spectrum $E(k, 0)$, as obtained from (3.51) and the initial velocity field, takes some prescribed form. The choice of initial spectral forms is another aspect which will be discussed in more detail in Chapter 8.

With $u_\alpha(\mathbf{k}, 0)$ prescribed, eqn (3.46) can be stepped forward in time to $t = 1$ (in arbitrary units). Clearly the evaluation of $W_\alpha(\mathbf{k}, 0)$, by means of the convolution sum (3.47), will require a large calculation. This operation can be very much reduced by using a pseudospectral method (Orszag 1971), which relies on the fact that the convolution in $\mathbf{k}$-space is the Fourier transform of a local product in $\mathbf{x}$-space.

The method can be explained as follows. Rewrite (3.47) for $W_\alpha(\mathbf{k}, t)$ as

$$W_\alpha(\mathbf{k}, t) = M_{\alpha\beta\gamma}(\mathbf{k})A_{\beta\gamma}(k, t) \tag{3.53}$$

where $A_{\beta\gamma}(\mathbf{k}, t)$ is just the convolution of u_β with u_α. By the convolution theorem (see Appendix D), its Fourier transform is

$$A_{\beta\gamma}(\mathbf{x}, t) = u_\beta(\mathbf{x}, t)u_\gamma(\mathbf{x}, t). \tag{3.54}$$

Now we proceed as follows.

(a) Fourier transform $u_\alpha(\mathbf{k}, t)$ to obtain $u_\alpha(\mathbf{x}, t)$. This is, of course, done numerically and requires a reciprocal lattice upon which $\mathbf{x}$ takes discrete values.

(b) Work out the product specified by (3.54) to obtain a value for $A_{\beta\gamma}(\mathbf{x}, t)$ for every point on the reciprocal lattice.

(c) Fourier transform $A_{\beta\gamma}(\mathbf{x}, t)$ to obtain $A_{\beta\gamma}(\mathbf{k}, t)$ and work out $W_\alpha(\mathbf{k}, t)$, using (3.53), for each point on the lattice in wavenumber space.

(d) Use a suitable finite-difference method to calculate $u_\alpha(\mathbf{k}, t = 1)$ from (3.46).

The whole process is then repeated to calculate $u_\alpha(\mathbf{k}, t = 2)$ and so on. In practice, the fast Fourier transform algorithm can be used to speed up steps (a) and (c).

The advantage of the pseudospectral method is that it reduces the number of arithmetic operations compared with the direct evaluation of the convolution sums. In the former case the number of operations is proportional to $(2K - 1)^6$, whereas it goes as $2K^3 \log_2(2K)$ if the convolution sums are worked out directly.

The snag with the method (there has, of course, to be one!) has to do with aliasing errors, due to the truncation of the Fourier sums, as specified by (3.50). A good discussion of this problem has been given by Rogallo (1981, p. 46), so we merely note here that it arises in cases where j and l add up to give values of $|\mathbf{k}| > (2\pi/L)K$. If we were evaluating the convolution sums directly, then we could simply set such invalid contributions equal to zero. But with the pseudospectral approach, the difficulty is in identifying these contributions. Orszag avoided this by performing the whole process on two grids, one of which is suitably shifted relative to the other, in order to break the symmetry. Then taking the two values of $W_\alpha(\mathbf{k}, t)$, as calculated on the two different grids, and averaging, eliminates the aliasing errors.

These methods were used by Orszag and Patterson (1972) to simulate decaying isotropic turbulence at a Taylor–Reynolds number $R_\lambda = 42$ with 32^3 grid points, and (incidentally) taking 1.5 hours of central processing unit (CPU) time on a CDC 6600 computer. Their shell-averaged statistical results agreed well with the results of laboratory experiments and with the predictions of analytic turbulence theories, and this work has come to be regarded as the pioneering turbulent simulation on a computer.

If we attempt to assess the significance of the Orszag–Patterson simulation in the context of practical problems, then we must begin by observing that isotropic turbulence is rather remote from the sort of turbulent flows that are typically involved in engineering applications. Moreover, the Reynolds number is quite low when converted to the form (1.5) appropriate to pipe flow. We may see this by taking (2.165) for R_λ and combining it with (2.64) for λ to obtain

$$R_\lambda = \frac{3.87u^2}{(v\varepsilon)^{1/2}} \tag{3.55}$$

where u is the r.m.s. fluctuating velocity (assumed isotropic).

Now it is usual to estimate the dissipation rate in the centre of a pipe by the formula (e.g. see Pond, Stewart, and Burling 1963; also compare eqn (3.37) of the present work)

$$\varepsilon = \frac{2.5u_\tau^3}{a}, \tag{3.56}$$

where a is the radius of the pipe and u_τ is the friction velocity as defined by (1.29). The fluctuating velocity can be estimated by putting $u = u_\tau$ (see Fig. 1.5) and the friction velocity related to the bulk mean velocity U through (1.48) in terms of the friction factor f. Then, invoking the well-known empirical relationship for f (Goldstein 1938, p. 339),

$$f = 0.067 \left(\frac{Ua}{v} \right)^{-1/4},$$

we can combine all these steps with (3.55) to obtain the relationship

$$R_\lambda = 0.95 R^{7/16} \tag{3.57}$$

where R is in terms of the bulk mean velocity and the pipe diameter, as given by (1.5).

From (3.57), it is readily found that the Taylor–Reynolds number $R_\lambda = 42$ of Orszag and Patterson (1972) is equivalent to about $R = 5 \times 10^3$ for pipe flow. This can be put in perspective by noting that the critical Reynolds number for transition is $R = 2 \times 10^3$. More recently Kerr (1985) used essentially the same algorithm to simulate isotropic turbulence at $R_\lambda = 83$, with 128^3 grid points and using 250 hours of CPU time on a CRAY 1S. In pipe flow terms this corresponds to $R = 2.5 \times 10^4$.

Of course this does not mean that Kerr's simulation is equivalent to reproducing pipe flow at this value of R, but merely that such a pipe flow would have a similar value of R_λ at its centre. A pipe or channel flow has the additional problems of inhomogeneity and anisotropy with which to contend. That is, instead of one scalar function $E(\mathbf{k}, t)$, dependent on the three scalar components of the wavevector $\mathbf{k}$, we now have to calculate E_{11}, E_{22}, E_{33}, and E_{12}, each of which additionally depends on the distance from the pipe or channel wall. At the time of writing, the most advanced full numerical simulation of a shear flow is that of Kim, Moin, and Moser (1987), who used 4×10^6 grid points to achieve a Reynolds number (based on channel semi-width and maximum mean velocity) of 3300. In our terms, this means $R = 5280$, on the usual assumption that $U = 0.8 U_C$.

In fact there is not all that much difference between these two simulations in terms of computer requirements, although Kim et al. would need to increase their Reynolds number by a factor of 10 to have the same Taylor–Reynolds number as in the isotropic simulation of Kerr. But the inexorable logic of eqn (3.42) is that the Reynolds number goes up more slowly than the square root of the number of grid points. Hence increases in computer power, although important, are not likely to make any great impact on the problem of simulating real turbulent flows in complicated geometries. Therefore we have to consider ways of reducing the number of degrees of freedom and this will be the subject of the rest of this section.

3.3.2   Large-eddy simulation (LES)

The basic idea of LES can be explained most simply in the context of the spectral simulations that we have just been considering. Suppose that we do

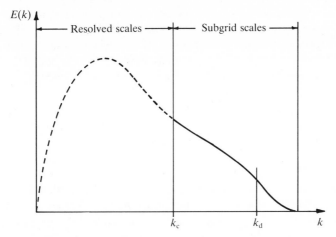

Fig. 3.12. The energy spectrum divided up into resolved scales ($k < k_c$) and subgrid scales ($k > k_c$), for the purposes of large-eddy simulation.

not attempt to simulate all the wavenumber modes up to the viscous cutoff ($> k_d$). Instead, we only simulate modes for which $k \leqslant k_C$, where the arbitrarily chosen cutoff wavenumber k_C satisfies the condition $k_C \ll k_d$. The situation we are envisaging is illustrated schematically in Fig. 3.12.

This means that we are simulating the Fourier-transformed Navier–Stokes equation with its wavenumber representation arbitrarily truncated to the interval $0 \leqslant k \leqslant k_C$. Referring to (2.76), we can readily appreciate the consequences of this. We note that the non-linear term on the r.h.s. couples all wavenumbers together, and that the overall effect is the net transfer of energy from any one wavenumber to higher wavenumbers. Hence, if we truncate at $k = k_C$, we are removing the mechanism by which energy is transferred from wavenumbers below k_C to those above. In practice, such a simulation would fail because energy would be cascaded down to $k = k_C$ and would pile up at the cut-off.

In fact this problem is essentially just the one considered by Heisenberg (see Section 2.8.1). Thus, if we introduce a Heisenberg-type effective viscosity $v(k|k_C)$ to represent the effect of transfers to $k > k_C$, then eqn (2.76) can be replaced by its truncated form

$$\left\{ \frac{\partial}{\partial t} + v_0 k^2 + v(k|k_C)k^2 \right\} u_\alpha(\mathbf{k}, t) = M_{\alpha\beta\gamma}(\mathbf{k}) \sum_{\mathbf{j}} u_\beta(\mathbf{j}, t) u_\gamma(\mathbf{k} - \mathbf{j}, t), \quad (3.58a)$$

where the wavevectors satisfy the condition

$$0 \leqslant |\mathbf{k}, \mathbf{j}, \mathbf{k} - \mathbf{j}| \leqslant k_C \quad (3.58b)$$

and v_0 denotes the molecular kinematic viscosity, where the subscript zero anticipates a later usage. From now on we shall use v or v_0 interchangeably, according to context, to represent the kinematic viscosity of the fluid.

This procedure is therefore an example of LES. That is, we simulate the eddies of size larger than $1/k_C$ by explicit numerical integration of the Navier–Stokes equation and account for the transfer of energy from the large scales (or eddies) to the small scales by the introduction of an effective viscosity which augments the molecular viscosity of the fluid.

We shall discuss the fundamental problems of obtaining a form for the effective viscosity for spectral simulations in a later chapter. Here we turn to the more general application of the LES technique to shear flows. We also return to configuration (**x**) space.

The original idea was due to Smagorinsky (1963), who calculated the general circulation of the atmosphere on a finite-difference grid and represented the drain of energy to turbulent scales smaller than the grid spacing h (the 'subgrid scales') by a subgrid model based on a Heisenberg-type effective viscosity.

Other meteorological simulations followed, but the first engineering LES was Deardorff's (1970) simulation of plane channel flow. Deardorff used Reynolds (spatial) averaging, applied to a unit cell of the finite-difference mesh, to define the large (or resolved) scales, and introduced the terminology 'filtered variables'. Although only 6720 grid points were used, the comparison with the laboratory experiments of Laufer (1954) was sufficiently favourable for the feasibility of the method to have been established.

Leonard (1974) was apparently the first to use the term 'large-eddy simulation'. He also introduced the idea of filtering as a formal convolution operation on the velocity field and gave the first general formulation of the method. We shall base our own discussion on Leonard's approach.

From (1.1) and (1.4), we have the continuity equation

$$\frac{\partial U_\beta(\mathbf{x}, t)}{\partial x_\beta} = 0,$$

and the Navier–Stokes equation

$$\frac{\partial U_\alpha}{\partial t} + \frac{\partial U_\alpha U_\beta}{\partial x_\beta} = -\frac{\partial P}{\partial x_\beta} + v_0 \nabla^2 U_\alpha, \tag{3.59}$$

where we have introduced kinematic units for the pressure, so that the density can be eliminated from (1.4).

We define the large scales by the general filtering operation

$$\tilde{U}_\alpha(\mathbf{x}) = \int G(\mathbf{x} - \mathbf{x}') U_\alpha(\mathbf{x}') \, d^3 \mathbf{x}', \tag{3.60}$$

where G is some arbitrarily chosen filter function and the tilde denotes the large-scale (or resolved) part of the velocity field. For convenience, this operation can be written in the contracted form

$$\tilde{U}_\alpha = G * U_\alpha. \tag{3.60a}$$

Also, we decompose the instantaneous velocity field into resolved and subgrid scales:

$$U_\alpha = \tilde{U}_\alpha + u'_\alpha \tag{3.61}$$

where u'_α is the subgrid part of the velocity field.

The continuity equation provides a relatively simple starting point in formulating the LES equations. Operating with G, according to (3.60a), on the l.h.s. of (1.1), we find

$$G * \frac{\partial U_\beta}{\partial x_\beta} = \frac{\partial \{G * U_\beta\}}{\partial x_\beta}. \tag{3.62}$$

This result can be proved by integration by parts, with respect to $\mathbf{x}'$, provided only that U_α vanishes on the boundaries. It then follows at once from (1.1), (3.60a), and (3.62) that the resolved scales satisfy the continuity equation in the form

$$\frac{\partial \tilde{U}_\beta(\mathbf{x}, t)}{\partial x_\beta} = 0. \tag{3.63}$$

The linear terms of the Navier–Stokes equation are readily treated in the same way but, as usual, the non-linearity requires some thought. If we denote the effect of the filter on the non-linear term by $A_{\alpha\beta}$, temporarily, then the operation of (3.60a) on (3.59) yields

$$\frac{\partial \tilde{U}_\alpha}{\partial t} + \frac{\partial A_{\alpha\beta}}{\partial x_\beta} = -\frac{\partial \tilde{P}}{\partial x_\alpha} + v_0 \nabla^2 \tilde{U}_\alpha, \tag{3.64}$$

where the filtered non-linear term is given by

$$\begin{aligned} A_{\alpha\beta} &= G * \{U_\alpha U_\beta\} \\ &= G * \{\tilde{U}_\alpha \tilde{U}_\beta + u'_\alpha \tilde{U}_\beta + \tilde{U}_\alpha u'_\beta + u'_\alpha u'_\beta\}, \end{aligned} \tag{3.65}$$

with the last step following from the substitution of (3.61).

The terms in (3.65) which involve u' can all be lumped together:

$$T_{\alpha\beta} = G * \{u'_\alpha \tilde{U}_\beta + \tilde{U}_\alpha u'_\beta + u'_\alpha u'_\beta\}. \tag{3.66}$$

As the simulation only calculates explicit values for the large scales, $T_{\alpha\beta}$ (which controls the subgrid drain of energy) must be modelled in some way. We shall return to this shortly.

The first term on the r.h.s. of (3.65) is in quite a different category, as it involves only the explicit scales and the filter function. Its computation might present some practical difficulties, but clearly there is no fundamental closure problem. Leonard (1974) wrote this term as

$$G^*\{\tilde{U}_\alpha \tilde{U}_\beta\} = \tilde{U}_\alpha \tilde{U}_\beta + L_{\alpha\beta}, \tag{3.67}$$

where

$$L_{\alpha\beta} = G^*\{\tilde{U}_\alpha \tilde{U}_\beta\} - \tilde{U}_\alpha \tilde{U}_\beta \tag{3.68}$$

and is usually referred to as the 'Leonard stress'.

In all then, we can now write eqn (3.64) for the filtered variables as

$$\frac{\partial \tilde{U}_\alpha}{\partial t} + \frac{\partial (\tilde{U}_\alpha \tilde{U}_\beta)}{\partial x_\beta} = -\frac{\partial \tilde{P}}{\partial x_\alpha} - \frac{\partial L_{\alpha\beta}}{\partial x_\beta} - \frac{\partial T_{\alpha\beta}}{\partial x_\beta} + v_0 \nabla^2 \tilde{U}_\alpha, \tag{3.69}$$

where $L_{\alpha\beta}$ is given by (3.68) and $T_{\alpha\beta}$ by (3.66).

It is of interest to compare the filtered equations of motion with the equations of mean motion (see Section 1.3.1). Clearly (3.63) is formally identical with (1.11) for continuity, if we replace $\tilde{U}_\alpha$ by U_α. There are also some similarities between the Reynolds-averaged and Leonard-filtered Navier–Stokes equations, as given by (1.12) and (3.69) respectively. The differences lie in the presence in (3.69) of the terms involving $T_{\alpha\beta}$ (the subgrid stress) and $L_{\alpha\beta}$ (the Leonard stress). Clearly, if we were to turn the filtering operation, as defined by (3.60), into a spatial average over the entire system volume, then the Leonard stress should vanish and the subgrid stress should reduce to the Reynolds stress. We will next consider how to deal with these two stress terms in a practical simulation.

The subgrid stress has to be modelled in some way and the original Smagorinsky model (Smagorinsky 1963) makes a good starting point. Smagorinsky employed the traditional analogy between turbulence effects and molecular properties. Thus, by analogy with eqn (1.3), the (kinematic) subgrid stress tensor is supposed to be expressible in terms of explicit scales by the relationship

$$T_{\alpha\beta} = -v_S \left(\frac{\partial \tilde{U}_\alpha}{\partial x_\beta} + \frac{\partial \tilde{U}_\beta}{\partial x_\alpha} \right), \tag{3.70}$$

where, on dimensional grounds, the subgrid effective viscosity takes the form

$$v_S = c^2 h^2 \tilde{S}^{1/2} \tag{3.71}$$

where c is a constant, h is a measure of the filter width or the mesh spacing, and $\tilde{S}$ is given by

$$\tilde{S} = \frac{\partial \tilde{U}_\alpha}{\partial x_\beta} \left(\frac{\partial \tilde{U}_\alpha}{\partial x_\beta} + \frac{\partial \tilde{U}_\beta}{\partial x_\alpha} \right). \tag{3.72}$$

Lilly (1967) showed that this form was consistent with the existence of the Kolmogorov energy spectrum (2.137) for $k \sim 1/h$, provided that the constant

c was given by

$$c = 0.23\alpha^{-3/4}, \tag{3.73}$$

where α is the constant of proportionality in the Kolmogorov spectrum.

Deardorff (1971) has quoted various values of c based on Lilly's estimates. The exact value chosen depends on various factors (not least, on the value assumed for the Kolmogorov constant), but they are generally of the order of $c = 0.2$. However, from a comparison with experimental results, Deardorff concluded that, if there was a mean rate of shear present, the constant in the Smagorinsky effective viscosity should be smaller than this, and a value of about $c = 0.10$ was found to be the best single value to apply right across a flow.

The Leonard stress depends only on the filter function chosen, and we follow Leonard (1974) in illustrating some typical filters (see Fig. 3.13). An important point to note is that $L_{\alpha\beta}$ can vanish identically if G is chosen appropriately. From (3.68), we see that this requires

$$G * \{\tilde{U}_\alpha \tilde{U}_\beta\} = \tilde{U}_\alpha \tilde{U}_\beta,$$

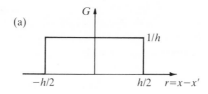

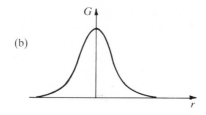

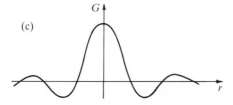

Fig. 3.13. Examples of filter functions used to define the resolved scales in configuration space: (a) top-hat or boxcar function; (b) Gaussian filter $G = (6/h\pi)^{1/2} \exp(-6r^2/h^2)$; (c) 'sinc' function $G = 2(\pi r)^{-1} \sin(\pi r/h)$.

or, in terms of an arbitrary test function F, we require the filter G to have the property

$$G*\tilde{F} = G*\{G*F\} = \tilde{F}. \qquad (3.74)$$

In order to find suitable forms of G, we Fourier transform G into $g(k)$ and F into $f(k)$. Then, from the convolution theorem, (3.74) becomes

$$g\tilde{f} = g^2 f = \tilde{f}; \qquad (3.75)$$

hence we require G to be such that $g^2 = g = 0$ or 1.

This condition can be satisfied in one of two ways. First, we can take G to be as shown in Fig. 3.13(a), and let h tend to infinity. Then g becomes a Dirac delta function and (3.75) holds. This is not really surprising as, under these circumstances, (3.60) just becomes a spatial average. The more interesting case of the two is when we choose G to be the filter shown in Fig. 3.13(c). The Fourier transform of this function is the unit top-hat function (like Fig. 3.13(a), only in **k**-space rather than **x**-space) and satisfies the condition $g(k) = 1$ or 0. In this case, g is often referred to as the 'spectrally sharp filter' and is the one normally used in spectral simulations.

In practice, the Leonard term can be swamped by the numerical errors inherent in the finite-difference representation and, on these grounds, it is often neglected. For the sake of completeness we should also mention that Leonard has given an approximate procedure, for estimating such errors, which is based on an expansion of the velocity fields in a Taylor series about $\mathbf{x} = \mathbf{x}'$: details can be found in the reference cited.

There is a rapidly growing literature on the subject of LES, and the interested reader will find that the review articles by Voke and Collins (1983) and Rogallo and Moin (1984) make a good starting point. Here we shall only mention that the pioneering simulation of channel flow by Deardorff (1970) was followed by the more advanced simulations of Schumann (1975) and Moin and Kim (1982).

The latter reference provides a striking instance of the virtues of LES as opposed to full simulation. Moin and Kim achieved a Reynolds number (based on centre-line velocity and channel half-width) of 1.4×10^4 with 0.52×10^6 grid points and 92 hours of CPU time on a ILLIAC-IV. The result was not only in quite good agreement with laboratory experiment, as far as velocity profiles and the like were concerned, but also in the successful reproduction of the coherent streaks of the wall region. In contrast, the full simulation of the same problem by Kim et al. (1987) only reached a Reynolds number of 3.3×10^3, despite using eight times as many grid points (roughly) and 250 hours of CPU time on a CRAY-XMP, a much more powerful machine than an ILLIAC-IV.

3.3.3 *The use of the Reynolds-averaged Navier–Stokes equations for practical applications*

We should also bear in mind that the Reynolds averaging procedure is in itself a way of reducing the number of degrees of freedom in the problem. Equation (1.12) for the mean velocity is formally identical to the original Navier–Stokes equation, provided that we absorb the Reynolds stress $\rho \langle u_\alpha u_\beta \rangle$ into some more general stress tensor, as in (1.13). Then, deferring for the moment the question of a turbulent constitutive relationship, we observe that the mean velocity can be expected to be a fairly smooth function of time and position, and hence the problem of computing eqn (1.12) should be very much reduced, when compared with the primitive Navier–Stokes equation, which for turbulence has solutions which depend randomly on space and time.

Further (continuing to defer the turbulent closure problem), we have to consider how to solve partial differential equations numerically for a particular geometric configuration and specified boundary conditions. And, although the Reynolds equations can sometimes be reduced to parabolic form, in general we are going to be faced with elliptical equations, which will mean an iterative method as the upstream solution will depend on downstream values.

First, of course, there is the question of the numerical representation: whether one should use finite differences, finite elements, or spectral methods, each of which has its adherents. However, in practice anyone wishing to make a turbulent flow calculation would be well advised to use the finite-difference method, which has been absolutely dominant in the development of the subject.

The second major question to be faced is that of which numerical procedure or algorithm should be used? It is at this stage that we are forced to the conclusion that CFD as a subject is as much art as science. By this I mean that, in practice, there are likely to be many problems to do with the stability, convergence or efficiency of numerical solutions and one's best approach is often to allow oneself to be guided by the past experience of others, even when the only justification for following a particular procedure or adopting a certain value for a parameter is that it has been found to work in the past! A noteworthy development was the SIMPLE algorithm (Patankar and Spalding 1972), which is an implicit method of handling elliptical problems. This algorithm has been widely used for engineering calculations (e.g. see the recent review by Ferziger 1987) and now exists in several refined versions (Van Doormaal and Raithby 1984).

We have previously touched on the fact that many engineering applications involve geometrical configurations which are very complicated. For completeness we should just mention that, as well as the traditional method of using coordinate transformations to map a difficult shape into a simple one (e.g. Carr and Forsey 1982), there is also a growing interest in the technique of adaptive gridding. That is, the finite-difference representation of the mean flow equation is formed on an orthogonal curvilinear grid, which is also constructed numerically, and is continually adjusted as the calculation proceeds (Dwyer 1984, Luchini 1987). This method can also be used in situations where there are free boundaries whose position will not usually be known in advance (e.g. Ryskin and Leal 1984).

As we pointed out initially, progress in all these areas—together with developments in computers—has been so rapid that the pressure is on for better turbulence models. This is a topic which we can defer no longer.

For engineering applications the general problem can be stated in the following way. We wish to solve eqn (1.12) for the mean velocity distribution $\bar{U}_\alpha(\mathbf{x}, t)$ in some specified physical location, with prescribed initial and boundary conditions. And, in order to do this, we need to know the Reynolds stress $\rho \langle u_\alpha u_\beta \rangle$.

The simplest prescription for the Reynolds stress is obtained from the mixing-length model which, as we saw in Section 1.5.2, relies on an analogy between the randomizing effects of turbulent eddies on the mean motion and the corresponding processes due to molecular motion in a gas. Thus eqn (1.54) for the Reynolds stress is obtained by applying the methods of gas kinetic theory to the macroscopic motions of the fluid continuum. This is the lowest level of turbulence model, and in fact it works quite well. Its snags are mainly that (apart from possessing the defects of the eddy-viscosity type of hypothesis) it lacks universality (i.e. the prescription of the mixing length varies from one type of flow to another), and in complicated flows it may be impossible to specify any form for the mixing length.

Above the mixing-length model there is a hierarchy of turbulent models based on transport equations for the fluctuating field. Such equations are derived—often heuristically rather than rigorously—from eqn (1.13) for the velocity fluctuation $u_\alpha(\mathbf{x}, t)$.

The next order of difficulty is the one-equation model, where the reference is to the single transport equation which has to be solved for the fluctuating field. Here we use (1.13) to derive an equation for the turbulent kinetic energy E (usually referred to as k in the CFD literature). From (1.20) for $\langle u_\alpha u_\beta \rangle$, and using (1.21) and (1.22), we obtain the energy equation in the form

$$\frac{\partial E}{\partial t} + \bar{U}_\beta \frac{\partial E}{\partial x_\beta} = \underbrace{-\langle u_\alpha u_\beta \rangle \frac{\partial \bar{U}_\alpha}{\partial x_\beta}}_{\text{(production)}} + \underbrace{\frac{\partial \{\langle u_\alpha^2 u_\beta \rangle / 2 + \langle u_\beta p \rangle / \rho \}}{\partial x_\beta}}_{\text{(diffusion)}} + \nu \nabla^2 E -$$

$$\underbrace{- \frac{\nu}{2} \left\langle \left(\frac{\partial u_\alpha}{\partial x_\beta} \right)^2 \right\rangle}_{\text{(dissipation)}} . \tag{3.76}$$

It should be noted that the last term on the r.h.s. of (3.76) is not strictly the dissipation, unless the turbulence is homogeneous. However, it is usual to argue that the small scales responsible for the dissipation are homogeneous and hence this is probably quite a good approximation, except possibly near a solid surface.

The diffusion term (see Section 1.6.2 for a discussion in the context of experimental results) involves (essentially) triple moments and a viscous diffusion effect which is negligible, except near the wall. The entire term is normally modelled as if non-linear inertial transfer were a gradient process proportional to the gradient of the kinetic energy E. It is important to realize that an assumption of this sort is a form of closure approximation, irrespective of any other assumptions still to be made.

With both these points in mind, eqn (3.76) can be written as

$$\frac{\partial E}{\partial t} + \bar{U}_\beta \frac{\partial E}{\partial x_\beta} = -\langle u_\alpha u_\beta \rangle \frac{\partial \bar{U}_\alpha}{\partial x_\beta} + \frac{\partial \{(\nu_T/\sigma_E)\partial E/\partial x_\beta\}}{\partial x_\beta} - \varepsilon, \qquad (3.77)$$

where ν_T is the turbulent eddy viscosity and σ_E is an empirical constant. The effective viscosity is still modelled by analogy with gas kinetic theory and accordingly we can write an expression for it as follows:

$$\nu_T = C'_\mu E^{1/2} L, \qquad (3.78)$$

where C'_μ is an empirical constant and L is an integral length scale. This is analogous to the relationship in kinetic theory between the viscosity, the particle energy, and the mean free path, and is often referred to as the Kolmogorov–Prandtl relationship (e.g. Rodi 1982). The dissipation rate can be modelled by

$$\varepsilon = C_D \frac{E^{3/2}}{L} \qquad (3.79)$$

which can be compared with equation (3.37) and where C_D is an empirical constant.

In practice this type of model is little advance on the mixing-length theory, because the length scale L still has to be fixed by simple empirical arguments. We have included it here because it illustrates the basis of the whole approach. The next level of complication is the introduction of a second transport equation, from which the length scale can be calculated. Models at this level are accordingly known as two-equation models and there are many of them. A critical examination of two-equation models was carried out by Chambers and Wilcox (1976), who concluded that there was little to choose between the various models. However, over the last decade, the k-ε model of Jones and Launder (1972) has been predominant in the literature and we shall discuss this model in some more detail in the next section.

Although the k-ε model has proved very successful in practice, it still suffers (as we shall see) from a reliance on the eddy-viscosity hypothesis. Thus it must inevitably break down where the concept of the eddy viscosity itself is invalid, and for such flows one must resort to turbulent stress modelling. This is the most general of all the single-point closures of the Navier–Stokes equation, and is based on the full equation for the Reynolds stress tensor (1.20). Again, individual terms in the equation have to be modelled in order to make the technique work. The best-known version of this kind of closure is given in the paper by Launder, Reece, and Rodi (1975).

Finally we have the proposal by Rodi (1976) that the various differential equations should be simplified in such a way that they reduce to purely algebraic expressions, while retaining most of their important features. These equations—known as algebraic stress models—can be applied to many of the flows which require turbulent stress models, but are much easier to compute.

General accounts of all these models, and of their performance when compared with experimental results, can be found in the book by Rodi (1980) and in the reviews by Markatos (1986) and Nallasamy (1987). Very detailed and comprehensive listings of equations and coefficients for various models are given in the reviews by Mellor and Herring (1973) and by Patel, Rodi, and Scheuerer (1984).

3.3.4 An example of a two-equation turbulence model

In his recent review, Nallasamy (1987) comments that the k-ε model has been used in a majority of all the two-dimensional (presumably, turbulent) calculations reported in the literature. Thus, at worst, we are certainly justified in choosing it for more detailed discussion as a representative example of a two-equation model.

We begin with the eddy-viscosity hypothesis in what is often known as the 'Boussinesq form' (Hinze 1975):

$$\langle u_\alpha u_\beta \rangle = \frac{2}{3} E \delta_{\alpha\beta} - v_\mathrm{T} \left(\frac{\partial \bar{U}_\alpha}{\partial x_\beta} + \frac{\partial \bar{U}_\beta}{\partial x_\alpha} \right) \tag{3.80}$$

where $\delta_{\alpha\beta}$ is the Kronecker delta. It is readily demonstrated that this reduces to (1.50) for a thin shear layer in which $\bar{U}_\alpha = \{\bar{U}_1(x_2), 0, 0\}$. It should be noted that (3.80) is an isotropic model for the (kinematic) Reynolds stress and assumes that the normal stresses are all equal: a discussion of this limitation and an attempt to improve upon it will be found in the paper by Speziale (1987).

The next stage in setting up the model is to combine equations (3.78) and (3.79) to obtain the relationship

$$v_\mathrm{T} = \frac{C_\mu E^2}{\varepsilon} \tag{3.81}$$

where $C_\mu = C'_\mu C_D$ is a new empirical constant. The kinetic energy can be calculated from its transport equation (3.77), and so a specification is needed for the dissipation rate ε. In principle, a rigorous equation can be derived from (1.13) for the fluctuating velocity, but in practice many empirical steps are needed to make it tractable. Hence, the simplest procedure is just to derive an equation for the dissipation rate by analogy with (3.77) for the kinetic energy:

$$\frac{\partial \varepsilon}{\partial t} + \bar{U}_\beta \frac{\partial \varepsilon}{\partial x_\beta} = -C_1 \frac{\varepsilon}{E} \langle u_\alpha u_\beta \rangle \frac{\partial \bar{U}_\alpha}{\partial x_\beta} + \frac{\partial \{(v_T/\sigma_\varepsilon)\partial \varepsilon/\partial x_\beta\}}{\partial x_\beta} - \frac{C_2 \varepsilon^2}{E} \quad (3.82)$$

where C_1 and C_2 are additional empirical constants.

The constants are determined by reference to the results of various laboratory experiments. For instance, if the model—comprising eqns (3.77), (3.80), (3.81), and (3.82)—is applied to the decay of turbulence behind a grid, then the production and dissipation terms vanish and the only unknown constant is C_2. From measurements of the decay of E behind a grid, it was found that

$$C_2 = 1.92.$$

The constant C_μ can be found by combining (3.81) with (3.77) and applying the result to a local-equilibrium shear layer, where the dissipation and production terms are approximately equal. This results in the relationship

$$C_\mu = \left(\frac{\langle u_1 u_2 \rangle}{E}\right)^2 = 0.09$$

where the numerical value comes from experimental measurements.

In the region near the wall, where the log profile applies, experimental results indicate that the convection of the dissipation is negligible, where the production and dissipation are in approximate local equilibrium. Application of the transport equation for the dissipation rate to these circumstances leads to

$$\frac{1}{\sigma_\varepsilon} = \frac{C_\mu^{1/2}(C_2 - C_1)}{K^2},$$

where K is the von Karman constant and takes the value $K = 0.4$. This relationship can be used to fix C_1 if σ_E and σ_ε are taken to be unity. Then, computer optimization of various flow calculations is used to obtain the values

$$C_1 = 1.44 \qquad \sigma_E = 1.0 \qquad \sigma_\varepsilon = 1.3.$$

These parameters are constant in the sense that they remain the same throughout any particular calculation. However, different physical situations may require values which are different from those given above. To be specific, the constants given will serve well for plane shear layers. However, axisymmetric flows would require modified values of C_μ and C_2 (Markatos 1986; although see the earlier paper by Pope 1978).

The many specific assumptions made in their derivation, and their consequent lack of universality, together rob models like k-ε of true fundamental status. However, anyone who has to carry out a practical flow calculation will have reason to be glad of their existence.

Lastly, we should mention the subject of 'wall functions'. These are empirical rules, based on the logarithmic 'law of the wall', which allow the calculation to be carried out with the last grid point on the edge of the viscous sublayer, rather than on the wall itself (Launder and Spalding 1974). In effect, they amount to a synthetic boundary condition and may be needed in order to save on computer time and storage or because a particular model is not valid in the region near the wall, where the effects of the viscosity are dominant. With developments in computer power, their use in conjunction with engineering models may no longer be necessary (Launder 1984).

3.4   Turbulent drag reduction by additives

The idea that the frictional drag exerted on a solid body by a flowing fluid can in some way be reduced is very widely held and is often associated in peoples' minds with the notion of a 'dolphin skin effect'. As a general subject, this area of investigation is rather fraught, and has quite a lot in common with studies of alleged paranormal phenomena, such as precognition, telepathy, ghosts, messages from the spirit world, and so on. In other words investigators feel that there is something in it, but tend to have difficulties in reproducing their positive results.

However, if we restrict our attention to one particular subclass of such (drag-reduction) phenomena—the reduction of turbulent drag obtained by dissolving small quantities of a chemical additive in a liquid—then we are (if we may run the risk of mixing our metaphors) on much more solid ground. The resulting effects are often dramatic and, for that matter, eminently repeatable.

In order to highlight the unambiguous nature of the phenomenon, we shall consider the very simple experiment which is illustrated schematically in Fig. 3.14. Here water flows from a constant-head tank through a long straight pipe. The pressure drop along the tube can be measured between two pressure tappings in the pipe wall, using a liquid-in-glass manometer. The rate of outflow in a given time can be readily obtained, thus yielding the volumetric flow rate and hence the bulk mean velocity U. Now suppose that we dissolve a synthetic resin called polyethyleneoxide (PEO) in the water and repeat the experiment. It turns out that the pressure drop needed to maintain the same volumetric flow rate is normally very much reduced by the polymer additive.

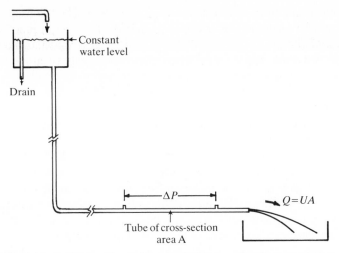

Fig. 3.14. Definition sketch for a typical drag-reduction experiment.

The amount by which the pressure drop will be reduced will depend to some extent on the Reynolds number. However, we can give a specific example, assuming that the Reynolds number is large enough for the flow to be turbulent. If the PEO is present in a concentration of about 10 wt ppm, then the pressure drop can be reduced by the order of 70–80 per cent, compared with that required for water alone. This remarkable result is made all the more remarkable by the fact that the laminar flow behaviour of such dilute solutions is very little different from that of the water on its own. That is, the density and viscosity of the polymer solutions are not very different from those for water.

Drag reduction by additives is not confined to aqueous solutions. For this reason, we give the formal definition of the amount of drag reduction DR, as a percentage, in terms of a general solvent. Let ΔP_S be the pressure drop required to produce a mean velocity U in the solvent alone, and ΔP_A the pressure drop required to produce the same value of mean velocity when the additive is present. Then the drag reduction DR, as a percentage, is defined through the relationship

$$DR = \left(\frac{\Delta P_S - \Delta P_A}{\Delta P_S}\right) \times 100 \text{ per cent} \qquad (3.83)$$

It was pointed out by Lumley (1969) that one must be careful to define drag reduction as the reduction of skin friction in turbulent flow, by means of additives, below the corresponding value which would be obtained with the solvent alone. This can avoid confusion with other effects where, for example, an additive increases the Reynolds number at which transition occurs ('delayed transition'). Such an effect would apparently reduce the drag because the flow would then be laminar rather turbulent. However, in practice, once the Reynolds number is increased to the point where transition now takes place, the resulting drag can be larger than in the solvent alone (Lumley 1969).

In the remainder of this section we shall give a brief account of the history of drag reduction, and then go on to discuss some of the experimental findings. We shall find it convenient to deal separately with the topics of polymer solutions and fibre suspensions.

However, before doing this, it may be helpful if we make a remark about the term 'drag'. Essentially it refers to the total force exerted by a fluid on a solid body. This total can be made up from several different kinds of force. For example, if you place your hand in the path of a jet from a hose pipe, then the force on your hand will be mainly due to its getting in the way of the momentum flux in the streamwise direction. Or, in the general case of flow around a bluff body, the main force may be due to the fact that the pressure at the upstream stagnation point is larger than the lower one—this is known as 'form drag'.

In the case of drag reduction by additives, we are concerned only with skin friction, which is due to the fluid shear stress at the solid surface. For this reason, we shall concentrate our attention mostly on simple pipe flows, where the drag is almost entirely due to skin friction alone. More general discussions fluid drag forces can be found in the book by Goldstein (1938).

3.4.1   Historical background

The discovery that chemical additives could reduce flow resistance has usually been credited to Toms (1948), who was actually interested in investigating the mechanical degradation of dissolved polymers. His results for polymethyl-methacrylate dissolved in monochlorobenzene were presented at an international conference on rheology (Toms 1948), in conjunction with a paper by Oldroyd (1948) who offered an explanation of the friction reduction in terms of an 'apparent slip' at the wall. Later on, Toms (1974), in an inaugural address to another international conference—this time on drag reduction—gave a personal account of the background to his accidental discovery and conveyed his surprise at the way in which—although initially ignored for ten years—it had led to the development of such a large and active research field. It is worth noting that, at the time he was speaking, research into drag reduction was at its peak, with a publication rate in excess of a hundred papers per year.

At one time there was a tendency to speak of the 'Toms effect', but it now seems to be generally preferred to refer to 'drag reduction by additives', in recognition of the great diversity of the subject and the complexity of its origins. For instance, in his review of the topic, Lumley (1969) has noted that Mysels, working under wartime security restrictions, may have been the first to observe drag reduction by additives. Apparently the first open publication of his work was a US patent concerning the flow of petrol which had been thickened by the addition of an aluminium soap (Mysels 1949). Later, short accounts of this research were given in the literature (Agoston *et al.* 1954) and at a conference on drag reduction (Mysels 1971).

In fact, as the subject has developed, it has become ever clearer that friction reduction due to additives is not really a new phenomenon at all. Hoyt (1972a) has pointed out that increased flow rates in silt-laden rivers were known as far back as the 1880s. He cites references for observations on the Mississippi in the 1880s and on the Nile in 1921.

The phenomenon of drag reduction by additives has also been invoked in order to explain inconsistent results—going back over many years—in tests of standard ship models in towing tanks. Fluctuations in measured drag are now believed to be due to long-chain polysaccharides in slimes produced by naturally occurring algae (Hawkridge and Gadd 1971).

Drag reduction which occurs naturally is one thing, but it was pointed out by Arunachalam and Fulford (1971, see also Rao 1970) that the first recorded case of the deliberate use of a drag-reducing additive was the injection of bile into water flows (Hele-Shaw 1897) in the hope of reproducing the effect of fish slimes. The results seem to have been inconclusive, but certainly fish slimes have since been shown to be effective drag-reducing additives (Kobets, Matjukhov, and Migrenko 1974).

Possibly the most influential development was the accidental discovery, by the oil industry, that the use of Guar gum as a lubricant during drilling operations also resulted in reductions in fluid friction. According to Hoyt (1972b), this led to the arousal of military interest and, in turn, to two very significant discoveries.

First, there was the discovery of the spectacular drag-reducing properties of PEO in water (Fabula, Hoyt, and Crawford 1963). Today PEO is still the single most effective additive (a concentration of 0.5 wt ppm can reduce turbulent drag in water by up to 40 per cent). It is almost certainly the most studied of all drag-reducing polymers.

Second, Ellis (1970) found that macroscopic asbestos fibres could, when in aqueous suspension, produce reductions in turbulent friction comparable with those of the best polymers. Prior to that, drag reduction in fibre suspensions was known to be a much less spectacular effect than that in polymer solutions, requiring much larger concentrations of the additive to give rather meagre reductions in drag. The difference was that Ellis's fibres were very long, thin, and flexible, with an enormous aspect ratio (i.e. length-to-diameter ratio) of about 10^5.

3.4.2   *Polymer properties*

From the numerous experimental studies it can be concluded that, in order to reduce turbulent drag, polymers should be (a) of very long chain structure, with little branching (correspondingly the molecular weight should be large, typically about 10^6), (b) flexible, and (c) well dissolved. It should perhaps be emphasized that the large molecular weight should be due to the molecule's possessing a large number of monomer units, rather than the individual monomers' being relatively massive.

Many polymer-solvent combinations which fit these requirements can be found in the literature, but the most effective in aqueous solution are undoubtedly the PEOs and the polyacrylamides. The commercial forms most frequently encountered are Polyox WSR 301, a PEO manufactured by Union Carbide, and Separan AP30, which is a polyacrylamide and is manufactured by the Dow Chemical Company. Most polymers as supplied are polydisperse (i.e. contain a range of molecular weights), but it is usual to take the mean molecular weight of Polyox WSR301 as 5.05×10^6, and that of Separan AP30 as 3.0×10^6.

An indication of how these synthetic polymers compare with natural drag reducers can be gained from a comparison of the concentrations needed to give DR = 67 per cent, at a Reynolds number of $R = 1.4 \times 10^4$, in pipe flow. According to Hoyt (1972b, Table 3), this requires 400 wt ppm of Guar gum compared with only 10 wt ppm of Polyox WSR301.

The drag-reducing properties of Polyox WSR301 and Separan AP30 have been extensively investigated. In Fig. 3.15 we compare the amounts by which each of them reduces drag in pipe flow for various additive concentrations. These particular results were obtained by injecting concentrated polymer solutions into water flowing through a pipe (McComb and Rabie 1982) and allowing the polymers to spread out until evenly mixed. They agree well with results obtained in pre-mixed dilute polymer solutions, and it is clear that Polyox WSR301 is slightly more effective than Separan AP30. However, by way of compensation, Separan AP30 has been found to be the more resistant of the two to degradation. This is the process whereby the polymer solutions progressively lose their ability to reduce drag, due (presumably) to the scission of the individual molecules under shear. Degradation is irreversible, except possibly in the case of certain soaps and association colloids (Little *et al.* 1975).

At this stage we should consider what is to be our criterion for deciding whether a polymer solution is concentrated or dilute. In practice it is usual to introduce a critical concentration C_S in the following way. Let us characterize the random-coiling molecules in solution by their mean radius of gyration G_M. Then, if we imagine each molecule to be replaced by a sphere with radius equal

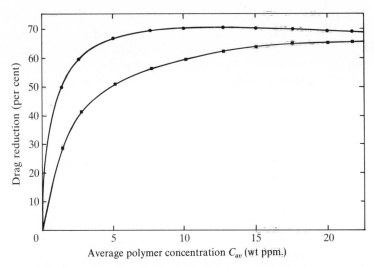

Fig. 3.15. Variation of drag reduction with polymer concentration in pipe flow at $R = 4.5 \times 10^4$ (McComb and Rabie 1982): ■, Separan AP30; ●, Polyox WSR301.

to G_M, we define C_S to be the concentration at which the imagined spheres would just touch (that is, when they would be close-packed).

The critical concentration supplies the necessary basis for our criterion. Suppose we have $C > C_S$, then it follows that molecules will tend to interact directly with each other, and we shall call such a solution concentrated. However, if $C < C_S$ then molecules will tend to interact indirectly, using the solvent as an intermediary, and naturally we refer to such solutions as being dilute (Merrill, Smith, Shin, and Mickley 1966).

Drag reduction can be obtained when the polymer concentration (by weight) is very much smaller than the critical value. However, Lumley (1973) has suggested that this can be somewhat misleading. If, instead of using weight-based concentrations, we work out the volume fraction occupied by the polymers—using the imaginary spheres with radius G_M referred to above—then this can take a much larger value.

A second misleading aspect of these dilute polymer solutions is the observation that they are almost indistinguishable from the pure solvent in terms of the values of physical properties like density and viscosity. In fact, these quantities are normally measured in simple shearing flows. If instead the flow is extensional (i.e. tending to stretch, rather than rotate, the fluid), then the associated (extensional) viscosity can be many orders of magnitude larger than the shear viscosity. This anomalous behaviour can be deduced in a qualitative way by dipping a finger into such a solution and noting that, as you withdraw your finger, the fluid tends to form strings. More qualitatively, Metzner and

Metzner (1970) have measured the response to extensional stress of polyacrylamide solutions at a concentration of 100 wt ppm, and found an extensional viscosity several times larger than the steady state shearing viscosity.

It would seem reasonable to associate the anomalous extensional viscosity of polymer solutions with their other form of anomalous behaviour, i.e. drag reduction. Indeed, such a notion forms the basis of the only convincing mechanism proposed for drag reduction by long-chain polymers. This is the idea—put forward in one form or another—that the bursting process which is responsible for turbulence production supposedly involves extensional motions which are resisted (or damped) by the very large extensional viscosity due to the dissolved polymers.

3.4.3 *The threshold effect*

The occurrence of drag reduction in a system shows up most clearly when we plot the friction factor f against the Reynolds number R. Referring back to eqns (1.47) and (1.48), it is clear that eqn (3.83) for the drag reduction in terms of the reduction in pressure drop can be written as

$$\text{DR} = \left(\frac{f_S - f_A}{f_S} \right) \times 100 \text{ per cent,} \qquad (3.84)$$

where f_S is the friction factor as measured in the solvent and f_A is the friction factor measured at the same flow rate when the additive is present.

In Fig. 3.16 we show some characteristic results, as a graph of friction factor against Reynolds number, for two different pipe diameters. Apart from the reduction in the friction factor, two other effects are clearly evident. First, the drag reduction is larger in the narrower of the two tubes. This is usually known as the 'diameter effect'. Second, the deviation from Newtonian behaviour begins at a different value of the Reynolds number in each of the two tubes. This is a form of what is known as the 'onset effect'.

The dependence on the diameter is readily explained if the basic mechanism of drag reduction is some sort of 'wall effect'. Clearly, the smaller the pipe diameter the greater the proportionate effect of the wall. However, this is probably not a very important aspect of the phenomenon, as it has been shown (e.g. Whitsitt, Harrington, and Crawford 1959; Paterson and Abernathy 1970) that the dependence on pipe diameter can be eliminated by plotting DR against the friction velocity, or, in other words, by making comparisons at constant wall shear stress.

This still leaves us with the onset effect, which can be regarded as a threshold value of the Reynolds number at which drag reduction will occur. Figure 3.17 illustrates this for Polyox WSR301 dissolved in water. Here the pipe diameter is now a constant, but the polymer concentration varies, and with it the value of the Reynolds number at onset.

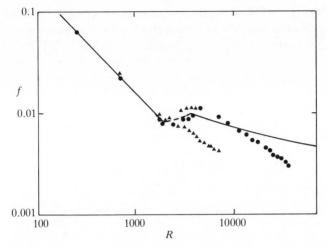

Fig. 3.16. Drag-reduction data of Toms (1948), as replotted by Savins (1961), showing the effect of different pipe diameters: ●, pipe diameter = 0.404 cm; ▲, pipe diameter = 0.129 cm.

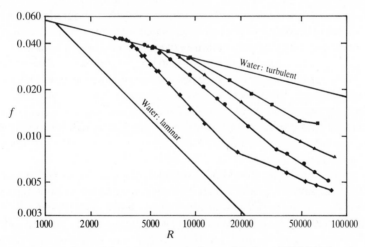

Fig. 3.17. Friction factors for aqueous solutions of Polyox WSR301 in pipe flow, showing the variation of onset with Reynolds number (Paterson and Abernathy 1970): ■, 1 wt ppm; ▲, 5 wt ppm; ●, 10 wt ppm; ◆, 50 wt ppm).

As we saw in the preceding chapter, as the Reynolds number is increased, smaller and smaller length and time scales are excited. Thus it is natural to seek an explanation of the onset effect in terms of some characteristic scale of the turbulence becoming small enough for the smallest eddies to interact strongly with the individual macromolecules.

Virk, Merrill, Mickley, Smith, and Mollo-Christensen (1967) put forward the suggestion that drag reduction occurred at a well-defined value of the wall shear stress, corresponding to the smallest turbulent length scale (the Kolmogorov scale) bearing a constant relationship to the mean diameter of the random-coiling polymers. Taking the former to be represented by the inner layer length scale (see (1.30) for a definition; it can be shown that this is of the same order as the Kolmogorov scale), and the latter to be equal to twice the radius of gyration, Virk et al. (1967) proposed that the onset criterion should be given by

$$\frac{2G_M u_\tau^*}{v} = C',$$
(3.85)

where the asterisk denotes the value at onset (and the onset shear stress is readily calculated from the onset friction velocity) and C' is the so-called onset constant. From an analysis of many experimental results, Virk et al. (1967) concluded that (3.85) held, with the onset constant given by

$$C' = 0.015 \pm 0.005.$$
(3.86)

From (3.85) and (3.86) it can be seen that the onset of drag reduction is assumed to occur when the smallest turbulent length is about fifty times larger than the mean diameter of the polymer, which must raise some doubts about there being any physical connection between the two length scales.

In fact, Fabula, Lumley, and Taylor (1965) had produced a more physically plausible criterion based on time-scales. The trouble was that all the evidence seemed to favour the hypothesis based on the relative length scales. For instance, Paterson and Abernathy (1970) concluded from an extensive series of tests that Virk's onset criterion was confirmed, although they attributed a greater variation in their values of the onset constant C' to the polydispersity of their samples.

To digress briefly, we should note that polydispersity of molecular weights is just one of several factors which can raise uncertainties about the microscopic nature of the polymer solutions being studied. The presence or otherwise of supermolecular aggregates, along with the sensitivity of these solutions to preparation methods and solution history, can all cause imponderable variations between one set of experiments and another.

For this reason, it is all the more surprising that the ingenious experiments of Berman and George (1974) were able to pronounce so decisively on the length versus time-scales controversy. These authors observed the onset of drag reduction produced by PEO in water alone, and with two different concentrations (25 per cent and 53 per cent) of glycerine. The glycerine changed the solvent viscosity dramatically but left the mean radius of the polymers virtually unchanged. Despite this, in the three experiments onset occurred at three very different values of the turbulent viscous length scale (i.e. $u_\tau/v = 940$, 671 and 181). In complete contrast, the turbulent time-scale at onset was constant to well within experimental error.

3.4.4 *Maximum drag reduction*

The existence of viscosity implies a lower bound on the rate of energy dissipation in a flowing fluid. From this fact we can infer an absolute upper bound on the amount by which skin friction can be reduced. That is, if we could eliminate all the turbulent motions, then the skin friction could be reduced to that for laminar flow at the same Reynolds number. Or, in other words, the specifically turbulent contribution to the fluid drag would have been eliminated.

According to Hoyt (1972b), at an early stage in the development of the subject it was appreciated that there appeared to be an upper bound on the amount of drag reduction which could be obtained. This upper limit was found to be independent of polymer type—different polymers simply required to be used in different concentrations—but it did depend on the Reynolds number. Studies carried out with many different kinds of polymer (Pruitt and Crawford 1965, cited as ref. 38 in Hoyt 1972b) showed that the maximum amount of the drag reduction that could be obtained was about 80 per cent of the theoretical maximum equivalent to suppressing the turbulence altogether.

Virk *et al.* (1967) concluded that an upper limit on drag reduction was a universal phenomenon, and introduced the concept of the ultimate drag reduction asymptote. Later, Virk, Mickley, and Smith (1970) produced an analytical representation of the ultimate drag reduction asymptote in the form

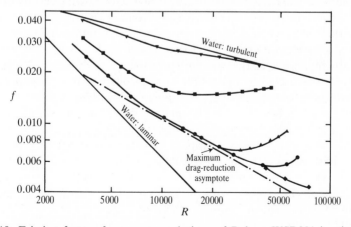

Fig. 3.18. Friction factors for aqueous solutions of Polyox WSR301 in pipe flow, showing the trend to the maximum drag reduction asymptote (Paterson and Abernathy 1970): ▼, 0.1 wt ppm; ■, 1 wt ppm; ▲, 5 wt ppm; ●, 10 wt ppm; ◆, 50 wt ppm).

$$\frac{1}{f^{1/2}} = 19.0 \log_{10}(Rf^{1/2}) - 32.4 \tag{3.87}$$

which can be seen to be a modification of the logarithmic resistance law for turbulent flow in pipes, as given by eqn (1.49).

In Fig. 3.18 we show how the results of Paterson and Abernathy (1970) are bounded from above by eqn (1.49), for the turbulent flow of water, and from below by the Virk maximum drag reduction asymptote, as given by (3.87).

3.4.5 Drag reduction in fibre suspensions

We have already briefly referred to the natural occurrence of drag reduction in solid–fluid systems. It takes only a moment's reflection to realize that this is likely to prove a very large field of study. As well as the obvious environmental (and, for that matter, physiological) flows, there are many industrial flow processes which involve droplet–gas or particle–gas or particle–liquid combinations.

The field is also rather confusing, partly because the effects are normally very much smaller than in the polymer case, and partly because investigations in this area are often undertaken as an adjunct of some industrial process and any possible drag reduction is a secondary consideration. Thus there has often been controversy over attempts (often long after the investigation was carried out, and often by others than the original researchers) to decide whether or not a particular flow showed any evidence of drag reduction.

The reader who wishes to pursue this general subject further will find the paper by Radin, Zakin, and Patterson (1975) helpful. These authors have provided a very extensive and analytical review of existing results in solid–fluid systems. They have also reported an extremely comprehensive investigation of their own. Their primary conclusion is that fibrous additives with a length-to-diameter l/d ratio greater than about 25 to 35 would always cause drag reduction provided that the concentration was high enough. Our interest here will be restricted to those fibre suspensions which are comparable in drag-reducing effectiveness to the best polymer solutions.

As we noted earlier, the scientific interest of this phenomenon was greatly enhanced by the discovery (Ellis 1970) that certain asbestos fibres (prepared by Turner Brothers Asbestos Company Ltd) could give high drag reductions at low additive concentrations. Typical figures are 40 per cent drag reduction at a fibre concentration in water of 100 wt ppm. Thus these fibres are quite comparable in effectiveness with the best polymers, a result which has been conformed by subsequent more extensive investigations (Hoyt 1972b; Radin et al. 1975).

If we summarize the desirable properties of fibres (much as we did for polymers at the beginning of Section 3.4.2), then experimental results suggest that fibres should be (a) very long and thin, (b) flexible, and (c) well dispersed (a surfactant may be necessary). Evidently there is a superficial resemblance between this list and the one for polymers. This might suggest that they are two aspects of the same phenomenon, that is, that they reduce drag by the same mechanism. In fact there is evidence that this is not the case and we shall return to this in Chapter 14.

An additional feature of considerable interest is the result of combining both macroscopic fibres and drag-reducing polymers in one suspension. Lee, Vaseleski, and Metzner (1974) found that when both types of additive were used together they reduced drag by amounts greater than the sum of the two independent effects. In particular, drag reductions greater than 95 per cent were obtained in suspensions containing TBA asbestos fibres and Separan AP30. They also found that these mixed suspensions were much less susceptible to degradation under shear. These results have been confirmed by Sharma, Seshadri, and Malhotra (1979), who reported similar behaviour when a TBA asbestos fibre suspension was injected into pipe flow of drag-reducing polymer solution, and by McComb and Chan (1985), who used laser-Doppler anemometry to measure turbulent structure in fibre suspensions.

3.5 Renormalization methods and the closure problem

The term renormalization belongs to quantum physics. It arises when the discrete formulation for particles is extended to include the case of continuous fields. Initially this is done by formulating the theory on a lattice (say a cube of side L, divided into a mesh with links of length l). When the continuum limit is taken, it is found that certain terms diverge. Renormalization is a standard method of removing the resulting singularities.

It is usual to refer to divergences as $L \to \infty$ as 'infra-red', and naturally divergences as $l \to 0$ are spoken of as being 'ultraviolet'. If singularities of either kind can be eliminated, then a particular theory is referred to as being 'renormalizable'. Specialist discussions of this subject can be found in Amit (1984) and Collins (1984). The former reference also considers the extension of these ideas, through the renormalization group, to critical phenomena.

In this section, we shall first give a brief discussion of other relevant areas of statistical physics, and consider the precedents for borrowing the term renormalization in order to describe a systematic way of taking interactions into account through a quasi-particle approach. We shall then consider the formulation of the turbulence problem as a branch of statistical physics, and show how it can be tackled using field-theoretical methods.

However, first we should sound a cautionary note. The analogy between incompressible fluid turbulence and quantum field theory should not be pushed too far. There are no intrinsic singularities in the Navier–Stokes equation, in either its x-space or its k-space forms. The so-called infrared divergence of turbulence theory arises (in a very artificial situation) when we try the Kolmogorov spectrum (2.137) as a solution (for all **k**) in certain theories of turbulence.

We shall discuss this problem in some detail in the following chapters, but for the moment we emphasize that there is only one, rather artificial, infrared singularity in turbulence, and there are no ultraviolet singularities at all! For this reason, we would argue that to use 'renormalizable' as a description of a theory is a specific technicality of quantum field theory and (unlike renormalization) is not appropriate to turbulence. We note some support for this view, in that Moiseev, Sagdeev, Tur, and Yanovskii (1984) state that the divergence associated with the Kolmogorov distribution is 'non-renormalizable', from the field-theoretical point of view.

3.5.1   *Renormalization methods in statistical physics*

Historically, the subject of statistical mechanics has been successful in treating the case of the perfect gas, where the constitutent particles do not interact with each other. The Hamiltonian H for such a system can be written as the sum over the degrees of freedom of single-particle Hamiltonians:

$$H = \sum_\alpha \frac{p_\alpha^2}{2m_\alpha} \tag{3.88}$$

where p_α and m_α are the momentum and mass respectively of the αth particle. The index α is summed up to N, for a system of N non-interacting particles.

In an interacting system, the total Hamiltonian can be expected to take a more complicated form:

$$H = \sum_\alpha H_\alpha + \sum_{\alpha,\beta} H_{\alpha\beta}, \tag{3.89}$$

where H_α is the single-particle Hamiltonian, as in (3.88), and $H_{\alpha\beta}$ may involve (for example) a potential which depends on the coordinates of several particles. A standard method of approach to many-body problems is to try to replace the interaction term by the average effect of all the other particles on the αth particle. The result can be a first approximation to (3.89) in the form

$$H = \sum_\alpha H_\alpha' \tag{3.90}$$

where H_α' is the effective Hamiltonian for the αth quasi-particle. Each of the N particles is therefore replaced by a quasi-particle, and each of these has a portion of the interaction energy added on to their single-particle form.

The immediate benefit of this method is that we can treat the system as a perfect gas of quasi-particles, with total Hamiltonian given approximately by (3.90). Thus statistical mechanics can be used to obtain the macroscopic properties of the system.

Of course, turbulence is dissipative, which means that some of specific points made about other, conservative, physical systems will not be directly applicable. Nevertheless, it may be instructive to consider possible analogies with other problems which involve interactions and, before returning to turbulence, we shall consider two examples of the successful use of the quasi-particle method.

Our first example is the motion of conduction electrons through a metal. Here it is possible to neglect the interaction between the electron and the periodic (lattice) potential by replacing each electron, of mass m, by a quasi-particle of effective mass m'. In general, m' will not be the same as the electron mass, and may even be negative. Ter Haar (1958) refers to m' as a renormalized mass.

As a second, and in some ways, more relevant example, let us consider an electron in either a plasma or an electrolyte. It was found by Debye and Hückel (as long ago as 1923) that the cloud of charge surrounding a single electron screens the Coulomb potential, so that the effective potential at a point becomes (in appropriate units)

$$V(r) = q_{\text{eff}}/r, \tag{3.91}$$

where the effective charge is given by

$$q_{\text{eff}} = q \exp\left(-\frac{r}{\lambda}\right). \tag{3.92}$$

Here q is the 'bare' electron charge and λ is the Debye–Hückel length, which depends on the number density of electrons and on the temperature. Thus the introduction of an effective charge allows the interaction potential to be replaced by a quasi-single-particle potential.

The Debye–Hückel result was obtained as a rather phenomenological theory. Nowadays this problem is treated systematically through perturbation theory, and the Debye–Hückel screened potential can be recovered as a first approximation of the more general method. We shall treat these matters in a little more detail in Chapter 5, but it is of interest to observe here that Balescu (1975) refers to the partial summation of the perturbation series that leads to the screened potential as a renormalization process, and remarks that the name stems from a loose analogy with the renormalization process of quantum field theory. A more extensive discussion of the same point has been given by Prigogine (1968).

This is the way in which we shall use the terminology in this book. We shall refer to any systematic procedure which replaces a bare quantity plus interactions by a 'dressed' quantity without interactions as a renormalization process. The dressed quantity will accordingly be regarded as the renormalized version of the bare quantity. At its simplest, in the context of turbulence, this will mean that the molecular viscosity of the fluid will be the bare quantity, and the effect of turbulent interactions will be taken into account through a renormalized viscosity. But, as we shall see, more general interpretations will be possible.

3.5.2 *Renormalized perturbation theory*

The application of renormalized perturbation methods to turbulence was begun by Kraichnan in a series of papers during the late 1950s, culminating in the direct interaction approximation (DIA) (Kraichnan 1959). At the time, the main claim of the DIA was that it did not suffer from the flaws of the quasi-normality hypothesis (see Section 2.8.2)—in particular, it did not predict negative spectra—and accordingly it was seen as a great advance.

We shall return to a detailed discussion of DIA, and other subsequent theories, in the next few chapters. Here we shall only give a rather simple introduction to the subject, and we shall base this, very loosely, on the formalism of Wyld (1961).

Let us first take a simple-minded look at the failure of quasi-normality. As we have seen, the essential hypothesis is that the third-order moment can be calculated on the assumption that the even-order moments are related as if the velocity had a Gaussian probability distribution. Yet, strictly, the existence of a non-zero third (or any odd-order) moment would be incompatible with such a distribution. On that basis, the failure of quasi-normality seems only to be expected (but, see the less simplistic discussion of this point in Section 2.8.1).

In a sense, this illustrates our underlying dilemma. We can only work out the general relationships between moments for a Gaussian distribution, yet the very essence of turbulence is its non-Gaussianity. The particular way in which this problem is resolved lies at the heart of any modern turbulence theory. In what follows, we shall consider one fairly general way of doing this.

We shall formulate our approach as a 'thought experiment', which we shall set up in the following way. We imagine a fluid stirred up into turbulence by some arbitrary forces. Initially the non-linear terms are to be thought of as 'switched off'. Then at some later time we switch them on and observe what happens.

First we have to specify our arbitrary stirring force. We represent its Fourier components by $f_\alpha(\mathbf{k}, t)$ and take it to satisfy the continuity condition in the form of eqn (2.74):

$$k_\alpha f_\alpha(\mathbf{k}, t) = 0. \tag{3.93}$$

We shall also take $f_\alpha(\mathbf{k}, t)$ to have a Gaussian probability distribution, with autocorrelation $w(\mathbf{k}, t)$ given by

$$\left(\frac{L}{2\pi}\right)^3 \langle f_\alpha(\mathbf{k}, t) f_\beta(-\mathbf{k}, t') \rangle = D_{\alpha\beta}(\mathbf{k}) w(k, t - t'). \tag{3.94}$$

It should be noted that we are also taking the random stirring forces to be both stationary and isotropic.

Now we consider the Navier–Stokes equation with the random force added on as an external driving force. That is, we write eqn (2.76) as

$$\left(\frac{\partial}{\partial t} + vk^2\right) u_\alpha(\mathbf{k}, t)$$

$$= f_\alpha(\mathbf{k}, t) + \lambda M_{\alpha\beta\gamma}(\mathbf{k}) \sum_{\mathbf{j}} u_\beta(\mathbf{j}, t) u_\gamma(\mathbf{k} - \mathbf{j}, t) \tag{3.95}$$

where λ is a bookkeeping parameter and is put equal to unity at the end of the calculation.

First, however, we examine the effect of putting $\lambda = 0$. Without the non-linear term, the response of the fluid is determined solely by the molecular viscosity (and of course by the continuity condition). Thus the stirring force could be expected to induce a velocity field $u_\alpha^{(0)}(\mathbf{k}, t)$ given by

$$u_\alpha^{(0)}(\mathbf{k}, t) = \int dt' \exp\{-vk^2(t - t')\} f_\alpha(\mathbf{k}, t) \tag{3.96}$$

where the superscript zero indicates that (3.96) is the solution of (3.95) with $\lambda = 0$.

Now for what follows, we shall use a very symbolic notation. We shall drop all labels, except those that are absolutely necessary. Thus (3.96) becomes

$$u^{(0)} = G^{(0)} f, \tag{3.97}$$

where $G^{(0)}$ stands for the Green function of eqn (3.95) with $\lambda = 0$, and it should be clear that the wavenumber label is the same for each of the three functions. Where we have combinatorial expressions involving many wavenumbers we shall just use integers to distinguish one from another. An example of this arises straightaway. It follows from (3.97) that the $u^{(0)}$ are also normally distributed and therefore we have the rigorous result (see Section 2.8.2)

$$\langle u^{(0)}(1) u^{(0)}(2) \ldots u^{(0)}(n) \rangle = 0, \text{ if } n \text{ is odd,} \tag{3.98}$$

and, combining the homogeneity requirement (in the form of eqn (2.83)) with the factorization properties of even-order moments, we also have

$$\langle u^{(0)}(1) u^{(0)}(2) \rangle = \delta_{12} Q^{(0)}(1) \tag{3.99}$$

$$\langle u^{(0)}(1) u^{(0)}(2) u^{(0)}(3) u^{(0)}(4) \rangle = \delta_{12} \delta_{34} Q^{(0)}(1) Q^{(0)}(3) +$$

$$+ \delta_{13} \delta_{24} Q^{(0)}(1) Q^{(0)}(2) +$$

$$+ \delta_{14} \delta_{23} Q^{(0)}(1) Q^{(0)}(2), \tag{3.100}$$

where the numbers $1, 2 \ldots n$ imply wavenumbers $k_1, k_2, \ldots, k_n$, and analogous formulae hold for all higher even-order moments.

Our thought experiment is completed by switching on the non-linear terms. The effect of this is to couple modes together. If we do this at $t = -\infty$, then any mode $u^{(0)}(k_1)$ will first be coupled to modes $u^{(0)}(k_2)$ and $u^{(0)}(k_3)$, and then to $u^{(0)}(k_2)$, $u^{(0)}(k_3)$, and $u^{(0)}(k_4)$, and so on. Ultimately the exact velocity field $u(k)$ will be generated and, because of the mode coupling, $u(k)$ will be perturbed away from a Gaussian distribution.

We can summarize this by writing down the perturbation series

$$u(k) = u^{(0)}(k) + u^{(1)}(k) + u^{(2)}(k) + \dots . \tag{3.101}$$

Here the fields $u^{(0)}(k)$ and $u(k)$ have a clear physical interpretation. In contrast, the higher-order terms on the r.h.s. of eqn (3.101) are merely combinations of the $u^{(0)}$ field at various wavenumbers, representing ever higher orders of mode couplings. We obtain $u^{(1)}$, $u^{(2)}$, ... by an iterative process, and this can be done as follows.

The operator on the l.h.s. of eqn (3.95) can be inverted in terms of the Green function, and the full equation of motion written in our present highly symbolic form as

$$u(k) = u^{(0)}(k) + G^{(0)}(k)M(k)u(j)u(k - j), \tag{3.102}$$

where the repeated variable j is, of course, summed. The expansion (3.101) is now substituted into both sides of eqn (3.102), and coefficients of the same order (i.e. power of λ) equated, with the result

$$u^{(1)} = G^{(0)}Mu^{(0)}u^{(0)}, \tag{3.103a}$$

$$u^{(2)} = 2G^{(0)}Mu^{(0)}u^{(1)}, \tag{3.103b}$$

$$u^{(3)} = 2G^{(0)}Mu^{(2)}u^{(0)} + G^{(0)}Mu^{(1)}u^{(1)}, \tag{3.103c}$$

and so on.

The coefficient $u^{(1)}$ is already in terms of $u^{(0)}$ only, and each higher-order coefficient can be similarly expressed by means of successive substitutions:

$$u^{(1)} = G^{(0)}Mu^{(0)}u^{(0)}, \tag{3.104a}$$

$$u^{(2)} = 2G^{(0)}Mu^{(0)}G^{(0)}Mu^{(0)}u^{(0)}, \tag{3.104b}$$

$$u^{(3)} = 4G^{(0)}Mu^{(0)}G^{(0)}Mu^{(0)}G^{(0)}Mu^{(0)}u^{(0)} +$$
$$+ G^{(0)}MG^{(0)}Mu^{(0)}u^{(0)}G^{(0)}Mu^{(0)}u^{(0)}, \tag{3.104c}$$

and so on. Clearly our symbolic notation is glossing over many complexities of wavenumber and other labels. We shall return to a properly detailed account of these methods in Chapter 5. Here we wish only to give a rather simple and rather schematic treatment, which will, we hope, allow the main ideas to stand out.

Now we use eqns (3.101) and (3.104) to obtain a perturbation series for the spectral density. We have (still in our symbolic notation)

$$Q(k) = \langle u(k)u(-k)\rangle \tag{3.105}$$

and, substituting from (3.101),

$$Q(k) = \langle u^{(0)}u^{(0)}\rangle + \langle u^{(0)}u^{(2)}\rangle + \langle u^{(1)}u^{(1)}\rangle + \langle u^{(2)}u^{(0)}\rangle +$$
$$+ O(\lambda^4). \tag{3.106}$$

In order to keep things simple, we have terminated the expansion at second order. We can readily deduce from eqns (3.98) and (3.104) that all odd orders in (3.106) vanish under the averaging process.

If we now substitute from (3.104) for $u^{(1)}$, $u^{(2)}$ and invoke (3.99) and (3.100), in order to work out the averages, then it should be clear, in a general way, that we shall end up with an equation for the spectral density in terms of zero-order quantities only. This will take the symbolic form

$$Q(k) = Q^{(0)} + G^{(0)}MMQ^{(0)}Q^{(0)} + O(\lambda^4), \tag{3.107}$$

where the second term on the r.h.s. will take a variety of forms according to how wavenumber (and other) labels are permuted.

Now, if λ were a small parameter, then we would have a conventional perturbation theory for $Q(k)$, and the expansion in (3.107) could be terminated at any given order, with an error that could be made as small as we wish. Unfortunately this is not the case here, and ultimately we will have to put $\lambda = 1$.

In the absence of a small parameter, Wyld's (1961) approach was to study the properties of the perturbation globally in terms of diagrams. We shall go into more detail in Chapter 5, but essentially Wyld found that diagrams representing the various terms of the perturbation series could be divided into two classes, according to their topology. One of these classes could be summed exactly, and the other could be partially summed in terms of irreducible diagrams, in which $Q^{(0)}$ and $G^{(0)}$ had been replaced by the exact forms G and Q.

This then is the process of renormalization as we shall understand it here. The primitive perturbation series, which is wildly divergent, has been replaced by a renormalized perturbation series or RPT.

The RPT has, in general, unknown properties, but, even if it did not converge, it might be some improvement on the primitive series. For example, it might take the form of an asymptotic expansion. This is something which will require investigation for each new theory, although, as we shall see, the difficulties of doing this are generally formidable, and normally one has to be content with a rather limited comparison with experimental results.

It may seem unsatisfactory that we have to rely on what may be regarded as matters of pure mathematics in order to obtain a theory of turbulence. However, it is worth remarking that turbulence is not alone in this, and in other problems of this type it is not unusual for workers in the field to rely on the topology of their diagrammatic representations.

At the same time, the renormalization procedures are not devoid of physical significance, a matter which we shall discuss in due course. For the moment it may be noted that eqn (3.106) can be interpreted as an expansion for the exact energy spectral density in powers of the Reynolds number. In order that turbulence can exist in the first place, the Reynolds number must be large. However, a process in which the molecular viscosity is replaced by an effective turbulent viscosity could result in a renormalized Reynolds number, with (as we have suggested above) the possibility that the renormalized series is either convergent or, failing that, asymptotic.

3.5.3   Renormalization group (RG) methods

This technique has recently had some success when applied to problems in critical phenomena. Such phenomena are characterized by fluctuations on all length scales, ranging from the molecular to the macroscopic. As the physical system moves towards a phase transition, it becomes increasingly dominated by the large-scale fluctuations. At the critical point, the length scale of the fluctuations is infinite, and accordingly the system becomes insensitive to changes of scale.

RG involves the progressive scaling away of the shortest wavelengths, whose effect on the larger wavelengths can be retained in an average form as a contribution to a transport coefficient. When the system becomes invariant under these scaling transformations, it is said to have reached a fixed point. The fixed point corresponds to a critical point.

Of course this method cannot be taken over from critical phenomena and applied in some prescriptive fashion to turbulence. As Wilson (1975) has pointed out, the difficulties faced initially in applying RG to a new problem may seem as formidable as those involved in applying any other technique. Thus, although various attempts have been made to adapt the method to the case of turbulence, there are many problems still to be overcome. This field of activity will be the subject of a later chapter. Here we shall only give a brief discussion of the general concept of RG applied to fluid turbulence.

We start with a notational change: that is, we replace the symbol v for molecular viscosity by v_0. This is in recognition of the fact that the molecular viscosity of the fluid is, in the context of turbulent interactions, the bare quantity which is to be renormalized. We now consider in outline how the Navier–Stokes equation can be used for this purpose.

Consider the velocity field as represented by the discrete set of components $u_\alpha(\mathbf{k}, t)$ on the interval $0 < k < k_0$, where k_0 is the largest wavenumber present and can be defined through the dissipation integral

$$\varepsilon = \int_0^\infty 2v_0 k^2 E(k)\, dk = \int_0^{k_0} 2v_0 k^2 E(k)\, dk, \qquad (3.108)$$

where ε is the dissipation rate and $E(k)$ is the energy spectrum.

We now choose a wavenumber cut-off k_1 such that $k_1 < k_0$. In practice, we would have both k_1 and k_0 of the same order of magnitude as the dissipation wavenumber $R_d^{(0)}$, where the superscript indicates that the dissipation wavenumber is based on the bare viscosity. Thus we have

$$k_1 \approx k_0 \approx k_d^{(0)} \qquad (3.109)$$

where the dissipation wavenumber is given by eqn (2.133) with v replaced by v_0.

In principle, RG then involves two stages.

(a) Solve the Navier–Stokes equation on the interval $k_1 \leqslant k \leqslant k_0$. Substitute that solution for the mean effect of the high k modes into the Navier–Stokes equation on $0 \leqslant k \leqslant k_1$. This results in an increment to the viscosity: $v_0 \to v_1 = v_0 + \delta v_0$.
(b) Rescale the basic variables so that the Navier–Stokes equation on $0 \leqslant k \leqslant k_1$ looks like the original Navier–Stokes equation on $0 \leqslant k \leqslant k_0$.

This procedure is then repeated for a wavenumber cut-off $k_2 \leqslant k_1 \leqslant k_0$, and so on.

The underlying physics of RG can be summarized as follows. In the viscous range of wavenumbers, it is reasonable to suppose that the turbulence is critically damped. That is, any more k in the band $k_1 \leqslant k \leqslant k_0$ is driven by energy transfer from modes $k < k_1$, and the injected energy is dissipated locally by the effects of the molecular viscosity. This offers the hope that the Navier–Stokes equation can be linearized, and solved, within the band $k_1 \leqslant k \leqslant k_0$, although inevitably this solution is coupled to modes $k < k_1$.

As a result of this procedure, the new Navier–Stokes equation has an increased effective viscosity in the reduced range of wavenumbers $0 \leqslant k \leqslant k_1$. If we then define an effective dissipation wavenumber $k_d^{(1)}$ for the new Navier–Stokes equation

$$k_d^{(1)} = \left(\frac{\varepsilon}{v_1^3}\right)^{1/4} < k_d^{(1)} = \left(\frac{\varepsilon}{v_0^3}\right)^{1/4}, \qquad (3.110)$$

we can repeat the whole procedure for $k_2 \approx k_d^{(1)}$ and we can again linearize the suitably scaled Navier–Stokes equation on the interval $k_2 \leqslant k \leqslant k_1$.

We shall discuss all these points in more detail later on, but it should not escape our attention that, from the computational point of view, the RG procedure is a systematic way of reducing the number of degrees of freedom in the problem.

References

AGOSTON, G. A., HARTE, W. H., HOTTEL, H. C., KLEMM, W. A., MYSELS, K. J., POMEROY, H. H., and THOMPSON, J. M. (1954). *Ind. eng. Chem.* **46**, 1017.

AMIT, D. J. (1984). *Field theory, the renormalisation group, and critical phenomena*, (2nd edn). World Scientific.

ANSELMET, F., GAGNE, Y., HOPFINGER, E. J., and ANTONIA, R. A. (1984). *J. fluid Mech.* **140**, 63.

ANTONIA, R. A., SATYAPRAKASH, B. R., and HUSSAIN, A. K. M. F. (1982). *J. fluid Mech.* **119**, 55.

ARUNACHALAM, Vr. and FULFORD, G. D. (1971). *Houille Blanche* **26**, 33.

BALESCU, R. (1975). *Equilibrium and nonequilibrium statistical mechanics.* John Wiley, New York.

BATCHELOR, G. K. (1971). *The theory of homogeneous turbulence* (2nd edn). Cambridge University Press, Cambridge.

—— and TOWNSEND, A. A. (1949). *Proc. R. Soc. A* **199**, 238.

BERMAN, N. S. and GEORGE, W. K. (1974). *Phys. Fluids* **17**, 250.

BLACKMAN, R. B. and TUKEY, J. W. (1958). *Measurement of power spectra.* Dover Publications, New York.

BROWN, G. L. and ROSHKO, A. (1974). *J. fluid Mech.* **64**, 775.

BUCHAVE, P., GEORGE, W. K., and LUMLEY, J. L. (1979). *Ann Rev. fluid Mech.* **11**, 443.

CARR, M. P. and FORSEY, C. R. (1982). In *Numerical methods in aerodynamics and fluid dynamics* (ed. P. L. ROE). Academic Press, New York.

CHAMBERS, T. L. and WILCOX, D. C. (1976) *AIAA Paper 76–352, AIAA 9th Fluid and Plasma Dynamics Conference, San Diego, Ca.*

COLLINS, J. C. (1984). *Renormalisation.* Cambridge University Press, Cambridge.

COMTE-BELLOT, G. *Ann. Rev. fluid Mech.* **8**, 209.

COOLEY, J. W. and TUKEY, J. W. (1965). *Math. Comput.* **19**, 297.

CORINO, E. R. and BRODKEY, R. S. (1969). *J. fluid Mech.* **37**, 1.

CORRSIN, S. (1962). *Phys. Fluids* **5**, 1301.

DEARDORFF, J. W. (1970). *J. fluid Mech.* **41**, 453.

—— (1971). *J. comput. Phys.* **7**, 120.

DRAIN, L. E. (1980). *The laser Doppler technique.* John Wiley, New York.

DURRANI, T. S. and GREATED, C. A. (1977). *Laser systems in flow measurement.* Plenum Press, London.

DURST, F., MELLING, A. and WHITELAW, J. H. (1976). *Theory and practice of laser Doppler anemometry.* Academic Press, New York.

DWYER, H. A. (1984). *AIAA J.* **22**, 1705.

ELLIS, H. D. (1970). *Nature, Lond.* **226**, 352.

FABULA, A. G., HOYT, J. W., and CRAWFORD, H. R. (1963). *Bull. Am. phys. Soc.* **8**, paper K5.

——, LUMLEY, J. L., and TAYLOR, W. D. (1965). In *Modern developments in the mechanics of continua* (ed. S. Eskinazi). Academic Press, New York.

FERZIGER, J. H. (1987). *J. comput. Phys.* **69**, 1.

FRENKIEL, F. N. and KLEBANOFF, P. S. (1975). *Boundary-layer Meteorol.* **8**, 173.

FRISCH, U., SULEM, P–L., and NELKIN, M. (1978). *J. fluid Mech.* **87**, 719.

FUJISAKA, H. and MORI, H. (1979). *Prog. theor. Phys.* **62**, 54.

GIBSON, C. H. and MASIELLO, P. J. (1972). *Proc. Symp. on Statistical Models and Turbulence* (eds M. ROSENBLATT and C. VAN ATTA). Springer, New York.

——, STEGEN, G. R. and McCONNELL, S. (1970). *Phys. Fluids* **13**, 2448.

GOLDSTEIN, S. (1938). *Modern developments in fluid dynamics.* Clarendon Press, Oxford. (Reprinted by Dover Publications, New York, 1965).

GOTTLIEB, D., HUSSAINI, M. Y., and ORSZAG, S. A. (1984). In *Spectral methods for partial differential equations* (eds R. G. VOIGHT, D. GOTTLIEB, and M. Y. HUSSAINI). SIAM, Philadelphia, PA.

GRANT, H. L., STEWART, R. W., and MOILLIET, A. (1962). *J. fluid Mech.* **12**, 241.

GURVICH, A. S. and YAGLOM, A. M. (1967). *Phys. Fluids Suppl.* **10**, S59.

HAWKRIDGE, H. R. J. and GADD, G. E. (1971). *Nature, Lond.* **230**, 253.

HELE-SHAW, H. S. (1897). *Trans. Inst. nav. Archit.* **39**, 145.

HINZE, J. O. (1975). *Turbulence* (2nd edn). McGraw-Hill, New York.

HOYT, J. W. (1972a). *J. Basic Eng.* **94**, 258.

—— (1972b). *Rep. TP 299. Naval Undersea Center*, San Diego, CA.

JONES, W. P. and LAUNDER, B. E. (1972). *Int. J. heat mass Transfer* **15**, 301.

KENNEDY, D. A. and CORRSIN, S. (1961). *J. fluid Mech.* **10**, 366.

KERR, R. M. (1985). *J. fluid Mech.* **153**, 31.

KIM, J., MOIN, P., and MOSER, R. (1987). *J. fluid Mech.* **177**, 133.

KING, L. V. (1914) *Proc. R. Soc. A* **214**, 373.

KLINE, S. J., REYNOLDS, W. C., SCHRAUB, F. A., and RUNSTADLER, P. W. (1967). *J. fluid Mech.* **30**, 741.

KOBETS, G. F., MATJUKHOV, A. P., and MIGRENKO, G. S. (1974). *Proc. 1st Int. Conf. on Drag Reduction, Cambridge*, Paper D4. BHRA, Bedford.

KOLMOGOROV, A. N. (1941a). *C.R. Acad. Sci. URSS* **30**, 301.

—— (1941b). *C.R. Acad. Sci. URSS* **32**, 16.

—— (1962). *J. fluid Mech.* **13**, 82.

KRAICHNAN, R. H. (1959). *J. fluid Mech.* **5**, 497.

—— (1974). *J. fluid Mech.* **62**, 305.

KUO, A. Y-S. and CORRSIN, S. (1971). *J. fluid Mech.* **50**, 285.

—— and —— (1972). *J. fluid Mech.* **56**, 447.

LANDAU, L. D. and LIFSHITZ, E. M. (1959). *Fluid mechanics.* Pergamon Press, London.

LAU, J. C., WHIFFEN, M. C., FISHER, M. J., and SMITH, D. M. (1981). *J. fluid Mech.* **102**, 353.

LAUFER, J. (1954). *NACA Tech. Rep. 1174.*

LAUNDER, B. E. (1984). *Int. J. heat mass Transfer* **27**, 1485.

—— and SPALDING, D. B. (1974). *Comput. Meth. appl. mech. Eng.* **3**, 269.

——, REECE, G. J., and RODI, W. (1975). *J. fluid Mech.* **68**, 537.

LAUTERBORN, W. and VOGEL, A. (1984). *Ann. Rev. fluid Mech.* **16**, 223.

LEE, W. K., VASELESKI, R. C. and METZNER, A. B. (1974). *AIChE J.* **20**, 128.

LEONARD, A. (1974). *Adv. Geophys. A* **18**, 237.

LILLY, D. K. (1967). *Proc. IBM Science and Computing Symp. on Environmental Science*, p. 195. Thomas J. Watson Research Center, Yorktown Heights.

LITTLE, R. C., HANSEN, R. J., HUNSTON, D. L., KIM, O-K., PATTERSON, R. L., and TING, R. Y. (1975). *Ind. eng. Fundam.* **14**, 283.

LOMAS, C. G. (1986). *Fundamentals of hot wire anemometry.* Cambridge University Press, Cambridge.

LUCHINI, P. (1987). *J. comput. Phys.* **68**, 283.

LUMLEY, J. L. (1969). *Ann. Rev. fluid Mech.* **1**, 367.

—— (1973). *J. polym. Sci. macromol. Rev.* **7**, 263.

McCOMB, W. D. and CHAN, K. T. J. (1985). *J. fluid Mech.* **152**, 455.

—— and RABIE, L. H. (1982). *AIChE J.* **28**, 547.

MANDELBROT, B. (1972). *Proc. Symp. on Statistical Models and Turbulence* (eds M. ROSENBLATT and C. VAN ATTA), Springer, New York.

—— (1974). *J. fluid Mech.* **62**, 331.

MARKATOS, N. C. (1986). *Appl. math. Modelling* **10**, 190.

MELINAND, J. P. and CHARNAY, G. (1978). *Proc. Dynamic Flow Conf.* (eds B. W. HANSEN and L. S. G. KOVASNAY). Marseille-Baltimore.

MELLOR, G. L. and HERRING, H. J. (1973). *AIAA J.* **11**, 590.

MERRILL, E. W., SMITH, K. A., SHIN, H., and MICKLEY, H. S. (1966). *Trans. Soc. Rheol.* **10**, 335.

METZNER, A. B. and METZNER, A. P. (1970). *Rheol. Acta* **9**, 174.

MOIN, P. and KIM, J. (1982). *J. fluid Mech.* **118**, 341.

MOISEEV, S. S., SAGDEEV, R. Z., TUR, A. V., and YANOVSKII, V. V. (1984). *Sov. Phys.—Dokl.* **28**, 643.

MONIN, A. S. and YAGLOM, A. M. (1975). *Statistical fluid mechanics*, Vol. 2, *Mechanics of turbulence*. MIT Press, Cambridge, Ma.

MOORE, D. W. and SAFFMAN, P. G. (1975). *J. fluid Mech.* **69**, 465.

MYSELS, K. J. (1949). *US Patent 2,492,173.*

—— (1971). *AIChE Chem. Eng. Prog. Symp. Ser.* **67**, 45.

NALLASAMY, M. (1987). *Comput. Fluids* **15**, 151.

NOVIKOV, E. A. and STEWART, R. W. (1964). *Izv. Akad. Nauk, Ser. geophys.* **3**, 408.

OBUKHOV, A. M. (1962). *J. fluid Mech.* **13**, 77.

OLDROYD, J. G. (1948). *Proc. 1st Int. Congr. on Rheology*, Vol. 2, p. 130. North-Holland, Amsterdam.

ORSZAG, S. A. (1971). *Stud. appl. Math.* **50**, 293.

—— and ISRAELI, M. (1974). *Ann. Rev. fluid Mech.* **6**, 281.

—— and PATTERSON, G. S. (1972). *Phys. rev. Lett.* **28**, 76.

OTNES, R. K. and ENOCHSON, L. D. (1972). *Digital time series analysis*. John Wiley, New York.

PATANKAR, S. V. and SPALDING, D. B. (1972). *Int. J. heat mass Transfer* **15**, 1787.

PATEL, V. C., RODI, W. and SCHEUERER, G. (1984). *AIAA J.* **23**, 1308.

PATERSON, R. W. and ABERNATHY, F. H. (1970). *J. fluid Mech.* **43**, 689.

PERRY, A. E. (1982). *Hot-wire anemometry*. Clarendon Press, Oxford.

PEYRET, R. and TAYLOR, T. D. (1983). *Computational methods for fluid flow*. Springer, New York.

POND, S., STEWART, R. W., and BURLING, R. W. (1963). *J. atmos. Sci.* **20**, 319.

POPE, S. B. (1978). *AIAA J.* **16**, 279.

PRIGOGINE, I. (1968). In *Topics in nonlinear physics* (ed. N. J. ZABUSKY). Springer, New York.

PRESS, W. H. (1981). *J. fluid Mech.* **107**, 455.

RADIN, I., ZAKIN, J. L., and PATTERSON, G. K. (1975). *AIChE J.* **21**, 358.

RAO, P. V. (1970). *Houille Blanche* **25**, 15.

ROACHE, P. J. (1982). *Computational fluid dynamics*. Hermosa, Albuquerque, NM.

RODI, W. (1976). *Z. Angew. Math. Mech.* **56**, T219.

—— (1980). *Turbulence models and their application in hydraulics*. IAHR.

—— (1982). *AIAA J.* **20**, 872.

ROGALLO, R. S. (1981). *NASA Tech. Note 81315.*

—— and MOIN, P. (1984). *Ann. Rev. fluid. Mech.* **16**, 99.

RYSKIN, G. and LEAL, L. G. (1984). *J. fluid Mech.* **148**, 1.

SAFFMAN, P. G. (1968). Lectures in homogeneous turbulence. In *Topics in nonlinear physics* (ed. N. ZABUSKY), Springer, New York.

SANDBORN, V. A. (1959). *J. fluid Mech.* **6**, 211.

SAVINS, J. G. (1961). *J. Inst. Petrol.* **47**, 329.

SCHUMANN, U. (1975). *J. comput. Phys.* **18**, 376.

SHARMA, R. S., SESHADRI, V., and MALHOTRA, R. C. (1979). *Chem. eng. Sci.* **34**, 703.

SHEIH, C. M., TENNEKES, H., and LUMLEY, J. L. (1971). *Phys. Fluids* **14**, 201.

SMAGORINSKY, J. (1963). *Mon. weath. Rev.* **91**, 99.

SPEZIALE, C. G. (1987). *J. fluid Mech.* **178**, 459.

TENNEKES, H. (1968). *Phys. Fluids* **11**, 669.

TER HAAR, D. (1958). *Introduction to the physics of many-body systems*. Interscience, London.

Toms, B. A. (1948). In *Proc. 1st Int. Congr. on Rheology*, Vol. 2, p. 135. North-Holland, Amsterdam.

—— (1974). Opening address. *Proc. 1st Int. Conf. on Drag Reduction*, Cambridge. BHRA, Bedford.

Townsend, A. A. (1948). *Aust. J. sci. Res. A*, **1**, 161.

Van Atta, C. W. and Antonia, R. A. (1980). *Phys. Fluids* **23**, 252.

—— and Chen, W. Y. (1970). *J. fluid Mech.* **44**, 145.

—— and Yeh, T. T. (1973). *J. fluid Mech.* **59**, 537.

Van Doormaal, J. P. and Raithby, G.D. (1984). *Num. Heat Transfer* **7**, 147.

Virk, P. S., Merrill, E. W., Mickley, H. S., Smith, K. A., and Mollo-Christensen, E. L. (1967). *J. fluid Mech.* **30**, 305.

——, Mickley, H. S., and Smith, K. A. (1970). *J. appl. Mech.* **37**, 488.

Voke, P. R. and Collins, M. C. (1983). *PCH PhysicoChem. Hydrodyn.* **4**, 119.

Watrasiewicz, B. M. and Rudd, M. J. (1977). *Laser Doppler measurements*. Butterworth, London.

Whitsitt, N. F., Harrington, L. J., and Crawford, H. R. (1969). In *Viscous drag reduction* (ed. C. S. Wells). Plenum Press, New York.

Wilson, K. G. (1975). *Rev. mod. Phys.* **47**, 773.

Winant, C. D. and Browand, F. K. (1974). *J. fluid Mech.* **63**, 237.

Wyld, H. W. (1961). *Ann. Phys.* (*NY*) **14**, 143.

Wyngaard, J. C. and Tennekes, H. (1970). *Phys. Fluids* **13**, 1962.

Yeh, H. and Cummins, H. Z. (1964). *Appl. phys. Lett.* **4**, 176.

4

STATISTICAL FORMULATION OF THE GENERAL PROBLEM

We have seen that fluid turbulence can be regarded as a problem involving a random field $u_\alpha(\mathbf{x}, t)$. Or, as we also saw in Section 3.3.1, Fourier transformation can provide us with a formulation in terms of many degrees of freedom: that is, the Fourier components $u_\alpha(\mathbf{k}, t)$. Naturally, this latter interpretation leads us to view the turbulence problem from the point of view of the subject of statistical mechanics. This will be the dominant theme of this chapter, although we shall not entirely neglect the 'field' interpretation (still valid in $\mathbf{k}$-space, especially if we take the limit $L \to \infty$).

In the first part of the chapter, we shall attempt to assess the turbulence problem in the context of statistical mechanics. This will be done against the background of a summary treatment of the classical N-body problem. Then we shall give the functional formalism for turbulence, followed by a discussion of the criteria to be satisfied by a theory of isotropic turbulence.

4.1 Turbulence in the context of classical statistical mechanics

Let us begin with some rather general ideas about the formulation of the statistical theory of turbulence. We shall want to introduce the concept of phase space, and so, paradoxically, we must work with the continuous field $u_\alpha(\mathbf{x}, t)$—although we are free, if we wish, to discretize the $\mathbf{x}$-coordinate on a lattice.

The idea of a phase is familiar from the subject of vibrations and waves. A simple harmonic oscillation takes the form $x = A\cos(\omega t + \phi)$, where $(\omega t + \phi)$ is the phase. Clearly one can specify the position of the particle executing simple harmonic motion by specifying the phase of its motion. In this sense the phase can be regarded as a coordinate. The idea can be extended to a space spanned by the velocity and position vectors of a particle, and for N particles free to move in all three directions we have a space of $6N$ coordinates. This is what is meant by phase space.

The time evolution of a dynamical system can be represented by the motion of its representative point in phase space. To take some specific examples, a free particle would have an orbit which was a straight line parallel to the x-axis and which extended to infinity in both directions. A simple pendulum would have a closed elliptical orbit in a two-dimensional space, and an ideal gas of N particles would be represented by a single point in a $6N$-dimensional space.

This representative point would have equal *a priori* probability of being found at any position in phase space.

If we refer to the N molecules of ideal gas in their container as a system, then we can generalize the concept of phase space to accommodate the idea of an ensemble. That is, we suppose that we have many such systems (the ensemble), each identical with, but completely independent of, the other. Then the ensemble can be pictured as a cloud of representative system points in a phase space of $6N$ dimensions.

We can use the ensemble to obtain (for example) the most probable of all the system distributions. To take a specific instance, for the case of the ideal gas in thermal equilibrium this would be the Maxwell–Boltzmann distribution. However, before turning to the generalities of classical statistical mechanics, we shall first consider how some of the relevant concepts can be taken over into the turbulent case.

The general idea of the ensemble is not too difficult in itself. If we take the case of isotropic turbulence in a cubical box of side L, then this forms our basic system and clearly an ensemble can be defined as consisting of many such (identical) boxes. Now let us consider the question: what is the probability that at a particular point x_1 and time t, the velocity field $u(x_1, t)$ takes a value which lies between v_1 and $v_1 + dv_1$?

(As an aside, we note that this is a formally correct statement, but that we shall in future often just refer to the probability that the velocity takes a particular value v_1. This should always be understood to be an abbreviation of the full statement, which amounts to the requirement that the velocity lies between two particular values, separated by an infinitesimal amount.)

The answer to the question can be given in terms of the one-point distribution function $p_1(x_1, v_1, t)$. That is, the probability that $u(x_1, t)$ lies between v_1 and $v_1 + dv_1$ is

$$p_1(x_1, v_1; t)\, dv = \langle \delta(u(x_1, t) - v_1) \rangle. \tag{4.1}$$

Here the delta function can be thought of as the distribution for any one system, whereas p_1 is the mean distribution which is obtained by averaging over all members of the ensemble.

Similarly, the two-point distribution can be written as

$$p_2(x_1, v_1, x_2, v_2; t) = \langle \delta(u(x_1, t) - v_1)\delta(u(x_2, t) - v_2) \rangle \tag{4.2}$$

and so on, to any order. It follows at once from the definitions above, that we can write down the following relationships:

$$\int p_1(u_1)\, du_1 = 1, \tag{4.3}$$

$$\int p_2(u_1, u_2)\, du_2 = p_1(u_1), \tag{4.4}$$

and, in general,

$$p_S(\mathbf{u}_1, \mathbf{u}_2, \ldots, \mathbf{u}_S) = \int d\mathbf{u}_{S+1} \ldots \int d\mathbf{u}_n p_n(\mathbf{u}_1, \mathbf{u}_2, \ldots, \mathbf{u}_s, \mathbf{u}_{S+1} \ldots \mathbf{u}_n) \quad (4.5)$$

where we have suppressed all $\mathbf{x}$ variables for conciseness.

Lastly, if we wish to study the continuous field $\mathbf{u}(\mathbf{x}, t)$, then we must consider the limit $n \to \infty$, where the discrete set of points $\{\mathbf{x}_1, \mathbf{x}_2, \ldots, \mathbf{x}_n\}$ is replaced by the continuous variable $\mathbf{x}$. Then we define the functional (i.e. function of a function) probability distribution of the velocity field as

$$P[\mathbf{u}(\mathbf{x}, t)] = \lim_{n \to \infty} p_n(\mathbf{u}_1, \mathbf{u}_2, \ldots \mathbf{u}_n). \quad (4.6)$$

This will be our standard notation for functionals. We shall return to the problem of obtaining the functional probability distribution for turbulence after a short digression in order to summarize some of the main results of classical statistical mechanics as applied to systems of particles.

4.1.1   Statistical mechanics of the classical N-particle system

The subject of statistical mechanics is normally restricted to those systems whose microscopic components obey Hamilton's equations. Naturally this also restricts the strict relevance of the subject to the study of fluid turbulence. Nevertheless, we shall find it helpful to review this background material, as a number of analogies can be drawn and it provides a frame of reference, so to speak, within which the statistical dynamics of turbulence can be analysed.

Let us consider a closed system containing N particles. The state of the system at any time can be specified in terms of the N-particle distribution function F_N. This can be regarded as a generalization of the distribution functions described above, where now we are interested in the probability that at time t, particle 1 is at the position $\mathbf{x}_1$ and has the velocity $\mathbf{u}_1$, and so on for all N particles. We can write this in abbreviated notation as

$$F_N(\mathbf{x}_1, \mathbf{x}_2, \ldots \mathbf{x}_n; \mathbf{u}_1, \mathbf{u}_2, \ldots \mathbf{u}_n; t) = F_N(\mathbf{x}, \mathbf{u}; t). \quad (4.7)$$

The traditional way of obtaining the governing equation for F_N is to argue that the number of systems in the ensemble is conserved and that F_N should satisfy the equation of continuity in phase space (compare the equivalent result for fluid continuity in real space, leading to (1.1)). Then (Woods 1975) we have

$$\frac{\partial F_N}{\partial t} + \mathbf{V} \cdot (F_N \mathbf{V}) = 0, \quad (4.8)$$

where the velocity in phase space takes the generalized form $\mathbf{V} = (\dot{\mathbf{x}}, \dot{\mathbf{u}})$. This can be written more compactly as

$$\frac{dF_N}{dt} = 0, \tag{4.9}$$

where the total derivative denotes the time rate of change as one follows a representative group of points through phase space. This is Liouville's equation and it is the central equation of statistical mechanics.

The same result can be reached by considering the time evolution of the microscopic constituents of the system, each evolving according to Hamilton's equations (see Goldstein 1953). First we make the change to generalized coordinates q_n, p_n, which are the canonically conjugate position and momentum coordinates of the nth particle. Then, introducing the abbreviated notation

$$(q_1, q_2, \ldots q_n; p_1, p_2, \ldots p_n) = (q, p), \tag{4.10}$$

we can write the time dependence of the system (Balescu 1975) in terms of the Hamiltonian H, which is the sum of all the single-particle contributions (see eqn (3.88):

$$F_N(q, p; t) = F_N(q, p; 0) \exp(Ht). \tag{4.11}$$

We note that the Hamiltonian H is usually (but not always) the total energy of the system.

Liouville's equation can be expressed in terms of the Hamiltonian either by transforming variables in eqn (4.8) (see Woods 1975, p. 112) or by direct differentiation of (4.11) with respect to time (see Balescu 1975) to obtain

$$\frac{\partial F_N}{\partial t} = \sum_{n=1}^{N} \left(\frac{\partial H}{\partial q_n} \frac{\partial F_N}{\partial p_n} - \frac{\partial H}{\partial p_n} \frac{\partial F_N}{\partial q_n} \right)$$

$$= [H(q, p), F_N(q, p)], \tag{4.12}$$

where Hamilton's equations have been invoked in order to eliminate the partial derivatives with respect to time on the r.h.s., and the compact representation of the r.h.s. using square brackets is known as a Poisson bracket. Formally, the r.h.s. of eqn (4.12) can be written in terms of the Liouvillian L, which is defined by the relationship

$$LF_N = [H, F_N], \tag{4.13}$$

and the time evolution of the system can now be written with equal formality as

$$F_N(q, p; t) = F_N(q, p) \exp(Lt). \tag{4.14}$$

For Hamiltonian systems, eqns (4.14) and (4.11) are equivalent. However, Balescu (1975) has pointed out that the Liouvillian is a good basis for generalizations, even for systems where the Hamiltonian does not exist. This is a point to which we shall return later, but for the moment we shall briefly consider the question of generalization to the case where the system is Hamiltonian but interactions are present (i.e. between individual molecules owing to long-range potentials).

We have already touched on interacting systems in Section 3.5. We shall use a slightly different notation here, but there should be no confusion as we are only interested in general formulations at this stage and not in renormalization or the quasi-particle picture.

In the interests of conciseness, we shall write the canonical position co-ordinates and momenta of the nth particle as

$$x_n = (\mathbf{q}_n, \mathbf{p}_n). \tag{4.15}$$

As in Section 3.2, we represent the Hamiltonian of the system as the sum of single-particle Hamiltonians plus an interaction term which we take to be a two-particle form. That is, the total Hamiltonian is assumed to be

$$H(x_1, \ldots, x_N) = \sum_{n=1}^{N} H(x_n) + \sum_{n<m=1}^{N} H'(x_n, x_m). \tag{4.16}$$

We could also include the case of an external field, but, from our present point of view, that would be an unnecessary complication and we shall not pursue it here.

It follows, by inspection of eqns (4.12), (4.13), and (4.16) that we can generalize the Liouvillian to the case of N interacting particles:

$$L(x_1, \ldots, x_n) = \sum_{n=1}^{N} L_n + \sum_{n<m=1}^{N} L'_{nm}, \tag{4.17}$$

where the functional dependence on the canonical variables is now indicated by subscripts. The corresponding generalization of Liouville's equation follows at once from (4.12) as

$$\frac{\partial F_N}{\partial t} = \sum_{n=1}^{N} L_n F_N + \sum_{n<m=1}^{N} L'_{nm} F_N. \tag{4.18}$$

Now $F_N(x_1, \ldots, x_N)$ contains all the information that in principle one could ever possess about the system. That is, one would literally know the position and momentum of every particle at all times. So, as in all statistical problems, we want to reduce a great mass of indigestible information to a more tractable (and more interesting) form. We shall close this section with a few remarks on the use of reduced distribution functions as an organized way of providing a more coarse-grained description of the system.

In the preceding section, we have already met the hierarchy of one-point, two-point, and, in general, S-point distribution functions for the turbulent velocity field. For the case of the N-body problem which we are discussing here, we know (unlike the turbulent case) that F_N provides all possible information about the system. Accordingly, all lower-order distributions can be obtained by integrating F_N over a particular set of the dependent variables $\{x_1, \ldots, x_N\}$. Thus the S-point distribution function becomes (as a generalization of (4.5) for the turbulent case)

$$f_S(x_1, \ldots, x_S) = \int dx_{S+1} \ldots \int dx_N F_N(x_1, \ldots, x_N). \tag{4.19}$$

As all the f_S (for all S) are obtained by reduction of F_N, they are (in this context) called reduced distribution functions.

Thus the state of the system can be specified by the set of reduced distribution functions

$$\mathbf{f} = \{f_0, f_1(x_1), f_2(x_1, x_2) \ldots f_N(x_1, \ldots, x_N)\}, \tag{4.20}$$

where $\mathbf{f}$ is known as the distribution vector of the system and $f_N \equiv F_N$.

As F_N satisfies the Liouville equation, we can readily derive an evolution equation for the reduced S-point distribution function. From (4.19) and (4.18), we can show that

$$\frac{\partial f_S}{\partial t} = \sum_{n=1}^{S} L_n f_S + \sum_{n<m=1}^{S} L'_{nm} f_S + \sum_{n=1}^{S} \int dx_{S+1} L'_{n,S+1} f_{S+1}. \tag{4.21}$$

The sting here is, appropriately enough, in the tail. The last term on the r.h.s. contains f_{S+1}. Thus, as with correlations in the Navier–Stokes case (see Sections 1.2 and 2.2.1), we are faced with an open hierarchy, and indeed one which can also be formulated in terms of correlations (Balescu 1975), although we shall not pursue that here.

This open set of equations is known as the Bogoliubov–Born–Green–Kirkwood–Yvon (BBGKY) hierarchy. It can be closed by making assumptions about the way in which distribution functions can be factorized, with obvious analogies to be drawn with the quasi-normality procedure in turbulence (see Section 2.8.2). However, from our present point of view, we shall be more interested in an alternative strategy: the development of kinetic equations. This will be the subject of the next section. We close this section with two observations, which we shall include in the interests of completeness.

First, we should note that Lundgren (1967) has given a BBGKY-type treatment of turbulence in which the multipoint distribution hierarchy is derived from the Navier–Stokes equation. A comparison with the molecular case suggests that the turbulence can be intepreted in terms of 'particles' which interact through non-central velocity-dependent force potentials. This is an interesting interpretation which might merit more attention.

Second, from time to time there is sporadic interest in the question of whether one should bypass the Navier–Stokes equation altogether, and formulate the turbulence problem directly from the molecular level. To most people, the fact that the smallest turbulence length scale is enormous when compared with molecular scales would seem to render this kind of approach rather academic. This is the view taken here, although we shall feel obliged to re-examine the question later on, when we discuss the possible application of lattice gas models to the numerical simulation of turbulence. In the meantime, we cite the work of Lewis (1975, 1977) and Montgomery (1976) as representative examples, purely for completeness.

4.1.2 *Kinetic equations in statistical mechanics*

If we consider the case of the classical gas of point particles in equilibrium, then the theoretical problems become rather trivial. For example, the single-particle distribution is independent of position coordinates and its dependence on momenta is given by the well-known Maxwellian distribution (i.e. $\exp(-p^2/2mkT)$), where m is the mass of the particle, k is the Boltzmann constant, and T is the absolute temperature. The N-particle distribution function F_N is just the product of N single-particle distributions, from which it follows that all reduced distribution functions factorize into products of single-particle functions.

The next level of difficulty is the non-equilibrium case (followed by the inclusion of long-range interactions, which will be treated in Chapter 5), where we are interested in the time dependence of the system. Nowadays, the whole subject of the time evolution of statistical systems is a very active field of research (Prigogine 1968) but we shall restrict our attention to the traditional *ad hoc* kinetic equations, such as the Boltzmann and Fokker–Planck equations. This is not because these equations are immediately applicable to the turbulence problem. Rather, it is because they suggest analogous approaches to turbulence, and (perhaps more than anything) we shall find it helpful to know the relevant terminology when we come to consider the actual turbulence theories.

We shall begin with the Boltzmann equation, and so we now consider the ideal gas away from equilibrium. We assume that particles interact through a repulsive potential which is effective at short range. In other words, particles interact through collisions.

Let $f(\mathbf{q}, \mathbf{v}; t)$ be the number of particles with position coordinate between $\mathbf{q}$ and $\mathbf{q} + d\mathbf{q}$ and with velocities in the range $[\mathbf{v} - d\mathbf{v}/2, \mathbf{v} + d\mathbf{v}/2]$. Evidently f is simply related to the single-particle distribution function $f_1(\mathbf{q}, \mathbf{p}; t)$. From eqn (4.19), with $S = 1$, we have a defining relation for f:

$$f_1(\mathbf{q}, \mathbf{p}; t) = m^3 f(\mathbf{q}, \mathbf{v}; t), \tag{4.22}$$

where we make the change from momentum to velocity (hence the factor m^3), as independent variable, in order to conform with convention. The evolution equation for f can be deduced from (4.21) and (4.22).

Firstly, where there are no interactions, it is trivial to show that f satisfies

$$\frac{\partial f}{\partial t} = Lf \qquad (4.23)$$

or in full, showing the convective derivative,

$$\frac{\partial f}{\partial t} + \mathbf{v} \cdot \nabla f(\mathbf{q}, \mathbf{v}; t) = 0. \qquad (4.24)$$

In the absence of long-range interactions, we consider the effect of point collisions. We can argue that two kinds of collision are of interest to us. First, those particles which already have velocity $\mathbf{v}$ may collide with other particles, with the result that their velocity changes. We would regard these collisions as being responsible for a 'loss' of particles from the set of particles of velocity $\mathbf{v}$. Conversely, particles whose velocity is not $\mathbf{v}$ may acquire that particular velocity through collisions. Hence, such collisions would cause a 'gain' of particles to the set with velocity $\mathbf{v}$.

In all therefore, one can generalize eqn (4.24) to include the effect of collisions:

$$\frac{\partial f}{\partial t} + \mathbf{v} \cdot \nabla f(\mathbf{q}, \mathbf{v}; t) = \text{rate of (gain} - \text{loss) due to collisions.} \qquad (4.24a)$$

The r.h.s. can be calculated from a mechanistic analysis of hard-sphere collisions, with some probabilistic assumptions thrown in. For instance, on grounds of relative probability, one can neglect three-body (and higher) orders of collision, in comparison with the two-body, or binary, collision.

Full details can be found in Woods (1975) or Balescu (1975); here we shall summarize only the main results of the calculation. Consider the probability that a particle with velocity $\mathbf{v}$ collides with a particle which has velocity $\mathbf{v}_1$, such that after the collision there are two particles with velocities $\mathbf{v}'$ and $\mathbf{v}'_1$, and denote the associated density by $p(\mathbf{v}, \mathbf{v}_1 | \mathbf{v}', \mathbf{v}'_1)$, where p must be symmetric under interchange of initial and final states. Further simplify matters by imposing a restriction to spatial homogeneity, with the implication that the dependence of f on $\mathbf{q}$ drops out, as does the convective term. Then the Boltzmann equation can be written as

$$\frac{\partial f(\mathbf{v}, t)}{\partial t} = \int d\mathbf{v}_1 \int d\omega p(\mathbf{v}, \mathbf{v}_1 | \mathbf{v}', \mathbf{v}'_1) \times$$

$$\times \{ f(\mathbf{v}')f(\mathbf{v}'_1) - f(\mathbf{v})f(\mathbf{v}_1) \}, \qquad (4.25)$$

where we have suppressed the explicit dependence on t on the r.h.s. in the interests of conciseness and the angular integration (with respect to ω) is over all possible scattering angles.

An important probabilistic argument is that the two-particle distribution function can be factored as

$$f_2 = f_1 f_1, \tag{4.26}$$

which can only be the case if correlations between the motions of the two particles are zero. In general this cannot always be true. However, the postulate that it is true led to a significant advance in the subject of non-equilibrium statistical mechanics, and is known as Boltzmann's *stossahlansatz*, or assumption of molecular chaos.

Our next topic is the more general one, where the time evolution is governed by a Markov process, that is, a stochastic process where memory effects are not important. We can then obtain the master equation, which can be solved (at least in principle) for the single-point probability distribution and which has many applications in chemistry, biophysics, and population dynamics, as well as physics. Here we shall find the specific application to Brownian motion particularly helpful to our later discussions of turbulence.

Let us return to our multipoint joint probability distribution function, as in (4.5) or (4.18). We are now interested in the generalization to the probability that a continuous variable X takes on specific values x_1 at time $t = t_1$, x_2 at time $t = t_2$, and, in general, x_n at time $t = t_n$, which we write as

$$f_n(x_1, t_1; x_2, t_2; \ldots x_n, t_n).$$

We shall introduce the conditional probability density (see Appendix B)

$$p(x_1, t_1 | x_2, t_2),$$

which is the probability density that $X = x_2$ when $t = t_2$, given that X had the value x_1 at time $t_1 < t_2$, and is defined by the identity

$$f_1(x_1, t) p(x_1, t_1 | x_2, t_2) = f_2(x_1, t_1; x_2, t_2). \tag{4.27}$$

Then, from the reduction property of the multipoint distributions (see (4.19)) and from (4.27) we can obtain a general relationship between the probabilities at different times as

$$f_1(x_2, t_2) = \int p(x_1, t_1 | x_2, t_2) f_1(x_1, t_1) \, dx_1. \tag{4.28}$$

Now we introduce the idea of a Markov process. We define such a process formally in terms of the conditional probabilities. If

$$p(x_1, t_1; x_2, t_2; \ldots x_{n-1}, t_{n-1} | x_n, t_n) = p(x_{n-1}, t_{n-1} | x_n, t_n), \tag{4.29}$$

then the current step only depends on the immediately preceding step and not on any preceding ones, and the process can be said to be Markovian.

It follows that the entire hierarchy of probability distributions can be constructed from the single-point form $f_1(x_1, t_1)$ and the conditional probability distribution $p(x_1, t_1 | x_2, t_2)$. This latter quantity is also known as the transitional probability. It can be shown to satisfy the Chapman–

Kolmogorov equation:

$$p(x_1, t_1 | x_3, t_3) = \int p(x_1, t_1 | x_2, t_2) p(x_2, t_2 | x_3, t_3) \, dx_2. \qquad (4.30)$$

We note that this equation tells us that the probability of two successive steps is just the product of the separate probabilities for each of the two steps. In other words, the two successive steps are statistically independent.

For a chain which has small steps between events, the integral relation of (4.30) can be turned into a differential equation known as the master equation. This is obtained (Reichl 1980) by expanding time dependences to first order in Taylor series, with the result

$$\frac{\partial f_1(x_2, t)}{\partial t} = \int dx_1 \{ W(x_1, x_2) f_1(x_1, t) - W(x_2, x_1) f_1(x_2, t) \}, \qquad (4.31)$$

where $W(x_1, x_2)$ is the rate (per unit time) at which transitions from state x_1 to state x_2 take place.

If, as in the turbulent cases which we shall be considering later, X is a continuous variable, then we can further derive the Fokker–Planck equation. That is, for the continuous case, eqn (4.31) reduces to

$$\frac{\partial f_1(x, t)}{\partial t} = -\frac{\partial \{ A(x) f_1(x, t) \}}{\partial x} + \frac{\frac{1}{2} \partial^2 \{ B(x) f_1(x, t) \}}{\partial x^2}. \qquad (4.32)$$

If we ignore the first term on the r.h.s., then (4.32) has the form of a diffusion equation with diffusion coefficient (or diffusivity) B, which would seem to suggest an underlying random walk process. Then the first term on the r.h.s. could be plausibly interpreted as a damping or friction effect, as normally encountered in dynamical systems. We can pursue these aspects—along with some other useful ideas—by briefly considering the application of the above formalism to the problem of Brownian motion. This is the irregular movement seen when small particles are suspended in a fluid and is due to the molecular agitation of the fluid.

On the microscopic scale, the effect of the molecular collisions on a Brownian particle can be represented by a random force $\mathscr{F}(t)$, which we shall take to have known statistics. On the macroscopic scale, the net effect of the surrounding fluid will be the viscous resistance to flow, which we assume to be linearly proportional to the velocity of the particle. Thus Newton's second law yields for the motion of a Brownian particle

$$\frac{\partial v}{\partial t} = -bv + \mathscr{F}(t), \qquad (4.33)$$

where v is the velocity and b is the friction coefficient. Equation (4.33) is known as the Langevin equation, and will turn up again, at a later stage, when we consider renormalization group approaches to turbulence.

Now we specify the random force $\mathscr{F}(t)$ in terms of its statistics. We begin by taking it to have zero mean when averaged over the ensemble of all the Brownian particles:

$$\langle \mathscr{F}(t) \rangle = 0. \tag{4.34}$$

Hence, if we average each term of (4.33), it follows from (4.34) that the macroscopic resistance law holds in the mean.

Second, we assume that $\mathscr{F}(t)$ has a Gaussian probability distribution and is, moreover, highly uncorrelated. That is, collisions at different times are statistically independent.

We express these properties in a rather extreme way by taking the auto-correlation of $\mathscr{F}(t)$ to be given by a Dirac delta function

$$\langle \mathscr{F}(t)\mathscr{F}(t') \rangle = h\delta(t - t'), \tag{4.35}$$

where h is defined more generally by the relationship

$$h = \int \langle \mathscr{F}^2(t) \rangle \, dt. \tag{4.36}$$

We now solve (4.33) as an initial value problem with given conditions

$$v = V \qquad \text{at} \qquad t = 0,$$

with the result that v at any time is given by

$$v(t) = V \exp(-bt) + \exp(-bt) \int_0^t dt' \, \exp(bt')\mathscr{F}(t'). \tag{4.37}$$

Then, squaring each side of (4.37) and averaging term by term, we obtain the variance of the random velocity as

$$\langle v^2 \rangle = V^2 \exp(-2bt) + \frac{h}{2b}\{1 - \exp(-2bt)\}, \tag{4.38}$$

where the use of eqn (4.34) ensures that terms linear in the random force vanish. We note that the dispersion of the Brownian particles is determined by their initial velocity at short times, but at longer times the variance of the velocity approaches the asymptotic value $h/2b$, which is entirely determined by the properties of the molecular collisions.

Application of the Fokker–Planck equation to the same problem allows us to identify the coefficients as

$$A(v) = -bv \tag{4.39}$$

and

$$B(v) = h, \tag{4.40}$$

where we have replaced x by v. The friction coefficient b can be obtained from macroscopic flow experiments, and we can use the fact the time evolution is towards thermal equilibrium to fix h in terms of the Boltzmann constant and the absolute temperature (Balescu 1975).

Obviously this expedient is not open to us in turbulence. However, as we shall see later, we can to some extent replace the concept of thermal equilibrium by statistical stationarity. Then the rate of doing work on the system (the analogue of h) can be equated to the energy dissipation rate.

4.1.3 The difficulties involved in generalizing statistical mechanics to include turbulence

So far we have discussed the classically successful areas of statistical mechanics, where systems are either in, or near, thermal equilibrium. In such cases the total energy of the system is constant, and we are interested in the way in which it is shared out among the many degrees of freedom. The fact that the sharing-out process may change with time (slightly non-equilibrium systems) need not cause any problems, as generally one can find a small parameter upon which to base a perturbation expansion. Usually an approximate treatment of this kind will depend on the existence of widely separated and distinct length scales or time-scales (Woods 1975). For example, in the dilute gas with weak interactions, the duration of an intermolecular collision is much shorter than the time between collisions, and the ratio of one to the other provides the necessary small expansion parameter.

In complete contrast with the above picture, fluid turbulence is highly dissipative and characteristically involves a flow of energy through the degrees of freedom. All length scales (or time-scales) can be seen as being of equal importance, and, for the case of high Reynolds number, there are very many such scales. In this respect, at least, the problem of well-developed turbulence would seem to have much in common with critical phenomena, and this is a topic which we shall pursue later in connection with the application of renormalization group methods.

In all, therefore, we can see that turbulence is very far from thermal equilibrium, which rules out treatments based on small departures from equilibrium such as fluctuation–dissipation theorems (e.g. Reichl 1980). It is clearly also the case that turbulence is unlikely to respond to weak perturbation methods of the kind used in slightly non-equilibrium situations. Indeed, in view of its strong non-linearity—which is of the very essence of the phenomenon—this can be seen as hardly surprising.

Another consequence of non-linearity is the non-Gaussian nature of the probability distributions. While it is true that the distribution of a single variable at a single point is approximately Gaussian, it is well known that experimentally measured joint probability distributions are of a more general type (Frenkiel and Klebanoff 1967, 1971, 1973; Van Atta and Chen 1968). As, broadly, the only functionals amenable to any general kind of manipulation are the Gaussian forms, it follows that the problems in the way of finding (and making use of) the general probability functional, as in eqn (4.6), are unlikely to be trivial.

One way of evading some of these difficulties is to consider the case of a fluid in which the viscosity is zero. In other words, we take the equation expressing conservation of momentum to be Euler's equation (e.g. see Batchelor 1967) rather than the Navier–Stokes equation. It was shown by Lee (1952) that one could then derive the Liouville equation.

In the absence of dissipation, other results of classical statistical mechanics are applicable to the macroscopic random velocity field of the inviscid fluid. For example, the system will be in thermal equilibrium and there will be energy equipartition. That is to say, the energy spectral density $q(k)$—as defined by eqn (2.97a)—will be constant, independent of wavenumber, and naturally the energy spectrum $E(k)$—as defined by (2.101)—will be proportional to the square of the wavenumber. Also, for small fluctuations about equilibrium, a classical fluctuation–dissipation theorem can be derived (Kraichnan 1959).

The concept of inviscid equilibrium ensembles must be purely hypothetical, as far as the macroscopic motion of real fluids is concerned. For instance, the mere fact of choosing the viscosity to be zero raises the difficulty that the viscous length scale (see eqn (2.131) also becomes zero, and we are faced with the problem of an infinite number of degrees of freedom. (It also follows that we must have an infinite amount of energy in the system.) In practice, we can easily circumvent this problem by truncating the Fourier representation of the velocity field at some upper cut-off wavenumber; Lee (1952) has suggested the inverse of atomic dimensions. The disadvantage is that this is only an expedient at best and, by its arbitrariness, is just as unphysical is the unmodified system.

(As an aside, we note that the presence of an infinite amount of energy in the system, as the upper cut-off wavenumber goes. to infinity, is not an ultraviolet catastrophe. Rather, it is a natural consequence of a hypothetical case where there is an infinite number of degrees of freedom. An ultraviolet catastrophe would be an infinite amount of energy contained in a finite number of degrees of freedom.)

Nevertheless, despite its apparent artificiality, Kraichnan (1964) has argued that the concept of the absolute equilibrium ensemble is important because it provides a basis from statistical mechanics for the energy cascade in real fluid turbulence. This is because the presence of viscous dissipation can be seen as frustrating the endeavours of the non-linear terms to achieve energy equipartition. More specifically, the local rate of dissipation increases with wavenumber, so that dissipation is (potentially) greatest in those wavenumbers which invariably have the weakest initial excitation. Thus high wavenumbers must behave as a sink, so that the non-linear interaction, in attempting to produce energy equipartition, will instead produce the energy cascade.

The study of inviscid fluid dynamics is a historic part of the subject, and many of the results are well known. From our present point of view, the topic is not worth pursuing very much further here. In the interests of completeness, we shall just mention a few modern uses of the Euler equation: Kraichnan (1964) regards the inviscid fluid as providing a special case which acts as a test of a general theory of the Navier–Stokes equations; Betchov and Larsen (1981) have shown, by integrating a truncated form of the Euler equation forward in time, that non-Gaussian probability distributions can develop, even in the absence of viscosity; Lee (1982) has shown, also by integrating the truncated Euler equation forward in time, that the inherent unpredictability of the non-linearity implies the development of mixing, and hence ergodicity.

Before turning to the general functional formalism of the next section, we should briefly mention one point. The idea of setting the viscosity equal to zero in order to produce an idealized fluid is not the same as taking the limit of infinite Reynolds number in a real fluid. This is a topic to which we shall return again in Section 6.2.7.

4.2   Functional formalisms for the turbulence problem

In eqn (4.6) we have the formal definition of the functional probability distribution $P[\mathbf{u}(\mathbf{x}, t)]$ of a turbulent system. In principle, this particular functional contains all the information one could ever possibly know about the system. Evidently the major objective of a statistical theory of turbulence can now be restated as the need to reduce this vast number of information to a tractable form. (An equivalent restatement would be need to eliminate many of the degrees of freedom.)

We have seen that the analogous problem in statistical mechanics is tackled by first obtaining the governing equation for the most general case (the Liouville equation) and then seeking approximation procedures which lead to a description of the system in terms of a vastly reduced number of degrees of freedom. For equilibrium systems one has the canonical distributions, whereas in the non-equilibrium case one has to resort to kinetic equations. The latter are only valid for slight departures from equilibrium, and so it must be appreciated that these particular equations are not directly applicable to turbulence. Our purpose in discussing them, however briefly, is that the procedures of classical statistical mechanics may offer us some general guidance, even if only by a rather loose analogy.

In this section we shall take the first steps along this route, in that we shall give the general formulation of the turbulence problem.

The first functional formulation of the turbulence problem was given by Hopf (1952), who considered the distribution $P[\mathbf{u}(\mathbf{x}), t]$ and derived its governing equation from the Navier–Stokes equation. Later, Lewis and Kraichnan (1962) obtained the equation for the evolution in time of the more general functional $P[\mathbf{u}(\mathbf{x}, t)]$. Paradoxically, their more general treatment is easier to follow, and so we shall ignore history and deal with the two pieces of work in the reverse of chronological order. First, however, we shall make a few remarks about the characteristic functionals upon which the theoretical approaches are based.

In Appendix B we deal with the characteristic function $m(k)$ of a distribution $p(x)$, and note its use as a generating function for the moments of $p(x)$. We can generalize the definition given there to the case of a distribution dependent on a vector argument:

$$m(\mathbf{k}) = \int p(\mathbf{x}) \exp\{i\mathbf{k} \cdot \mathbf{x}\} \, d\mathbf{x}$$

$$= \langle \exp\{i\mathbf{k} \cdot \mathbf{x}\} \rangle. \tag{4.41}$$

The scalar product of three-dimensional vectors in the exponent can be extended to n dimensions. Let $\mathbf{f}$ and $\mathbf{g}$ be two vectors in an n-dimensional space. We can write each of them as an ordered set of real numbers in the form

$$\mathbf{f} = \{f_1, f_2, \ldots, f_n\} \qquad \mathbf{g} = \{g_1, g_2, \ldots, g_n\},$$

and the extended scalar (or inner) product of the vectors $\mathbf{f}$ and $\mathbf{g}$ follows at once as

$$[\mathbf{f} \cdot \mathbf{g}] = f_1 g_1 + f_2 g_2 + \cdots + f_n g_n. \tag{4.42}$$

If we then take the case where n tends to infinity, such that the vectors take a limiting form as continuous functions of a variable x, then we can replace $\mathbf{f}$ and $\mathbf{g}$ by $f(x)$ and $g(x)$, and the summation in (4.42) by an integration, to obtain a general form of inner product

$$[\mathbf{f} \cdot \mathbf{g}] = \int f(x) g(x) \, dx. \tag{4.43}$$

In order to deal with the functionals in the turbulence problem, we must further make the obvious extension of (4.43) to the case of vector arguments. In order to do this, we introduce a vector field $\mathbf{Z}(\mathbf{x}, t)$ which has dimensions of inverse velocity and which satisfies the condition $Z \to 0$ as $|\mathbf{x}| \to \infty$. Clearly, the straightforward extension of (4.43) to vector functions with vector arguments is just

$$[\mathbf{Z} \cdot \mathbf{u}] = \int Z_\alpha(\mathbf{x}, t) u_\alpha(\mathbf{x}, t) \, d\mathbf{x} \, dt, \tag{4.44}$$

from which it is clear that the characteristic functional can be introduced by analogy with the definition of the characteristic function given in eqn (4.41). This takes the form

$$M[\mathbf{Z}(\mathbf{x}, t)] = \int P[\mathbf{u}(\mathbf{x}, t)] \exp\{i[\mathbf{Z} \cdot \mathbf{u}]\} \, d\mathbf{u}(\mathbf{x}, t)$$

$$= \langle \exp\{i[\mathbf{Z} \cdot \mathbf{u}]\} \rangle \tag{4.45}$$

where $M[\mathbf{Z}(\mathbf{x}, t)]$ is the characteristic functional of the distribution $P[\mathbf{u}(\mathbf{x}, t)]$.

The first r.h.s. of eqn (4.45) is equivalent to saying that the characteristic functional M is the functional Fourier transform of P. If we persist in this interpretation, then we shall have to reckon with the intricacies of functional integration. However interesting that might be, it would really lead us too far astray from the problem of fluid turbulence. A better alternative is to interpret M as the expectation value of the imaginary exponential $\exp\{i[\mathbf{Z}\cdot\mathbf{u}]\}$. Then the functional integration becomes a purely formal operation and the expansion of the exponential in the second r.h.s. of eqn (4.45) generates the moments of the distribution $P[\mathbf{u}(\mathbf{x}, t)]$ (see also Appendix B).

This brings us to the reason for working with M rather than P. Quite simply, it allows us to forget (broadly speaking) the niceties of functional calculus. As far as integration is concerned, we never have to do anything more than formally interpret an integration against P as an expectation value (and it can be shown that this is always well behaved, provided that P is properly normalized). Differentiation only requires us to remember that a functional (in the present context, at least) is just a function of a function. The general operation is simply illustrated by the formula for the nth-order derivative of the characteristic functional:

$$\frac{\delta^n M}{\delta\mathbf{Z}(\mathbf{x}_1, t_1)\delta\mathbf{Z}(\mathbf{x}_2, t_2)\ldots\delta\mathbf{Z}(\mathbf{x}_n, t_n)} = i^n\langle\mathbf{u}(\mathbf{x}_1, t_1)\mathbf{u}(\mathbf{x}_2, t_2)\ldots(\mathbf{x}_n, t_n)\exp\{i[\mathbf{Z}\cdot u]\}\rangle$$

(4.46)

where $\delta/\delta\mathbf{Z}$ stands for the functional differentiation.

The general n-point n-time correlation is obtained by evaluating both sides of eqn (4.46) at $Z = 0$. The result is

$$\langle\mathbf{u}(\mathbf{x}_1, t_1)\mathbf{u}(\mathbf{x}_2, t_2)\ldots\mathbf{u}(\mathbf{x}_n, t_n)\rangle = \frac{i^{-n}\delta^n M}{\delta\mathbf{Z}(\mathbf{x}_1, t_1)\delta\mathbf{Z}(\mathbf{x}_2, t_2)\ldots\delta\mathbf{Z}(\mathbf{x}_n, t_n)}\bigg|_{z=0}.$$

(4.47)

Thus, if we can derive a governing equation for $M[\mathbf{Z}(\mathbf{x}, t)]$, eqn (4.47) can be used to reformulate the basic turbulence problem back into terms of the moment hierarchy.

4.2.1 The space–time functional formalism

We shall base this section on the work of Lewis and Kraichnan (1962). We should begin by making a clear distinction between their approach and that of Hopf (1952). Lewis and Kraichnan start from the probability $P[\mathbf{u}(\mathbf{x}, t)]$ that the velocity has the value $\mathbf{u}(\mathbf{x}, t)$. In contrast, Hopf's theory involves the probability $P[\mathbf{u}(\mathbf{x}), t]$ that the velocity field is $\mathbf{u}(\mathbf{x})$ at time t. The former allows us to obtain correlations at many times (see eqn (4.47) above), whereas the latter—which we shall discuss in Section 4.2.3—restricts us to correlations of many space points evaluated at a single time.

We are interested in the characteristic functional $M[\mathbf{Z}(\mathbf{x}, t)]$, as defined by (4.45). We wish to obtain the equation for the evolution in time for this quantity, and it should be clear that such an equation must be based on the Navier–Stokes equations. These equations govern the behaviour of the velocity field $\mathbf{u}(\mathbf{x}, t)$; and hence, through eqn (4.46), that of the characteristic functional.

Let us take the incompressibility condition as an easy introductory example. This must be satisfied by the velocity field in the form shown by eqn (1.1). What is the corresponding condition on $M[\mathbf{Z}(\mathbf{x}, t)]$?

If we put $n = 1$ in eqn (4.46), then we obtain the first-order functional derivative of M as

$$\frac{\delta M}{\delta Z_\alpha(\mathbf{x}, t)} = i\langle u_\alpha(\mathbf{x}, t)\exp\{i[\mathbf{Z} \cdot u]\}\rangle. \qquad (4.48)$$

Now, from (4.44), $[\mathbf{Z} \cdot \mathbf{u}]$ is a constant, and so differentiating both sides of (4.48) with respect to $\mathbf{x}$ leads to

$$\frac{\partial\{\partial M/\delta Z_\beta(\mathbf{x}, t)\}}{\partial x_\beta} = 0, \qquad (4.49)$$

where we have used eqn (1.1) to eliminate the divergence of the r.h.s. of (4.48).

Equation (4.49) provides a constraint on the characteristic functional. In order to obtain the evolution equation we employ exactly the same approach to the Navier–Stokes equation. Writing (1.6) in terms of the fluctuating fields with zero mean, $u_\alpha(\mathbf{x}, t)$ and $p(\mathbf{x}, t)$, and rearranging, we have

$$\frac{\partial u_\alpha}{\partial t} = -\frac{\partial\{u_\alpha u_\beta\}}{\partial x_\beta} - \frac{\partial p}{\partial x_\alpha} + v\frac{\partial^2 u_\alpha}{\partial x_\beta \partial x_\beta}. \qquad (4.50)$$

Now, we derive the time evolution equation for the characteristic functional by differentiating both sides of (4.48) with respect to the time:

$$\frac{\partial\{\delta M/\delta Z_\alpha(\mathbf{x}, t)\}}{\partial t} = i\left\langle\left(\frac{\partial u_\alpha}{\partial t}\right)\exp\{i\{\mathbf{Z} \cdot \mathbf{u}\}\}\right\rangle$$

$$= i\left\langle\left(-\frac{\partial\{u_\alpha u_\beta\}}{\partial x_\beta} - \frac{\partial p}{\partial x_\alpha} + v\frac{\partial^2 u_\alpha}{\partial x_\beta \partial x_\beta}\right) \times\right.$$

$$\left.\times \exp\{i[\mathbf{Z} \cdot \mathbf{u}]\}\right\rangle, \qquad (4.51)$$

where we have borne in mind that $[\mathbf{Z} \cdot \mathbf{u}]$ is constant and eqn (4.50) has been used to substitute for $\partial u_\alpha/\partial t$.

As we have seen in Section 2.1, the pressure term can be eliminated, along with the continuity equation. Thus we need to obtain the two velocity terms on the r.h.s. of (4.51) in terms of the characteristic functional.

The non-linearity is dealt with by differentiating $M[\mathbf{Z}(\mathbf{x}, t)]$ twice. That is, we set $n = 2$ in eqn (4.46) to obtain

$$\frac{\delta^2 M}{\delta Z_\alpha(\mathbf{x}, t)\delta Z_\beta(\mathbf{x}, t)} = -\langle u_\alpha(\mathbf{x}, t)u_\beta(\mathbf{x}, t)\exp\{i[\mathbf{Z}\cdot\mathbf{u}]\}\rangle. \qquad (4.52)$$

Then, differentiating both sides with respect to x_β, and rearranging, we have for the non-linear term on the r.h.s. of (4.51),

$$-\left\langle\frac{\partial\{u_\alpha u_\beta\}}{\partial x_\beta}\exp\{i[\mathbf{Z}\cdot\mathbf{u}]\}\right\rangle = \frac{\partial\{\delta^2 M/\delta Z_\alpha(\mathbf{x}, t)\delta Z_\beta(\mathbf{x}, t)\}}{\partial x_\beta}. \qquad (4.53)$$

Note that at this stage we are transforming the problem from one that is non-linear in $\mathbf{u}$ to one that is linear in $M[\mathbf{u}]$.

The viscous term is easily found. Differentiate both sides of (4.48) twice with respect to x_β, and we have the explicit form

$$i\left\langle\left(\frac{\partial^2 u_\alpha}{\partial x_\beta\partial x_\beta}\right)\exp\{i[\mathbf{Z}\cdot\mathbf{u}]\}\right\rangle = \frac{\partial^2\{\delta M/\delta Z_\alpha(\mathbf{x}, t)\}}{\partial x_\beta\partial x_\beta}, \qquad (4.54)$$

which just leaves the pressure term in (4.51). As we know this can be eliminated, we can just leave this in its implicit form and, for compactness, write

$$\Pi = \langle ip(\mathbf{x}, t)\exp\{i[\mathbf{Z}\cdot\mathbf{u}]\}. \qquad (4.55)$$

Then, with (4.53), (4.54), and (4.55), eqn (4.51) becomes

$$\frac{\partial\{\delta M/\delta Z_\alpha(\mathbf{x}, t)\}}{\partial t} = \frac{i\partial\{\delta^2 M/\delta Z_\alpha(\mathbf{x}, t)\delta Z_\beta(\mathbf{x}, t)\}}{\partial x_\beta} + \frac{\nu\partial^2\{\delta M/\delta Z_\alpha(\mathbf{x}, t)\}}{\partial x_\beta\partial x_\beta} - \frac{\partial\Pi}{\partial x_\alpha}. \qquad (4.56)$$

The pressure could have been eliminated previously by using continuity, along with the Navier–Stokes equation, just as we did in Chapter 2. Indeed, this is what we shall do in the next section, in connection with the spectral version of the formalism. But here we shall follow Lewis and Kraichnan (1962) and introduce a testing field $Y_\alpha(\mathbf{x}, t)$, which satisfies the continuity equation, such that

$$\frac{\partial Y_\beta(\mathbf{x}, t)}{\partial x_\beta} = 0, \qquad (4.57)$$

and $Y_\alpha(\mathbf{x}, t) \to 0$ as $|\mathbf{x}| \to \infty$. Then, using the formula for integrating by parts, we have

$$\int Y_\alpha\left\{\frac{\partial\Pi}{\partial x_\alpha}\right\}d\mathbf{x} = [Y_\alpha\Pi]_{-\infty}^\infty - \int\left\{\frac{\partial Y_\alpha}{\partial x_\alpha}\right\}\Pi\, d\mathbf{x}$$

$$= 0, \qquad (4.58)$$

where the last step follows from (4.57) and the boundary condition on Y_α at infinity.

The elimination of the pressure term can now be accomplished if we form the inner product of $Y_\alpha(\mathbf{x}, t)$ with each term of (4.56) and invoke (4.58) to end up with

$$\int Y_\alpha(\mathbf{x}, t) \left[\frac{\partial \{\delta M / \delta Z_\alpha(\mathbf{x}, t)\}}{\partial t} - \frac{i\partial \{\delta^2 M / \delta Z_\alpha(\mathbf{x}, t)\delta Z_\beta(\mathbf{x}, t)\}}{\partial x_\beta} - \frac{v\partial^2 \{\delta M / \delta Z_\alpha(\mathbf{x}, t)\}}{\partial x_\beta \partial x_\beta} \right] d\mathbf{x}\, dt = 0. \tag{4.59}$$

This is the general functional differential equation for the characteristic functional $M[\mathbf{Z}]$, and must be satisfied for all testing fields Y_α which satisfy (4.57). The most important thing about it is that it is linear, a property which it shares with the analogous Liouville equation of classical statistical mechanics.

4.2.2 The k-space–time functional formalism

In view of the emphasis on spectral methods in many parts of the present book, it would seem appropriate to discuss the functional formalism in $\mathbf{k}$-space and time. Although Lewis and Kraichnan (1962) have given the appropriate generalization in their paper, we shall adopt a different approach here. In particular, we shall present an approach which allows us to draw freely on the earlier discussion, and avoids undue repetition. We shall also eliminate the pressure term at an earlier stage by working with the solenoidal form of the Navier–Stokes equation.

We begin by introducing the Fourier transformations

$$Z_\alpha(\mathbf{k}, t) = \int Z_\alpha(\mathbf{x}, t) \exp\{i\mathbf{k} \cdot \mathbf{x}\}\, d\mathbf{x} \tag{4.60}$$

and

$$Z_\alpha(\mathbf{x}, t) = \left(\frac{1}{2\pi}\right)^3 \int Z_\alpha(\mathbf{k}, t) \exp\{-i\mathbf{k} \cdot \mathbf{x}\}\, d\mathbf{k}. \tag{4.61}$$

Note that this Fourier pair bears an inverse relation to our previous definitions for the velocity field. This is apparent from a comparison with either eqns (2.89) and (2.90) or eqns (D.21) and (D.22).

Defining the Fourier transformation of the $\mathbf{Z}$ field as in eqns (4.60) and (4.61) may seem quite a natural choice when one considers that $\mathbf{Z}$ is Fourier conjugate to $\mathbf{u}$. However, a compelling reason for choosing the arbitrary normalizations and signs of phases to be as in (4.60) and (4.61) is that it leads to the important simplification that the functional inner product $[\mathbf{Z} \cdot \mathbf{u}]$ is form invariant under Fourier transformation.

Substituting (4.61) for $Z_\alpha(\mathbf{x}, t)$ and (D.21) for the Fourier transformation of the velocity field $u_\alpha(\mathbf{x}, t)$ into the r.h.s. of eqn (4.44), we find

$$[\mathbf{Z} \cdot \mathbf{u}] = \left(\frac{1}{2\pi}\right)^3 \int dt \, d\mathbf{x} \, d\mathbf{k} \, d\mathbf{k}' Z_\alpha(\mathbf{k}', t) u_\alpha(\mathbf{k}, t) \exp\{i(\mathbf{k} - \mathbf{k}') \cdot \mathbf{x}\}. \qquad (4.62)$$

Now we have the general result (e.g. see Appendix D)

$$\left(\frac{1}{2\pi}\right)^3 \int \exp\{i(\mathbf{k} - \mathbf{k}') \cdot \mathbf{x}\} \, d\mathbf{x} = \delta(\mathbf{k} - \mathbf{k}'), \qquad (4.63)$$

and so, integrating over $\mathbf{x}$ and $\mathbf{k}'$, we obtain

$$[\mathbf{Z} \cdot \mathbf{u}] = \int Z_\alpha(\mathbf{k}, t) u_\alpha(\mathbf{k}, t) \, dt \, d\mathbf{k}, \qquad (4.64)$$

which may be seen to be identical with the original definition (4.44), with $\mathbf{x}$ replaced by $\mathbf{k}$.

This is an important result, as it allows us to take over much of the formalism of the preceding section, merely changing $\mathbf{x}$ to $\mathbf{k}$, as appropriate. For instance, the definition of the characteristic functional embodied in eqn (4.45) can be extended to the case of $Z(\mathbf{k}, t)$ as

$$M[Z(\mathbf{k}, t)] = \langle \exp\{i[\mathbf{Z} \cdot \mathbf{u}]\} \rangle, \qquad (4.65)$$

where the functional inner product $[\mathbf{Z} \cdot \mathbf{u}]$ is now understood to be given by (4.64) rather than (4.44). Then we can take over the general moment-generating property of eqn (4.46), differentiating with respect to $Z(\mathbf{k}, t)$, to generate the moment hierarchy in $\mathbf{k}$-space:

$$\frac{\delta^n M}{\delta Z(\mathbf{k}_1, t_1) \delta Z(\mathbf{k}_2, t_2) \ldots \delta Z(\mathbf{k}_n, t_n)}$$
$$= i^n \langle \mathbf{u}(\mathbf{k}_1, t_1) \mathbf{u}(\mathbf{k}_2, t_2) \ldots \mathbf{u}(\mathbf{k}_n, t_n) \exp\{i[\mathbf{Z} \cdot \mathbf{u}]\} \rangle. \qquad (4.66)$$

As before, we use the flow equations to obtain the governing equations for the characteristic functional. Again, as an easy example, we obtain the continuity condition first. We begin by setting $n = 1$ on both sides of eqn (4.66), with the result

$$\frac{\delta M}{\delta Z_\alpha(\mathbf{k}, t)} = i \langle u_\alpha(\mathbf{k}, t) \exp\{i[\mathbf{Z} \cdot \mathbf{u}]\} \rangle, \qquad (4.67)$$

where we have also put $\mathbf{k}_1 = \mathbf{k}$ and $t_1 = t$ without loss of generality. The equation of continuity for the velocity field in $\mathbf{k}$-space is given by (2.74). It follows that, if we scalar multiply both sides of (4.67) by $\mathbf{k}$, the r.h.s of eqn (4.67) vanishes and we have

$$k_\beta \frac{\delta M}{\delta Z_\beta(\mathbf{k}, t)} = 0 \tag{4.68}$$

as the required condition on the characteristic functional.

The Navier–Stokes equation in $\mathbf{k}$-space is given by (2.76). We note that the pressure term has already been eliminated, and rewrite this equation slightly as

$$\frac{\partial u_\alpha(\mathbf{k}, t)}{\partial t} = -\nu k^2 u_\alpha(\mathbf{k}, t) + M_{\alpha\beta\gamma}(\mathbf{k}) \sum_{\mathbf{j}} u_\beta(\mathbf{j}, t) u_\gamma(\mathbf{k} - \mathbf{j}, t). \tag{4.69}$$

We can derive the time evolution for $M[\mathbf{Z}(\mathbf{k}, t)]$ by using the same general procedure as in the preceding section. That is, we differentiate both sides of (4.67) with respect to t, and substitute for $\partial u_\alpha(\mathbf{k}, t)/\partial t$ from eqn (4.69) to obtain

$$\frac{\partial \{\delta M/\delta Z_\alpha(\mathbf{k}, t)\}}{\partial t} = i \left\langle \left[\frac{\partial u_\alpha(\mathbf{k}, t)}{\partial t} \right] \exp\{i[\mathbf{Z} \cdot \mathbf{u}]\} \right\rangle$$

$$= i \left\langle \exp\{i[\mathbf{Z} \cdot \mathbf{u}]\} \{-\nu k^2 u_\alpha(\mathbf{k}, t) + \right.$$

$$\left. + M_{\alpha\beta\gamma}(\mathbf{k}) \sum_{\mathbf{j}} u_\beta(\mathbf{j}, t) u_\gamma(\mathbf{k} - \mathbf{j}, t)\} \right\rangle. \tag{4.70}$$

We now need to express the r.h.s. of (4.70) solely in terms of the characteristic functional. For the linear term this is trivial and we simply substitute from (4.67). The appropriate substitution for the non-linear term is found by setting $n = 2$ on both sides of eqn (4.66), along with $\mathbf{k}_1 = \mathbf{j}$, $\mathbf{k}_2 = \mathbf{k} - \mathbf{j}$, and $t_1 = t_2 = t$. In this way eqn (4.70) can be reduced to the form

$$\left\{ \frac{\partial}{\partial t} + \nu k^2 \right\} \frac{\delta M}{\delta Z_\alpha(\mathbf{k}, t)} = i M_{\alpha\beta\gamma}(\mathbf{k}) \sum_{\mathbf{j}} \frac{\delta^2 M}{\delta Z_\beta(\mathbf{j}, t) \delta Z_\gamma(\mathbf{k} - \mathbf{j}, t)}, \tag{4.71}$$

which is the required equation for the characteristic functional in $\mathbf{k}$-space.

4.2.3 *The Hopf equation for the characteristic functional*

The functional formalism for turbulence was originated by Hopf (1952), who considered the probability distribution $P[\mathbf{u}(\mathbf{x}), t]$ and the characteristic functional $M[\mathbf{Z}(\mathbf{x}), t]$. The relationship between the two is given in the usual way by

$$M[\mathbf{Z}(\mathbf{x}), t] = \int P[\mathbf{u}(\mathbf{x}), t] \exp\{i[\mathbf{Z} \cdot \mathbf{u}]\} \, d\mathbf{u}(\mathbf{x}), \tag{4.72}$$

where this time the functional inner product is

$$[\mathbf{Z} \cdot \mathbf{u}] = \int Z_\beta(\mathbf{x}) u_\beta(\mathbf{x}) \, d\mathbf{x}. \tag{4.73}$$

The problem now is that $u_\alpha(\mathbf{x})$ does not depend on the time. This effectively rules out our previous procedure, where we would have differentiated both sides of (4.72) with respect to $Z_\alpha(\mathbf{x})$, and then differentiated with respect to time. In dealing with this difficulty, we have to rely on the identity (Hopf 1952)

$$\int P[\mathbf{u}(\mathbf{x}), t] \exp\{i[\mathbf{Z} \cdot \mathbf{u}]\} \, d\mathbf{u}(\mathbf{x}) = \int P[\mathbf{u}(\mathbf{x}, 0), 0] \exp\{i[\mathbf{Z} \cdot \mathbf{u}(t)]\} \, d\mathbf{u}(\mathbf{x}, 0),$$

$$\tag{4.74}$$

where the functional inner product, as defined by eqn (4.73), is slightly generalized to include the case

$$[\mathbf{Z} \cdot \mathbf{u}(t)] = \int Z_\beta(\mathbf{x}) u_\beta(\mathbf{x}, t) \, d\mathbf{x}. \tag{4.75}$$

Thus, from (4.72) and (4.74), we can write the requisite characteristic functional as

$$M[\mathbf{Z}(\mathbf{x}), t] = \langle \exp\{i[\mathbf{Z} \cdot \mathbf{u}(t)]\} \rangle. \tag{4.76}$$

It should be noted that we can use $M[\mathbf{Z}(\mathbf{x}), t]$ as a moment-generating functional, just as we did with $M[\mathbf{Z}(\mathbf{x}, t)]$ in Section 4.2.1. The only proviso is that we must now differentiate with respect to $\mathbf{Z}(\mathbf{x})$, rather than $\mathbf{Z}(\mathbf{x}, t)$, and with this modification eqn (4.46) can be taken over into the present case. As an example of this process, we note that the continuity condition still applies to $M[\mathbf{Z}(\mathbf{x}), t]$ and eqn (4.49), appropriately modified, now becomes

$$\frac{\partial \{\delta M / \delta Z_\beta(\mathbf{x})\}}{\partial x_\beta} = 0. \tag{4.77}$$

With these preliminaries out of the way, the derivation of the time evolution equation is now quite straightforward. We note that the phase on the r.h.s. of eqn (4.76) is time dependent and accordingly we can differentiate straightaway with respect to time. (This can be contrasted with the space–time case, where we had first to perform the functional differentiation with respect to $\mathbf{Z}$.) The result is a genuine first-order time evolution equation for M:

$$\frac{\partial M}{\partial t} = \left\langle i\mathbf{Z} \cdot \left(\frac{\partial \mathbf{u}}{\partial t} \right) \exp\{i[\mathbf{Z} \cdot \mathbf{u}(t)]\} \right\rangle$$

$$= \left\langle i\mathbf{Z} \cdot \left(-\frac{\partial \mathbf{u} u_\beta}{\partial x_\beta} - \nabla p + \nu \nabla^2 \mathbf{u} \right) \exp\{i[\mathbf{Z} \cdot \mathbf{u}(t)]\} \right\rangle \tag{4.78}$$

where, as before, we have substituted from the Navier–Stokes equation, in the form of eqn (4.50), for $\partial \mathbf{u}/\partial t$. Again, as before, we obtain expressions for the linear and non-linear velocity terms in (4.78) by functional differentiation of M, but this time with respect to $\mathbf{Z}(\mathbf{x})$. The final result is

$$\frac{\partial M}{\partial t} = \mathrm{i} \int \mathrm{d}\mathbf{x} Z_\alpha(\mathbf{x}) \left[\frac{\partial \{\delta^2 M/\delta Z_\alpha(\mathbf{x})\delta Z_\beta(\mathbf{x})\}}{\partial x_\beta} + \right.$$
$$\left. + \frac{\nu \partial^2 \{\delta M/\delta Z_\alpha(\mathbf{x})\}}{\partial x_\beta \partial x_\beta} - \frac{\partial \Pi}{\partial x_\alpha} \right], \tag{4.79}$$

where Π is as defined by eqn (4.55) and the functional inner product has been written out in full.

We conclude by noting that the pressure term can be eliminated by a permissible arbitrary generalization of $\mathbf{Z}$ (Hopf 1952) or by using the solenoidal Navier–Stokes equation, in the form of eqn (2.15), and the procedures of Section 4.2.2.

4.2.4   *General remarks on functional formalisms*

The functional approaches to turbulence can be seen as elegant and rigorous formulations of the general problem. In this respect, comparisons can be made with the functional formalism of quantum field theory, in the hope that useful analogies may be drawn (Monin and Yaglom 1975, p. 757). Unfortunately, the field-theoretical approach seems to have proved rather barren in the case of turbulence (see Beran 1968; Monin and Yaglom 1975), although purely deductive approaches may serve to remind us that the functional and moment formulations are fully equivalent (Rosen 1967)!

Another view of the functional formalisms is that they may present the general problem in a different light, so that one may be led to general approximation schemes which would not be apparent in the moment formulation (Lewis and Kraichnan 1962). This is close to our point of view in the present chapter; we hope to follow (by analogy) the route taken in statistical mechanics, in which one starts with a rigorous differential equation (the Liouville equation) and makes approximations on physical grounds in order to derive an equation which is approximate, but solvable (e.g. the Boltzmann equation or the Fokker–Planck equation). We shall return to these points in Chapter 5.

4.3   **Test problems in isotropic turbulence**

Theorists study isotropic turbulence for the good reason that it appears to be the simplest non-trivial version of the turbulence problem. From this point of view, considerations of its 'relevance' to practical situations are, in themselves, irrelevant. Likewise, questions to the effect (often encountered in review articles), 'does isotropic turbulence really exist?', are missing the point.

We would suggest that there are only two general questions about which we need worry. First, can we formulate the problem (or, perhaps, problems)? Second, can we make some empirical check of our theoretical calculations? The purpose of this section is to suggest that the answer to both these questions is reasonably reassuring.

However, having stated our general position, we should then freely concede that the concept of isotropic turbulence is rather artificial in that it requires a property, which is naturally present to some extent in all turbulent flows, to be the dominant or characteristic property of at least some turbulence fields. As a consequence, one price we pay for the relative analytical simplicity of isotropic turbulence is a certain amount of difficulty in checking our results.

Before turning to the formulation of specific test problems in isotropic turbulence, let us first consider the more typical everyday situation, which is represented by flow through a pipe. In principle, we would want to solve a closed form of the Reynolds equation for the mean velocity, subject to the boundary condition that the velocity vanishes at the wall of the pipe.

The fact that the flow occurs at all is due to an imposed pressure gradient, and so naturally we must specify this gradient. For simplicity, we might take it to be steady, so that the flow itself would be independent of time, and of course this is quite easily realizable in practice. Also, for a really thorough approach, we might prescribe the mean velocity distribution at the entrance to the pipe. A simple assumption would be to take the velocity to be zero everywhere in a cross-section at the entrance to the pipe. This is also physically realizable, at least to a good approximation, if we allow the pipe to protrude from the side of a very large reservoir, although a simpler and more usual method would be to ignore an 'entrance length' and begin the calculation at some downstream distance where the mean velocity distribution would be independent of the initial conditions (i.e. universal in form).

The point of all this is that the classical problem of unidirectional flow in a pipe can be formulated in a very simple and direct way. Also, the prediction of any theory, i.e. the mean velocity distribution for a given pressure gradient, can be tested experimentally in a fairly uncontroversial and rigorous way. This is also true of most practical fluid flow configurations, but it is not true of isotropic turbulence, where the lowest-order statistical quantity is the energy spectrum which is much more difficult to measure and interpret than the mean velocity distribution.

4.3.1   *Free decay of turbulence*

Consider the energy balance for isotropic turbulence, as given by eqn (2.118):

$$\left\{\frac{d}{dt} + 2\nu k^2\right\} E(k, t) = T(k, t),$$

where the inertial transfer term $T(k, t)$ is related to the non-linearity of the Navier–Stokes equation through (2.119) and (2.91). Formally, the turbulence closure problem can be seen in this context as the need to obtain an expression relating $T(k, t)$ to the energy spectrum $E(k, t)$.

Assuming that in fact we have such an expression, we note that (2.118) is first order in time, and so we are faced with an initial value problem. That is, we wish to integrate (2.118) forward in time, given the initial energy spectrum

$$E(k, 0) = e(k) \qquad (4.80)$$

where $e(k)$ is some function which determines the initial values of the total energy and the rate of energy dissipation. Note that both quantities are referred to unit mass of fluid; this will be our convention throughout, but we shall issue the occasional reminder.

However, these considerations aside, the important feature of $e(k)$ is that it is completely arbitrary. It is in this respect that the isotropic problem differs from pipe flow, as discussed above. Thus we are forced to rely on the possibility of universal behaviour developing as time goes on. In other words, we envisage the turbulence as having been generated in some arbitrary way at $t = 0$, but having forgotten about its conditions of formation as time passes, to the point where the current generation of eddies has come entirely from non-linear interactions. These interactions are, of course, common to all situations, as they are a property of the equations of motion only, and so can be expected to have some universal effect.

In practice, this turns out to be the case: self-similar behaviour in time (also known as self-preserving behaviour) has been verified experimentally. We shall not go into detail here, but good discussions of these points will be found in Batchelor (1971), Hinze (1975) and Monin and Yaglom (1975), particularly for the case of grid-generated turbulence (see also Section 2.4).

However, from our present point of view, the really important thing is not just that grid-generated turbulence is (to a good approximation) isotropic, but also that it embodies the above characteristics to the extent that it can be generated in an arbitrary fashion (i.e. anyone can design their own grid and make their own choice of grid parameters) but decays to a form which is reasonably independent of those arbitrary choices. As we pointed out in Chapter 2, the free decay of grid turbulence with downstream distance can be converted into a free decay with time by the simple expedient of transforming to a frame of reference moving with the free-stream velocity.

Historically, the study of isotropic turbulence was crucially dependent on grid-generated turbulence. However, nowadays the existence of computer simulations gives us a powerful new method of assessing analytical theories. An especially valuable aspect of these computer experiments is the ability to choose the initial spectrum to match that of the theoretical calculation. We shall return to this and other points in Chapter 8, when we shall consider the assessment of various analytical theories in some detail.

4.3.2 *Stationary isotropic turbulence*

We have already pointed out that it is quite easy to formulate the problem of stationary turbulence for the case of flow through a pipe. Moreover, experimental achievement of stationary flow is a trivial task and this is true of most practical flow situations.

It is not, however, true of isotropic turbulence, except as a local approximation in flows of large physical extent, such as in the atmosphere or the oceans. One may hope that the computer simulations mentioned at the end of the previous section may put this right in time. But, at the moment, the assessment of stationary theories of turbulence is not without its difficulties. We shall touch on these presently, but first we consider the formulation of the theoretical problem.

The formation of a stationary isotropic turbulence field requires some input of energy to compensate for the losses due to viscous dissipation. It is customary to introduce the concept of stirring forces, which are random in nature and which produce a random velocity field by their direct action on the fluid. This is actually a step with some profound implications—as we shall see in Chapter 6, the prescription of the stirring forces essentially amounts to a specification of the turbulent ensemble. But, for our present purposes, it will suffice to note that we must choose the stirring forces with some care, so that we end up with a turbulent field which is characteristic of the Navier–Stokes equations and not just of the arbitrarily chosen input.

Let us consider the equation of motion generalized by the addition of a random force with Fourier components $f_\alpha(\mathbf{k}, t)$. Equation (2.76) can be written as

$$\left\{\frac{\partial}{\partial t} + vk^2\right\} u_\alpha(\mathbf{k}, t) = M_{\alpha\beta\gamma}(\mathbf{k}) \sum_{\mathbf{j}} u_\beta(\mathbf{j}, t) u_\gamma(\mathbf{k} - \mathbf{j}, t) + f_\alpha(\mathbf{k}, t) \qquad (4.81)$$

where, in order to maintain the incompressibility of the velocity field, the stirring forces should satisfy a generalization of eqn (2.74):

$$k_\beta f_\beta(\mathbf{k}, t) = 0. \qquad (4.82)$$

Note also that the stirring forces (like all the other terms of (4.81)) must actually be forces per unit mass, and therefore have the dimensions of acceleration. Nevertheless, we shall conform to convention and continue to refer to them as forces.

We now use (4.81) to derive the energy-balance equation, just as in Chapter 2. That is, we multiply each term by $u_\alpha(-\mathbf{k}, t)$ and average. Then we sum over

α, multiply through by $2\pi k^2$, and use (2.101) for the energy spectrum to obtain

$$\left\{\frac{d}{dt} + 2\nu k^2\right\} E(k, t) = T(k, t) + 2\pi k^2 \langle f_\alpha(\mathbf{k}, t) u_\alpha(-\mathbf{k}, t)\rangle, \qquad (4.83)$$

which differs from (2.118) only by the presence of the input term on the r.h.s.. In order to find an explicit form for this, we now have to specify the nature of the random stirring forces.

We can approach this task rather as we did in the problem of Brownian motion in Section 4.1.2. We take the probability distribution of the forces to be normal (or Gaussian) and we choose their autocorrelation to be given by

$$\langle f_\alpha(\mathbf{k}, t) f_\beta(-\mathbf{k}, t')\rangle = D_{\alpha\beta}(\mathbf{k}) w(k, t - t'), \qquad (4.84)$$

where $D_{\alpha\beta}(\mathbf{k})$ is given by (2.78) and $w(k, t - t')$ remains to be specified. Note that the form of the r.h.s. of (4.84) has been chosen to give a force correlation which is homogeneous, isotropic, and stationary; the reader may find it instructive to make a comparison with the corresponding results for the velocity, as shown in eqns (2.93) and (2.97).

We may be guided further by the case of Brownian motion and choose the random stirring forces to be highly uncorrelated in time. This means that the function $w(k, t - t')$ should be very much peaked near $t = t'$, and we again approximate such behaviour by a delta function:

$$w(k; t - t') = W(k)\delta(t - t'). \qquad (4.85)$$

This assumption has the virtue that if there are no initial time correlations, then any subsequent correlations will be due to the non-linear coupling of the Navier–Stokes equation.

In order to specify $W(k)$, we have to evaluate the last term on the r.h.s. of eqn (4.83). We shall give a rather simple treatment of this here, although more general discussions are given elsewhere in the book.[1] Assume that the system response, for small time intervals $|t - t'|$, is given by a Green function $g(k, t - t')$, such that

$$u_\alpha(\mathbf{k}, t) = \int g(\mathbf{k}, t - t') f_\alpha(\mathbf{k}, t') \, dt', \qquad (4.86)$$

where g is statistically sharp, so that $\langle g\rangle = g$, and has the properties

$$g(\mathbf{k}, t - t') = \begin{cases} 0 & \text{for } t < t' \\ 1 & \text{for } t = t' \end{cases}. \qquad (4.87)$$

The first of the two conditions (4.87) indicates that the effect cannot precede the cause, and the second reflects the fact that at zero time separation the velocity is just given by the time integral of the acceleration. These are in fact the only properties of g that we shall need, and so, substituting (4.86) for

$u_\alpha(-\mathbf{k}, t)$, we can write the input term of (4.83) as

$$2\pi k^2 \langle f_\alpha(\mathbf{k}, t) u_\alpha(-\mathbf{k}, t) \rangle = 2\pi k^2 \int g(\mathbf{k}, t - t') \langle f_\alpha(\mathbf{k}, t) f_\alpha(-\mathbf{k}, t') \rangle$$

$$= 4\pi k^2 W(k), \tag{4.88}$$

where we have substituted from (4.84) for the force autocorrelation and used $\mathrm{tr}\, D_{\alpha\beta}(\mathbf{k}) = 2$.

With (4.88), eqn (4.83) for the energy balance becomes

$$\left\{ \frac{d}{dt} + 2vk^2 \right\} E(k, t) = T(k, t) + 4\pi k^2 W(k). \tag{4.89}$$

The stationary state is achieved when the input term (which represents the rate at which the random stirring forces do work on the fluid) is exactly the same as the rate of energy dissipation by viscosity. Under these circumstances, the time derivative vanishes, and it is of interest to integrate each remaining term over all values of wavenumber k. Recalling, from eqn (2.126), that the integral of $T(k, t)$ over all k is zero—reflecting the fact that the non-linear terms are conservative and do no net work on the system—we obtain

$$\int_0^\infty 2vk^2 E(k)\, dk = \int_0^\infty 4\pi k^2 W(k)\, dk$$

$$= \varepsilon, \tag{4.90}$$

where the last step follows from eqn (2.121). Thus, under steady state conditions, the autocorrelation of the random forces can be expressed in terms of the energy dissipation rate.

Ideally we would complete the specification of a test problem by stating the required solution, or, at least, by saying how it could be checked. Unfortunately, this takes us into a rather controversial area. For some years it was generally accepted that, for high enough Reynolds numbers, the inertial-range solution of equation (4.89) was the famous Kolmogorov 5/3 law. And, indeed, for the case of infinite Reynolds number, the Kolmogorov spectrum—as given by eqn (2.137)—would be valid for all wavenumbers.

However, as we saw in Section 3.2.2, it has been widely conjectured that intermittency effects might change the exponent in the Kolmogorov power law (even if only very slightly). For this reason, many theorists have become sceptical about the use of the Kolmogorov spectrum as a test for turbulence theories. We shall postpone a detailed discussion of this point to Chapter 8, where we consider the general assessment of theories. For the moment, we can fall back on the pragmatic view that any deviation from (2.137) seems to be too small to detect experimentally.

4.4 Further reading

This chapter does not purport to offer a course in statistical mechanics. Our (restricted) interest in the subject has been twofold. First, we wanted to put the statistical theory of turbulence in some sort of perspective within statistical physics. Second, many of the theories of turbulence have been influenced by analogies drawn between turbulent problems and problems such as Brownian motion. For instance, we have just made use of such an analogy when we were formulating the problem of stationary isotropic turbulence.

In order to achieve our objectives, all we needed was a survey of some relevant highlights of the subject. The interested reader, who would like to fill in some of the details, should be able to find a wide choice of classic texts in any good science library. The books which we found helpful in preparing this chapter (and the next as well) were Woods (1975; conservation laws and the utility of distinct time-scales as a basis for expansions), Balescu (1975; many-body systems), and Reichl (1980; despite the apparent quantum bias, good on stochastic theory and classical dynamical systems). In each case, the aspect of the book found particularly helpful is mentioned in the parentheses.

We have also treated functional calculus rather cavalierly. More general discussions of functional differentiation and integration will be found in Beran (1968), while discussions of the relationship to functional integration in quantum field theory have been treated by Gel'fand and Yaglom (1960) and Monin and Yaglom (1975).

Note

1. Equation (4.88) for the cross-correlation of the stirring force with the resulting velocity field is valid for stirring forces with autocorrelations which are delta functions in time. The more general formulations of this term in Sections 5.5.4, 6.1.3, 6.2.4, and Appendix H all reduce to eqn (4.88) when the stirring forces are specialized to have delta function autocorrelations.

References

BALESCU, R. (1975). *Equilibrium and nonequilibrium statistical mechanics.* Wiley, New York.

BATCHELOR, G. K. (1967). *An introduction to fluid dynamics.* Cambridge University Press, Cambridge.

—— (1971). *The theory of homogeneous turbulence* (2nd edn). Cambridge University Press, Cambridge.

BERAN, M. J. (1968). *Statistical continuum theories.* Wiley, New York.

BETCHOV, R. and LARSEN, P. S. (1981). *Phys. Fluids* **24**, 1602.

FRENKIEL, F. N. and KLEBANOFF, P. S. (1967). *Phys. Fluids* **10**, 507.

—— and —— (1971). *J. fluid Mech.* **48**, 183.

—— and —— (1973). *Phys. Fluids* **16**, 725.

GEL'FAND, I. M. and YAGLOM, A. M. (1960). *J. math. Phys.* **1**, 48.

GOLDSTEIN, H. (1953). *Classical mechanics.* Addison-Wesley, Cambridge, Ma.

HINZE, J. O. (1975). *Turbulence* (2nd edn). McGraw-Hill, New York.

HOPF, E. (1952). *J. ratl mech. Anal.* **1**, 87.

KRAICHNAN, R. H. (1959). *Phys. Rev.* **113**, 1181.

—— (1964). *Phys. Fluids* **7**, 1030.

LEE, J. (1982). *J. fluid mech.* **120**, 155.

LEE, T. D. (1952). *Q. appl. Math.* **10**, 69.

LEWIS, M. B. (1975). *Phys. Fluids* **18**, 313.

—— (1977). *Phys. Fluids* **20**, 1058.

LEWIS, R. M. and KRAICHNAN, R. H. (1962). *Commun. pure appl. Math.* **15**, 397.

LUNDGREN, T. S. (1967). *Phys. Fluids* **10**, 969.

MONIN, A. S. and YAGLOM, A. M. (1975). *Statistical fluid mechanics: mechanics of turbulence*, Vol. 2. MIT Press, Cambridge, Ma.

MONTGOMERY, D. (1976). *Phys. Fluids* **19**, 802.

PRIGOGINE, I. (1968). In *Topics in nonlinear physics* (ed. N. J. ZABUSKY). Springer, New York.

REICHL, L. E. (1980). *A modern course in statistical physics.* Edward Arnold, London.

ROSEN, G. (1967). *Phys. Fluids* **10**, 2614.

VAN ATTA, C. W. and CHEN, W. Y. (1968). *J. fluid Mech.* **34**, 497.

WOODS, L. C. (1975). *The thermodynamics of fluid systems.* Clarendon Press, Oxford.

5

RENORMALIZED PERTURBATION THEORY AND THE TURBULENCE CLOSURE PROBLEM

The application of renormalization methods to the problem of closing the turbulent moment hierarchy was begun by Kraichnan (1959). He was soon followed by other pioneers (Wyld 1961; Edwards 1964; Herring 1965), and since then there have been many similar approaches. It is this body of work which forms the principal topic of the present book, and we shall refer to any theoretical approach of this general type as a 'renormalized perturbation theory' (RPT). This gives us a convenient generic term and, at the same time, allows us to distinguish between the subject of the next few chapters and the renormalization group (RG), which will be discussed later in Chapter 9. Introductory discussions of these topics can be found in Section 3.5.

In Chapters 6 and 7, we shall divide the various RPTs into two classes, which are (respectively) those which do not yield the Kolmogorov spectrum as a solution, and those which do, and discuss particular theories in some detail. In this chapter our aim is to expose the basic ideas behind RPTs in general. We shall try to achieve this by introducing RPT in other many-body problems (particularly in certain situations where we can be confident about the soundness of this method), and then by considering its application to turbulence through the Wyld (1961) formulation.

We begin by discussing some background problems and, in the process, defining some of the terminology, which we shall need later on.

5.1 Time evolution and propagators

We have briefly mentioned the time dependence of the many-particle system in the previous chapter (see eqns (4.11) and (4.14)). Here we shall consider how that kind of evolution equation arises. We begin with a simple dynamical system, characterized by a variable $X(t)$, which satisfies Hamilton's equation

$$\frac{\mathrm{d}X}{\mathrm{d}t} = [X, H], \tag{5.1}$$

where the Poisson bracket [ ] is defined in eqn (4.12) and H is the Hamiltonian of the system. For instance, we might be considering a one-dimensional oscillator, where X is the displacement from equilibrium at any time and H is the sum of the kinetic and potential energies.

Now, in general we would like to be able to use (5.1) to find $X(t)$, given $X(0)$. Assuming that derivatives of X exist to all orders, we can write $X(t)$ in terms of X and its derivatives at $t = 0$ by means of the Taylor series

$$X(t) = X(0) + t\dot{X}(0) + \frac{t^2}{2}\ddot{X}(0) + \cdots, \tag{5.2}$$

where we now use a dot to represent differentiation with respect to time. The derivatives of X, to any order, can be obtained from eqn (5.1). Let us denote the operation of taking a Poisson bracket by

$$[H]X = [X, H]. \tag{5.3}$$

It then follows that the derivatives of X can be written as

$$\dot{X} = [X, H] = [H]X, \tag{5.4a}$$

$$\ddot{X} = [\dot{X}, H] = [[X, H], H] = [H^2]X, \tag{5.4b}$$

and so on, to any order n.

With the appropriate substitutions from (5.4a), (5.4b), and so on, eqn (5.2) can readily be obtained in the form

$$X(t) = \sum_{n=0}^{\infty} \frac{t^n[H^n]X(0)}{n!}$$

$$= G(t)X(0). \tag{5.5}$$

The function $G(t)$ is known as a propagator, and is given by

$$G(t) = \exp\{[H]t\}. \tag{5.6}$$

For stationary systems, which depend only on the differences between times and not on their absolute value, eqn (5.5) has the immediate generalization

$$X(t) = G(t - t')X(t'). \tag{5.7}$$

The term propagator comes from quantum field theory (March, Young, and Sampanthar 1967; Mattuck 1976), where it has the technical significance that a particular probability amplitude corresponds to the propagation of a particle. We do not need to worry about this in any great detail, as the intuitive significance of the term seems clear enough in the context of an equation like (5.7). Evidently we can think of a propagator as an evolution in time or, for that matter, in space.

Formally, in quantum mechanics, one deals in probability amplitudes. That is, the propagator can be defined as

$g(\mathbf{x}, \mathbf{x}'; t, t') =$ probability amplitude that, if at time t' we add a particle at $\mathbf{x}'$ to a system in its ground state, the particle will be found at $\mathbf{x}$ for some time $t > t'$, with the system still in its ground state

$\qquad\qquad = 0, \quad$ if $t < t'$. $\hfill(5.8)$

The associated probability density $P(\mathbf{x}, \mathbf{x}'; t, t')$ is given by

$$P(\mathbf{x}, \mathbf{x}'; t, t') = g(\mathbf{x}, \mathbf{x}'; t, t')g^*(\mathbf{x}, \mathbf{x}'; t, t'), \qquad (5.9)$$

which is the usual relationship in quantum mechanics, where g is a solution of the Schrödinger equation and g^* is its complex conjugate. In fact, to be more specific, g is actually the Green function of the Schrödinger equation, and this equivalence of the Green function and the propagator will be of interest to us later on.

In order to take the idea over into the classical many-body problem, we interpret the propagator directly in terms of a probability density. Thus, denoting the classical propagator by $G(\mathbf{x}, \mathbf{x}'; t, t')$, we have the appropriate definition

$G(\mathbf{x}, \mathbf{x}'; t, t') =$ probability density that if a particle is at rest at a
point $\mathbf{x}'$ at time t' then at a later time t it will be
found at the point x

$$= 0, \quad \text{if } t < t'. \qquad (5.10)$$

If there are no interactions with the existing particles of the many-body system, then the test particle moves from $\mathbf{x}'$ at time t' to $\mathbf{x}$ at time t in a purely deterministic way, and we put $G = G_0$, the free propagator.

5.2   Perturbation methods using Feynman-type diagrams

In this section we aim to provide a very simple introduction to the methods used in many-body theory in order to generalize the traditional perturbation series to the case of strong interactions. Formal treatments of this subject, along with the introduction of diagrammatic representations of the mathematical terms in the perturbation expansions, can be found in March et al. (1967) and (from the point of view of statistical mechanics) in Balescu (1975). A distinctly informal treatment of the subject has been given by Mattuck (1976), and this is recommended to the reader who has little or no background in quantum theory.

Let us consider a specific classical (i.e. non-quantum) problem. A test particle at point $\mathbf{x}'$ at time t' moves through a medium containing other particles or scattering centres. What is the probability that the test particle will reach the point $\mathbf{x}$ at time t? In order to simplify things as much as possible, we take the scattering centres to be fixed in space and the scattering forces to be short range. We also drop the time arguments.

Suppose that the scattering centres are labelled by the set of letters $\{A, B, C, \ldots\}$, that they are located at spatial positions $\{\mathbf{x}_A, \mathbf{x}_B, \mathbf{x}_C, \ldots\}$, and that they have (respectively) the probability of scattering the test particle $\{P(A), P(B), P(C), \ldots\}$. Then the total probability of the test particle going from $\mathbf{x}'$ to $\mathbf{x}$ is found by adding up all the probabilities of all possible paths through the system. We can expect this overall probability (i.e. the propagator) to be in the form of an expansion, with the order of each term being governed by the number of interactions with scattering centres. Thus, in all we would expect a form like

$$G(\mathbf{x}, \mathbf{x}') = G_0(\mathbf{x}, \mathbf{x}') + G_1(\mathbf{x}, \mathbf{x}') + G_2(\mathbf{x}, \mathbf{x}') \dots, \qquad (5.11)$$

where $G_0(\mathbf{x}, \mathbf{x}')$ is the free propagator corresponding to the case where the test particle travels from $\mathbf{x}'$ to $\mathbf{x}$ without being scattered by any of the system particles.

The first-order term is due to processes where the test particle is scattered once only, while travelling from $\mathbf{x}'$ to $\mathbf{x}$. If we take scattering centre A as an example, then we have the probability that the particle goes from $\mathbf{x}'$ to $\mathbf{x}_A$ (i.e. $G_0(\mathbf{x}', \mathbf{x}_A)$), the probability that the particle is scattered by A (i.e. $P(A)$), and the subsequent probability that the scattered particle goes from $\mathbf{x}_A$ to $\mathbf{x}$ (i.e. $G_0(\mathbf{x}, \mathbf{x}_A)$). The resulting probability for the process is the product of the three separate probabilities just enumerated, and, with similar results for the other scattering centres, we have the first-order contribution to the r.h.s. of eqn (5.11)

$$G_1(\mathbf{x}, \mathbf{x}') = G_0(\mathbf{x}, \mathbf{x}_A) P(A) G_0(\mathbf{x}_A, \mathbf{x}') + G_0(\mathbf{x}, \mathbf{x}_B) P(B) G_0(\mathbf{x}_B, \mathbf{x}') + \cdots. \quad (5.12)$$

The second order is more complicated because we now have the possibility of two types of interaction. First, the test particle can be scattered twice from any given centre. Second, the test particle can be scattered from one centre and then from any other. Thus, the second-order term on the r.h.s. of (5.11) can be written as

$$G_2(\mathbf{x}, \mathbf{x}') = G_0(\mathbf{x}, \mathbf{x}_A) P(A) G_0(\mathbf{x}_A, \mathbf{x}_A) P(A) G_0(\mathbf{x}_A, \mathbf{x}') +$$
$$+ G_0(\mathbf{x}, \mathbf{x}_A) P(A) G_0(\mathbf{x}_A, \mathbf{x}_B) P(B) G_0(\mathbf{x}_B, \mathbf{x}') + \cdots, \qquad (5.13)$$

where the series can be completed by writing down all the possible pairs from $\{A, B, C, \dots\}$, including repetitions.

In this way, we can obtain terms in the expansion of $G(\mathbf{x}, \mathbf{x}')$ to all orders, with each order itself being given by a sum over all scattering centres. Two points are fairly obvious. First, if we take a simple view of things, then the expansion for G is essentially just a power series in the free propagator G_0. Second, the combinatorial structure of each order means that the complexity of the expansion increases rapidly with order. We can make the former (simpler) view prevail by adopting a diagrammatic representation for the expansion. That is, we replace the mathematical symbols by pictures which are topologically equivalent, but which allow one to comprehend the overall structure of the mathematics rather more readily.

We can introduce the diagram technique by noting that, once we substitute from (5.12) and (5.13), the expansion in eqn (5.11) is built up from terms which

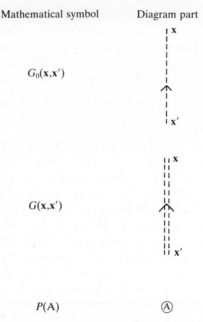

Mathematical symbol	Diagram part
$G_0(\mathbf{x},\mathbf{x}')$	
$G(\mathbf{x},\mathbf{x}')$	
$P(A)$	Ⓐ

Fig. 5.1. Equivalence of mathematical symbols and diagram parts for the perturbation expansion, as defined by eqns (5.11), (5.12), and (5.13).

are combinations of the free propagator G_0 and the scattering probability P, each having labels selected from the set $\{A, B, C, \ldots\}$. In Fig. 5.1, we replace each of these basic symbols by a pictorial representation, which we call a diagram part. Thus, with combinations of these diagram parts, we can build up more elaborate pictorial representations, eventually depicting the perturbation series as shown in Fig. 5.2.

The diagrammatic representation centainly makes the perturbation series look very much simpler. This is part of its value, but the main reason for making the change is that one can classify terms more easily by their topological properties. We shall make use of this facility at appropriate points later on to carry out partial summations, and it will be seen that this is the basis of our renormalization programme. Here we shall confine ourselves to a few preliminary remarks.

The probabilities G_0 and P all take values somewhere between zero and unity. Suppose that all the free propagators are about unity in magnitude. Then, if all the P factors are very much smaller than unity, the perturbation series will converge very rapidly and we can obtain rational approximations, truncating at first, or higher, order to obtain any desired accuracy. This is, of course, the classical application of perturbation theory.

However, suppose that instead of all the scattering probabilities being small,

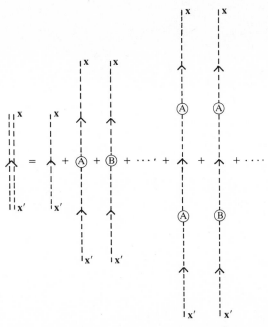

Fig. 5.2. The perturbation expansion in diagrams: eqns (5.11), (5.12), and (5.13) have been rewritten by changing symbols into diagrams, according to the prescription given in Fig. 5.1.

we were faced with the case where one scattering probability, $P(A)$ say, had associated factors which were very much larger than unity. Then $P(A)$ would be dominant, and the perturbation expansion could be approximated as shown in Fig. 5.3. In general we could expect this series to converge only slowly, or even to diverge. This is the general problem of strong interactions. But, despite the lack of convergence, all is not lost. From the simple regularity of the expansion in Fig. 5.3, it is clear that it can be expressed in terms of a geometric progression. Or, rewriting it in analytic form, we can decompose and factorize the r.h.s. to obtain

$$G(\mathbf{x}, \mathbf{x}') = G_0(\mathbf{x}, \mathbf{x}') + G_0(\mathbf{x}, \mathbf{x}_A)P(A)G_0(\mathbf{x}_A, \mathbf{x}') \times$$
$$\times \{1 + P(A)G_0(\mathbf{x}_A, \mathbf{x}_A) + [P(A)G_0(\mathbf{x}_A, \mathbf{x}_A)]^2 + \cdots\} \quad (5.14)$$

and, summing the geometric series,

$$G(\mathbf{x}, \mathbf{x}') = G_0(\mathbf{x}, \mathbf{x}') + G_0(\mathbf{x}, \mathbf{x}_A)P(A)G_0(\mathbf{x}_A, \mathbf{x}')\frac{1}{1 - P(A)G_0(\mathbf{x}_A, \mathbf{x}_A)}. \quad (5.15)$$

This technique is known as 'partial summation', and, as we have previously remarked, it will be the basis of our renormalization programme. However,

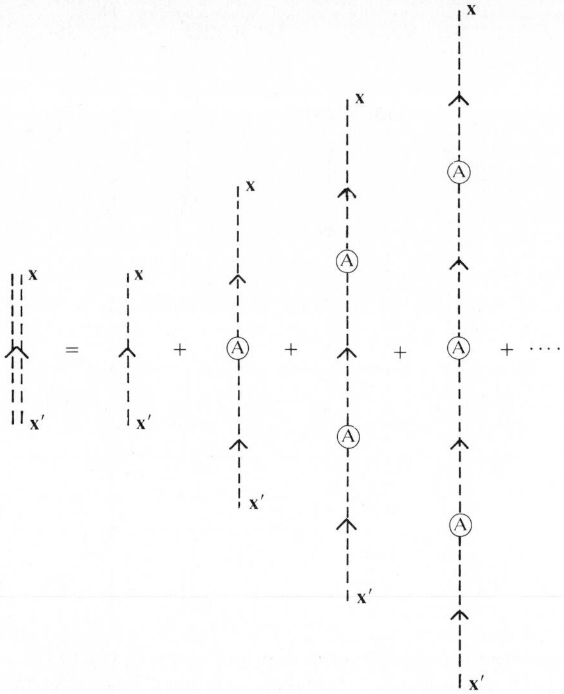

Fig. 5.3. Perturbation series for the case where $P(A)$ is large, and all the other scattering probabilities are negligible in comparison.

it is worth noting that even the simple example given here as a demonstration of the method is by no means trivial. For instance, very similar expansions occur in modern treatments of Brownian motion (Van Beijeren and Dorfman 1979).

Lastly, it should perhaps be pointed out that the method of diagrams, and partial summation, constitutes a general mathematical method. For instance, it can be used in the iterative solution of integral equations (Mathews and Walker 1965, p. 288). Or, as we shall see in the next section, it can be applied to statistical mechanical problems where there are long-range interactions.

5.3   Equilibrium system with weak interactions: an introduction to renormalized perturbation theory

In this section we shall consider the N-particle system again, but this time we shall be interested in the effect of interactions. Basically, we shall take our N-particle system to be an ideal gas (non-interacting hard spheres) with some rather weak interactions added on. We shall examine, first, short-range interactions, as found in a gas of neutral molecules, and then, in Section 5.4, long-range interactions, as found in an electron gas or high temperature classical plasma.

However, we should first underline that the present discussion has little direct application to turbulence. In the N-particle problem, all N particles have identical properties and are collectively in thermal equilibrium. We also impose the restriction to weak interactions. In contrast, the turbulent Fourier modes have different levels of excitation (i.e. the turbulent spectrum does not correspond to equipartition), and the dominant dynamical feature is the flow (unidirectional, on average) of energy through the modes. Also, the interactions in turbulence (essentially measured by the Reynolds number) are far from weak.

Accordingly, our purpose is to introduce the RPT technique in the context of a problem which is simpler than that of fluid turbulence. This approach has two advantages. First, the greater conceptual simplicity allows the main points of the method to stand out more clearly. Second, the fact that the RPT method can be seen to work and indeed to have a clear physical interpretation must surely be an advantage when later we turn to the turbulence problem.

5.3.1   Interactions in dilute systems and the connection with macroscopic thermodynamics

For N hard spheres, interacting weakly through two-body potentials, we can adapt equation (4.16) for the total Hamiltonian to

$$H = \sum_{i=1}^{N} H(\mathbf{p}_i) + \lambda \sum_{i<j=1}^{N} H'(\mathbf{q}_i, \mathbf{q}_j) \tag{5.16}$$

where, for particles of mass m,

$$H(\mathbf{p}_i) = \frac{p_i^2}{2m}, \tag{5.17}$$

and the parameter λ is inserted in front of H' as a bookkeeping parameter in order to remind us that H' is a small quantity. We shall therefore treat λ itself as a small quantity.

The interaction Hamiltonian depends only on the intermolecular potentials, and, representing these by W, we can write

$$H'(\mathbf{q}_i, \mathbf{q}_j) = W(\mathbf{q}_i, \mathbf{q}_j). \tag{5.18}$$

In general, we can expect the potentials to depend only on the distance between any pair of molecules and not on either absolute position or orientation. Thus the pair potential can be simplified to

$$W(\mathbf{q}_i, \mathbf{q}_j) = W(|\mathbf{q}_i - \mathbf{q}_j|) = W(r_{ij}), \tag{5.19}$$

where the relative coordinate is

$$r_{ij} = |\mathbf{q}_i - \mathbf{q}_j|. \tag{5.20}$$

In the interests of conciseness, we shall often use the abbreviated notation

$$W_{ij} = W(r_{ij}). \tag{5.21}$$

At this stage we should consider the form of the intermolecular potentials. A general discussion of this whole subject will be found in Hirschfelder, Curtiss, and Bird (1954). For our present purposes it will be sufficient to have forms which possess the general features of interest to us here. To begin with, we have assumed the molecules to be hard spheres, which implies that they do not penetrate each other. If the molecular radius is b then the hard-sphere repulsive potential can be specified by

$$W'(r) = \begin{cases} \infty & \text{if } r < 2b \\ 0 & \text{if } r > 2b \end{cases} \tag{5.22}$$

where r is the radial distance between any pair of molecules.

For molecules which are electrically neutral, a realistic potential which combines a steep repulsion at very short range, with a weak attraction at slightly longer ranges, is given by the Lennard-Jones form

$$W(r) = 4E\left[\left(\frac{b}{r}\right)^{12} - \left(\frac{b}{r}\right)^{6} \right], \tag{5.23}$$

where E has the units of energy and b is a measure of the molecular radius.

In Section 5.4, we shall need the intermolecular potential for the case where the gas is ionized. A reasonable model is obtained by combining the hard-sphere potential, as given by (5.22), with the Coulomb potential:

$$W_{ij} = W'_{ij} + \frac{e^2}{r_{ij}} \tag{5.24}$$

where e is the electronic charge. This model is completed by regarding eqn (5.24) as giving the potential between any two electrons in a medium in which the positive charges are supposed to be smeared out in order to provide a uniform background positive charge, thus ensuring the overall electrical neutrality of the system.

Having now set up some of the microscopic concepts, we conclude with a brief consideration of the way in which we can use these to obtain values for macroscopic variables such as the system pressure. In fact the crucial quantity for making the connection between microscopic statistical mechanics and macroscopic thermodynamics is the partition function Z. With some over-simplification this can be defined as the normalization of the probability density distribution of the system:

$$Z = \int d\mathbf{p}\, d\mathbf{q} \exp(-\beta H), \tag{5.25}$$

where β is given by

$$\beta = \frac{1}{k_B T}, \tag{5.26}$$

and k_B is the Boltzmann constant and T is the absolute temperature.

Technical problems over the definition of the partition function need not concern us here as, noting that eqn (5.16) indicates that H can be written as the sum of a part depending on $\mathbf{p}$ only and a part depending on $\mathbf{q}$ only, it is clear that the integration over $\mathbf{p}$ can be factored out (e.g. see Balescu 1975, p. 186). Thus it follows that the full partition function can ifself be factored as

$$Z = Z_0 Q, \tag{5.27}$$

where Z_0 is the partition function for the ideal gas and Q is known as the configuration integral. It is given by

$$Q = V^{-N} \int d^N \mathbf{q} \exp\left\{ -\beta\lambda \sum_{i<j=1}^{N} W_{ij} \right\}. \tag{5.28}$$

Here V is the volume of the system and its presence is due to the way in which Z_0 is defined.

Clearly the evaluation of the configuration integral Q is the essence of the problem where interactions are involved. Correspondingly, we note that as the free energy of the system is given by the natural logarithm of the partition function, we can introduce the configurational free energy A through the relationship

$$A = -\beta^{-1} \ln Q. \tag{5.29}$$

Our problem now is to use perturbation theory to obtain approximate solutions for Q and A.

5.3.2 Primitive perturbation expansion for the configuration integral

Using the series representation of the exponential function, the expansion of the configuration integral can be written down directly from (5.28) as

$$Q = V^{-N} \int d^N \mathbf{q} \sum_{n=0}^{\infty} (n!)^{-1} (-\beta\lambda)^n \left(\sum_{i<j=1}^{N} W_{ij} \right). \tag{5.30}$$

As this amounts to a power series in a small quantity (λ), we might expect to need to retain only a few orders to achieve any desired accuracy. However, this depends on the integrals being well-behaved. We can therefore anticipate later discussions, to some extent, by noting that we are likely to encounter problems with the hard-sphere repulsive potentials, which may cause divergences at $r = 0$. Likewise, in the case of the electron gas, we can expect the long-range Coulomb potential to cause divergences as r goes to infinity.

The complications of dealing with eqn (5.30) increase rapidly with order n. We shall find it convenient to proceed inductively, considering increasing orders in turn. We begin by rewriting (5.30) as

$$Q = \sum_{n=0}^{\infty} (n!)^{-1}(-\beta\lambda)^n Q_n, \tag{5.31}$$

and then we consider Q_0, Q_1, Q_2, and so on.

Zero order is trivial, and we just have

$$Q_0 = V^{-N} \int d^N \mathbf{q} = 1.$$

The case of $n = 1$ is a lot more interesting, with Q_1 given by

$$Q_1 = V^{-N} \int d^N \mathbf{q} \left(\sum_{i<j=1}^{N} W_{ij} \right)$$

$$= V^{-N} \int d^N \mathbf{q}(W_{12} + W_{13} + W_{23} + W_{14} + W_{24} + \cdots)$$

$$= V^{-2} \left\{ \int d\mathbf{q}_1 \, d\mathbf{q}_2 \, W_{12} + \int d\mathbf{q}_1 \, d\mathbf{q}_3 \, W_{13} + \cdots \right\}. \tag{5.32}$$

We have expanded out the double sum in order to make it easy to see that Q_1 is made up from many identical terms, each of which is a double spatial integral over the potential interaction between a pair of particles. Evidently we can work out the value of Q_1 merely by evaluating the first integral (say) on the r.h.s. of eqn (5.32), and multiplying by the number of pairs which can be selected from N particles (i.e. $N(N - 1)/2$), with the result

$$Q_1 = \frac{N(N - 1)}{2} I_1, \tag{5.33}$$

where the integral I_1 is given by

$$I_1 = V^{-1} \int dr_{12} \, W_{12}. \tag{5.34}$$

Note that in I_1 we have made the change of variables from $\mathbf{q}_1, \mathbf{q}_2$ to $\mathbf{q}_1, r_{12}$, where r_{12} is as defined by eqn (5.20). This leaves a free integration with respect to $\mathbf{q}_1$ over the system volume, which just gives a factor V.

For $n = 2$, we can carry out the same procedure to obtain

$$Q_2 = V^{-N} \int d^N \mathbf{q} \{ W_{12}^2 + W_{13}^2 + W_{23}^2 + \cdots + 2W_{12}W_{13} +$$

$$+ 2W_{12}W_{23} + \cdots + 2W_{12}W_{34} + 2W_{12}W_{35} + \cdots \}, \tag{5.35}$$

where we show explicitly just a few of the terms that arise when we write out the double sum and square the result. The resulting products have been set out on three lines in order to draw attention to the fact that there are three types of product. These are as follows:

(a) products with two indices in common, e.g. W_{12}^2;
(b) products with one index in common, e.g. $W_{12} W_{13}$;
(c) products with no indices in common, e.g. $W_{12} W_{34}$.

The differences between these three types will turn out to be very important indeed.

As a first step, we note that we can apply the techniques just used for Q_1, but we must do it separately for each of the three classes of term making up Q_2. That is, we divide up Q as follows:

$$Q_2 = AI_{2a} + BI_{2b} + CI_{2c}, \qquad (5.36)$$

where A, B, and C are the numbers (respectively) of type A, type B and type C terms. The integrals I_{2a}, I_{2b}, and I_{2c} are defined analogously to I_1, and we now consider each of these in turn.

We begin with type A. Taking the first term on the first line of the r.h.s. of eqn (5.35), we have

$$I_{2a} = V^{-N} \int d^N \mathbf{q} \, W_{12}^2$$

$$= V^{-2} \int d\mathbf{q}_1 \, d\mathbf{q}_2 \, W_{12}^2$$

$$= V^{-1} \int dr_{12} \, W_{12}^2, \qquad (5.37)$$

where we have made the change of variables to $\mathbf{q}$, r_{12}, and performed the free integration to eliminate one factor V.

Type B is quite straightforward:

$$I_{2b} = 2V^{-3} \int d\mathbf{q}_1 \, d\mathbf{q}_2 \, d\mathbf{q}_3 \, W_{12} W_{13}$$

$$= 2V^{-2} \int dr_{12} \, dr_{13} \, W_{12} W_{13}$$

$$= 2V^{-2} \left\{ \int dr_{12} \, W_{12} \right\} \left\{ \int dr_{12} \, W_{12} \right\}$$

$$= 2I_1^2. \qquad (5.38)$$

Type C is also straightforward:

$$I_{2c} = 2V^{-4} \int d\mathbf{q}_1 \, d\mathbf{q}_2 \, d\mathbf{q}_3 \, d\mathbf{q}_4 \, W_{12} W_{34}$$

$$= 2V^{-2} \int dr_{12} \, dr_{34} \, W_{12} W_{34} = 2I_1^2. \tag{5.39}$$

Clearly, therefore, one very important distinction between the three types of term occurring at second order is the ability of types B and C to be factored.

In order to complete our calculation of Q_2, by substituting from (5.37), (5.38), and (5.39) into eqn (5.36), we would need to know values for A, B, and C. The first of these is easy, as A is just the number of pairs which can be selected from N particles. Hence A is given by $N(N-1)/2$, as in the first-order case. The values for B and C are rather more difficult, and a consideration of these matters would take us into more detail than would be appropriate to our purposes. The interested reader will find general discussions of technical points of this kind, along with a treatment of the nth-order term, in Balescu (1975). From our present point of view, we have now gone far enough with the perturbation expansion to provide a basis for the introduction to the graphical method.

Let us consider the first-order contribution to the configuration integral. Essentially we require a graphical representation of I_1. We represent the particles by large dots (usually referred to as 'vertices', because they normally occur in diagrams where lines intersect) and the potential with respect to the coordinate joining the particles by a line (bond). The result is as shown in Fig. 5.4.

At second order, the type A interaction is a straightforward extension of the first-order case. We still have two particles but now we have the product of two potentials. We take account of the second potential by means of a second line joining the two vertices. Again, the result is shown in Fig. 5.4.

Type B interactions must have the physical interpretation that three particles are involved, which implies three vertices in a diagram. Similarly, type C interactions imply four vertices in the corresponding diagram. Then the restriction to two bonds dictates the only two remaining possibilities, which are as shown in Fig. 5.4.

At order n, we can expect all possible diagrams which are topologically distinct, and which possess n bonds and a number of vertices which can range from a minimum of 2 to a maximum of $2n$. A procedure for generating diagrams of order n from those of order $n-1$ can be stated in terms of the various ways on can add an extra line to the lower-order graphs.

(1) The extra bond is connected by both its ends to all possible pairs of vertices in the lower-order graph.

(2) The extra bond is connected by one of its ends to every vertex in the lower-order graph. Its other end is terminated by a new vertex.

$n=1$ •————• $\longleftrightarrow$ $I_1 = V^{-1} \int dr_{12} W_{12}$
       1   2

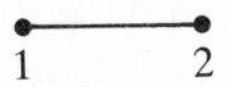

$n=2$ ⬭ $\longleftrightarrow$ $I_{2a} = V^{-2} \int dq_1 dq_2 W_{12}^2$
       1   2

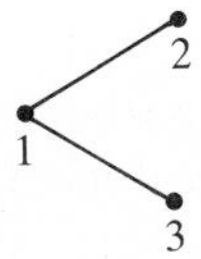

$\longleftrightarrow$ $I_{2b} = 2V^{-2} \{\int dr_{12} W_{12}\}^2$
$= 2I_1^2$

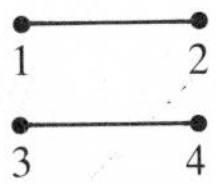

$\longleftrightarrow$ $I_{2b} = 2V^{-2} \{\int dr_{12} W_{12}\}\{dr_{34} W_{34}\}$
$= 2I_1^2$

Fig. 5.4. First and second orders of the perturbation expansion for the configuration integral, showing the constituent integrations and their graphical equivalents.

(3) The extra bond is terminated at both ends by new vertices and added without connection to the lower-order graph.

It can readily be verified that the second-order diagrams of Fig. 5.4 are generated from the first-order diagram by applying the above rules in the order given.

We now require a graphical technique which is the equivalent of evaluating Q_1, Q_2, and so on. As before, we start with first order. We know that I_1, as defined by eqn (5.34), is one of many similar integrals, each differing from the other merely by the differences in the labelling of the potentials. As all these integrals have the same numerical value, Q_1 can be obtained by multiplying I_1 by the total number of integrals.

The graphical equivalent is to remove the labels (1, 2) from the first-order graph and call the result an unlabelled graph. Then we take the unlabelled graph and proceed to label its vertices with all integers (i, j), such that $i < j$ and $j < N$. This process is illustrated in Fig. 5.5. The value of Q_1 is plainly just equal to the value of the unlabelled graph multiplied by the number of labelled graphs which can generated from it.

Of course this is rather trivial. But even at second order, complications accrue, and as the order increases, so do the complications (e.g. see Balescu 1975). Therefore the value of the method emerges at the higher orders. How-

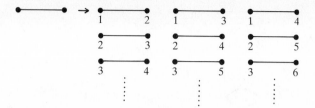

Fig. 5.5. Some of the many (in this case $N(N - 1)/2$) ways of labelling an unlabelled graph.

ever, we shall not go into further detail about higher orders and shall merely state the general principles by which the graphical method can be used to obtain the configuration integral. These can be summarized as follows:

(1) Use the rules above to draw the unlabelled graphs for each order of the expansion. These graphs must be topologically distinct (i.e. no duplicates).

(2) Perform the integrations necessary to obtain a numerical value for each unlabelled graph.

(3) Label each graph in all possible ways, as illustrated in Fig. 5.5.

(4) Multiply the result of (3) by that of (2), and add up over all topologically distinct graphs for a particular order.

(Readers who consult one of the more specialized works on this subject may find it helpful to note that we have included the multinomial coefficient in our definition of the I_n.)

We conclude this section by setting out the rules for making a general topological classification of the perturbation diagrams into one of three types. This classification is based on the three types of diagram which appear at second order (see Fig. 5.4). We shall take types A, B, and C of Fig. 5.4 in reverse order, as we generalize their properties to diagrams of any order.

Disconnected graphs: these have two or more bonds with no particle in common. Type C of Fig. 5.4 is an example.

Reducibly connected graphs: these have two or more bonds with only one particle in common, so that cutting the diagram at the vertex corresponding to the common particle results in two or more disconnected diagrams. Type B of Fig. 5.4 is an example.

Irreducibly connected graphs: all bonds have two particles in common and hence cannot be reduced to disconnected diagrams with only one cut. Type A of Fig. 5.4 is an example, as is the sole diagram at first order.

Diagrams in the last category are often referred to more simply as **irreducible graphs**. As we shall see in the next section, this type of classification is crucial to a successful reduction of the perturbation series.

5.3.3 *Renormalized expansion for the free energy*

The critical test for theories in statistical mechanics is their behaviour in the thermodynamic limit, when we let N and V become indefinitely large such that N/V tends to a finite constant limit ρ—the density. In fact the expansion for the configuration integral Q fails this test, as its terms of a given order in λ have a variety of different dependences on the density. In other words, the expansion is inhomogeneous.

Somewhat analogous problems arise in turbulence, and an important lesson to be learned at the outset is that we should regard the primitive perturbation series as so much raw material. That is, we can use it as the basis of a fairly general and systematic approach to the fundamental problems of the subject. But we should not look on it as any kind of answer in itself.

Reverting now to the statistical mechanical problem, the inhomogeneity of the expansion does not present any serious problems: we can get round it by using the free energy A, as this is the quantity with physical meaning. We can do this rather easily, by means of a trick, as follows. If we think of the potentials W_{ij} as random variables, then we can interpret the configuration integral Q as a characteristic function. This is discussed in Appendix B, where it is shown that the characteristic function $m(k)$ of a distribution $p(x)$ is used to generate the moments of the distribution. We can make the connection by identifying x with the potentials and putting $k = -\beta\lambda$. Then the expansion (5.31) defines the moments Q_n.

We can also introduce the cumulant generating function $\ln\{m(k)\}$: its expansion generates cumulants K_n, as defined in Appendix B where we also give equations expressing the cumulants in terms of the moments. Clearly, if the configuration integral is a moment-generating function, it follows from (5.29) that the free energy A is a cumulant-generating function. Hence, with some rearrangement, eqn (5.29) takes the form

$$-\beta A = \ln Q = \ln\{m(-\beta\lambda)\}$$

$$= \sum_{n=0}^{\infty} (n!)^{-1}(-\beta\lambda)^n K_n. \tag{5.40}$$

Thus, as we now know the moments to any order, the cumulants to any order follow from the well-known relationships connecting the two quantities.

We shall not go into detail, but we quote the first two such relationships in order to explain a most important point. We have (see Appendix B) the equations

$$K_1 = Q_1 \quad \text{and} \quad K_2 = Q_2 - Q_1^2.$$

Now, the first of these is trivial but the second is quite crucial. It can be shown that, if we substitute for Q_1 and Q_2, the contributions from the integrals (5.38) and (5.39) are cancelled by the counter-term Q_1^2. This leaves only the contribution from (5.37) at second order. In other words, only the irreducible diagram contributes and this turns out to be true for all orders. In Fig. 5.6 the irreducible diagrams contributing to the free energy are shown up to fifth order.

In Section 5.2, we discussed the technique of infinite partial summation. An interesting example of this arises in the case of the free energy. As we mentioned earlier, we can expect problems with the convergence of the perturbation expansion owing to the divergence of integrals over the potentials as r goes to zero. We can avoid this situation by considering dilute systems, which means that a power series in the density $\rho = N/V$ can be truncated at low order.

In practice this amounts to a rearrangement of the perturbation series such that one finds an infinite series of diagrams associated with ρ, ρ^2, ρ^3, etc. Each of these infinite series must be summed to give a coefficient in our new expansion in powers of the density.

We shall indicate the general form of this approach at the first order in the density only. Select all diagrams, which have only two vertices, from the irreducible graphs shown in Fig. 5.6; such graphs are of order V^{-1} and hence of order ρ. Using (5.34), (5.37), and (5.40), with the n-order term (in λ) following by induction, we can write the first order in the density expansion in the form

$$-\beta A \simeq V^{-1} \int dr_{12} \left[-\beta\lambda W_{12} + \frac{(-\beta\lambda W_{12})^2}{2!} + \cdots + \right.$$

$$\left. + \frac{(-\beta\lambda W_{12})^n}{n!} + \cdots \right] + O(V^{-2}). \tag{5.41}$$

At this stage, we introduce the Mayer function f_{ij}, which is named after Mayer (1937) who introduced the method of graphs to statistical mechanics.[1] Mayer functions are defined by

$$f_{ij} = \exp(-\beta\lambda W_{ij}) - 1. \tag{5.42}$$

They have the useful property that, as $r \to 0$, and $W_{ij} \to \infty$, $f_{ij} \to -1$. They also allow the expansion on the r.h.s. of (5.41) to be summed:

$$-\beta A \simeq V^{-1} \int dr_{12} f_{12} + O(V^{-2}). \tag{5.43}$$

This procedure can be carried out for higher powers of the density, although the algebra becomes very complicated. At each order of ρ, the infinite series in powers of λ can be summed using Mayer functions. The general procedure is to take the lowest-order diagram in the series for a particular power of the density and replace the potential term $(-\beta\lambda W_{12})$ by f_{12}. This can be shown pictorially by replacing each single bond by double lines in order to denote partial summation or renormalization.

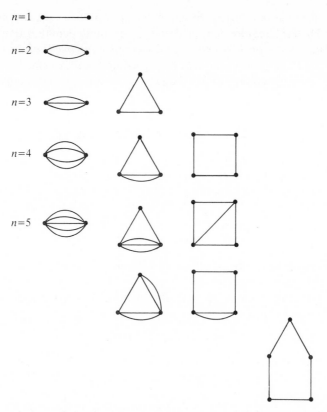

$n=1$

$n=2$

$n=3$

$n=4$

$n=5$

Fig. 5.6. Irreducible diagrams in the expansion for the free energy up to fifth order.

5.4   The electron gas: an example with long-range forces

By now it should be clear that, even when the basic (primitive) expansion is divergent, the situation for perturbation theory is by no means hopeless. We have seen that the primitive series can be rearranged and partially summed in various ways, or that terms in the expansion can be classified according to their topological properties. Then, if certain classes of terms can be neglected, the problem can be much reduced.

Of course, the difficulty in tackling any new physical situation is to know just how to achieve these simplifications. As an illustration of how a successful phenomenological theory has provided the necessary guidance for the implementation of many-body theory, we consider the case where the particles of the system are charged. In this case, the classical phenomenological theory was first developed for electrolytes (Debye and Hückel 1923), but can also be applied to the classical plasma or to conduction electrons in a metal.

The basic model is set up by assuming that all N particles are, say, negatively charged. Then the requirement of electrical neutrality demands that we have an equal number of positive charges. However, we can avoid dealing with the dynamics of these positive charges by regarding them as being smeared out uniformly over the volume of the system, thus creating a continuous medium or background charge density.

5.4.1 *Phenomenological theory: the screened potential*

Of the N particles we shall consider just one pair in isolation. We take them to be separated by distance r_{12} and to interact through the Coulomb potential

$$W_{12} = e^2/r_{12},$$

or,

$$W(r) = e^2/r, \tag{5.44}$$

where the subscripts can be dropped without confusion, as we shall only consider this one pair specifically.

In thermal equilibrium the probability density of finding particle 2 a distance r from particle 1 is just

$$p(r) = \rho \exp\{-\beta W(r)\}, \tag{5.45}$$

where ρ is the (number) density of charged particles. Now suppose that we can switch on the effect of all the other charged particles. The long-range nature of the interactions ensures that they will all affect the potential $W(r)$, changing it to, say, $W'(r)$. Thus eqn (5.45) would be modified to

$$p'(r) = \rho \exp\{-\beta W'(r)\}, \tag{5.46}$$

where we shall find it convenient to express the collective potential energy in terms of a collective electrostatic potential $\phi(r)$:

$$W'(r) = e\phi(r). \tag{5.47}$$

The potential $\phi(r)$ is due to all N electrons, and Debye and Hückel (1923) proposed that it should be determined by the Poisson equation, just as if the distribution of N charges could be regarded as a continuous medium, with the charge density determined self-consistently by the probability $p'(r)$ as given by eqn (5.46). In the terminology of modern many-body theory, this proposal amounts to the method of the self-consistent field.

We can put this into practice by writing the Poisson equation as

$$\begin{aligned}
\nabla^2 \phi &= -4\pi[ep'(r) - e\rho] \\
&= -4\pi[e\rho \exp\{-\beta W'(r)\} - e\rho] \\
&= -4\pi e\rho[\exp\{-\beta W'(r) - 1], \tag{5.48}
\end{aligned}$$

where eqn (5.46) has been used to calculate the charge density due to the electrons and the second term in the square bracket represents the background (positive) charge density, which is constant over the system.

We restrict our attention to the case of weak interactions, for which $W' \ll k_B T$ and hence $\beta W' \ll 1$. Under these circumstances, the exponential on the r.h.s. of eqn (5.48) can be expanded to first order. Cancelling the constant terms, we have the simplified equation

$$\nabla^2 \phi = 4\pi e \rho \beta W' = 4\pi e^2 \rho \beta \phi, \tag{5.49}$$

which has the spherically symmetric solution

$$\phi = \frac{e}{r} \exp\left(-\frac{r}{l_D}\right), \tag{5.50}$$

where l_D is the Debye length and is given by

$$l_D = [4\pi e^2 \rho \beta]^{-1/2}. \tag{5.51}$$

For completeness, we note that the self-consistent potential energy is given by

$$W'(r) = \frac{e^2}{r} \exp\left(-\frac{r}{l_D}\right). \tag{5.52}$$

We should note that the effect of including the collective interactions is to cut off the long '$1/r$' tail in the Coulomb potential so that its effective range is reduced to, roughly, the Debye length. This is why it is often referred to as a screened potential. Physically, we can interpret the situation as a cloud of electrons in the immediate neighbourhood of one electron screening it from the effect of all the other electrons in the system. Then it follows that the Debye length can be interpreted as the approximate radius of the screening cloud about any one electron.

It is also the case that we can interpret the calculation which led to eqns (5.50) and (5.51) as a process of charge renormalization, in which the 'bare' charge has been replaced by an effective charge which depends not only on particle density and system temperature, but also on the spatial position coordinate.

Lastly, the continuum approximation which underlies the introduction of the Poisson equation can be expected to be valid provided that the distance between particles is much smaller than the Debye length. That is, we would require $l_D \gg \rho^{-1/3}$ or $l_D^3 \gg \rho^{-1}$. With the substitution of (5.51) for l_D, this criterion can be written as

$$8\pi^{3/2} e^3 \rho^{1/2} \beta^{1/3} \ll 1, \tag{5.53}$$

giving us a characteristic small parameter for the plasma problem.

5.4.2 *Perturbation calculation of the free energy*

We assume that the two-body interaction potential is given by the combination of the hard-sphere and Coulomb potentials, as shown in eqn (5.24).

Renormalization of the hard-sphere part of W is readily carried out using the Mayer functions, as in Section 5.3.3, and will not be repeated here. However, when we try to carry out the same programme for the Coulomb part of W, we run straight into the problems associated with long-range forces.

The expansion which is summed in terms of the Mayer functions is given by eqn (5.41). With the substitution of the Coulomb potential, the integrals at first and second order become respectively

$$I_1 = \int_0^\infty 4\pi r^2 W_{12}(r)\,dr$$

$$= 4\pi e^2 \int_0^\infty r\,dr, \tag{5.54}$$

where we can drop the subscripts on r without loss of generality, and

$$I_2 = 4\pi e^4 \int_0^\infty dr. \tag{5.55}$$

Clearly these integrals are divergent, the problem being that the Coulomb potential does not fall off fast enough as r tends to infinity.

In practice, problems associated with the divergent integrals are easily circumvented. To begin with, the first-order integral is cancelled by the background and therefore presents only a trivial problem. Higher orders are dealt with by cutting off each integral at some arbitrary value of r, and then summing to all orders in the interaction parameter (in this case $\lambda = e^2$). This was first done by Mayer (1950), who modified the potentials to the Debye–Hückel form, as given by eqn (5.52), and let l_D tend to infinity at the end of the calculation. However, the use of a straightforward cut-off in the integration works just as well. That is, we evaluate all the integrals over the range $0 < r < R$, add up all the individual terms, and then let R tend to infinity in the final sum.

The problem of identifying the appropriate class of diagrams to sum can be solved—for this particular problem—by making a guess about which particular first-order approximation would lead to the Debye–Hückel screened potential. The appropriate diagrams (in the irreducible set; see Fig. 5.6) turn out to be those with no internal bonds. There is only one such diagram in each order and we show these, up to sixth order, in Fig. 5.7. (Note that the first-order diagram is not included, as it is cancelled out by the positive background charge.)

Mayer (1950) referred to these as 'cycle' graphs, but more recently, and in

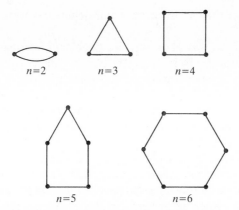

Fig. 5.7. Ring diagrams (up to sixth order) which contribute to the perturbation calculation of the Debye–Hückel screened potential.

other contexts, they have come to be known as 'ring' graphs (March *et al.* 1967, Balescu 1975, Mattuck 1976, Reichl 1980). From our present point of view, an interesting development was the reinterpretation of Mayer's methods by Salpeter (1958), who argued that the complicated combinatorial arguments could be short-circuited by topological methods.

Lastly, although we shall not consider the effect of including higher-order terms in the renormalized perturbation series, it should be noted that the perturbation method can, unlike the phenomenological theory, be carried on to higher orders of accuracy.

5.5 Perturbation expansion of the Navier–Stokes equation

We now wish to examine the properties of the Navier–Stokes equation in the same spirit as we approached the other many-body problems in the preceding sections of this chapter. Our first step in doing so is to recognize the position of the turbulence problem in the hierarchy of statistical problems. That is to say, it can be stated as a problem of many bodies (the Fourier modes) making up a non-equilibrium system with strong interactions.

It should be clear that each of the two properties just stated amounts to a barrier in the way of any straightforward extension of the existing many-body theories to the case of fluid turbulence. Let us consider them in turn.

First, we must understand that the absolutely characteristic feature of turbulence is the flow of energy through the modes, as exemplified by the energy cascade. Thus we are not faced with the slight departure from equilibrium seen in Section 4.1.2, where we discussed kinetic equations. Instead we have to consider a situation which is far from energy equilibrium. This automatically rules out the methods of equilibrium thermodynamics, along with perturbation theory based on small departures from the equilibrium state.

Second, there is no way of evading the problem of strong interactions. We can express this formally by rescaling the Navier–Stokes equation into dimensionless variables, in which case the Reynolds number R will appear as a measure of the ratio of the non-linear term to the viscous (linear) term. Even if we choose length and velocity scales which are such that R takes a small numerical value, the actual non-linear term would then be numerically larger than the viscous term. Of course the mere fact that we have a turbulent flow at all tells us that the non-linear term in the Navier–Stokes equation is much larger than the viscous term (and in principle this ratio could be varied up to infinity). Thus we inevitably have the problem of a wildly divergent primitive perturbation series when we expand the Navier–Stokes equation.

In this section we shall see how the techniques of rearrangement and partial summation (as discussed in the preceding sections) can be applied, in a general way, to the primitive perturbation expansion of the Navier–Stokes equaton. This was first done by Wyld (1961), who overcame technical difficulties involving the double counting of diagrams by using methods which were only valid for a simplified scalar analogue of the full Navier–Stokes equation. Later, Lee (1965) extended Wyld's analysis to include hydromagnetic turbulence, and introduced a new way of overcoming the double-counting problem which was valid for the Navier–Stokes equation. We shall base our present treatment very largely on the work reported in these two references.

However, before we begin, it will be as well to emphasize that the procedure due to Wyld is not itself a theory (any more than the analogous Dyson equations in quantum field theory (March *et al.* 1967; Reichl 1980)). Our purpose in treating it here is really to emphasize its kinship (as it were) with many-body theories known to be successful in other areas. In this sense it should be regarded as being (like the other material in this chapter) merely prefatory to the actual turbulence theories to be presented in the following two chapters. Indeed, we shall close the work of this section by showing how the pioneering direct-interaction approximation (Kraichnan 1959) can be recovered from the general diagrammatic analysis of the primitive expansion.

5.5.1 *The zero-order isotropic propagators*

Our starting point is the solenoidal Navier–Stokes equation in wavenumber space, with the addition of a stirring force. We write eqn (4.81) in the slightly modified form as

$$\left(\frac{\partial}{\partial t} + \nu k^2\right) u_\alpha(\mathbf{k}, t) = \lambda M_{\alpha\beta\gamma}(\mathbf{k}) \sum_j u_\beta(\mathbf{j}, t) u_\gamma(\mathbf{k} - \mathbf{j}, t) + D_{\alpha\beta}(\mathbf{k}) f_\beta(\mathbf{k}, t), \quad (5.56)$$

where we have added λ as a bookkeeping parameter in front of the non-linear term. The factor $D_{\alpha\beta}(\mathbf{k})$, as defined by eqn (2.78), ensures that the stirring forces satisfy the continuity condition (2.74) for abitrary $\mathbf{f}(\mathbf{k}, t)$. The stirring forces (per unit mass of fluid $\mathbf{f}(\mathbf{k}, t)$ are taken to be random and to have a Gaussian distribution with zero mean (see Section 4.3.2, for a discussion).

If we switch the non-linear term off, we can introduce the linear response function $G_0(k; t, t')$ for isotropic fields. This is essentially a Green function (see Appendix D), and can be introduced through the zero-order Green tensor which satisfies the linearized Navier–Stokes equation:

$$\left(\frac{\partial}{\partial t} + vk^2\right) G_{\alpha\beta}^{(0)}(\mathbf{k}; t, t') = D_{\alpha\beta}(\mathbf{k})\delta(t - t'), \tag{5.57}$$

where the $D_{\alpha\beta}(\mathbf{k})$ ensures the correct solenoidal structure for $G_{\alpha\beta}^{(0)}(\mathbf{k}; t, t')$ such that

$$G_{\alpha\beta}^{(0)}(\mathbf{k}; t, t') = D_{\alpha\beta}(\mathbf{k})G_0(k; t, t'). \tag{5.58}$$

(Comparison with eqn (2.97) for the energy spectrum tensor shows that (5.58) is the correct form for an isotropic solenoidal tensor.)

With the substitution of (5.58), eqn (5.57) takes the simpler form

$$\left(\frac{\partial}{\partial t} + vk^2\right) G_0(k; t, t') = \delta(t - t'), \tag{5.59}$$

which, for future use, we can also write in the more symbolic fashion

$$L_0 G_0(k; t, t') = \delta(t - t'). \tag{5.60}$$

It then follows that the solution of eqn (5.56), with the non-linear term put equal to zero and hence the response of the fluid system determined solely by viscous effects, is

$$u_\alpha^{(0)}(\mathbf{k}, t) = \int^t dt' G_{\alpha\beta}^{(0)}(\mathbf{k}; t, t') f_\beta(\mathbf{k}, t')$$

$$= D_{\alpha\beta}(\mathbf{k}) \int^t dt' G_0(k; t, t') f_\beta(\mathbf{k}, t'). \tag{5.61}$$

We can also use this function to relate the velocity field, associated with mode $\mathbf{k}$, to itself at two different times, provided that such variations are due to the action of the fluid viscosity. We do this by setting $\lambda = 0$ and $\mathbf{f}(\mathbf{k}) = 0$ in equation (5.56), and solving for $u_\alpha(\mathbf{k}, t)$, given $u_\alpha(\mathbf{k}, t')$ for $t > t'$, to obtain

$$u_\alpha(\mathbf{k}, t) = G_0(k; t, t')u_\alpha(\mathbf{k}, t')$$

$$= \exp\{-vk^2(t - t')\}u_\alpha(\mathbf{k}, t'). \tag{5.62}$$

Evidently this relationship justifies us referring to $G_0(k; t, t')$ as the zero-order or 'viscous' propagator.

Lastly, we can also extend the Fourier transform from the wavenumber–time domain (see eqn (2.71)) to the wavenumber–frequency domain by using

$$u_\alpha(\mathbf{x}, t) = \sum_{\mathbf{k}} \sum_{\omega} u_\alpha(\mathbf{k}, \omega) \exp\{i\mathbf{k} \cdot \mathbf{x} + i\omega t\} \tag{5.63}$$

where ω is the angular frequency. If we substitute (5.63), instead of (2.71), for $u_\alpha(\mathbf{x}, t)$ into eqn (2.75), then the Navier–Stokes equation becomes

$$(i\omega + vk^2)u_\alpha(\mathbf{k}, \omega) = \lambda M_{\alpha\beta\gamma}(\mathbf{k}) \sum_{\mathbf{j}} \sum_{\omega'} u_\beta(\mathbf{j}, \omega')u_\gamma(\mathbf{k} - \mathbf{j}, \omega - \omega') +$$

$$+ D_{\alpha\beta}(\mathbf{k})f_\beta(\mathbf{k}, \omega), \tag{5.64}$$

where $f_\beta(\mathbf{k}, \omega)$ is the Fourier transform of $f_\beta(\mathbf{k}, t)$, with respect to t, and is defined by a straightforward extension of (5.63). Clearly, if we carry out the same procedure for eqn (5.59), while remembering that the Fourier transform of the delta function is unity, we can easily see that the Fourier transform of the zero-order propagator is given by

$$G_0(\mathbf{k}, \omega) = \frac{1}{i\omega + vk^2}. \tag{5.65}$$

What we are now interested in is the renormalization of $G_0(k, \omega)$ to the exact form $G(k, \omega)$, which takes into account the collective interactions among the modes of the system when the non-linear term is switched on. This will be the subject of the next three subsections.

5.5.2   The primitive perturbation expansion

The underlying physics of the perturbation expansion of the Navier–Stokes equation, when subject to a random force, has been discussed in Section 3.5.2. Accordingly, we shall concentrate here on treating the mathematical formalities. We begin by considering the specification of the stirring force autocorrelation for non-stationary forces. The appropriate generalization of eqn (3.94) can readily be seen to be

$$\left(\frac{L}{2\pi}\right)^3 \langle f_\alpha(\mathbf{k}, t)f_\beta(-\mathbf{k}, t')\rangle = D_{\alpha\beta}(\mathbf{k})w(k; t, t'). \tag{5.66}$$

The Fourier transformation from $(\mathbf{k}, t)$- to $(\mathbf{k}, \omega)$-space can then be accomplished by a suitable extension of eqn (5.63), yielding for (5.66)

$$\left(\frac{L}{2\pi}\right)^3 \left(\frac{T}{2\pi}\right) \langle f_\alpha(\mathbf{k}, \omega)f_\beta(-\mathbf{k}, \omega')\rangle = D_{\alpha\beta}(\mathbf{k})w(k; \omega, \omega'). \tag{5.67}$$

Note that T is the time period for Fourier series, and tends to infinity in order to allow us to make the transition to Fourier transforms. As we shall only use this representation for analytical convenience, and will ultimately change back to the $(\mathbf{k}, t)$ representation, we do not need to worry too much about normalizing factors and from now on we shall take the factor $T/2\pi$ to be unity. At the same time, we should note that we cannot be so cavalier about the space dependence, as we will remain in k-space and will wish to make comparisons between theoretical and experimental results for wavenumber spectra.

Also, using the same Fourier transformation and recalling the convolution theorem, we can write eqn (5.61) for the zero-order velocity field in the form

$$u_\alpha^{(0)}(\mathbf{k}, \omega) = D_{\alpha\beta}(\mathbf{k})G_0(k, \omega)f_\beta(\mathbf{k}, \omega), \qquad (5.68)$$

which then provides the zero-order term in the perturbation expansion

$$u_\alpha(\mathbf{k}, \omega) = u_\alpha^{(0)}(\mathbf{k}, \omega) + \lambda u_\alpha^{(1)}(\mathbf{k}, \omega) + \lambda^2 u_\alpha^{(2)}(\mathbf{k}, \omega) + \cdots +$$
$$+ \lambda^n u_\alpha^{(n)}(\mathbf{k}, \omega) + \cdots \qquad (5.69)$$

As we have seen in Section 3.5.2, we can solve the Navier–Stokes equation iteratively to express the various higher-order coefficients on the r.h.s. of (5.69) in terms of $u_\alpha^{(0)}(\mathbf{k}, \omega)$. In practice, we shall treat the problem statistically, and therefore we will be interested in equivalent problem of expressing the exact correlation $Q_{\alpha\beta}(\mathbf{k}; \omega, \omega')$ in terms of the zero-order result given (after substitution from (5.68)) by

$$\langle u_\alpha^{(0)}(\mathbf{k}, \omega)u_\beta^{(0)}(-\mathbf{k}, \omega')\rangle$$
$$= D_{\alpha\gamma}(\mathbf{k})D_{\beta\sigma}(-\mathbf{k})G_0(k, \omega)G_0(-k, \omega')\langle f_\gamma(\mathbf{k}, \omega)f_\sigma(-\mathbf{k}, \omega')\rangle$$
$$= D_{\alpha\gamma}(\mathbf{k})D_{\beta\sigma}(\mathbf{k})D_{\gamma\sigma}(\mathbf{k})G_0(k, \omega)G_0(k, \omega')\left(\frac{2\pi}{L}\right)^3 w(k; \omega, \omega')$$
$$= \left(\frac{2\pi}{L}\right)^3 D_{\alpha\beta}(\mathbf{k})G_0(k, \omega)G_0(k, \omega')w(k; \omega, \omega'), \qquad (5.70)$$

where we have used the invariance of $D_{\alpha\beta}(\mathbf{k})$ and $G_0(k, \omega)$ under interchange of $\mathbf{k}$ and $-\mathbf{k}$, along with the contraction property

$$D_{\alpha\gamma}(\mathbf{k})D_{\gamma\beta}(\mathbf{k}) = D_{\alpha\beta}(\mathbf{k}).$$

The zero-order spectral density function Q_0 can be defined by analogy with eqns (2.88) and (2.97) for the exact spectral density Q for homogeneous isotropic turbulence:

$$\left(\frac{L}{2\pi}\right)^3 \langle u_\alpha^{(0)}(\mathbf{k}, \omega)u_\beta^{(0)}(-\mathbf{k}, \omega')\rangle = D_{\alpha\beta}(\mathbf{k})Q_0(k; \omega, \omega'). \qquad (5.71)$$

Combining this definition with eqn (5.70), we obtain the following simple relationship between the spectral density function of the zero-order velocity and the corresponding spectral function for the random stirring forces:

$$Q_0(k; \omega, \omega') = G_0(k, \omega)G_0(k, \omega')w(k; \omega, \omega'). \qquad (5.72)$$

Our next step is to set up the perturbation expansion which will relate the exact Q to the zero-order form as given by eqn (5.72). We do this by taking eqns (2.88) and (2.97), making the obvious extension to $(\mathbf{k}, \omega)$ space, and substituting eqn (5.69) for the velocity field to obtain

$$D_{\alpha\beta}(\mathbf{k})Q(k;\omega,\omega') = \left(\frac{2\pi}{L}\right)^3 \langle u_\alpha(\mathbf{k},\omega)u_\beta(-\mathbf{k},\omega')\rangle$$

$$= \left(\frac{2\pi}{L}\right)^3 \{\langle u_\alpha^{(0)}(\mathbf{k},\omega)u_\beta^{(0)}(-\mathbf{k},\omega')\rangle +$$

$$+ \lambda^2[\langle u_\alpha^{(0)}u_\beta^{(2)}\rangle + \langle u_\alpha^{(1)}u_\beta^{(1)}\rangle + \langle u_\alpha^{(2)}u_\beta^{(0)}\rangle] + O(\lambda^4)\}$$

$$= D_{\alpha\beta}(\mathbf{k})Q_0(k;\omega,\omega') + \left(\frac{2\pi}{L}\right)^3 \{\lambda^2[\langle u_\alpha^{(0)}u_\beta^{(2)}\rangle +$$

$$+ \langle u_\alpha^{(1)}u_\beta^{(1)}\rangle + \langle u_\alpha^{(2)}u_\beta^{(0)}\rangle] + O(\lambda^4)\}, \qquad (5.73)$$

where some frequency variables have been suppressed in the interests of conciseness and, for the same reason, we only give explicit results for second order. Note that it follows from eqn (5.68) that the $u^{(0)}$ are (like the stirring forces) normally distributed with zero mean. Hence, terms like $\langle u^{(0)}u^{(1)}\rangle$ vanish, as they are odd functionals in the $u^{(0)}$ (see eqns (3.103a)–(3.103c)).

Now we need the relationships between the various coefficients of the perturbation expansion. As reference to Section 3.5.2 will show, we obtain these by substituting the perturbation expansion into the equation of motion and equating coefficients at each order.

We shall now consider this procedure in detail (up to second order) as follows. First, we take eqn (5.64) and invert the linear operator on the l.h.s. by using (5.65). Then, substituting from (5.68) for the zero-order velocity field, we can write

$$u_\alpha(\mathbf{k},\omega) = u_\alpha^{(0)}(\mathbf{k},\omega) + \lambda G_0(k,\omega)M_{\alpha\beta\gamma}(\mathbf{k}) \sum_{\mathbf{j}}\sum_{\omega'} u_\beta(\mathbf{j},\omega')u_\gamma(\mathbf{k}-\mathbf{j},\omega-\omega')$$

$$(5.74)$$

as the equation of motion.

We now substitute the perturbation expansion, as given by (5.69), for the exact velocity field on each side of eqn (5.74) to obtain

$$u_\alpha^{(0)}(\mathbf{k},\omega) + \lambda u_\alpha^{(1)}(\mathbf{k},\omega) + \lambda^2 u_\alpha^{(2)}(\mathbf{k},\omega) + \cdots$$

$$= u_\alpha^{(0)}(\mathbf{k},\omega) + \lambda G_0(k,\omega)M_{\alpha\beta\gamma}(\mathbf{k}) \times$$

$$\times \sum_{\mathbf{j}}\sum_{\omega'} \{u_\beta^{(0)}(\mathbf{j},\omega')u_\gamma^{(0)}(\mathbf{k}-\mathbf{j},\omega-\omega') +$$

$$+ \lambda[u_\beta^{(0)}(\mathbf{j},\omega')u_\gamma^{(1)}(\mathbf{k}-\mathbf{j},\omega-\omega') +$$

$$+ u_\beta^{(1)}(\mathbf{j},\omega')u_\gamma^{(0)}(\mathbf{k}-\mathbf{j},\omega-\omega')] + O(\lambda^2)\}. \qquad (5.75)$$

Then, equating coefficients at each order of λ, we find

$$u_\alpha^{(1)}(\mathbf{k}, \omega) = G_0(k, \omega) M_{\alpha\beta\gamma}(\mathbf{k}) \sum_{\mathbf{j}} \sum_{\omega'} u_\beta^{(0)}(\mathbf{j}, \omega') u_\gamma^{(0)}(\mathbf{k} - \mathbf{j}, \omega - \omega'), \quad (5.76a)$$

$$u_\alpha^{(2)}(\mathbf{k}, \omega) = 2G_0(k, \omega) M_{\alpha\beta\gamma}(\mathbf{k}) \sum_{\mathbf{j}} \sum_{\omega'} u_\beta^{(1)}(\mathbf{j}, \omega') u_\gamma^{(0)}(\mathbf{k} - \mathbf{j}, \omega - \omega'), \quad (5.76b)$$

and so on. It should be noted that the factor of 2 in (5.76b) arises from the fact that the terms $u^{(0)}u^{(1)}$ and $u^{(1)}u^{(0)}$ on the r.h.s. of eqn (5.75) can be added together and their sum put equal to $2u^{(1)}u^{(0)}$, with variables being renamed as appropriate. This step relies on the fact that both $\mathbf{j}$ and $\mathbf{k} - \mathbf{j}$ are dummy variables. Note that, if this is not immediately obvious, then $\mathbf{k} - \mathbf{j}$ can be replaced by $\mathbf{l}$, along with the condition that $\mathbf{j} + \mathbf{l} = \mathbf{k}$.

The coefficient $u^{(2)}$ can be expressed in terms of $u^{(0)}$, through equation (5.76a). An intermediate stage is to rewrite the expression for $u^{(1)}$ as

$$u_\beta^{(1)}(j, \omega') = G_0(j, \omega') M_{\beta\rho\sigma}(\mathbf{j}) \sum_{\mathbf{l}} \sum_{\omega''} u_\rho^{(0)}(\mathbf{l}, \omega'') u_\sigma^{(0)}(\mathbf{j} - \mathbf{l}, \omega' - \omega''), \quad (5.77)$$

followed by substitution into (5.76b) with the result

$$u_\alpha^{(2)}(\mathbf{k}, \omega) = 2G_0(k, \omega) M_{\alpha\beta\gamma}(\mathbf{k}) \sum_{\mathbf{j}} \sum_{\omega'} G_0(j, \omega') M_{\beta\rho\sigma}(\mathbf{j}) \times$$

$$\times \sum_{\mathbf{l}} \sum_{\omega''} u_\rho^{(0)}(\mathbf{l}, \omega'') u_\sigma^{(0)}(\mathbf{j} - \mathbf{l}, \omega' - \omega'') u_\gamma^{(0)}(\mathbf{k} - \mathbf{j}, \omega - \omega'). \quad (5.78)$$

Similar procedures can be used for all higher orders, although here we shall restrict ourselves to second order in the expansion parameter λ.

Using eqns (5.76a) and (5.78), the perturbation expansion (5.73) for the exact correlation $Q(k, \omega)$ can be worked out explicitly in terms of the zero-order correlation $Q_0(k, \omega)$. Details are given in Appendix G, but the general ideas behind the derivation can be followed quite readily here.

First, we should note that the only non-trivial operation is to evaluate the second-order terms on the r.h.s. of (5.73) as fourth-order correlations of the zero-order velocity field. Unlike the quasi-normality hypothesis discussed in Section 2.8.2, this is a completely rigorous procedure. The random stirring forces are specified to have a Gaussian distribution and, as we have pointed out previously, it follows from eqn (5.68) that the zero-order velocity field must also be normally distributed. So, just as in Section 2.8.2, each $\langle u^0 u^0 u^0 u^0 \rangle$ can be factored into three products of pair correlations $\langle u^0 u^0 \rangle \langle u^0 u^0 \rangle$, one of which contributes zero. The other two can be written as $2\langle u^0 u^0 \rangle \langle u^0 u^0 \rangle$, with appropriate renaming of dummy variables.

The other main points to watch out for are as follows.

(a) The homogeneity property will lead (as in eqn (2.93) applied to the zero-order field) to a delta function, which can then be eliminated, along with one of the summations, from the r.h.s. of (5.73).

(b) The formation of the product $Q_0 Q_0$ will result (see eqn (5.71)) in the introduction of two factors of $(2\pi/L)^3$ on the r.h.s. of (5.73). One of these factors will cancel with the corresponding factor on the l.h.s. The other will be absorbed, along with the remaining summation, into an integration when we take the infinite system limit.

With all these considerations in mind (or in detail from Appendix G), the expansion for the exact correlation function in terms of zero-order correlation functions can be written down, to second order, as

$$D_{\alpha\beta}(\mathbf{k})Q(k;\omega,\omega')$$

$$= D_{\alpha\beta}(\mathbf{k})Q_0(k;\omega,\omega') +$$

$$+ \lambda^2 \Bigg[4G_0(k,\omega')M_{\beta\delta\gamma}(-\mathbf{k}) \int d^3j \int d\omega'' \int d\omega''' G_0(j,\omega'') \times$$

$$\times M_{\delta\rho\sigma}(\mathbf{j})D_{\alpha\rho}(\mathbf{k})D_{\gamma\sigma}(\mathbf{k}+\mathbf{j}) \times$$

$$\times Q_0(k;\omega,\omega''')Q_0(|\mathbf{k}+\mathbf{j}|;\omega'-\omega'',\omega''-\omega''') +$$

$$+ 2G_0(k,\omega)M_{\alpha\delta\gamma}(\mathbf{k}) \int d^3j \int d\omega'' \int d\omega''' G_0(k,\omega') \times$$

$$\times M_{\beta\rho\sigma}(-\mathbf{k})D_{\delta\sigma}(\mathbf{j})D_{\gamma\rho}(\mathbf{k}-\mathbf{j}) \times$$

$$\times Q_0(|\mathbf{k}-\mathbf{j}|;\omega-\omega'',\omega''')Q_0(j;\omega'',\omega'-\omega''') +$$

$$+ 4G_0(k,\omega)M_{\alpha\delta\gamma}(\mathbf{k}) \int d^3j \int d\omega'' \int d\omega''' G_0(j,\omega'') \times$$

$$\times M_{\delta\rho\sigma}(\mathbf{j})D_{\beta\rho}(\mathbf{k})D_{\sigma\gamma}(\mathbf{k}-\mathbf{j}) \times$$

$$\times Q_0(k;\omega',\omega''')Q_0(|\mathbf{k}-\mathbf{j}|;\omega-\omega'',\omega''-\omega''') \Bigg] +$$

$$+ O(\lambda^4). \tag{5.79}$$

Two general points should be noted about this equation. First, it is still formally an expansion for the correlation tensor $Q_{\alpha\beta}(\mathbf{k};\omega,\omega')$, albeit for the isotropic case. However, it is easy to convert it to an expression for the correlation function (or spectral density function) $Q(k;\omega,\omega')$, merely by putting $\alpha = \beta$, summing over the repeated subscript, and dividing both sides of eqn (5.79) by $\operatorname{tr} D_{\alpha\beta}(\mathbf{k}) = 2$. Thus we can regard it as also being an equation for the correlation function.

Second, we have changed from summations to integrations over the angular frequency ω. This is a purely formal step, and has been taken in the interests of consistency with the way in which wavenumbers are treated. It has no real significance as, when we deal with the application of theories in later chapters, we shall invariably transform back into the time domain.

5.5.3   *Graphical representation of the perturbation series*

Equation (5.79) represents the lowest non-trivial order of the perturbation expansion for the exact pair correlation of fluctuating velocities. Clearly we can carry on the procedures given, so that we can obtain terms of any order. Thus the exact correlation $Q(k; \omega, \omega')$ can be rigorously specified in terms of the zero-order (or 'viscous') propagator $G_0(k, \omega)$ and the zero-order pair correlation $Q_0(k; \omega, \omega')$, albeit by means of a divergent series.

Our purpose now is to consider ways in which we can extract some useful information from this divergent series, and we begin by introducing diagrams, as in the preceding sections on many-body problems. We do this by representing the three main constituents of the expansion, $u^{(0)}$, G_0, and $M_{\alpha\beta\gamma}(\mathbf{k})$ as follows:

$$\text{full line} \quad \leftrightarrow u^{(0)}$$

$$\text{broken line} \quad \leftrightarrow G_0$$

$$\text{point (vertex)} \leftrightarrow M.$$

These basic representations will be referred to as 'elements'.

We now draw diagrams which will represent the terms of the perturbation series, as given by eqn (5.69), to all orders. The result is shown (to third order) in Fig. 5.8.

Beginning with $u^{(1)}$, we can usefully compare the diagram, as shown in Fig. 5.8, with its equivalent algebraic form, as given by eqn (5.76a). This is, in fact, the prototype non-linear term of the Navier–Stokes equation, as expressed in zero-order quantities. Note the general characteristic that three elements join up at a vertex, and that these three elements always show wavenumber conservation. That is, the wavenumber of the element to the left of the vertex is equal to the sum of the two wavenumbers of the elements on the right. In this particular case, we have $\mathbf{k}$, $\mathbf{j}$, and $\mathbf{k} - \mathbf{j}$.

Equation (5.78) reveals that $u^{(2)}$ contains two M factors and hence the corresponding diagram in Fig. 5.8 has two point vertices. The factor of 2 is also shown explicitly. The three elements at each vertex show wavenumber conservation, the first being $(\mathbf{k}, \mathbf{j}, \mathbf{k} - \mathbf{j})$ and the second being $(\mathbf{j}, \mathbf{l}, \mathbf{j} - \mathbf{l})$. Note that $G_0(j, \omega')$ is the linking element between the two vertices.

Superficially, the order of each term in the expansion is the same as the number of M factors, which is the same as the number of vertices in the corresponding diagram. Thus, although we only have algebraic results to second order, it would be quite easy to devise general rules for drawing the diagrams of any order. For instance, at third order there are three vertices and clearly there are two ways in which we might connect them up (i.e. in series or in parallel; see Fig. 5.8).

Factors of 2 arise firstly because of the expansion of the original quadratic terms of any order to the zero-order velocity field. The general rule is that we

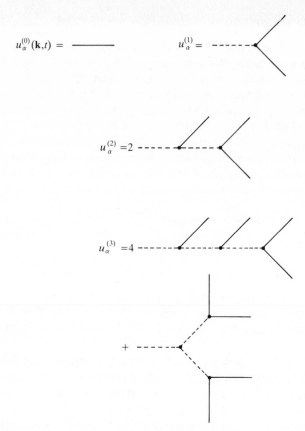

Fig. 5.8. Diagrams corresponding to terms in the perturbation series (5.69) for the velocity field.

multiply a diagram by 2 for (a) each vertex which has only one element $u^{(0)}$ connected to it and (b) each asymmetric branching of the diagram. Note that the first third-order diagram of Fig. 5.8 has two such vertices but the second diagram has none. Also, the branching in the second third-order diagram is symmetrical.

Our next step is to use these diagram elements to set up a graphical representation for the primitive perturbation expansion of the correlation function. Just as the series for $\mathbf{u}$ is made up from $\mathbf{u}^{(0)}$, G_0, and M, the series for Q will be made up from Q_0, G_0, and M. Clearly what we need is the graphical equivalent of the way in which we factored a fourth-order moment into two non-vanishing products of pair-correlations in order to derive eqn (5.79).

The general procedure is obtained by taking any pair of diagrams and

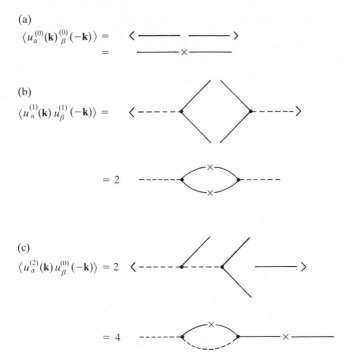

(a)
$$\langle u_\alpha^{(0)}(\mathbf{k})\, u_\beta^{(0)}(-\mathbf{k})\rangle =$$

(b)
$$\langle u_\alpha^{(1)}(\mathbf{k})\, u_\beta^{(1)}(-\mathbf{k})\rangle =$$

(c)
$$\langle u_\alpha^{(2)}(\mathbf{k})\, u_\beta^{(0)}(-\mathbf{k})\rangle = 2$$

Fig. 5.9. Examples of the averaging process on the zero-order field, leading to diagrams in the expansion for $Q_{\alpha\beta}(\mathbf{k};\omega,\omega')$.

placing them with their branches facing each other. Then we join up the emergent full lines (i.e. the $u^{(0)}$) in all possible ways. This is the same as selecting all n possible pairs of velocities from the set making up the moment of order $2n$, and we indicate that two velocities are correlated by placing a cross at their junction point. The numerical factors are obtained by multiplying each diagram by the number of different ways in which the full lines can be joined up to make duplicate diagrams.

In order to illustrate this process, we first consider the trivial example of the diagram for Q_0, which is shown in Fig. 5.9(a). Evidently, we replace the angular brackets, which denote the operation of taking an average, by the cross at the junction point, indicating that an average has been taken and that the velocities are now correlated. It should also be noted that each emergent line (full or broken) is labelled by $\mathbf{k}$, if to the left of the diagram, and by $-\mathbf{k}$, if to the right.

At second order, we show two illustrative examples in Fig. 5.9. The first of these (Fig. 5.9(b)) is the graphical representation of the middle second-order term on the r.h.s. of eqn (5.79). Therefore we can see that it stands for (in a very symbolic notation)

$$Q_{\alpha\beta}(\mathbf{k};\omega,\omega')=$$

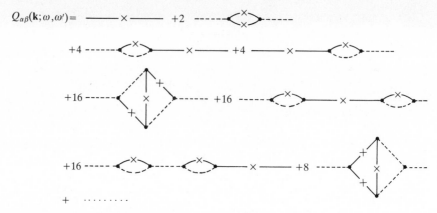

Fig. 5.10. Diagrams in the primitive perturbation expansion for $Q_{\alpha\beta}(\mathbf{k};\omega,\omega')$ showing four of the 29 fourth-order diagrams.

$$2G_0(k)M(k)Q_0(j)Q_0(|\mathbf{k}-\mathbf{j}|)M(-\mathbf{k})G_0(|-\mathbf{k}|).$$

Similarly, Fig. 5.9(c) stands for the last second-order term on the r.h.s. of eqn (5.79), and can be written symbolically as

$$4G_0(k)M(k)Q_0(|\mathbf{k}-\mathbf{j}|)G_0(j)M(\mathbf{j})Q_0(|-\mathbf{k}|),$$

where again we suppress the integral signs along with the variables and indices (with the exception of the wavenumber arguments).

The third of the second-order diagrams is simply a mirror image of Fig. 5.9(c), and is shown in Fig. 5.10 where we write down the series for the exact correlation function to fourth order. It should be noted that we only include a few of the fourth-order diagrams. We do this so that we can present the very simple ideas involved in renormalizing the perturbation series without being distracted by complications. Any reader who wishes to check the detailed bookkeeping will find full sets of diagrams, up to sixth order, in Lee (1965).

We now face the problem of summing the perturbation series for $Q_{\alpha\beta}(\mathbf{k};\omega, \omega')$ to all orders. As in previous sections, this will involve techniques of partial summation and, as before, we shall find it convenient to introduce new diagram elements:

thick full line $\leftrightarrow u$ (exact velocity field)

thick broken line $\leftrightarrow G$ (renormalized propagator)

open circle $\leftrightarrow$ (renormalized vertex).

It follows that when two thick full lines are joined by a cross, the resulting element stands for the exact pair-correlation of the fluctuating velocity field. The latter two elements will be explained as they arise.

Lastly, we shall find it helpful to divide the problem into two parts. That is, we shall write the correlation tensor as

$$Q_{\alpha\beta}(\mathbf{k};\omega,\omega') = Q_{\alpha\beta}(\mathbf{k};\omega,\omega')_{\mathrm{A}} + Q_{\alpha\beta}(\mathbf{k};\omega,\omega')_{\mathrm{B}}, \qquad (5.80)$$

where $Q_{\alpha\beta}(\mathbf{k};\omega,\omega')_{\mathrm{A}}$ is the sum of certain diagrams which we shall call class A, and $Q_{\alpha\beta}(\mathbf{k};\omega,\omega')_{\mathrm{B}}$ is the sum of all the other diagrams. We shall introduce class A diagrams in the next subsection.

5.5.4 Class A diagrams: the renormalized propagator

We follow Wyld (1961) in defining class A diagrams as those diagrams which can be split into two pieces by cutting a single Q_0 line. In Fig. 5.10, they consist of Q_0 itself, the second and third diagrams of the second order, and the second and third diagrams of the fourth order. There are, of course, many more fourth-order diagrams which we have not shown.

In order to see why this classification is significant, we first examine the zero-order case. From eqn (5.72), we see that Q_0 can be expressed in terms of two zero-order propagators acting on the spectrum of the stirring forces $w(k;\omega,\omega')$. We show this graphically in Fig. 5.11(a).

Next take the second-order class A diagrams. The first of these can have the Q_0 line emerging to the right factored into $G_0 G_0 w$, and, for the second, we can do the same thing to the emergent Q_0 line on the left. The result for both diagrams is shown in Fig. 5.11(b).

Now, let us summarize what we have done so far. At zero order, we obtain w with a G_0 on each side. At second order, w has a G_0 on one side and a diagram which connects like a G_0 on the other. We can carry this procedure on to fourth and higher orders. The result is that we can write the total contribution from class A diagrams to the correlation function as a generalization of eqn (5.72):

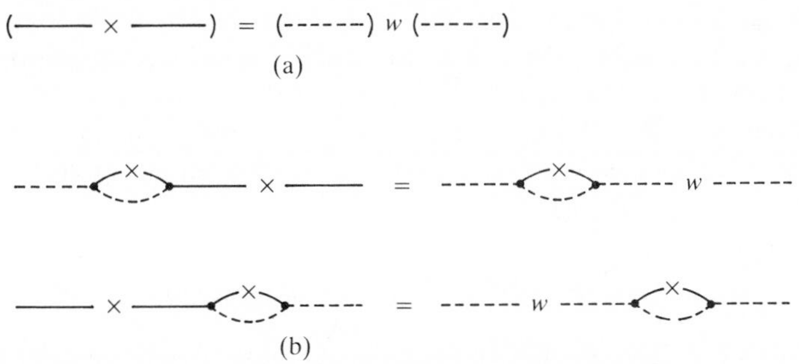

$$(\text{———} \times \text{———}) = (\text{------}) \, w \, (\text{------})$$
$$(a)$$

$$\text{-----}\langle\times\rangle\text{———} \times \text{———} = \text{-----}\langle\times\rangle\text{-----} \, w \, \text{------}$$

$$\text{———} \times \text{———}\langle\times\rangle\text{-----} = \text{------} \, w \, \text{-----}\langle\times\rangle\text{-----}$$
$$(b)$$

Fig. 5.11. Decomposition of class A diagrams into 'propagator-like' parts acting on the stirring force spectrum $h(k)$: (a) zero order; (b) second order.

$$Q_{\alpha\beta}(k;\omega,\omega')_A = \quad \text{------} \; w(k) \; \text{------}$$

(a)

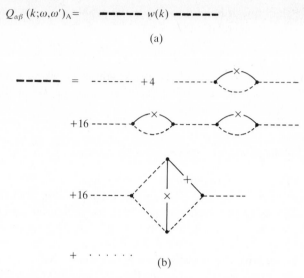

(b)

Fig. 5.12. Summation of class A diagrams: (a) sum of contributions from class A diagrams to $Q_{\alpha\beta}(\mathbf{k};\omega,\omega')$ in terms of the exact propagator $G(k,\omega)$; (b) expansion of the exact propagator $G(k,\omega)$ showing two of the nine fourth-order diagrams.

$$Q(k;\omega,\omega')_A = G(k,\omega)G(k,\omega')w(k;\omega,\omega'), \qquad (5.81)$$

where $G(k,\omega)$ is the renormalized propagator. This is shown graphically in Fig. 5.12(a), with the first few terms of the infinite series for G being given in Fig. 5.12(b).

In other words, substitute from Fig. 5.12(b) for the thick broken lines in Fig. 5.12(a), and all the class A diagrams are generated. That is, replacing G_0 by G amounts to a partial summation of the diagrams.

We should note that eqn (5.81) not only gives the direct contribution of the stirring forces to the correlation of velocities, but also (in effect) defines the exact propagator G. The same result was first obtained by Kraichnan as part of the direct-interaction approximation, which is based a on definition of the propagator (or response function) in terms of the relationship between the velocity field and the stirring forces. A discussion of Kraichnan's theory will be found in Chapter 6, where the basic ansatz is given as eqn (6.3) and the equivalent of eqn (5.81) is eqn (6.32).

5.5.5   Class B diagrams: renormalized perturbation series

Referring to Fig. 5.10, the class B diagrams are those which cannot be divided into two parts merely by cutting a single Q_0 line, that is, the first second-order diagram, the first, and fourth (and many other) fourth-order diagrams, and so on to higher orders.

Here we extend the ideas used to tackle the class A diagrams. There we saw that certain diagram parts were 'propagator like', that is, they connected like G_0. From purely topological considerations, if we can renormalize G_0 by adding up all diagrams which connect like G_0, it follows that the 'bare' vertex can also be renormalized by adding up all diagrams which connect like a vertex.

As an example, let us consider the fourth of the fourth-order diagrams shown in Fig. 5.10:

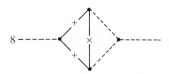

The part

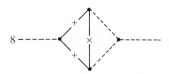

connects just like a point vertex. That is, we can connect three lines to it. If we replaced this part by a point vertex[2], then this particular fourth-order diagram would be reduced to the first second-order diagram in Fig. 5.10.

An alternative way of putting this is to define an expansion for the modified (renormalized) vertex as follows:

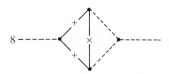

Then, (as an example) suppose that we replace one vertex of the class B second-order diagram by the renormalized vertex and substitute the above expansion, the result is that many of the higher-order diagrams would be generated this way:

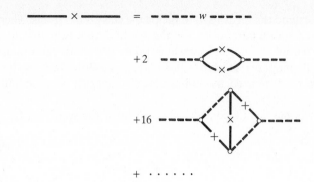

Fig. 5.13. Diagrams corresponding to an integral equation for $Q(k; \omega, \omega')$.

Therefore the key to the class B diagrams is as follows.

(1) Find those diagrams which cannot be reduced to a lower order by replacing diagram parts.
(2) Call these the irreducible diagrams.
(3) Replace all elements in the irreducible diagrams by their exact or renormalized forms.
(4) Write down all these modified diagrams in order, thus generating a 'renormalized' perturbation expansion.

The result for $Q(k; \omega, \omega')$ is shown in Fig. 5.13, where we have also included the sum of class A diagrams, as given by eqn (5.81) or, graphically by Fig. 5.12.

The same procedure is used for obtaining the renormalized vertex and propagator expansions. That is, we take the 'bare' series for each quantity and replace zero-order elements by renormalized elements, thus generating a renormalized expansion. The results for the renormalized vertex are shown in Fig. 5.14, and those for the renormalized propagator in Fig. 5.15.

We should draw attention to one peculiarity of the diagrams for the renormalized propagator in Fig. 5.15; that is, the propagator emerging to the left of the diagram is unrenormalized. This is a modification, which was introduced by Lee (1965), and is designed to overcome a problem involving the double counting of bare diagrams in the primitive perturbation series (Wyld 1961). It may seem a rather arbitrary modification of the general techniques that we have just been discussing, but this is not really so. If we consider the equation of motion in the (rather symbolic) form

$$L_0 u(k) = \lambda M(k) u(j) u(k - j),$$

where the linear operator $L_0 = \{\partial/\partial t + \nu k^2\}$, then we could carry out all the above procedures on the r.h.s. This would mean that the r.h.s. could be represented by a renormalized perturbation series like that given in Fig. 5.15,

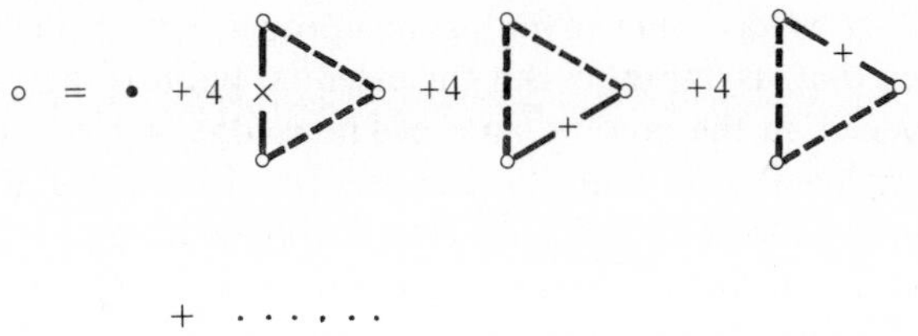

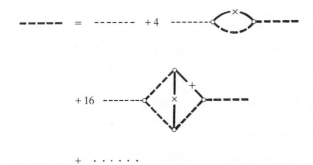

$$+ \cdots \cdots$$

Fig. 5.14. Diagrams corresponding to an integral equation for the renormalized vertex.

$$+ \cdots \cdots$$

Fig. 5.15. Diagrams corresponding to an integral equation for the renormalized propagator $G(k, \omega)$.

but with no emergent propagator to the l.h.s. of the diagrams. However, subsequent inversion of L_0 would result in a zero-order propagator on the l.h.s. of each diagram, hence giving Lee's result.

It should, perhaps, be emphasized that this particular problem need not have arisen in the first place. In Kraichnan's approach, he dealt with the governing equation for the correlation tensor rather than with the correlation tensor itself (see eqn (6.22)). Evidently the l.h.s. is exact in unrenormalized form, and the perturbation series is only used to expand the triple moment on the r.h.s.

5.5.6   Second-order closures

Figures 5.13–5.15, when compared with the bare perturbation series, show a substantial reduction in the number of diagrams needed to describe turbulence (all the more impressive when one consults Wyld (1961) or Lee (1965), where all diagrams are given to fourth and sixth order respectively). The general neatness and self-consistency of the method are certainly very appealing.

However, at the same time we must reiterate that this method is not in itself a theory. We have merely replaced a series which is widely divergent, in any case of interest, by one of unknown properties. Naturally we hope that the renormalized series has some advantageous properties—it might be convergent or, failing that, asymptotic. But the essential point is that we simply do not know. Indeed, in the present state of knowledge, it seems that the only reasonable approach is to truncate the renormalized series at some order where the complications are still tolerable, and then attempt to compute the correlation (or spectrum) and make a comparison with experimental results.

We finish this chapter with two examples of this approach, applied to the Wyld formulation, in which well-known theories can be recovered (Wyld 1961). First we truncate the renormalized perturbation series, as follows.

Figure 5.13: truncate at second order (in number of vertices).
Figure 5.14: truncate at first order (unrenormalized vertex).
Figure 5.15: truncate at zero order ($G \to G_0$, unrenormalized propagator).

This procedure leads to Chandrasekhar's (1955) theory, which is essentially much the same as quasi-normality (see Section 2.8.2) but as applied to two-time (rather than single-time) correlations.

A higher order of approximation is gained by also renormalizing the propagator to second order. That is, we truncate expansions, as follows.

Figure 5.13: truncate at second order.
Figure 5.14: truncate at first order.
Figure 5.15: truncate at second order.

The result is the pioneering direct-interaction approximation (DIA) (Kraichnan 1959) and the relevant diagrams are shown in Fig. 5.16.

We shall discuss the DIA (and various other renormalized perturbation theories) in the following chapters. We conclude here by noting that the DIA is technically (in terms of the Wyld formalism) a second-order closure with line renormalization. There is no vertex renormalization.

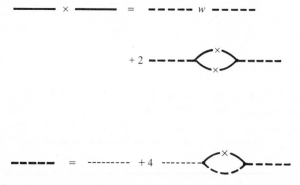

Fig. 5.16. Diagrams corresponding to a second-order closure of the statistical hierarchy, the DIA (Kraichnan 1959).

In contrast, Chandrasekhar's theory has no renormalization at all: the retention of renormalized lines for the correlations in Fig. 5.13 is meaningless in the absence of any partial summations.

Notes

1. In the literature, this reference (Mayer 1937) is normally cited for the introduction of graphs. In fact, they were introduced in the second part of that work (Mayer and Ackermann 1937). A quick look through that paper would miss the graphs altogether, as they are very small and were completely incorporated into the text. It is only in Mayer and Mayer (1940) that something like the familiar modern form of diagram emerges.
2. Intuitively, this step seems arbitrary, if one gives any credence to topological considerations. In fact, as well as vertex parts containing combinations of G_0, G_0, and Q_0, one might also expect vertex parts with both combinations of G_0, Q_0, and Q_0, along with the vertex based on $Q_0 Q_0 Q_0$. The need for three kinds of vertex function was first pointed out by Kraichnan (1972), who noted that the line-renormalized expansion for the non-linear term conserved energy. That is, if we substitute the line-renormalized expansion for $T(k, t)$ on the r.h.s. of eqn (2.126), then this equation would hold, irrespective of the order at which we truncated the expansion. However, according to Kraichnan, the vertex-renormalized expansions of Wyld and Lee do not satisfy this consistency requirement.

   As a result, a more general procedure is needed in which one expands both the correlation and response equations (i.e. eqns (6.22) and (6.21)). Apparently the full analysis has still not been published, but Martin, Siggia, and Rose (1973) have independently noted the need for additional vertex functions.

References

BALESCU, R. (1975). *Equilibrium and nonequilibrium statistical mechanics*. Wiley, New York.

CHANDRASEKHAR, S. (1955). *Proc. R. Soc. A* **233**, 322.

DEBYE, P. and HÜCKEL, E. (1923). *Z. Phys.* **24**, 185.

EDWARDS, S. F. (1964). *J. fluid Mech.* **18**, 239.

HERRING, J. R. (1965). *Phys. Fluids* **8**, 2219.

HIRSCHFELDER, J. O., CURTISS, C. F., and BIRD, R. B. (1954). *Molecular theory of gases and liquids*. Wiley, New York.

KRAICHNAN, R. K. (1959). *J. fluid Mech.* **5**, 497.

—— (1972). In *Statistical mechanics: new concepts, new problems, new approaches* (eds S. A. RICE, K. F. FREED and J. C. LIGHT), p. 208. University of Chicago Press, Chicago, IL.

LEE, L. L. (1965). *Ann. Phys. (NY)* **32**, 292.

MARCH, N. H., YOUNG, W. H. and SAMPANTHAR, S. (1967). *The many-body problem in quantum mechanics*. Cambridge University Press, Cambridge.

MARTIN, P. C., SIGGIA, E. D., and ROSE, H. A. (1973). *Phys. Rev. A* **8**, 423.

MATHEWS, J. and WALKER, R. L. (1965). *Mathematical methods of physics*. Benjamin, New York.

MATTUCK, R. D. (1976). *A guide to Feynman diagrams in the many-body problem*. McGraw-Hill, New York.

MAYER, J. E. (1937). *J. chem. Phys.* **5**, 67.
—— (1950). *J. chem. Phys.* **18**, 1426.
—— and ACKERMANN, P. G. (1937). *J. chem. Phys.* **5**, 74.
—— and MAYER, M. G. (1940). *Statistical mechanics.* Wiley, New York.
REICHL, L. E. (1980). *A modern course in statistical physics.* Edward Arnold, London.
SALPETER, E. E. (1958). *Ann. Phys. (NY)* **5**, 183.
VAN BEIJEREN, H. and DORFMAN, J. R. (1979). *Phys. Rev. A* **19**, 416.
WYLD, H. W. (1961). *Ann. Phys. (NY)* **14**, 143.

6

RENORMALIZED PERTURBATION THEORIES OF THE FIRST KIND

In discussing renormalized perturbation theories (RPTs), we shall divide them into two categories. We shall then devote a separate chapter to each of these categories. Firstly, in this chapter we will concentrate on those theories which are not compatible with the Kolmogorov inertial-range spectrum. These can be thought of as the older theories although, in point of strict chronological order, this is not true of every case. Secondly, in Chapter 7, after first turning our attention to the problems involved in finding an analytical theory which does have the Kolmogorov spectrum as its solution, we then consider those theories which can claim to have achieved that result. Questions concerning the significance of one or other category will be deferred until Chapter 8, where we shall attempt to make an overall assessment of the various theories.

The theories which we are just about to discuss in the present chapter can again be subdivided, but this time on the basis of the general approach to the problem. That is, there are those theories which are based on a direct manipulation of the equations of motion and there are those which work with the probability distribution of the fluctuating velocities. Loosely speaking, we could refer to the first as a 'Chapter 5' type of theory, and the second as a 'Chapter 4' type of theory. We shall discuss the work of Kraichnan (1959), as representing the first type, and the work of Edwards (1964), as representing the second.

Another theory which qualifies as an RPT of the first kind, and which has some particularly interesting features, is the self-consistent field theory (SCF) of Herring (1965, 1966). We discuss this theory in some detail and then conclude the chapter with a brief account of other self-consistent theories (Phythian 1969; Balescu and Senatorski 1970).

6.1 The direct-interaction approximation (DIA)

The original statement of DIA is normally taken to be that of Kraichnan (1959), although details of some derivations can be found in Kraichnan (1958). Our presentation here will differ somewhat from that given originally by Kraichnan as, for pedagogical reasons, we will base our approach on the preliminary discussions of the preceding chapter.

At this stage, it may be helpful if we restate the problem. That is to say, we wish to solve the Fourier-transformed solenoidal Navier–Stokes equation (2.76) for the statistical (mean) properties of a given turbulence field. In fact all our procedures will be sufficiently general to apply to any solenoidal field, but we can reduce the algebraic complexity by restricting our attention to homogeneous isotropic fields. This we shall continue to do, and the solution of the test problems discussed in Section 4.3 remains our immediate objective: we want to solve for the correlation function $Q(k; t, t')$ in terms of prescribed initial conditions or inputs.

It may also be helpful to anticipate the coming discussion and state that the DIA consists of two simultaneous equations for the correlation function and a new quantity, the response function $G(k; t, t')$. The latter quantity is, in the language of Chapter 5, a renormalized propagator. But it was introduced from quite a different point of view, and represented a major step forward for turbulence theory. We begin our treatment of DIA by discussing the introduction of the associated response tensor.

6.1.1 The infinitesimal response tensor

We begin with the Navier–Stokes equation in the form (5.56), where we have added solenoidal stirring forces to the r.h.s. We also recall that the ordering parameter λ is equal to unity. Now consider the effect of a small change in the stirring forces:

$$f_\alpha(\mathbf{k}, t) \to f_\alpha(\mathbf{k}, t) + \delta f_\alpha(\mathbf{k}, t). \tag{6.1}$$

In turn, this can be expected to produce a small change in the velocity field:

$$u_\alpha(\mathbf{k}, t) \to u_\alpha(\mathbf{k}, t) + \delta u_\alpha(\mathbf{k}, t). \tag{6.2}$$

Kraichnan introduced the relationship between these two infinitesimal changes as

$$\delta u_\alpha(\mathbf{k}, t) = \int_{-\infty}^{t} \hat{G}_{\alpha\beta}(\mathbf{k}; t, t') \delta f_\beta(\mathbf{k}, t') \, dt', \tag{6.3}$$

where $\hat{G}_{\alpha\beta}(\mathbf{k}; t, t')$ is the infinitesimal response tensor. We shall usually abbreviate this to be 'response tensor'. Note also that we not using the notation of Kraichnan (1958, 1959), where this concept was introduced, but rather the later notation used by Kraichnan when reworking the DIA in a Lagrangian-history form: this will be one of the subjects of the next chapter.

The importance of the infinitesimal change in the stirring force is that we can linearize the Navier–Stokes equation in order to calculate the response (i.e. the corresponding infinitesimal change in the velocity field). Let us make the substitutions (6.1) and (6.2) in eqn (5.56) to obtain

$$\left(\frac{\partial}{\partial t} + vk^2 \right) \{ u_\alpha(\mathbf{k}, t) + \delta u_\alpha(\mathbf{k}, t) \}$$

$$= \lambda M_{\alpha\beta\gamma}(\mathbf{k}) \sum_j \{ u_\beta(\mathbf{j}, t) u_\gamma(\mathbf{k} - \mathbf{j}, t) + u_\beta(\mathbf{j}, t) \delta u_\gamma(\mathbf{k} - \mathbf{j}, t) +$$

$$+ \delta u_\beta(\mathbf{j}, t) u_\gamma(\mathbf{k} - \mathbf{j}, t) + \delta u_\beta(\mathbf{j}) \delta u_\gamma(\mathbf{k} - \mathbf{j}) \} +$$

$$+ D_{\alpha\beta}(\mathbf{k}) \{ f_\beta(\mathbf{k}, t) + \delta f_\beta(\mathbf{k}, t) \}. \tag{6.4}$$

Then the equation for $\delta u_\alpha(\mathbf{k}, t)$ is obtained by subtracting eqn (5.56) from (6.4), interchanging the dummy variables $\mathbf{j}$ and $\mathbf{k} - \mathbf{j}$, and using the invariance $M_{\alpha\beta\gamma}(\mathbf{k}) = M_{\alpha\gamma\beta}(\mathbf{k})$ to obtain

$$\left(\frac{\partial}{\partial t} + vk^2\right)\delta u_\alpha(\mathbf{k}, t) - 2\lambda M_{\alpha\beta\gamma}(\mathbf{k}) \sum_{\mathbf{j}} u_\beta(\mathbf{j}, t)\delta u_\gamma(\mathbf{k} - \mathbf{j}, t)$$

$$= D_{\alpha\beta}(\mathbf{k})\delta f_\beta(\mathbf{k}, t) + O(\delta u)^2, \tag{6.5}$$

where we have neglected terms which are quadratic in the infinitesimal velocity perturbation.

Formally, eqn (6.5) is linear in $\delta u_\alpha(\mathbf{k}, t)$, which means that its solution can be written in terms of a Green function. And, of course, the relevant solution is none other than eqn (6.3), where $\hat{G}_{\alpha\beta}(\mathbf{k}; t, t')$ can now be interpreted as the Green tensor which satisfies

$$\left(\frac{\partial}{\partial t} + vk^2\right)\hat{G}_{\alpha\sigma}(\mathbf{k}; t, t') - 2\lambda M_{\alpha\beta\gamma}(\mathbf{k}) \sum_{\mathbf{j}} u_\beta(\mathbf{j}, t)\hat{G}_{\gamma\sigma}(\mathbf{k} - \mathbf{j}; t, t')$$

$$= D_{\alpha\sigma}(\mathbf{k})\delta(t - t'). \tag{6.6}$$

The presence of the term involving the random velocity field $u_\alpha(\mathbf{k}, t)$ suggests that $\hat{G}_{\alpha\beta}(\mathbf{k}; t, t')$ may well fluctuate from one realization to another. As we shall see shortly, this can be taken care of by working in terms of the ensemble-averaged response function denoted by

$$G_{\alpha\beta}(\mathbf{k}; t, t') = \langle \hat{G}_{\alpha\beta}(\mathbf{k}; t, t') \rangle. \tag{6.7}$$

6.1.2 Perturbation expansion of the velocity field

Our immediate aim now is to use (6.6) and (5.56) in order to obtain governing equations for the correlation tensor $Q_{\alpha\beta}(\mathbf{k}; t, t')$ and the mean response tensor $G_{\alpha\beta}(\mathbf{k}; t, t')$. We follow the procedures discussed in the previous chapter, and expand $u_\alpha(\mathbf{k}, t)$ and $\hat{G}_{\alpha\beta}(\mathbf{k}; t, t')$ as perturbation series in the ordering parameter λ. We shall only need to do this to first order:

$$u_\alpha(\mathbf{k}, t) = u_\alpha^{(0)}(\mathbf{k}, t) + \lambda u_\alpha^{(1)}(\mathbf{k}, t) + O(\lambda^2) \tag{6.8}$$

$$\hat{G}_{\alpha\beta}(\mathbf{k}; t, t') = \hat{G}_{\alpha\beta}^{(0)}(\mathbf{k}; t, t') + \lambda \hat{G}_{\alpha\beta}^{(0)}(\mathbf{k}; t, t') + O(\lambda^2). \tag{6.9}$$

It should be noted that (6.8) is just the same expansion as that given in eqn (5.69); the only difference is that this time we are working in $(\mathbf{k}, t)$-space rather than $(\mathbf{k}, \omega)$-space. As before, the zero-order term $u_\alpha^{(0)}(\mathbf{k}, t)$ is the solution of the Navier–Stokes equation (5.56), with $\lambda = 0$, and is given by (5.61). Similarly, we can make the identification

$$\hat{G}_{\alpha\beta}^{(0)}(\mathbf{k}; t, t') = G_{\alpha\beta}^{(0)}(\mathbf{k}; t, t'), \tag{6.10}$$

where $G_{\alpha\beta}^{(0)}(\mathbf{k}; t, t')$ satisfies eqn (6.4), with $\lambda = 0$, or, what is the same thing, is the solution of eqn (5.57). At the same time, we should also note that $G_{\alpha\beta}^{(0)}(\mathbf{k}; t, t')$

is statistically sharp (i.e. it is invariant under averaging) and from now on we shall simply drop the circumflex symbol, as far as the zero-order Green tensor is concerned.

Now substitute from (6.8) and (6.9) into the Navier–Stokes equation (5.56), and equate coefficients of powers of λ. We have already identified the zero-order term above and, in addition to this, we shall only need the first-order term, which is readily shown to be given by

$$\left(\frac{\partial}{\partial t} + vk^2\right) u_\alpha^{(1)}(\mathbf{k}, t) = M_{\alpha\beta\gamma}(\mathbf{k}) \sum_{\mathbf{j}} u_\beta^{(0)}(\mathbf{j}, t) u_\gamma^{(0)}(\mathbf{k} - \mathbf{j}, t), \tag{6.11}$$

or, inverting the operator on the l.h.s., in terms of the zero-order Green tensor as defined by eqn (5.57),

$$u_\alpha^{(1)}(\mathbf{k}, t) = \int_{-\infty}^{t} ds\, G_{\alpha\sigma}^{(0)}(\mathbf{k}; t, s) M_{\sigma\beta\gamma}(\mathbf{k}) \sum_{\mathbf{j}} u_\beta^{(0)}(\mathbf{j}, s) u_\gamma^{(0)}(\mathbf{k} - \mathbf{j}, s). \tag{6.12}$$

Similarly, by substituting (6.8) and (6.9) into eqn (6.6) and equating coefficients at each order, we find the equation for the first-order term in the expansion for the response tensor to be

$$\left(\frac{\partial}{\partial t} + vk^2\right) \hat{G}_{\alpha\sigma}^{(1)}(\mathbf{k}; t, t') = 2M_{\alpha\beta\gamma}(\mathbf{k}) \sum_{\mathbf{j}} u_\beta^{(0)}(\mathbf{j}, t) G_{\gamma\sigma}^{(0)}(\mathbf{k} - \mathbf{j}; t, t'), \tag{6.13}$$

or, inverting the operator on the r.h.s., as we did for eqn (6.11), we can write the first-order Green tensor as

$$\hat{G}_{\alpha\sigma}^{(1)}(\mathbf{k}; t, t') = 2 \int_{-\infty}^{t} ds\, G_{\alpha\rho}^{(0)}(\mathbf{k}; t, s) M_{\rho\beta\gamma}(\mathbf{k}) \sum_{\mathbf{j}} u_\beta^{(0)}(\mathbf{j}, s) G_{\gamma\sigma}^{(0)}(\mathbf{k} - \mathbf{j}; s, t')$$

$$= 2 \int_{t'}^{t} ds\, G_{\alpha\rho}^{(0)}(\mathbf{k}; t, s) M_{\rho\beta\gamma}(\mathbf{k}) \sum_{\mathbf{j}} u_\beta^{(0)}(\mathbf{j}, s) G_{\gamma\sigma}^{(0)}(\mathbf{k} - \mathbf{j}; s, t'), \tag{6.14}$$

where the lower limit on the integral with respect to time has been changed to take account of the fact that $G_{\gamma\sigma}^{(0)}(\mathbf{k} - \mathbf{j}; s, t') = 0$ for $s < t'$.

6.1.3 Perturbation series for the mean-response and correlation tensors

Formally, we obtain an equation for the mean-response tensor—as defined by (6.7)—by averaging each term of (6.6):

$$\left(\frac{\partial}{\partial t} + vk^2\right) G_{\alpha\sigma}(\mathbf{k}; t, t') - 2\lambda M_{\alpha\beta\gamma}(\mathbf{k}) \sum_{\mathbf{j}} \langle u_\gamma(\mathbf{k} - \mathbf{j}, t) \hat{G}_{\beta\sigma}(\mathbf{j}; t, t') \rangle$$

$$= D_{\alpha\sigma}(\mathbf{k}) \delta(t - t'), \tag{6.15}$$

where the dummy variables $\mathbf{j}$ and $\mathbf{k} - \mathbf{j}$ have been interchanged to fit in with a usage which we shall meet later on.

In order to evaluate the average $\langle u\hat{G}\rangle$, we substitute the perturbation series (6.8) and (6.9) for the velocity field and the response tensor respectively to obtain

$$\left(\frac{\partial}{\partial t} + vk^2\right)G_{\alpha\sigma}(\mathbf{k}; t, t') -$$

$$- 2\lambda M_{\alpha\beta\gamma}(\mathbf{k}) \sum_{\mathbf{j}} \{\langle u_\gamma^{(0)}(\mathbf{k} - \mathbf{j}, t)G_{\beta\sigma}^{(0)}(\mathbf{j}; t, t')\rangle +$$

$$+ \lambda\langle u_\gamma^{(1)}(\mathbf{k} - \mathbf{j}, t)G_{\beta\sigma}^{(0)}(\mathbf{j}; t, t')\rangle +$$

$$+ \lambda\langle u_\gamma^{(0)}(\mathbf{k} - \mathbf{j}, t)\hat{G}_{\beta\sigma}^{(1)}(\mathbf{j}; t, t')\rangle + O(\lambda^2)\}$$

$$= D_{\alpha\sigma}(\mathbf{k})\delta(t - t'), \tag{6.16}$$

where we recall from eqn (6.10) that the zero-order response function is statistically sharp.

Now consider the terms under the summation sign. The first of these is (symbolically) $\langle u^{(0)}G^{(0)}\rangle = \langle u^{(0)}\rangle G^{(0)} = 0$, as the $u^{(0)}$ are prescribed to have zero mean, and $G^{(0)}$ is of course statistically sharp and hence invariant under the averaging process. The second term also vanishes, but this is a little more difficult to see. Evidently we have $\langle u^{(1)}G^{(0)}\rangle = \langle u^{(1)}\rangle G^{(0)}$. It turns out that the average of the first-order velocity coefficient vanishes. We can see that this is so, as follows. From eqn (6.12) we have

$$\langle u_\alpha^{(1)}(\mathbf{k}, t)\rangle = \int_{-\infty}^t ds\, G_{\alpha\sigma}^{(0)}(\mathbf{k}; t, s)M_{\sigma\beta\gamma}(\mathbf{k}) \sum_{\mathbf{j}} \langle u_\beta^{(0)}(\mathbf{j}, s)u_\gamma^{(0)}(\mathbf{k} - \mathbf{j}, s)\rangle$$

$$= \int_{-\infty}^t ds\, G_{\alpha\sigma}^{(0)}(\mathbf{k}; t, s)M_{\sigma\beta\gamma}(\mathbf{k}) \sum_{\mathbf{j}} Q_{\beta\gamma}^{(0)}(\mathbf{j}; s, s)\delta(\mathbf{j} + \mathbf{k} - \mathbf{j})$$

$$= 0, \tag{6.17}$$

as $\delta(\mathbf{k})M_{\alpha\beta\gamma}(\mathbf{k}) = M_{\alpha\beta\gamma}(0) = 0$, which follows from eqn (2.77). Note that the homogeneity requirement, as embodied in eqn (2.93), also applies to the zero-order velocity field and so can be used (as above) when taking the step $\langle u^{(0)}u^{(0)}\rangle = Q^{(0)}$.

This then leaves us with the third term within the curly brackets and we deal with it by substituting from (6.14) for $\hat{G}^{(1)}$. We obtain the appropriate expression from eqn (6.14) by changing the labels $\mathbf{k}$, α to $\mathbf{j}$, β. Less obvious perhaps is the need to change dummy variables in order to avoid confusion when we substitute back into (6.16). Thus in (6.14) we also rename the dummy variables $\mathbf{j}$, β, γ as $\mathbf{l}$, δ, ε. The result is

$$\hat{G}_{\beta\sigma}^{(1)}(\mathbf{j}; t, t') = 2 \int_{t'}^t ds\, G_{\beta\rho}^{(0)}(\mathbf{j}; t, s)M_{\rho\delta\varepsilon}(\mathbf{j}) \sum_{\mathbf{l}} u_\delta^{(0)}(\mathbf{l}, s)G_{\varepsilon\sigma}^{(0)}(\mathbf{j} - \mathbf{l}; s, t'), \tag{6.18}$$

which we immediately substitute into (6.16) to obtain

$$\left(\frac{\partial}{\partial t} + vk^2\right)G_{\alpha\sigma}(\mathbf{k}; t, t') - 4\lambda^2 M_{\alpha\beta\gamma}(\mathbf{k}) \sum_{\mathbf{j}} \int_{t'}^{t} \mathrm{d}s\, G_{\beta\rho}^{(0)}(\mathbf{j}; t, s) M_{\rho\delta\epsilon}(\mathbf{j}) \times$$

$$\times \sum_{\mathbf{l}} G_{\epsilon\sigma}^{(0)}(\mathbf{j} - \mathbf{l}; s, t') \langle u_\gamma^{(0)}(\mathbf{k} - \mathbf{j}, t) u_\delta^{(0)}(\mathbf{l}, s)\rangle + O(\lambda^3)$$

$$= D_{\alpha\sigma}(\mathbf{k})\delta(t - t'). \tag{6.19}$$

The next step is to evaluate the pair correlation of the zero-order fluctuating velocities. Combining eqns (2.83) and (2.93), we have

$$\langle u_\gamma^{(0)}(\mathbf{k} - \mathbf{j}, t) u_\delta^{(0)}(\mathbf{l}, s)\rangle = \left(\frac{2\pi}{L}\right)^3 \delta_{\mathbf{k} - \mathbf{j} + \mathbf{l}, 0} Q_{\gamma\delta}^{(0)}(\mathbf{k} - \mathbf{j}; t, s). \tag{6.20}$$

Inserting this result into eqn (6.19), we can eliminate the summation over $\mathbf{l}$ along with the Kronecker delta. The other summation can be converted to an integral with respect to $\mathbf{j}$, as we take the infinite system limit

$$\lim_{L \to \infty} \left(\frac{2\pi}{L}\right)^3 \sum_{\mathbf{j}} \to \int \mathrm{d}^3 j,$$

and (6.19) for the response tensor becomes

$$\left(\frac{\partial}{\partial t} + vk^2\right)G_{\alpha\sigma}(\mathbf{k}; t, t') - 4\lambda^2 M_{\alpha\beta\gamma}(\mathbf{k}) - \int \mathrm{d}^3 j \int_{t'}^{t} \mathrm{d}s\, G_{\beta\rho}^{(0)}(\mathbf{j}; t, s) M_{\rho\delta\epsilon}(\mathbf{j}) \times$$

$$\times G_{\epsilon\sigma}^{(0)}(\mathbf{k}; s, t') Q_{\gamma\delta}^{(0)}(\mathbf{k} - \mathbf{j}; t, s) + O(\lambda^3)$$

$$= D_{\alpha\sigma}(\mathbf{k})\delta(t - t'). \tag{6.21}$$

The corresponding equation for the correlation tensor is obtained from the Navier–Stokes equation in the form of (5.56). The stirring forces are as prescribed in Section 3.5.2, with the autocorrelation of the forces given by (3.64). We multiply each term of eqn (5.56) by $u_\sigma(-\mathbf{k}, t')$ and average. Then, invoking eqn (2.88), we obtain

$$\left(\frac{\partial}{\partial t} + vk^2\right)Q_{\alpha\sigma}(\mathbf{k}; t, t') - \lambda M_{\alpha\beta\gamma}(\mathbf{k})\left(\frac{L}{2\pi}\right)^3 \sum_{\mathbf{j}} \langle u_\beta(\mathbf{j}, t) u_\gamma(\mathbf{k} - \mathbf{j}, t) u_\sigma(-\mathbf{k}, t')\rangle$$

$$= I_{\alpha\sigma}(\mathbf{k}; t, t'), \tag{6.22}$$

where the input term $I_{\alpha\sigma}$ is given by

$$I_{\alpha\sigma}(\mathbf{k}; t, t') = \left(\frac{L}{2\pi}\right)^3 D_{\alpha\beta}(\mathbf{k}) \langle f_\beta(\mathbf{k}, t) u_\sigma(-\mathbf{k}, t')\rangle. \tag{6.23}$$

Note that the correlation between the stirring force and the velocity field has already been treated in Section 4.3.2. We shall leave the input in the above non-specific form for the present. In the next section we shall specialize the equations to isotropic turbulence, and state the explicit form for $I_{\alpha\sigma}$, which is obtained as an exact result of the DIA theory. This result—see eqn (6.32)—is more general than, and reduces to, eqn (4.88).

The triple correlation in (6.22) can be expanded out using the perturbation series (6.8) for the velocity field. In an abbreviated notation we have

$$\langle \mathbf{u}(\mathbf{j})\mathbf{u}(\mathbf{k}-\mathbf{j})\mathbf{u}(-\mathbf{k})\rangle = \langle \mathbf{u}^{(0)}(\mathbf{j})\mathbf{u}^{(0)}(\mathbf{k}-\mathbf{j})\mathbf{u}^{(0)}(-\mathbf{k})\rangle +$$

$$+ \lambda\{\langle \mathbf{u}^{(0)}(\mathbf{j})\mathbf{u}^{(0)}(\mathbf{k}-\mathbf{j})\mathbf{u}^{(1)}(-\mathbf{k})\rangle +$$

$$+ \langle \mathbf{u}^{(0)}(\mathbf{j})\mathbf{u}^{(1)}(\mathbf{k}-\mathbf{j})\mathbf{u}^{(0)}(-\mathbf{k})\rangle +$$

$$+ \langle \mathbf{u}^{(1)}(\mathbf{j})\mathbf{u}^{(0)}(\mathbf{k}-\mathbf{j})\mathbf{u}^{(0)}(-\mathbf{k})\rangle + O(\lambda^2). \quad (6.24)$$

Now it follows at once from the fact that the zero-order velocity field is normally distributed (in itself a consequence of our prescription of the arbitrary stirring forces) that the zero-order term on the r.h.s. of (6.24) vanishes when we carry out the average. Next, we note that, of the first-order terms, the second and third are identical. This follows from the interchange of the dummy variables $\mathbf{j}$ and $\mathbf{k}-\mathbf{j}$, along with the property

$$M_{\alpha\beta\gamma}(\mathbf{k}) = M_{\alpha\gamma\beta}(\mathbf{k}).$$

Thus we can simplify matters by replacing the second and third first-order terms by twice the third term, and eqn (6.22) can be written as

$$\left(\frac{\partial}{\partial t} + vk^2\right) Q_{\alpha\sigma}(\mathbf{k}; t, t') -$$

$$- \lambda^2 M_{\alpha\beta\gamma}(\mathbf{k})\left(\frac{L}{2\pi}\right)^3 \sum \{\langle u_\beta^{(0)}(\mathbf{j}, t)u_\gamma^{(0)}(\mathbf{k}-\mathbf{j}, t)u_\sigma^{(1)}(-\mathbf{k}, t')\rangle +$$

$$+ 2\langle u_\beta^{(1)}(\mathbf{j}, t)u_\gamma^{(0)}(\mathbf{k}-\mathbf{j}, t)u_\sigma^{(0)}(-\mathbf{k}, t')\rangle\} + O(\lambda^3)$$

$$= I_{\alpha\sigma}(\mathbf{k}; t, t'). \quad (6.25)$$

Now we substitute from (6.12) for $\mathbf{u}^{(1)}$. First we rename the dummy variables in (6.12) in order to avoid confusion with the names of the existing dummy variables in (6.25). That is, the variables $\mathbf{j}$, σ, β, and γ are renamed $\mathbf{l}$, ρ, δ, and ε. In addition, for $u_\sigma^{(1)}(-\mathbf{k}, t')$ we change labels as appropriate, and it follows that

$$u_\sigma^{(1)}(-\mathbf{k}, t') = \int_{-\infty}^{t'} ds\, G_{\sigma\rho}^{(0)}(-\mathbf{k}; t', s)M_{\rho\delta\varepsilon}(-\mathbf{k}) \sum_{\mathbf{l}} u_\delta^{(0)}(\mathbf{l}, s)u_\varepsilon^{(0)}(-\mathbf{k}-\mathbf{l}, s),$$

$$(6.12)$$

and similarly for $u_\beta^{(1)}(\mathbf{j}, t)$ we obtain

$$u_\beta^{(1)}(\mathbf{j}, t) = \int_{-\infty}^{t} ds\, G_{\beta\rho}^{(0)}(\mathbf{j}; t, s)M_{\rho\delta\varepsilon}(\mathbf{j}) \sum_{\mathbf{l}} u_\delta^{(0)}(\mathbf{l}, s)u_\varepsilon^{(0)}(\mathbf{j}-\mathbf{l}, s). \quad (6.12b)$$

Once we substitute from (6.12a) and (6.12b) into (6.25), we then have to factor the two fourth-order moments into pairs of second-order moments using the relationships appropriate to a normal distribution. We have already carried out this procedure in Section 2.8.2, and reference should be made there, or to Appendix E, for the details. As before, we use the rules given by eqns (2.83) and (2.90) for homogeneous moments. Also as before, we eliminate one summation, along with the Kronecker delta which results from the homogeneity condition, and turn the other summation over $\mathbf{j}$ into an integration, as we take the infinite system limit. In all, the result is that eqn (6.25) becomes

$$
\left(\frac{\partial}{\partial t} + vk^2 \right) Q_{\alpha\sigma}(\mathbf{k}; t, t') -
$$

$$
- \lambda^2 M_{\alpha\beta\gamma}(\mathbf{k}) \int d^3 j \left\{ \int_{-\infty}^{t'} ds\, G_{\sigma\rho}^{(0)}(-\mathbf{k}; t', s) M_{\rho\delta\varepsilon}(-\mathbf{k}) \times \right.
$$

$$
\times 2 Q_{\beta\delta}^{(0)}(\mathbf{j}; t, s) Q_{\varepsilon\gamma}^{(0)}(\mathbf{k} - \mathbf{j}; t, s) +
$$

$$
+ \int_{-\infty}^{t} ds\, G_{\beta\rho}^{(0)}(\mathbf{j}; t, s) M_{\rho\delta\varepsilon}(\mathbf{j}) \times
$$

$$
\left. \times 4 Q_{\delta\gamma}^{(0)}(\mathbf{k} - \mathbf{j}; t, s) Q_{\varepsilon\sigma}^{(0)}(-\mathbf{k}; t', s) \right\} + O(\lambda^3)
$$

$$
= I_{\alpha\sigma}(\mathbf{k}; t, t'). \tag{6.26}
$$

Equations (6.21) and (6.26) are shown arbitrarily truncated at second order in the bookkeeping parameter λ, but of course they are actually dependent on an infinite number of terms of increasing order in $Q^{(0)}$ and $G^{(0)}$. The two equations are analogous to the graphical representations of Figs 5.13 and 5.15 respectively, and would be brought into an exact correspondence if we were to invert the l.h.s. of (6.21) and (6.26) in terms of $G^{(0)}$, One consequence of the present way of doing things is that we can see how natural it is to leave the left-hand propagator unrenormalized (see the last paragraph of Section 5.5.5).

6.1.4 Second-order equations for the isotropic response and correlation functions

The perturbation expansion which underlies eqns (6.21) and (6.26) can be renormalized by simultaneously replacing each $Q^{(0)}$ and $G^{(0)}$ by the exact Q and G. We have seen in Chapter 5 that this is really equivalent to a partial summation of the terms of the primitive perturbation series. Also, in terms of the diagrams given in Chapter 5, this is a process of line renormalization; there is no vertex renormalization.

It can be seen that the lowest non-trivial term (of order λ^2) involves the interaction of the three wavenumbers $(\mathbf{k}, \mathbf{j}, \mathbf{k} - \mathbf{j})$ which characterize the exact non-linearity of the Navier–Stokes equation. This is the direct interaction. In contrast, terms of higher order in the perturbation series (either primitive or renormalized) involve interactions via intermediate wavenumbers, with these mode couplings becoming ever more complicated as the order (i.e. the power of λ) increases. These are the indirect interactions.

The recipe for Kraichnan's (1958, 1959) DIA can now be stated as follows:

(a) replace $Q^{(0)}$ and $G^{(0)}$ by Q and G in eqns (6.21) and (6.26);

(b) truncate the perturbation series at second order (i.e. at the direct-interaction terms).

(c) put the bookkeeping parameter λ equal to unity.

While doing this, we shall also take the opportunity to make the simplifying restriction to isotropic turbulence. We use the general result for a second-order isotropic tensor (2.97), with trivial extensions to time dependence and to tensors other than the correlation tensor, to write

$$Q_{\alpha\beta}(\mathbf{k}; t, t') = D_{\alpha\beta}(\mathbf{k})Q(k; t, t') \tag{6.27a}$$

$$G_{\alpha\beta}(\mathbf{k}; t, t') = D_{\alpha\beta}(\mathbf{k})G(k; t, t') \tag{6.27b}$$

$$I_{\alpha\beta}(\mathbf{k}; t, t') = D_{\alpha\beta}(\mathbf{k})I(k; t, t'), \tag{6.27c}$$

where $Q(k; t, t')$ is the correlation function, $G(k; t, t')$ is the response function, and $I(k; t, t')$ is the input term, which is only of temporary significance, as we shall shortly express it in terms of the forcing spectrum. Note that $Q(k; t, t')$ satisfies

$$Q(k; t, t) = q(k, t),$$

as defined by eqn (2.163).

Following the above procedures, and substituting as appropriate from (6.27a) and (6.27b), we can write eqn (6.21) as

$$\left(\frac{\partial}{\partial t} + \nu k^2\right) D_{\alpha\sigma}(\mathbf{k}) G(k; t, t') -$$

$$- 4M_{\alpha\beta\gamma}(\mathbf{k}) \int d^3j \int_{t'}^{t} ds \, D_{\beta\rho}(\mathbf{j}) G(j; t, s) M_{\rho\delta\varepsilon}(\mathbf{j}) \times$$

$$\times D_{\varepsilon\sigma}(\mathbf{k}) G(k; s, t') D_{\gamma\delta}(\mathbf{k} - \mathbf{j}) Q(|\mathbf{k} - \mathbf{j}|; t, s)$$

$$= D_{\alpha\sigma}(\mathbf{k})\delta(t - t'). \tag{6.28}$$

This may seem even more complicated than the previous form, but this is illusory. A dramatic simplification is now possible. Set $\alpha = \sigma$, sum over α, and then cancel the factor $\sum_\alpha D_{\alpha\alpha}(\mathbf{k}) = 2$ across. The result is

$$\left(\frac{\partial}{\partial t} + vk^2\right) G(k; t, t') +$$

$$+ \int d^3j \int_{t'}^{t} ds\, L(\mathbf{k}, \mathbf{j}) G(j; t, s) G(k; s, t') Q(|\mathbf{k} - \mathbf{j}|; t, s)$$

$$= \delta(t - t'), \tag{6.29}$$

as the equation governing the response function $G(k; t, t')$. The coefficient $L(\mathbf{k}, \mathbf{j})$ has previously been encounted in connection with the quasi-normality theory in Section 2.8.2. From comparison of eqns (6.28) and (6.29) it is

$$L(\mathbf{k}, \mathbf{j}) = -2M_{\alpha\beta\gamma}(\mathbf{k}) M_{\rho\delta\varepsilon}(\mathbf{j}) D_{\beta\rho}(\mathbf{j}) D_{\varepsilon\alpha}(\mathbf{k}) D_{\gamma\delta}(\mathbf{k} - \mathbf{j})$$

$$= -2M_{\varepsilon\beta\gamma}(\mathbf{k}) M_{\beta\delta\varepsilon}(\mathbf{j}) D_{\gamma\delta}(\mathbf{k} - \mathbf{j}), \tag{6.30}$$

where the last line is the same as the form given previously in eqn (2.162) and follows from the contraction properties

$$D_{\alpha\varepsilon}(\mathbf{k}) D_{\varepsilon\rho}(\mathbf{k}) = D_{\alpha\rho}(\mathbf{k})$$

$$D_{\alpha\varepsilon}(\mathbf{k}) M_{\varepsilon\beta\gamma}(\mathbf{k}) = M_{\alpha\beta\gamma}(\mathbf{k})$$

which, in turn, are easily deduced from eqns (2.77) and (2.78).

The equation for the correlation function can be obtained, using exactly the same procedures, from (6.26). The only additional feature is the substitution of (6.27c) for the input term:

$$\left(\frac{\partial}{\partial t} + vk^2\right) Q(k; t, t') -$$

$$- \int d^3j\, L(\mathbf{k}, \mathbf{j}) \left\{ \int_{-\infty}^{t'} ds\, G(k; t', s) Q(j; t, s) Q(|\mathbf{k} - \mathbf{j}|; t, s) - \right.$$

$$\left. - \int_{-\infty}^{t} ds\, G(j; t, s) Q(|\mathbf{k} - \mathbf{j}|; t, s) Q(k; t', s) \right\}$$

$$= I(k; t, t'), \tag{6.31}$$

where the coefficient $L(\mathbf{k}, j)$ takes the same form as given in eqn (2.162); additional details can be found in Appendix E.

The input term $I(k; t, t')$, as defined by (6.27c) and (6.23), can be regarded as a cross-correlation of the random stirring force with the velocity field. It can be shown, by a generalization of the treatment for stationary turbulence in Kraichnan (1958), that the input term can be expressed in terms of the stirring spectrum $w(k; t, t')$:

$$I(k; t, t') = \int_{-\infty}^{t'} G(k; t', s) w(k; t, s)\, ds \tag{6.32}$$

where $w(k; t, t')$ is defined by eqn (3.94).

Equations (6.28), (6.31), and (6.32) comprise the original direct-interaction approximation (Kraichnan 1958, 1959). In the next section, we shall briefly consider the implications of DIA for energy transfer and the spectrum in the inertial range of wavenumbers. This treatment will merely be a prelude to a much more comprehensive investigation of the properties of the DIA (including its generalization to Lagrangian-history formulations) in Chapters 7 and 8.

6.1.5 *Spectral transport of energy: the inertial range*

In this section we shall consider some of the properties of the DIA, especially in the inertial range of wavenumbers. We shall begin by rewriting the relevant equations, so that they take on a somewhat simpler appearance. Thus, starting with (6.31) for the correlation function, we now write this as

$$\left(\frac{\partial}{\partial t} + vk^2\right) Q(k; t, t') = P(k; t, t') + I(k; t, t'), \tag{6.33}$$

where the inertial transfer term is given by

$$P(k; t, t') = \int d^3j \, L(\mathbf{k}, \mathbf{j}) \left\{ \int_{-\infty}^{t'} ds \, G(k; t', s) Q(j; t, s) Q(|\mathbf{k} - \mathbf{j}|; t, s) - \right.$$
$$\left. - \int_{-\infty}^{t} ds \, G(j; t, s) Q(k; t', s) Q(|\mathbf{k} - \mathbf{j}|; t, s) \right\}. \tag{6.34}$$

On the time diagonal, where $t = t'$, we have the special form

$$\left(\frac{\partial}{\partial t} + 2vk^2\right) Q(k; t, t) = 2P(k; t, t) + 2I(k; t, t), \tag{6.35}$$

where the factors of 2 arise when one multiplies $\partial u_\alpha(\mathbf{k}, t)/\partial t$ in the Navier–Stokes equation by $u_\alpha(\mathbf{k}, t)$ before averaging (see the derivation of eqn (2.115) for the single-time correlation).

The energy spectrum $E(k, t)$ can be introduced by generalizing eqn (2.101) to the time-dependent case:

$$E(k, t) = 4\pi k^2 Q(k, t),$$

where

$$Q(k, t) = Q(k; t, t)$$

and, for consistency with earlier notation,

$$Q(k, 0) = q(k).$$

Then, multiplying each term of (6.35) by $4\pi k^2$, we obtain the DIA equation for the energy spectrum:

$$\left(\frac{d}{dt} + 2vk^2\right) E(k, t) = T(k, t) + 4\pi k^2 w(k, t), \tag{6.36}$$

where the transfer spectrum $T(k, t)$ is defined by

$$T(k, t) = 8\pi k^2 P(k; t, t) \tag{6.37}$$

and $P(k; t, t)$ is given by (6.34) with $t = t'$.

In Section 2.7.1 we discussed the general energy-balance equation and the requirement that the non-linear terms should conserve energy. In practice this means that the transfer spectrum $T(k, t)$ must vanish when integrated over all wavenumbers (see eqn (2.126)). It is easily shown that the DIA form of $T(k, t)$ passes this test. Let us write the transfer spectrum as

$$T(k, t) = \int d^3j \, A(k, j), \tag{6.38}$$

where it can readily be deduced from eqns (6.34) and (6.37) that $A(k, j)$ is given by

$$A(k, j) = 8\pi k^2 L(\mathbf{k}, \mathbf{j}) \int_{-\infty}^{t} ds \{ G(k; t, s) Q(j; t, s) Q(|\mathbf{k} - \mathbf{j}|; t, s) -$$

$$- G(j; t, s) Q(k; t, s) Q(|\mathbf{k} - \mathbf{j}|; t, s) \}. \tag{6.39}$$

Now it is clear that, if we interchange k and j on the r.h.s. of (6.39), we merely change the second term into the first term and vice versa. Thus we see that

$$A(j, k) = -A(k, j), \tag{6.40}$$

meaning that $A(k, j)$ is antisymmetric under the interchange of k and j. Hence it follows that

$$\int_0^\infty T(k, t) \, dk = \int d^3k \int d^3j \, A(k, j) = 0, \tag{6.41}$$

as required.

Kraichnan (1959) also introduced the transport power $\Pi(k, t)$, which is defined to be the rate at which energy is transferred from modes $k' < k$ to modes $k' > k$. The formal expression (actually, in the later form to be found in Kraichnan (1964a)) is

$$\Pi(k, t) = \int_k^\infty T(k', t) \, dk' = -\int_0^k T(k', t) \, dk' \tag{6.42}$$

where the second equality follows from eqn (6.41).

In Section 2.7.1 the antisymmetry of the integrand involving the triple moment was established as a general result. It is interesting to note how this property has been achieved in the DIA. Clearly, from (6.39), the antisymmetry arises because there are two separate terms, one of the opposite sign to the other. The physical significance of these two terms can be explained most clearly by writing the energy-balance equation out in words.

First we rearrange equation (6.36) as

$$\frac{dE(k,t)}{dt} = 4\pi k^2 w(k,t) - 2\nu k^2 E(k,t) + T(k,t).$$

This can be restated in words as follows:

(rate of change of energy in mode **k**) equals (rate at which stirring forces do work on mode **k**) less (rate at which viscous dissipation turns energy in mode **k** into heat) plus {(rate at which energy is transferred into mode **k** from modes **k**' < **k**) − (rate at which energy is transferred out of mode **k** to modes **k**' > **k**)}.

Note that the second term making up $T(k,t)$ is the one proportional to $Q(k; t, s)$. This can be seen to be physically reasonable, as follows. Suppose that the turbulence is in a steady state under the combined effects of stirring forces and viscous dissipation. Now suppose that we inject some additional energy directly into mode **k**. According to the above, the output term of $A(k,j)$ would increase in magnitude owing to the presence of $Q(k)$, whereas the input term containing $Q(j)Q(|\mathbf{k} - \mathbf{j}|)$ would initially be unaffected. Hence we could expect the original steady state to be restored by such a process.

The precise effect of the transfer spectrum $T(k,t)$ depends on the physical situation and the value of the wavenumber being considered. An interesting case arises when we satisfy the conditions for the existence of an inertial range, as discussed in Section 2.7.2. Essentially we require the Reynolds number to be large, so that we can have a range of wavenumbers which is independent of the viscosity. Also, the turbulence should be at least approximately steady. Then, with these restrictions and integrating the various terms of (6.36) from some value k up to infinity (compare eqns (2.145) and (2.156)), we find the condition for an inertial range

$$\Pi(k) = \varepsilon, \tag{6.43}$$

where ε is the constant rate at which the inertial forces transfer energy from low to high wavenumbers. It is, of course, numerically equal to the viscous dissipation rate (see Section 2.7.2)

6.1.6 *The DIA energy spectrum in the inertial range*

It was shown by Kraichnan (1959) that the DIA predicted a $-3/2$ power law for the energy spectrum in the inertial range. At the time, the experimental accuracy so far achieved was not sufficient to distinguish between this result and the Kolmogorov $-5/3$ power law. Nowadays we know that the Kolmogorov spectrum is the more nearly correct, yet the analysis leading to the DIA $-3/2$ law is of considerable interest.

Following Kraichnan (1959), we begin with the equation for the response function. Now, restricting our attention to stationary turbulence, we rewrite equation (6.29) as

$$\left(\frac{\partial}{\partial t} + vk^2\right) G(k; t - t')$$

$$= -\int d^3j \int_{t'}^{t} dt'' L(\mathbf{k}, \mathbf{k} - \mathbf{j}) \times$$

$$\times G(|\mathbf{k} - \mathbf{j}|; t - t'')G(k; t'' - t')Q(j; t - t''), \qquad (6.44)$$

where we have interchanged the dummy variables $\mathbf{j}$ and $\mathbf{k} - \mathbf{j}$, and $L(\mathbf{k}, \mathbf{k} - \mathbf{j})$ is easily obtained by making the same interchange in eqn (6.30) for $L(\mathbf{k}, \mathbf{j})$. It is given by (see Appendix E)

$$L(\mathbf{k}, \mathbf{k} - \mathbf{j}) = (k^4 - 2k^3 j\mu + kj^3\mu)(1 - \mu^2)|\mathbf{k} - \mathbf{j}|^{-2}, \qquad (6.45)$$

where μ is the cosine of the angle between the vectors $\mathbf{k}$ and $\mathbf{j}$. For the case $j \ll k$, it becomes

$$L(\mathbf{k}, \mathbf{k} - \mathbf{j}) = k^2(1 - \mu^2) + O\left(\frac{j^2}{k^2}\right), \qquad (6.46)$$

and this is the case which will concern us here.

It should also be noted that we have dropped the delta function, which is on the r.h.s. of (6.29), as this only represents the discontinuity in the first derivative of $G(k; t, t')$ at $t = t'$. It can be avoided by keeping $t - t' > 0$.

Now make the substitutions

$$\tau = t - t' \qquad s = t'' - t',$$

and (6.44) takes the form

$$\dot{G}(k, \tau) + vk^2 G(k, \tau)$$

$$= -\int d^3j \int_0^\tau ds \, L(\mathbf{k}, \mathbf{k} - \mathbf{j})G(|\mathbf{k} - \mathbf{j}|, \tau - s)G(k, s)Q(j, \tau - s)$$

$$(6.47)$$

where the dot denotes differentiation with respect to τ. This equation can be simplified to the point where analytical solution is possible, if we consider the case where k is very much larger than some k_e defined by

$$\int_0^{k_e} E(j)\,dj \simeq \int_0^\infty E(j)\,dj = 3\frac{v_0^2}{2} \qquad (6.48)$$

where v_0 is the r.m.s. velocity of the turbulence. Hence k_e effectively marks the top of the energy-containing range of wavenumbers. Under these circumstances, we can argue that the presence of $4\pi j^2 Q(j, \tau - s) = E(j, \tau - s)$ ensures that the integration on the r.h.s. of (6.47) is dominated by wavenumbers $j < k_e \ll k$. Hence the triangle condition leads to $|\mathbf{k} - \mathbf{j}| \simeq k$. Therefore putting

$$G(|\mathbf{k} - \mathbf{j}|, \tau - s) \simeq G(k, \tau - s),$$

invoking eqn (6.46), and performing the integration over j, we obtain the simplified version of (6.47) as

$$\dot{G}(k, \tau) + \nu k^2 G(k, \tau) = -v_0^2 k^2 \int_0^\tau G(k, \tau - s) G(k, s) \, ds. \tag{6.49}$$

This can be solved subject to the boundary condition

$$G(k, 0+) = 1,$$

which follows from the definition of G. The result is (Kraichnan 1959)

$$G(k, \tau) = \frac{\exp(-\nu k^2 \tau) J_1(2v_0 k \tau)}{v_0 k \tau} \tag{6.50}$$

where J_1 is a first-order Bessel function of the first kind.

Various points about this solution are noteworthy. For instance, the oscillatory behaviour of the Bessel function is unlikely to be correct: a monotonic decline with increasing time separation τ seems more likely. We shall return to this sort of consideration in Chapter 8, when we make a quantitative assessment of the RPTs. For the moment we are more interested in the qualitative features of (6.50). Of these, the most striking is the presence of the viscous time-scale $(\nu k^2)^{-1}$ and the energy-containing range time-scale $(v_0 k)^{-1}$, but not the inertial range time-scale. In fact it is obvious, with the assumptions made above, that the viscous time-scale will be very large (i.e. viscous processes are slow for wavenumbers in the inertial range) and the exponential factor approximately equal to unity.

Thus we have surprising result that the response function in the inertial range of wavenumbers scales on the time-scale associated with the energy-containing range of wavenumbers. As $(v_0 k)^{-1}$ is the characteristic time one would associate with uniform convection of a periodic pattern with spatial frequency k past a fixed point with velocity v_0, it is often referred to as the convective or sweeping time-scale.

According to Kraichnan (1959), it can be concluded from a similar analysis that the two-time correlation function $Q(k, t - t')$ also scales on the convective time of the large eddies.

Lastly, the above results for the response and time-correlation functions lead to the inertial range spectrum

$$E(k) = f(0)(\varepsilon v_0)^{1/2} k^{-3/2}, \tag{6.51}$$

where $f(0)$ is a numerical constant. The details can be found in Kraichnan (1959). Here we shall merely note that (6.51) follows from the condition (6.43) for the existence of an inertial range, along with the imposition of the requirement that the transport power $\Pi(k)$ is determined locally by wavenumbers k', j and $|\mathbf{k} - \mathbf{j}|$, all in the neighbourhood of k.

6.1.7   *Alternative derivation of DIA by the method of reversion of power series*

Another method of deriving renormalized perturbation series has been given by Kraichnan (1977). The real significance of the new method seems to be that it provides a more powerful and systematic technique for deriving new theories—in particular, Lagrangian renormalized expansions, a topic which we shall discuss in the next chapter. Nevertheless, there seems to be some tendency to regard it as, in some way, improving the status of DIA, even in its Eulerian version. For this reason, and also for completeness, we shall briefly consider the topic here.

The method of reversion (or, sometimes, inversion) of power series has been used in other comparable problems in theoretical physics. For example, the renormalization of the expansion for the free energy, which we discussed in Section 5.3.3, can be treated in this way. To be precise, the expansion of the configuration integral in terms of cluster functions can be reverted into a power series in the Mayer functions (Reichl 1980, p. 357).

The general method can readily be explained, as follows. Consider a pair of real variables x and y, which are connected by the power series

$$y = ax + bx^2 + cx^3 + dx^4 + \cdots \qquad (6.52)$$

We now wish to invert this relationship and express x in terms of y. We begin by supposing that x is small enough for us to neglect its square. Then we have the immediate result

$$y = ax \qquad \text{or} \qquad x = y/a.$$

Evidently this is the lowest-order approximation to the general result which we are seeking. It can be made the basis of an iteration. We anticipate the required result by writing

$$x = Ay + By^2 + Cy^3 + Dy^4 + \cdots, \qquad (6.53)$$

where it follows immediately that the first unknown coefficient is given by

$$A = 1/a. \qquad (6.54)$$

The second coefficient is found by going to second order. That is, we assume that x is not quite so small and that we need to include its square; thus

$$\begin{aligned}
Y = ax + bx^2 &= a(Ay + By^2) + b(Ay + By^2)^2 \\
&= aAy + (aB + A^2b)y^2 + O(y^3) \\
&= y + (aB + A^2b)y^2 + O(y^3), \qquad (6.55)
\end{aligned}$$

where we have substituted from (6.53) for x and, in the first term on the r.h.s., from (6.54) for A. Consistency then requires that the term of order y^2 on the r.h.s. of (6.55) vanishes and hence

$$B = -b/a^3. \tag{6.56}$$

Clearly this iteration can be carried on to any order but we shall not pursue that here. The method can also be extended to functional power series (Kraichnan 1977, Appendix) and specifically to the present renormalization problem as follows.

(a) We begin with the primitive power series for Q and G in terms of Q_0 and G_0.

(b) We revert these primitive expansions to obtain Q_0 and G_0 as power series in Q and G.

(c) Then substitute these new expansions for each Q_0 and G_0 factor in the primitive expansions for the triple moments.

(d) Lastly, multiply out and collect terms of each order.

The result of all this is line renormalization of the primitive perturbation series, which we have just obtained from other methods, in the process of deriving eqns (6.29) and (6.31). According to Kraichnan (1977), it is also possible to carry out a further reversion which leads to vertex renormalization. We shall return to some additional consideration of these points in the next chapter, in connection with Lagrangian-history theories.

6.1.8 *Concluding remarks*

We have presented the DIA theory of turbulence very much in the context of other renormalization procedures, in other areas of physics, and also very much without frills. Our hope is that the general basis of the whole approach can be made rather clearer in this way. However, DIA was put forward originally as a complete theory, buttressed by physical hypotheses of 'maximal randomness' and 'weak dependence'. We are not ignoring these concepts, but are merely deferring them to Chapter 8, where we shall make an overall assessment of the various theories.

An important claim made for DIA is that its physical realizability is guaranteed by the fact that when applied to certain model dynamical equations it gives exact solutions. In view of the fact that the failure of the quasi-normality hypothesis occurred in the form of unphysical negative energy spectra (see Section 2.8.2), this was seen as a major strength of the theory. We shall not go into details here, but a full account can be found in Kraichnan (1961). This topic has been taken further by Frisch and Bourret (1970).

6.2 The Edwards–Fokker–Planck theory

The theory derived by Edwards (1964, 1965) can be seen as drawing on certain analogies with the development of kinetic equations in statistical mechanics.

We have discussed this subject in a rather concise way in Chapter 4, and the work of Edwards essentially follows a similar route, beginning with the derivation of the Liouville equation governing the turbulent velocity field. At this point the theory is equivalent to the Hopf formalism, but Edwards broke away from sterile formalism by approximating his Liouville equation into the Fokker–Planck form. This allows a solution to be obtained for the probability distribution of the fluctuating velocity field, and hence a closed equation for the energy spectrum.

6.2.1   The derivation of the Liouville equation

We begin by considering the turbulent fluid to be in a box, such that the Fourier modes $\mathbf{u}(\mathbf{k})$ are denumerable. Later, we shall take the limit as the system size goes to infinity, although we shall not draw any distinction between the two cases in the notation used. The probability F that the velocity field $\mathbf{u}(\mathbf{k}, t)$ takes the particular set of values $\mathbf{u}(\mathbf{k})$ at time t is given by

$$F = \prod_{\mathbf{k}} \delta[\mathbf{u}(\mathbf{k}, t) - \mathbf{u}(\mathbf{k})]. \tag{6.57}$$

We have previously discussed this kind of distribution in Section 4.1. An equation for the evolution of F with time can then be derived as follows. Differentiating both sides of (6.57) and using the chain rule for differentiating products, we find

$$\begin{aligned}
\frac{\partial F}{\partial t} &= \sum_{\mathbf{k}} \frac{\partial \mathbf{u}(\mathbf{k}, t)}{\partial t} \frac{\partial}{\partial \mathbf{u}(\mathbf{k}, t)} \left\{ \prod_{\mathbf{j}} \delta[\mathbf{u}(\mathbf{j}, t) - \mathbf{u}(\mathbf{j})] \right\} \\
&= -\sum_{\mathbf{k}} \frac{\partial \mathbf{u}(\mathbf{k}, t)}{\partial t} \frac{\partial}{\partial \mathbf{u}(\mathbf{k})} \left\{ \prod_{\mathbf{j}} \delta[\mathbf{u}(\mathbf{j}, t) - \mathbf{u}(\mathbf{j})] \right\} \\
&= -\sum_{\mathbf{k}} \frac{\partial}{\partial \mathbf{u}(\mathbf{k})} \frac{\partial \mathbf{u}(\mathbf{k}, t)}{\partial t} \left\{ \prod_{\mathbf{j}} \delta[\mathbf{u}(\mathbf{j}, t) - \mathbf{u}(\mathbf{j})] \right\},
\end{aligned} \tag{6.58}$$

where the second line uses the relation

$$\frac{\partial f(x - y)}{\partial x} = -\frac{\partial f(x - y)}{\partial y}$$

and the third line relies on the fact that $\partial \mathbf{u}(\mathbf{k}, t)/\partial t$ is not a function of $\mathbf{u}(\mathbf{k})$.

Next we substitute for $\partial \mathbf{u}(\mathbf{k}, t)/\partial t$ in (6.58) from the Navier–Stokes equation (4.81) to obtain

$$\begin{aligned}
\frac{\partial F}{\partial t} &= -\sum_{\mathbf{k}} \frac{\partial}{\partial u_\alpha(\mathbf{k})} \left\{ -vk^2 u_\alpha(\mathbf{k}, t) + M_{\alpha\beta\gamma}(\mathbf{k}) \sum_{\mathbf{j}} u_\beta(\mathbf{j}, t) u_\gamma(\mathbf{k} - \mathbf{j}, t) + f_\alpha(\mathbf{k}, t) \right\} \times \\
&\quad \times \left\{ \prod_{\mathbf{j}} \delta[\mathbf{u}(\mathbf{j}, t) - \mathbf{u}(\mathbf{j})] \right\}.
\end{aligned} \tag{6.59}$$

Now we wish to obtain the probability distribution and, following the example discussed in the sequence of eqns (4.1)–(4.6), we introduce the ensemble-averaged distribution function

$$P[\mathbf{u(k)}, t] = \langle F \rangle \tag{6.60}$$

where P becomes a functional as the box size tends to infinity and $\mathbf{u(k)}$ becomes a continuous function of $\mathbf{k}$.

The ensemble can be specified by the introduction of the stirring forces. These have already been discussed in Sections 3.5.2 and 4.3.2, but now we follow Edwards (1964) and formally specify their probability distribution:

$$J[\mathbf{f(k}, t)] = N \exp \left\{ -\sum_{\mathbf{k}} \int dt \int dt'\, f(\mathbf{k}, t) w^{-1}(k, t - t') f(-\mathbf{k}, t') \right\}, \tag{6.61}$$

where N is an appropriate normalization such that integration of J over the space of the functions $\mathbf{f(k}, t)$ gives unity and $w^{-1}(k, t - t')$ is the functional inverse of the force autocorrelation, i.e.

$$w(k, t - t'') w^{-1}(k, t'' - t')\, dt'' = \delta(t - t'). \tag{6.62}$$

The force autocorrelation $w(k, t - t')$ is given by eqn (3.94) which we shall repeat here for completeness:

$$\left(\frac{L}{2\pi} \right)^3 \langle f_\alpha(\mathbf{k}, t) f_\beta(-\mathbf{k}, t') \rangle = D_{\alpha\beta}(\mathbf{k}) w(k, t - t').$$

Also, for the reasons we discussed in Chapter 4 in connection with Brownian motion, we shall specialize the autocorrelation to a delta function in time, and again, for completeness, we reproduce the relevant result (eqn (4.85)) here:

$$w(k, t - t') = W(k) \delta(t - t').$$

Having specified our turbulent ensemble in terms of eqns (6.61), (6.62), (3.94), and (4.85), we can usefully rewrite (6.60) for the distribution of fluctuating velocities in a more specific form as

$$P[\mathbf{u(k)}, t] = \int FJ[\mathbf{f}]\delta\mathbf{f}. \tag{6.63}$$

Then we can obtain the evolution equation for P by multiplying each term of (6.59) by $J[\mathbf{f}]$ and integrating over the variables $\mathbf{f(k}, t)$. We can write out the intermediate stage explicitly as

$$\int \left[\frac{\partial F}{\partial t} \right] J[\mathbf{f}]\delta\mathbf{f} = \int \sum_{\mathbf{k}} \frac{\partial}{\partial u_\alpha(\mathbf{k})} \left\{ vk^2 u_\alpha(\mathbf{k}, t) - \right.$$

$$\left. - M_{\alpha\beta\gamma}(\mathbf{k}) \sum_{\mathbf{j}} u_\beta(\mathbf{j}, t) u_\gamma(\mathbf{k} - \mathbf{j}, t) + f_\alpha(\mathbf{k}, t) \right\} \times$$

$$\times \left\{ \prod_{\mathbf{j}} \delta[\mathbf{u(j}, t) - \mathbf{u(j)}] \right\} J[\mathbf{f}]\delta\mathbf{f}. \tag{6.59a}$$

We shall now work out an explicit form for each term individually, beginning with the l.h.s. The time derivative is unaffected by the functional integration and we readily obtain

$$\int \frac{\partial F}{\partial t} J[\mathbf{f}] \delta \mathbf{f} = \frac{\partial}{\partial t} \int F J[\mathbf{f}] \delta \mathbf{f} = \frac{\partial P}{\partial t}. \qquad (6.64)$$

The first term on the r.h.s. is just

$$\int \sum_{\mathbf{k}} \frac{\partial}{\partial u_\alpha(\mathbf{k})} \{ v k^2 u_\alpha(\mathbf{k}, t) \} \left\{ \prod_{\mathbf{j}} \delta[\mathbf{u}(\mathbf{j}, t) - \mathbf{u}(\mathbf{j})] \right\} J[\mathbf{f}] \delta \mathbf{f}$$

$$= \sum_{\mathbf{k}} \frac{\partial}{\partial u_\alpha(\mathbf{k})} \{ v k^2 u_\alpha(\mathbf{k}) \} \int F J[\mathbf{f}(\mathbf{k})] \delta \mathbf{f}$$

$$= \sum_{\mathbf{k}} \frac{\partial}{\partial u_\alpha(\mathbf{k})} \{ v k^2 u_\alpha(\mathbf{k}) P \}, \qquad (6.65)$$

and the non-linear term follows similarly.

A problem arises when we consider the last term on the r.h.s. of eqn (6.59a). This contains the stirring force $\mathbf{f}(\mathbf{k}, t)$ explicitly, and of course $\mathbf{f}(\mathbf{k}, t)$ is the independent variable when we carry out the functional integration. We have already solved an analogous problem when we obtained eqn (4.88) for the cross-correlation $\langle f_\alpha(\mathbf{k}, t) u_\beta(-\mathbf{k}, t) \rangle$. The corresponding term in the equation for P is very much more complicated and was first derived by Edwards, who drew on an analogy with the theory of Brownian motion. The details are complicated and the reader who wishes to pursue this further should consult the original paper (Edwards 1964). Alternatively, it can be obtained from the later functional formalism of Novikov (see Appendix H). The result is

$$\int \sum_{\mathbf{k}} \frac{\partial}{\partial u_\alpha(\mathbf{k})} \left\{ \prod_{\mathbf{j}} \delta[\mathbf{u}(\mathbf{j}, t) - \mathbf{u}(\mathbf{j})] \right\} \{ f_\alpha(\mathbf{k}, t) J[\mathbf{f}] \delta \mathbf{f} \}$$

$$= -\sum_{\mathbf{k}} \frac{\partial}{\partial u_\alpha(\mathbf{k})} \left\{ W(k) \frac{\partial}{\partial u_\alpha(-\mathbf{k})} \right\} P. \qquad (6.66)$$

Formally then, we start with eqn (6.59) for F, multiply each term by $J[\mathbf{f}]$, and integrate according to eqn (6.63):

$$\frac{\partial P}{\partial t} = -\sum_{\mathbf{k}} \frac{\partial}{\partial u_\alpha}(\mathbf{k}) \left\{ -v k^2 u_\alpha(\mathbf{k}) + M_{\alpha\beta\gamma}(\mathbf{k}) u_\beta(\mathbf{j}) u_\gamma(\mathbf{k} - \mathbf{j}) - W(k) \frac{\partial}{\partial u_\alpha(-\mathbf{k})} \right\} P, \qquad (6.67)$$

where we have also used equations (6.64), (6.65), and (6.66). This is the required equation for the probability distribution functional of the turbulent velocities $\mathbf{u}(\mathbf{k})$ at time t. For the special case of random stirring forces with Gaussian (multivariate normal) statistics and autocorrelations which are delta functions in time, it is an exact equation. In succeeding sections we shall consider how it can be solved as an approximation.

6.2.2   The Edwards–Fokker–Planck equation

We are considering a formulation with prescribed stirring forces, and hence we can expect the system to attain a steady state in which the rate at which the stirring forces do work is exactly balanced by the rate at which the viscous forces dissipate turbulent kinetic into heat. We can simplify matters some-what by anticipating this eventuality and just assume that the turbulence is statistically stationary. Thus the time derivative in eqn (6.67) can be equal to zero and, with some rearrangement, the equation for P becomes

$$\sum_{\mathbf{k}} \frac{\partial}{\partial u_\alpha(\mathbf{k})} \left\{ W(k) \frac{\partial}{\partial u_\alpha(-\mathbf{k})} + \nu k^2 u_\alpha(\mathbf{k}) \right\} P -$$

$$-\sum_{\mathbf{k}} \sum_{\mathbf{j}} M_{\alpha\beta\gamma}(\mathbf{k}) u_\beta(\mathbf{j}) u_\gamma(\mathbf{k} - \mathbf{j}) \frac{\partial P}{\partial u_\alpha(\mathbf{k})} = 0. \qquad (6.68)$$

Note that in the last term, the differential only acts on one of the $u(\mathbf{j})$ or $u(\mathbf{k} - \mathbf{j})$ if $\mathbf{k} = \mathbf{j}$ or $\mathbf{k} = \mathbf{k} - \mathbf{j}$. In either case, the triangle condition ensures that the other wavevector is zero and hence the whole term vanishes because of the boundary condition on the velocities, which is

$$\mathbf{u}(\mathbf{k}) = 0 \qquad \text{for } \mathbf{k} = 0. \qquad (6.69)$$

We also note that a velocity mode with $\mathbf{k} = 0$ would correspond to a uniform translation of the whole system and so would have no dynamical significance.

We now introduce a very symbolic notation (rather like that of Herring (1965), but there are some important differences). We write eqn (6.68) as

$$LP - VP = 0, \qquad (6.70)$$

where the operators L and V are defined by

$$L = \sum_{\mathbf{k}} \frac{\partial}{\partial u_\alpha(\mathbf{k})} \left\{ \nu k^2 u_\alpha(\mathbf{k}) + W(k) \frac{\partial}{\partial u_\alpha(-\mathbf{k})} \right\} \qquad (6.71)$$

and

$$V = \sum_{\mathbf{k}} \sum_{\mathbf{j}} M_{\alpha\beta\gamma}(\mathbf{k}) u_\beta(\mathbf{j}) u_\gamma(\mathbf{k} - \mathbf{j}) \frac{\partial}{\partial u_\alpha(\mathbf{k})}$$

$$= -\sum_{\mathbf{k}} \sum_{\mathbf{j}} M_{\alpha\beta\gamma}(\mathbf{k}) u_\beta(\mathbf{j}) u_\gamma(-\mathbf{k} - \mathbf{j}) \frac{\partial}{\partial u_\alpha(-\mathbf{k})} \qquad (6.72)$$

where we have replaced $\mathbf{k}$ by $-\mathbf{k}$ and used the relation $M_{\alpha\beta\gamma}(\mathbf{k}) = -M_{\alpha\beta\gamma}(-\mathbf{k})$.

Now we look for a solution of (6.70) as an expansion in terms of a book-keeping parameter λ. This is the same approach as before, and again λ is assumed to be superficially of the same order as $M_{\alpha\beta\gamma}(\mathbf{k})$. The essential difference this time is that, instead of expanding the velocity field, we are expanding the probability distribution P as follows:

$$P = P_0 + \lambda P_1 + \lambda^2 P_2 + \cdots \qquad (6.73)$$

where the zero-order coefficient satisfies

$$L_0 P_0 = 0, \qquad (6.74)$$

for some operator L_0, the form of which remains to be determined.

For this method to work, clearly we have to find a form for L_0 such that $P_0 \simeq P$. Then we can simultaneously add and subtract $L_0 P$ from eqn (6.70) to obtain

$$L_0 P - VP - (L_0 - L)P = 0. \qquad (6.75)$$

Further, we can associate the first order in λ with VP, as this essentially represents the non-linearity of the Navier–Stokes equation, and the second order in λ with the term $(L_0 - L)P$, which represents the correction to our perturbation procedure (i.e. it vanishes when $P = P_0$). Thus we can base our perturbation expansion on (6.75), rewritten as

$$L_0 P - \lambda VP - \lambda^2 (L_0 - L)P = 0. \qquad (6.75a)$$

At a later stage we shall show that these assignments of a particular order of λ to each term are in fact self-consistent.

Now we substitute (6.73) for P in eqn (6.75a), and equate coefficients of powers of λ. The result for zero order is just eqn (6.74), while at first order we have

$$L_0 P_1 = VP_0, \qquad (6.76)$$

and at second order

$$L_0 P_2 = VP_1 + (L_0 - L)P_0, \qquad (6.77)$$

or, using eqn (6.73),

$$L_0 P_2 = VP_1 - LP_0. \qquad (6.78)$$

This process can, of course, be carried on to any order, but the second order will be sufficient for our purposes.

Our next step is to obtain a form for L_0. As this is intended to approximate the operator L, we begin by considering this quantity. There are two things which we should note about it. First, it is the sum over $\mathbf{k}$ of single-mode operators. Second, each of the single-mode operators is of the Fokker–Planck form. This can be seen by comparison with the stationary case of (4.32). We replace the displacement **x** of Brownian motion with the velocity **u** for turbulence, and set the variable t equal to zero, for stationarity.

Neither of these two properties is particularly surprising. The operator L only involves linear effects, and the turbulence problem has indeed been formulated by us in terms of a close analogy with Brownian motion. However, once the non-linearity comes into play, we know that mode-coupling effects will dominate the dynamics and so we really ought not to expect that a renormalized version of L will retain the single-mode form.

Nevertheless, we have previously seen that many-body theories tend to rely on mean-field approaches, in which the interaction picture can be replaced by an effective field or a renormalized single-particle picture. In particular, we have seen how a renormalization programme is carried out in the DIA by replacing zero-order quantities by exact quantities, while retaining the general form of the equations unchanged. The Edwards–Fokker–Planck (EFP) theory (which is very different from DIA) effectively makes this kind of assumption at an earlier stage (and then proceeds to a quite different *post hoc* justification). Edwards assumed that the force spectrum $W(k)$ and the modal decay rate due to viscosity vk^2 could be replaced by more general forms $d(k)$ and $\omega(k)$:

$$d(k) = W(k) + s(k) \qquad (6.79)$$

$$\omega(k) = vk^2 + r(k) \qquad (6.80)$$

where $r(k)$ and $s(k)$ represent the effects of the non-linearity. As we shall see shortly, these two quantities are not independent of each other.

In physical terms, it can be argued that any mode **k** will experience an input of energy from the stirring forces plus a contribution from non-linear transfer from modes less than **k**. Similarly, the energy loss from mode **k** will be partially due to non-linear transfer to modes greater than **k**, along with some direct viscous dissipation of energy in mode **k**. Hence, in this way, we have a very straightforward interpretation of $d(k)$ and $\omega(k)$.

Equations (6.79) and (6.80) represent, in effect, the simultaneous renormalization of the stirring forces and the viscosity. Correspondingly, the renormalized version of L is readily obtained when one adds on terms containing $s(k)$ and $r(k)$ to obtain

$$L_0 = \sum_{\mathbf{k}} \frac{\partial}{\partial u_\alpha(\mathbf{k})} \left\{ \omega(k) u_\alpha(\mathbf{k}) + d(k) \frac{\partial}{\partial u_\alpha(-\mathbf{k})} \right\}. \qquad (6.81)$$

and the correction operator $L_0 - L$, where we subtract terms containing $s(k)$ and $r(k)$, takes the form

$$L_0 - L = \sum_{\mathbf{k}} \frac{\partial}{\partial u_\alpha(\mathbf{k})} \left\{ r(k) u_\alpha(k) + s(k) \frac{\partial}{\partial u_\alpha(-\mathbf{k})} \right\}. \qquad (6.82)$$

With L_0 still made up of single-mode operators, each of Fokker–Planck form, it is natural to interpret $\omega(k)$ as a dynamical friction and $d(k)$ as an eddy diffusivity (see Section 4.1.2). It is easily seen that the solution of (6.74) is now given by

$$P_0 = N \exp\left\{ -\sum_k \frac{u_\alpha(\mathbf{k})u_\alpha(-\mathbf{k})}{q(k)} \right\} \tag{6.83}$$

where N is, as usual, an appropriate normalization, and the dynamical friction and the diffusion coefficient must satisfy the condition

$$2\omega(k)q(k) = d(k), \tag{6.84}$$

where the second moment of the velocity field is given by

$$\left(\frac{2\pi}{L}\right)^3 \langle u_\alpha(k)u_\beta(-\mathbf{k}) \rangle = \int P_0[\mathbf{u}(k)]u_\alpha(\mathbf{k})u_\beta(-\mathbf{k})\delta\mathbf{u}(\mathbf{k})$$

$$= D_{\alpha\beta}(\mathbf{k})q(k). \tag{6.85}$$

Note that this is equivalent to requiring that the homogeneous isotropic second-order moment (see eqns (2.88) and (2.97)) is given by an expectation value evaluated against the zero-order probability distribution P_0. This requirement is the major hypothesis of the EFP theory.

6.2.3   Evaluation of the coefficients in the expansion for the probability distribution of velocities

In order to solve eqns (6.76) and (6.78), we have to invert the operator L_0. Now, it is well known that most inhomogeneous differential equations in mathematical physics can be solved by eigenfunction expansion, provided that the associated differential operator is linear. For instance, let us consider the inhomogeneous equation

$$Hy(x) = p(x),$$

where $p(x)$ is a given function and $y(x)$ satisfies some prescribed boundary conditions. The eigenfunctions of the operator H satisfy the equation

$$Hf_n(x) = e_n f_n(x),$$

where the e_n are the eigenvalues and the eigenfunctions $\{f_0, f_1, \ldots f_n, \ldots\}$ are assumed to form a complete orthonormal set. Note that the index n is not summed in the above equation. The solution to the inhomogeneous equation is then found by expanding both $y(x)$ and $p(x)$ in terms of the eigenfunctions of H, with the result

$$y(x) = \sum_n \left\{ \int f_n(x')p(x')\,dx' \right\} \frac{f_n(x)}{e_n}.$$

Details of this procedure can be found in any book on mathematical methods. (e.g. Mathews and Walker 1965); from our point of view the important fact is that the inversion of the operator can be accomplished by the purely arithmetic inversion of the eigenvalues.

We now follow Edwards (1964) and introduce the eigenfunctions of L_0 First we write L_0 as the sum of its constituent single-mode operators $L_0(k)$:

$$L_0 = \sum_k L_0(\mathbf{k}). \qquad (6.86)$$

Then we introduce the eigenfunctions of $L_0(\mathbf{k})$ in the usual way:

$$L_0(\mathbf{k})f_n[\mathbf{u}(\mathbf{k})] = e_n(k)f_n[\mathbf{u}(\mathbf{k})], \qquad (6.87)$$

where the eigenfunctions are given by

$$f_n[\mathbf{u}(\mathbf{k})] = H_n[\mathbf{u}(\mathbf{k})]P_0[\mathbf{u}(\mathbf{k})] \qquad (6.88)$$

and the eigenvalues are given by

$$e_n(k) = n\omega(k). \qquad (6.89)$$

Here, the H_n are Hermite polynomials. These are special cases of the confluent hypergeometric function, and there are recursion relations or generating functions which allow one to find the Hermite polynomial of any order n (e.g. see Matthews and Walker 1965). We shall only need the first three polynomials, which are

$$H_0[\mathbf{u}(\mathbf{k})] = 1,$$

$$H_1[\mathbf{u}(\mathbf{k})] = \frac{\mathbf{u}(\mathbf{k})}{\{q(\mathbf{k})\}^{1/2}}$$

$$H_2[\mathbf{u}(\mathbf{k})] = \frac{\mathbf{u}(\mathbf{k}) \cdot \mathbf{u}(\mathbf{k}) - q(\mathbf{k})}{2^{1/2}q(\mathbf{k})}. \qquad (6.90)$$

Now, inverting L_0 on the l.h.s., we can write eqn (6.76) as

$$P_1 = L_0^{-1}\{VP_0\}$$

$$= -L_0^{-1}\left\{\sum_k \sum_j M_{\alpha\beta\gamma}(\mathbf{k})u_\beta(\mathbf{j})u_\gamma(\mathbf{k}-\mathbf{j})\frac{\partial P_0}{\partial u_\alpha(-\mathbf{k})}\right\}$$

$$= L_0^{-1}\left\{\sum_k \sum_j \frac{M_{\alpha\beta\gamma}(\mathbf{k})u_\beta(\mathbf{j})u_\gamma(\mathbf{k}-\mathbf{j})u_\alpha(\mathbf{k})P_0}{q(k)}\right\}, \qquad (6.91)$$

where the last line follows from the operation of the functional derivative upon P_0 as given be (6.83).

Comparison of the r.h.s. of (6.91) with eqn (6.90) for the Hermite polynomials shows that P_1 contains the three first-order functions $H_1(\mathbf{j})$, $H_1(\mathbf{k}-\mathbf{j})$, and $H_1(\mathbf{k})$. Thus L_0^{-1} can be represented in terms of the first-order eigenvalues and, with some rearrangement, P_1 becomes

$$P_1 = \sum_{\mathbf{k}} \sum_{\mathbf{j}} M_{\alpha\beta\gamma}(\mathbf{k}) u_\alpha(\mathbf{k}) u_\beta(\mathbf{j}) u_\gamma(\mathbf{k} - \mathbf{j}) \times$$

$$\times \frac{P_0}{q(k)\{\omega(k) + \omega(j) + \omega(|\mathbf{k} - \mathbf{j}|)\}}. \qquad (6.92)$$

We should note two points here. First, P_1 is of the form of an operator acting on P_0. This is true of all P_n. Second, the operator V acting on P_0 is first order in the eigenfunction expansion which is consistent with our assignment of the bookkeeping parameter λ in equation (6.75a).

The second-order coefficient P_2 is given by eqn (6.78), and we write this equation more explicitly by substituting from (6.72) and (6.92) to obtain the first term on the r.h.s., and from (6.71) and (6.83) to obtain the second:

$$L_0 P_2 = -\sum_{\mathbf{l}} \sum_{\mathbf{p}} \sum_{\mathbf{k}} \sum_{\mathbf{j}} M_{\delta\rho\varepsilon}(\mathbf{l}) M_{\alpha\beta\gamma}(\mathbf{k}) u_\rho(\mathbf{p}) u_\varepsilon(\mathbf{l} - \mathbf{p}) \times$$

$$\times \frac{\partial \{u_\alpha(\mathbf{k}) u_\beta(\mathbf{j}) u_\gamma(\mathbf{k} - \mathbf{j})\}}{\partial u_\delta(-\mathbf{l})} \frac{P_0}{q(k)\{\omega(k) + \omega(j) + \omega(|\mathbf{k} - \mathbf{j}|)\}} -$$

$$-\sum_{\mathbf{k}} \left\{ vk^2 - \frac{W(k)}{q(k)} \right\} \frac{\{q(k) - \mathbf{u}(\mathbf{k}) \cdot \mathbf{u}(\mathbf{k})\} P_0}{q(k)}. \qquad (6.93)$$

We shall not formally carry out the inversion of the L_0 on the l.h.s., and so we merely note from the last term on the r.h.s. that P_2 contains the second-order Hermite function, which is consistent with our putting the last term on the r.h.s. of eqn (6.75) as order λ^2.

A general procedure for writing down terms of any order is given in Edwards (1964), but we shall only work to second order here.

6.2.4   The energy-balance equation

In eqn (6.85) we have the requirement that the velocity field correlation should be given by an expectation average evaluated against P_0 rather than P. This implies a constraint on the higher terms in the expansion for P:

$$\int u_\alpha(\mathbf{k}) u_\alpha(-\mathbf{k}) \{P_2 + P_4 + P_6 + \cdots\} \delta\mathbf{u}(\mathbf{k}) = 0. \qquad (6.94)$$

Note that terms P_1, P_3, and so on contain odd powers of the velocity and hence vanish automatically when integrated over all function space of the velocities. Hence, if we satisfy this requirement to second order, we have the relation

$$\int u_\alpha(\mathbf{k}) u_\alpha(-\mathbf{k}) P_2[\mathbf{u}(\mathbf{k})] \delta\mathbf{u}(\mathbf{k}) = 0, \qquad (6.95)$$

and it turns out that this gives us the required energy-balance equation for $q(k)$, and hence, through (2.101), the energy spectrum $E(k)$.

The task of working this out in detail is very much simplified by the observation, due to Leslie (1973), that we do not need to solve (6.93) explicitly for P_2. Instead we can use the fact that (6.95) is equivalent to

$$\int u_\alpha(\mathbf{k})u_\alpha(-\mathbf{k})L_0P_2[\mathbf{u}(\mathbf{k})]\delta\mathbf{u}(\mathbf{k}) = 0. \tag{6.96}$$

Hence, multiplying through eqn (6.78) for $P_2[\mathbf{u}(\mathbf{k})]$ by $u_\alpha(-\mathbf{k})$, and integrating with respect to $\mathbf{u}(\mathbf{k})$, we obtain

$$\int u_\alpha(\mathbf{k})u_\alpha(-\mathbf{k})LP_0\delta\mathbf{u} = \int u_\alpha(\mathbf{k})u_\alpha(-\mathbf{k})VP_1\delta\mathbf{u}. \tag{6.97}$$

Now, substituting from (6.71) for L, we can write the l.h.s. of (6.97) as

$$\int u_\alpha(\mathbf{k})u_\alpha(-\mathbf{k})\sum_j\left\{\frac{\partial}{\partial u_\beta(\mathbf{j})}\right\}\left\{vj^2u_\beta(\mathbf{j}) + W(j)\frac{\partial}{\partial u_\beta(\mathbf{j})}\right\}P_0\delta\mathbf{u}$$

$$= \int u_\alpha(\mathbf{k})u_\alpha(-\mathbf{k})\sum_j\frac{\partial}{\partial u_\beta(\mathbf{j})}\left\{vj^2u_\beta(\mathbf{j}) - \frac{W(j)u_\beta(\mathbf{j})}{q(j)}\right\}P_0\delta\mathbf{u}$$

$$= -\sum_j\int\delta(\mathbf{k}-\mathbf{j})\delta_{\alpha\beta}u_\alpha(-\mathbf{k})\left\{vj^2u_\beta(\mathbf{j}) - \frac{W(j)u_\beta(\mathbf{j})}{q(j)}\right\}P_0\delta\mathbf{u}$$

$$= -2vk^2q(k) + W(k), \tag{6.98}$$

where the penultimate line is obtained by integrating by parts and using the boundary condition that P_0 becomes exponentially small as $\mathbf{u}$ tends to infinity. The last line is obtained straightforwardly with the use of (6.85).

With this result, eqn (6.97) can now be written as

$$W(k) - 2vk^2q(k) = \int\sum_l\sum_p\sum_m\sum_j M_{\delta\rho\varepsilon}(\mathbf{l})M_{\alpha\beta\gamma}(\mathbf{m})u_\sigma(\mathbf{k})u_\sigma(-\mathbf{k}) \times$$

$$\times u_\rho(\mathbf{p})u_\varepsilon(\mathbf{l}-\mathbf{p})\left\{\frac{\partial}{\partial u_\delta(-\mathbf{l})}\right\}u_\alpha(\mathbf{m})u_\beta(\mathbf{j})u_\gamma(\mathbf{m}-\mathbf{j}) \times$$

$$\times \frac{1}{q(m)\{\omega(m) + \omega(j) + \omega(|\mathbf{m}-\mathbf{j}|\}}P_0[\mathbf{u}]\delta\mathbf{u}, \tag{6.99}$$

where the l.h.s. was obtained by substitution of (6.72) for V and (6.92) for P_1. The functional derivative can be eliminated using partial integration, just as in the manipulation leading to eqn (6.98). Then the wavenumber summation over $\mathbf{l}$ can be carried out to eliminate the delta function $\delta(\mathbf{k}-\mathbf{l})$, and, replacing the functional integration against $P_0[\mathbf{u}]$ by the Dirac brackets $\langle\ \rangle$, we have

$$W(k) - 2vk^2 q(k) = \sum_{\mathbf{p}} \sum_{\mathbf{m}} \sum_{\mathbf{j}} M_{\sigma\rho\varepsilon}(\mathbf{k}) M_{\alpha\beta\gamma}(\mathbf{m}) \times$$

$$\times \langle u_\sigma(\mathbf{k}) u_\rho(\mathbf{p}) u_\varepsilon(\mathbf{l} - \mathbf{p}) u_\alpha(\mathbf{m}) u_\beta(\mathbf{j}) u_\gamma(\mathbf{m} - \mathbf{j}) \rangle \times$$

$$\times \frac{1}{q(m)\{\omega(m) + \omega(j) + \omega(|\mathbf{m} - \mathbf{j}|)\}}. \quad (6.100)$$

The correlation of six velocities can be evaluated in the same way as we have done previously for four velocities. That is we make the factorization

$$\langle 123456 \rangle = \langle 12 \rangle \langle 34 \rangle \langle 56 \rangle + \text{all combinations.}$$

Note that although this will give us three q factors in the numerator of (6.100), the factor q in the denominator will give cancellations. Details will be found in Appendix E, and we merely quote the result here. In the limit of infinite system volume, eqn (6.100) becomes

$$W(k) - 2vk^2 q(k) = -2 \int d^3j \, L(\mathbf{k}, \mathbf{j}) q(|\mathbf{k} - \mathbf{j}|) \{q(j) - q(k)\} \times$$

$$\times \frac{1}{\{\omega(k) + \omega(j) + \omega(|\mathbf{k} - \mathbf{j}|)\}} \quad (6.101)$$

which is the required equation for $q(k)$. We can confirm its physical interpretation as an energy-balance equation by working out the expectation value of the triple moment against the probability distribution functional $P[\mathbf{u(k)}]$. To second order, this can be written as

$$\left\langle M_{\alpha\beta\gamma}(\mathbf{k}) \sum_{\mathbf{j}} u_\beta(\mathbf{j}) u_\gamma(\mathbf{k} - \mathbf{j}) u_\alpha(-\mathbf{k}) \right\rangle = \int M_{\alpha\beta\gamma}(\mathbf{k}) \sum_{\mathbf{j}} u_\beta(\mathbf{j}) u_\gamma(\mathbf{k} - \mathbf{j}) u_\alpha(-\mathbf{k}) P_1 \delta u.$$

$$(6.102)$$

It is simple to show, and is therefore left as an exercise for the reader, that substituting (6.92) for P_1 reduces (6.102) to the r.h.s. of eqn (6.99).

6.2.5 The response equation

At this stage we have three unknown functions $q(k)$, $d(k)$ and $\omega(k)$, to be determined from eqns (6.101) and (6.84). Or, equivalently, from eqns (6.79) and (6.80), we can replace $d(k)$ and $\omega(k)$ by $s(k)$ and $r(k)$ respectively as the functions to be determined. However, either way, it is evident that another condition is needed to provide the requisite third equation.

Edwards (1964) argued (in effect) that $d(k)$ came about as a renormalization of the random stirring forces $\mathbf{f}(\mathbf{k}, t)$ to take account of the non-linear term in the Navier–Stokes equation. Hence, one could think of $d(k)$ as being made up from the spectrum of the stirring forces $W(k)$ plus the spectrum of the random forces due to the non-linear term which could be interpreted as $s(k)$.

That is to say, we are assuming that the effect of the non-linear term on the velocity field in mode $\mathbf{k}$ can be represented by a random force $\hat{f}_\alpha(\mathbf{k}, t)$ which is given by

$$\hat{f}_\alpha(\mathbf{k}, t) = \sum_j M_{\alpha\beta\gamma}(\mathbf{k}) u_\beta(\mathbf{j}, t) u_\gamma(\mathbf{k} - \mathbf{j}, t). \tag{6.103}$$

This is, of course, rigorously true. However, to go further and assume that $\mathbf{f}$ is, like the stirring forces, Gaussian and uncorrelated with $\mathbf{u}(\mathbf{k})$ seems rather imponderable. Nevertheless, such an assumption leads to quite reasonable answers. If we define $s(k)$ by analogy with (3.94) and (4.85) for $W(k)$, then it can be shown (Edwards 1964) that the assumption

$$s(k) = \int \langle \hat{f}_\alpha(\mathbf{k}, t) \hat{f}_\alpha(-\mathbf{k}, t) \rangle \, \mathrm{d}t \tag{6.104}$$

leads to the relationship

$$s(k) = 2 \int \frac{\mathrm{d}^3 j \, L(\mathbf{k}, \mathbf{j}) q(|\mathbf{k} - \mathbf{j}|q(j)}{\omega(k) + \omega(j) + \omega(|\mathbf{k} - \mathbf{j}|)}. \tag{6.105}$$

Now we can use (6.84) to obtain a relationship for the dynamical friction. From eqns (6.79) and (6.80) we rewrite (6.84) as

$$2\{vk^2 + r(k)\} q(k) = W(k) + s(k).$$

Then, we rearrange the energy-balance equation (6.101) to take the above form, using (6.105) for $s(k)$:

$$2 \left\{ vk^2 + \int \frac{\mathrm{d}^3 j \, L(\mathbf{k}, \mathbf{j}) q(|\mathbf{k} - \mathbf{j}|)}{\omega(k) + \omega(j) + \omega(|\mathbf{k} - \mathbf{j}|)} \right\} q(k)$$
$$= W(k) + 2 \int \frac{\mathrm{d}^3 j \, L(\mathbf{k}, \mathbf{j}) q(|\mathbf{k} - \mathbf{j}|) q(j)}{\omega(k) + \omega(j) + \omega(|\mathbf{k} - \mathbf{j}|)}, \tag{6.106}$$

from which it follows that we can identify $\omega(k)$ as

$$\omega(k) = vk^2 + \int \frac{\mathrm{d}^3 j \, L(\mathbf{k}, \mathbf{j}) q(|\mathbf{k} - \mathbf{j}|)}{\omega(k) + \omega(j) + \omega(|\mathbf{k} - \mathbf{j}|)}. \tag{6.107}$$

As we shall see in the next section, when we make a comparison with DIA, eqn (6.107) can be interpreted as the response equation of the EFP theory.

6.2.6 Comparison with the DIA

The final form of the EFP theory consists of eqns (6.101) and (6.107) for $q(k)$ and $\omega(k)$ respectively. Clearly the most obvious difference between EFP and DIA is that the former is independent of the time, although a consideration of the Green function (propagator) of the Liouvillian suggests (Edwards 1964) that the time-dependent case could be handled by the relationship

$$Q(k, t - t') = q(k) \exp\{-\omega(k)|t - t'|\}. \tag{6.108}$$

In order to make a detailed comparison of the two theories, we first rewrite the DIA energy balance equation (6.36) as

$$\left(\frac{d}{dt} + 2vk^2\right) Q(k, t) = 2P(k, t) + w(k, t), \tag{6.109}$$

where we have divided both sides by $4\pi k^2$ and $P(k, t)$ is given by (6.34).

For the response equation, we take (6.44) and interchange the dummy variables j and $|\mathbf{k} - \mathbf{j}|$ to obtain

$$\left(\frac{d}{dt} + vk^2\right) G(k, t - t')$$

$$= -\int d^3 j \int_{t'}^t dt'' \, L(\mathbf{k}, \mathbf{j}) G(j, t - t'') G(k, t' - t'') Q(|\mathbf{k} - \mathbf{j}|, t - t''). \tag{6.110}$$

Now, in order to deal with the time-dependences in (6.109) and (6.110), we follow Kraichnan (1964b) and arbitrarily introduce exponential time dependences. That is, we assume that the two-time correlation is given by (6.108), even though this cannot be true for small values of $t - t'$ for the requirement that there should be invariance under interchange of t and t' implies that the time derivative of Q must vanish at $t = t'$, a condition not satisfied by the exponential form.

Similarly, a characteristic time can be associated with the response function:

$$\frac{1}{\eta(k)} = \int_0^\infty G(k, t) \, dt, \tag{6.111}$$

with the corresponding exponential form

$$G(k, t - t') = \begin{cases} \exp\{-\eta(k)(t - t')\} & \text{for } t > t' \\ 0 & \text{for } t < t'. \end{cases} \tag{6.112}$$

Kraichnan showed that substituting (6.108) and (6.112) into the energy-balance equation (6.109), and performing integrations over intermediate times, led to the EFP energy equation, as given by (6.101), provided that we take the additional step of putting $\eta(k) = \omega(k)$. A similar procedure applied to the response equation (6.110) led to the following equation:

$$\omega(k) = vk^2 + \int \frac{d^3 j \, L(\mathbf{k}, \mathbf{j}) q(|\mathbf{k} - \mathbf{j}|)}{\omega(j) + \omega(|\mathbf{k} - \mathbf{j}|)}, \tag{6.113}$$

which differs from the EFP form (6.107) only by the presence of two eigen-values, rather than three, in the denominator.

6.2.7 The limit of infinite Reynolds number

We have previously seen that conservation of energy requires the non-linear transfer term to vanish when integrated over all **k**-space (see eqn (2.125)). We have also seen that the DIA energy equation achieves this result by virtue of two cancelling terms (see eqn (6.42)). Similar considerations can be seen to apply to the EFP theory.

Consider the energy-balance equation (6.101) rewritten as

$$2r(k)q(k) - s(k) = W(k) - 2vk^2 q(k), \qquad (6.114)$$

where we have taken the non-linear term over to the l.h.s. (and vice versa) and substituted from eqns (6.105) and (6.107) for $s(k)$ and $r(k)$. In this particular form, the division of the non-linear term into input (i.e. $s(k)$) and output (i.e. $2r(k)q(k)$) parts is rather clear. We have already discussed the physical signifi-cance of this division into two terms for the DIA energy equation in Section 6.1.5. It should be noted that that discussion also applies to the EFP theory, but not to the time-independent form of DIA, as represented by (6.113) for the response equation. The problem is that the response integral has only two eigenvalues in the denominator, and therefore cannot be written as part of the energy integral which has three such eigenvalues in its denominator.

However, conservation of energy can be formally shown for EFP by inte-grating each individual term of (6.114) to obtain

$$\int 2r(k)q(k)\,\mathrm{d}^3k - \int s(k)\,\mathrm{d}^3k = \varepsilon - \varepsilon = 0, \qquad (6.115)$$

where the r.h.s. vanishes because the two integrals are identical and the l.h.s. vanishes because the two terms are each equal to the dissipation rate (see eqn (4.90)).

Edwards (1965) has shown that the conservation property of the energy integral leads to some rather interesting results in the limit of infinite Reynolds number. Let us suppose that we allow the viscosity to shrink to zero in such a way that the dissipation rate remains constant. It should be noted that this is not the same thing as assuming that the fluid is ideal (i.e. inviscid and hence governed by the Euler equation), which corresponds to an artificial case of zero dissipation. We may indeed think of it as a 'thought experiment' in Einstein's sense; that is, one which may be physically impossible to carry out yet which does not violate any known physical principle. Then it is reasonable to conclude that the dissipation becomes concentrated into ever higher wave-numbers (see eqn (2.133) for the dissipation wavenumber), and, in the limit, occurs only at infinite wavenumbers.

That is, we take the limit of zero viscosity

$$\lim_{v \to 0} 8\pi v k^2 q(k) = \varepsilon \delta(k - \infty), \tag{6.116}$$

which clearly has the necessary property

$$\varepsilon \int_0^\infty \delta(k - \infty)\,\mathrm{d}k = \varepsilon. \tag{6.117}$$

The input $W(k)$ is, of course, arbitrary but should be chosen to be peaked near the origin in order that we can obtain universal behaviour of the energy spectrum at higher wavenumbers. As we decrease the viscosity, it is easier to excite low wavenumbers, and it is arguable that we can excite the system at $k = 0$ in the limit of vanishing viscosity. Thus we take for the input

$$\lim_{v \to 0} 4\pi k^2 W(k) = \varepsilon \delta(k). \tag{6.118}$$

Under these circumstances, the entire **k**-space should satisfy the conditions for an inertial range, and eqn (2.141) for the Kolmogorov spectrum should take the form

$$E(k) = \alpha \varepsilon^{2/3} k^{-5/3} \tag{6.119}$$

for all k. The corresponding form for the spectral density function $q(k)$ is

$$q(k) = \left(\frac{\alpha}{4\pi}\right) \varepsilon^{2/3} k^{-11/3}, \tag{6.120}$$

which again holds for all k.

If we substitute (6.120) into the energy-balance equation (6.101), then dimensional considerations indicate that the modal decay rate $\omega(k)$ must take the form

$$\omega(k) = \beta \varepsilon^{1/3} k^{2/3} \tag{6.121}$$

where β, like α, is a constant.

Formally we can write eqn (6.101), in the limit of infinite Reynolds number, as

$$8\pi k^2 \int \mathrm{d}^3 j\, L(\mathbf{k},\mathbf{j}) q(|\mathbf{k} - \mathbf{j}|)\{q(k) - q(j)\} \frac{1}{\{\omega(k) + \omega(j) + \omega(|\mathbf{k} - \mathbf{j}|)\}}$$

$$= \varepsilon \delta(k) - \varepsilon \delta(k - \infty), \tag{6.122}$$

where we have multiplied both sides by $4\pi k^2$ and invoked eqns (6.116) and (6.118) for the particular case of the infinite Reynolds number limit. Then, if we substitute (6.120) and (6.121), we find

$$8\pi k^2 \left(\frac{\alpha}{4\pi}\right)^2 \beta^{-1} \int d^3j\, L(\mathbf{k},\mathbf{j})|\mathbf{k}-\mathbf{j}|^{-11/3} \frac{k^{-11/3}-j^{-11/3}}{k^{2/3}+j^{2/3}+|\mathbf{k}-\mathbf{j}|^{2/3}}$$

$$= \delta(k) - \delta(k-\infty). \tag{6.123}$$

It may seem surprising that the integral actually gives delta functions at the origin and at infinity. In fact this reflects a form of non-uniform convergence, which is due to the integral's being made up of two terms which are individually divergent but which cancel sufficiently rapidly for their difference to be integrable. A discussion is given in Edwards (1965) in terms of a simple one-dimensional model.

Let us consider the integral of both sides of (6.122) over k. If we integrate up to $k = K$, where K is not equal to zero or infinity but is otherwise arbitrary, then the integration over j will cancel by symmetry for the range $0 < j \leqslant K$. Hence the value of the integral with respect to k will be independent of its upper limit, a characteristic of a delta function. The integral has been evaluated for the arbitrary choice $K = 1$. That is, if we write

$$2\alpha^2\beta^{-1} \int_0^1 dk \int_1^\infty dj\, k^2 j^2 L(\mathbf{k},\mathbf{j})|\mathbf{k}-\mathbf{j}|^{-11/3} \frac{k^{-11/3}-j^{-11/3}}{k^{2/3}+j^{2/3}+|\mathbf{k}-\mathbf{j}|^{2/3}} = 1 \tag{6.124}$$

as

$$\frac{\alpha^2 C}{\beta} = 1, \tag{6.125}$$

then numerical integration (Edwards 1965) gives $C = 0.19$.

It should be noted that we still need a second equation in order to solve simultaneously with (6.125) for the Kolmogorov constants of proportionality. The obvious candidate would seem to be the response equation (6.107). Unfortunately the above procedures do not go through in this case and the response integral turns out to be divergent. This is the reason why EFP theory does not give the Kolmogorov spectrum, despite the behaviour of the energy equation. We shall return to this point again, when making an overall assessment of RPT theories.

6.3 Self-consistent field theory

The method of the self-consistent field has long been used in the quantum many-body problem. An exposition can be found in almost any text on quantum mechanics, and so we shall only give a very brief description here.

Suppose that we wish to calculate the wavefunction for an electron in a hydrogen atom. In principle, at least, there is no problem. The electron moves in the Coulomb potential of the nucleus and one simply solves the Schrödinger equation for its wavefunction. However, if we extend the problem to multi-electron atoms, then the interaction with the repulsive potential of all the other electrons must be considered.

Consider helium as an example. We take one electron to move in a potential V which is due partly to the nucleus and partly to the other electron treated as a distribution of charge density proportional to its wavefunction. A method of successive approximations is then used. Assume the wavefunction for the electron, calculate the effective potential V from Poisson's equation, and use V in the Schrödinger equation to calculate the new waveform. Self-consistency then demands that the initial and final wavefunctions should be the same.

The method is admirably pragmatic and easily understood in principle. Its introduction to turbulence theory was due to Herring (1965), who took as his starting point the Liouville equation derived by Edwards. In order to discuss the self-consistent field (SCF) theory, we shall change our notation to that of Herring and rewrite eqn (6.70) as

$$L^0 P - VP = 0, \tag{6.126}$$

where the change from L to L_0 is significant and reflects the fact that we shall be looking at a formal renormalization programme in which L_0 is the bare or viscous operator. In this sense, SCF will be seen to resemble DIA rather than EFP, although there are in fact many procedural similarities between the theories of Herring and Edwards.

Herring began by assuming that the basis operator for his expansion method was the sum of single-mode operators. The effect of mode coupling was then taken into account through an expansion in single-mode interactions. In this respect, SCF can be seen as less phenomenological than EFP, in which the renormalized single-mode (i.e. quasi-particle) form is assumed from the outset.

We begin with some definitions relating to single-mode forms. First, we can generalize the concept of reduced distributions (e.g. see (4.19)) in particle statistical mechanics to single-mode distributions in the turbulent case. Let us write (4.6) in **k**-space as

$$P[\mathbf{u}(\mathbf{k})] = \lim_{n \to \infty} P_n[\mathbf{u}(\mathbf{k}_1), \mathbf{u}(\mathbf{k}_2), \ldots, \mathbf{u}(\mathbf{k}_n)], \tag{6.127}$$

where the limit of infinite system volume (corresponding to $n \to \infty$) will not be taken until the end of the calculation. Here the notation $\mathbf{k}_n$ is shorthand for the l.h.s. of (2.72), which is our definition of a wavevector in a finite system. A particular value of n then corresponds to a particular set $\{n_1, n_2, n_3\}$.

In order to obtain a single-mode distribution, we integrate out all variables except (say) $\mathbf{u}(\mathbf{k}_n)$; thus

$$P_n[\mathbf{u}(\mathbf{k}_n)] = \int P[\mathbf{u}(\mathbf{k})]\, d\mathbf{u}(\mathbf{k}_1)\, d\mathbf{u}(\mathbf{k}_2)\ldots d\mathbf{u}(\mathbf{k}_{n-1})\, d\mathbf{u}(\mathbf{k}_{n+1})\ldots, \quad (6.128)$$

which defines the single-mode distribution $P_n[\mathbf{u}(\mathbf{k}_n)]$.

Now define the single-mode projection of an operator A by

$$\langle A \rangle = \sum_n \langle A \rangle_n, \qquad (6.129)$$

where

$$\langle A \rangle_n = \int A \prod_{m \neq n} P[\mathbf{u}(\mathbf{k}_m)]\, d\mathbf{u}(\mathbf{k}_1)\, d\mathbf{u}(\mathbf{k}_2)\ldots d\mathbf{u}(\mathbf{k}_{n-1})\, d\mathbf{u}(\mathbf{k}_{n+1})\ldots. \quad (6.130)$$

Note that the notation $\langle\ \rangle$ can be used here for this specialized purpose, as in this section there should be no confusion with its more usual meaning of ensemble average.

Then a single-mode operator can be defined as one which is invariant under the above averaging process, or one which satisfies the relation

$$\langle A \rangle = A. \qquad (6.131)$$

Now consider eqn (6.126). If $V = 0$, then it is clear that there will be no non-linear mixing and the solution P will be a product of single-mode distributions. However, if V is not zero, then non-linear mixing must result in some mode coupling and we should write the general solution as

$$P = \prod_n P_n + R, \qquad (6.132)$$

where the P_n are the exact single-mode distributions and R is a remainder term to take account of the mode couplings induced by V. It follows that R must satisfy the condition

$$\langle R \rangle_n = 0, \qquad (6.133)$$

where the operation $\langle\ \rangle_n$ is defined by eqn (6.130), and the general perturbation method then consists of expanding R about $V = 0$.

In order to do this, it is convenient to introduce a new operator L, which should satisfy two requirements. First it should be of single-mode form, or

$$\langle L \rangle = \sum_n L_n = L, \qquad (6.134)$$

and second, in order to conserve probability, it should satisfy

$$\int L_n F[\mathbf{u}(\mathbf{k}_n)]\, d\mathbf{u}(\mathbf{k}_n) = 0 \qquad (6.135)$$

for any well-behaved test function $F[\mathbf{u}(\mathbf{k}_n)]$. We also introduce the associated function P' which is the solution of

$$LP' = 0 \tag{6.136}$$

such that

$$P' = \prod_n P'_n. \tag{6.137}$$

In other words, P' is the product of single-mode operators. Lastly, as V tends to zero, consistency requires that $L \rightarrow L_0$, and $P' \rightarrow P$.

We now add the operator L to both sides of eqn (6.126), thereby leaving it unchanged, to obtain

$$LP = (L - L_0 + V)P. \tag{6.138}$$

The formal solution of this equation can be written as

$$P = P' + R', \tag{6.139}$$

and we then must solve (6.138) perturbatively for R' in terms of L, L_0, V, and P_n. Our aim is now to find an operator L which (a) satisfies eqns (6.134)–(6.136) and (b) is such that $P'_n = P_n$: this is the self-consistency criterion.

Our procedure is as follows. Noting (6.136), we rewrite eqn (6.138) as

$$LP = LP' + (\Delta + V)P, \tag{6.140}$$

where

$$\Delta = L - L_0. \tag{6.141}$$

Then, inverting the operator on the l.h.s., we obtain

$$P = P' + L^{-1}(\Delta + V)P. \tag{6.142}$$

At this stage there are two points worth mentioning. First there is a technical point here which we are glossing over, in that we should invert the non-singular part of L; see Herring (1965) for a discussion. Second, comparison of eqns (6.142) and (6.139) indicates that

$$R' = L^{-1}(\Delta + V)P.$$

Our next step is to solve (6.142) iteratively, with the result[1]

$$P = P' + L^{-1}TP', \tag{6.143}$$

where T is given by

$$T = (L - L_0 + V) + (L - L_0 + V)L^{-1}T. \tag{6.144}$$

Now, the probability distribution given by (6.143) is in the following form: guessed probability + corrections. For self-consistency we require that the effect of corrections will vanish, such that $P'_n = P_n$. This implies—along with equation (6.132)—that we must have

$$\langle P \rangle_n = \langle \prod_n P_n \rangle + \langle R \rangle_n = \prod_n P_n \tag{6.145}$$

where we have used (6.131) and (6.133). Also, from (6.131) and (6.137), we note that

$$\langle P' \rangle_n = \prod_n P'_n. \tag{6.146}$$

Hence, if we operate on both sides of (6.143) with the single-mode expectation value, we have

$$\prod_n P_n = \prod_n P'_n + \langle L^{-1} T P' \rangle_n$$

and, from the self-consistency condition,

$$\langle L^{-1} T P' \rangle_n = 0$$

or

$$\langle L^{-1} T \rangle_n P' = 0 \tag{6.147}$$

as P' is the product of single-mode distributions and therefore is unaffected by the operation $\langle \ \rangle_n$.

This result can also be written as

$$[L_n - L_{0,n} + \langle V L^{-1} T \rangle] P' = 0, \tag{6.147a}$$

where we have used eqn (6.144) and $L_{0,n}$ is defined by

$$L_0 = \sum_n L_{0,n}. \tag{6.148}$$

Herring has shown that the self-consistency condition can be satisfied, along with (6.132) and (6.135), by choosing L_n such that

$$L_n = L_{0,n} - \langle V L^{-1} T \rangle_n, \tag{6.149}$$

and eqn (6.136) is retained as a separate constraint;

$$L_n P_n = 0 \tag{6.150}$$

where we have put $P'_n = P_n$ in (6.136), in accordance with our self-consistency requirement.

We can obtain the perturbation expansion for $\Delta = L - L_0$ by substituting for T from (6.144) into (6.149) and then iterating about $\Delta = 0$. The result is

$$L = L_0 - \langle V L^{-1} V \rangle + O(V^4). \tag{6.151}$$

Similarly, we can obtain the perturbation expansion for P by substituting (6.151) into (6.143) and expanding T:

$$P = P' + L^{-1} V P' + L^{-1} [V L^{-1} V - \langle V L^{-1} V \rangle] P' + \cdots \tag{6.152}$$

It should be noted that truncation of the perturbation series at any order actually retains infinite powers of V. This is because of the non-linearity of (6.140). It is at this point that we see the underlying resemblance to the renormalized perturbation theory of the DIA.

On the other hand, Herring uses the same eigenfunction expansion as Edwards in order to invert the basis operator, and certainly (6.142) is very similar (notational changes apart) to the EFP form (6.73) when one substitutes, as appropriate, from (6.76) and (6.78). The differences essentially lie in the requirement in SCF that a particular operator should vanish:

$$\{L_n - L_n^{(0)} + \langle VL^{-1}T\rangle_n\} = 0,$$

where EFP would require that the product of the same operator with the P_n should vanish. Also, SCF involves a weaker assumption that the basis operator should be the product of single-mode operators, without specifying the form these operators should take. In contrast, EFP imposes the additional requirement that the basis operator should have the Fokker–Planck form. It should be noted that the basis operator in SCF is found not to be of this form.

Herring's SCF leads to coupled equations for $q(k)$ and $\omega(k)$. The energy equation is identical with the EFP equation (6.101), which is itself the same as the DIA equation, when exponential time dependences are substituted. The equation for $\omega(k)$ is the same as the time-independent DIA response equation, as given by (6.113).

We shall deal briefly with the time-dependent SCF in the next section.

6.3.1   Time-dependent SCF

The first step in extending SCF to non-stationary turbulence (Herring 1966) is to restore the time dependence to the Liouville equation. That is, we rewrite (6.126) as

$$\frac{\partial P}{\partial t} + L_0 P = VP, \tag{6.153}$$

where L_0 is still given by (6.71)—although we must bear in mind our change of notation from L to L_0—and V is given by (6.72). Then we generalize (6.144) to the form

$$\frac{\partial P}{\partial t} + \int_0^\infty L(t,t')P(t')\,\mathrm{d}t' = \int_0^\infty \Omega(t,t')P(t')\,\mathrm{d}t', \tag{6.154}$$

where

$$\Omega(t,t') = L(t,t') - \delta(t-t')(V-L_0) \tag{6.155}$$

such that

$$L(t, t') = 0 \qquad \text{for } t' > t. \tag{6.156}$$

It should be noted that this is all fully analogous to the procedure, which led to eqn (6.138) for the stationary case and was based on the introduction of the operator L, as defined by (6.134).

Now, in order to obtain an iterative solution of (6.154), Herring introduced a new operator $U(k; t, t')$. The interested reader will find the details of these procedures in the original papers (Herring 1965, 1966); we shall only quote the defining relationship here for completeness, as we shall need it in order to write down the final governing equations. Accordingly, we have

$$U(k; t, t') = \begin{cases} 1 - \displaystyle\int_{t'}^{t} ds \int_{0}^{\infty} ds' \, L(s, s') U(k; s', t') & \text{for } t > t' \\ 0 & \text{for } t < t'. \end{cases} \tag{6.157}$$

Then the equation for the single-time correlation $Q(k, t)$ is found to be

$$\left(\frac{d}{dt} + 2\nu k^2 \right) Q(k, t)$$

$$= 2 \int d^3 j \, L(\mathbf{k}, \mathbf{j}) \left\{ \int_0^t g(k; t, s) M(j; t, s) M(|\mathbf{k} - \mathbf{j}|; t, s) \, ds - \right.$$

$$\left. - \int_0^t g(j; t, s) M(k; t, s) M(|\mathbf{k} - \mathbf{j}|; t, s) \, ds \right\}, \tag{6.158}$$

where $g(k; t, t')$ can be interpreted as the response function for mode $\mathbf{k}$, and $M(k; t, t')$ plays the part of a two-time velocity covariance function. The two quantities arise, in fact, as non-trivial functional integrals, and can be written in a rather stylized notation in the form

$$g(k; t, t') = - \int \delta\mathbf{u}(\mathbf{k}, t) \mathbf{u}(\mathbf{k}, t) U(k; t, t') \frac{\partial P[\mathbf{u}(\mathbf{k}, t')]}{\partial \mathbf{u}(\mathbf{k}, t)} \tag{6.159}$$

and

$$M(k; t, t') = \int \delta\mathbf{u}(\mathbf{k}, t) \mathbf{u}(\mathbf{k}, t) U(k; t, t') \mathbf{u}(\mathbf{k}, t) P[\mathbf{u}(\mathbf{k}, t)], \tag{6.160}$$

and satisfy the conditions

$$g(k; t, t) = 1 \tag{6.161}$$

$$M(k; t, t) = Q(k, t). \tag{6.162}$$

Herring found governing equations for $g(k; t, t')$ and $M(k; t, t')$ which took the following forms:

$$\left(\frac{d}{dt} + vk^2\right)g(k; t, t')$$

$$= -\int d^3j\, L(\mathbf{k}, \mathbf{j}) \int_{t'}^{t} ds\, g(j; t, s) M(|\mathbf{k} - \mathbf{j}|; t, s) g(k; s, t') \qquad (6.163)$$

$$\left(\frac{d}{dt} + vk^2\right)M(k; t, t')$$

$$= -\int d^3j\, L(\mathbf{k}, \mathbf{j}) \int_{t'}^{t} ds\, g(j; t, s) M(|\mathbf{k} - \mathbf{j}|; t, s) M(k; s, t'), \qquad (6.164)$$

where, as in eqn (6.158), the coefficient $L(\mathbf{k}, \mathbf{j})$ is as defined by (6.30). Its analytical form is given by equation (2.162), and further details can be found in Appendix E.

It can be deduced by inspection of eqns (6.163) and (6.164) that the following relationship holds:

$$M(k; t, t') = g(k; t, t')Q(k, t'). \qquad (6.165)$$

This allows us to make contact with the DIA equations for $G(k; t, t')$ and $Q(k; t, t')$, and $Q(k; t, t) = Q(k, t)$. It is readily verified that, if we replace the DIA equation (6.31) for the two-time correlations $Q(k; t, t')$ by (6.165), then eqns (6.29) and (6.35) for $G(k; t, t')$ and $Q(k, t)$ become identical with the SCF equations (6.163) and (6.158) for $g(k; t, t')$ and $Q(k, t)$.

Lastly, it is interesting to note that the time-independent SCF equations for $q(k)$ and $\omega(k)$ can be recovered from eqns (6.158) and (6.163) by making an exponential approximation for $g(k; t, t')$ of the kind shown in eqn (6.112).

6.3.2   Other self-consistent methods

The Liouville equation has also been taken as a starting point by Balescu and Senatorski (1970), who attempt to combine methods applied to two different fields of non-equilibrium statistical mechanics (i.e. statistical electrodynamics and Heisenberg spins on a lattice), arguing that in this way they can take account of the main features of the turbulence problem, i.e. an infinite number of degrees of freedom and strong coupling. The equations of motion are transformed by the introduction of action-angle variables and the basic statistical problem is reformulated in terms of a master equation. This is treated perturbatively and the summation of certain classes of diagrams leads to the recovery of Herring's SCF theory.

Self-consistent methods have also been applied directly to the equation of motion (rather than the Liouville equation) by Phythian (1969). Strictly speaking, Phythian works with the one-dimensional Burgers model equation, but the analysis also goes through for the Navier–Stokes equation. Although the methods used rather resemble those of Edwards and Herring, Phythian does formally introduce an infinitesimal response function which is identical in its basic definition with the corresponding DIA quantity as defined by eqn (6.3). It is perhaps not altogether surprising, therefore, that Phythian ends up with the DIA equations.

However, the importance of this work has turned out to be more than just another way of deriving the DIA theory. Kraichnan has found that Phythian's theory leads to a new model representation for the DIA equations and this in turn leads on to the development of 'almost Markovian' theories. This is a subject which we shall discuss in the next chapter.

Note

1. Equation (6.142) has to be solved by iteration. We write it in the form

$$P = P' + XP$$

where $X = L^{-1}(\Delta + V)$. Then we proceed as follows:

zero-order approximation $\quad P_0 = P' + XP'$,

first-order approximation $\quad P_1 = P' + XP_0$

$$= P' + X(P' + XP')$$

$$= P' + XP' + X^2P',$$

second-order approximation $\quad P_2 = P' + XP'$

$$= P' + X(P' + XP' + X^2P')$$

$$= P' + XP' + X^2P' + X^3P',$$

and so on.
The exact distribution P is given by

$$P = P' + X(1 + X + X^2 + \cdots)P'$$

$$= P' + X\left(\frac{1}{1 - X}\right)P'$$

provided that $\{1/(1 - X)\}P'$ is non-singular. Then, substituting for X, we find

$$P = P' + L^{-1}TP',$$

which is just eqn (6.143), and the operator T is given by

$$T = \frac{(\Delta + V)}{1 - L^{-1}(\Delta + V)},$$

which can be shown to reduce to (6.144).

References

BALESCU, R. and SENATORSKI, A. (1970). *Ann. Phys.* (NY) **58**, 587.
EDWARDS, S. F. (1964). *J. fluid Mech.* **18**, 239.

—— (1965). *Int. Conf. on Plasma Physics, Trieste*, p. 595. IAEA, Vienna.

FRISCH, U. and BOURRET, R. (1970). *J. math. Phys.* **11**, 364.

HERRING, J. R. (1965). *Phys. Fluids* **8**, 2219.

—— (1966). *Phys. Fluids* **9**, 2106.

KRAICHNAN, R. H. (1958). *Phys. Rev.* 109, 1407; errata, 111, 1747.

—— (1959). *J. fluid Mech.* **5**, 497.

—— (1961). *J. math. Phys.* **2**, 124.

—— (1964a). *Phys. Fluids* **7**, 1030.

—— (1964b). *Phys. Fluids* **7**, 1163.

—— (1977). *J. fluid Mech* **83**, 349.

LESLIE, D. C. (1973). *Developments in the theory of turbulence*. Clarendon Press, Oxford.

MATHEWS, J. and WALKER, R. L. (1965). *Mathematical methods of physics*. W. A. Benjamin, New York.

PHYTHIAN, R. (1969). *J. Phys. A* **2**, 181.

REICHL, L. E. (1980). *A modern course in statistical physics*. Edward Arnold, London.

7

RENORMALIZED PERTURBATION
THEORIES OF THE SECOND KIND

In the preceding chapter, we discussed the pioneering theories of turbulence, in which systematic renormalization methods were introduced to the subject for the first time, but we did not dwell to any extent on the virtues (or otherwise) of these theories. We shall continue to defer such considerations, for the most part, to Chapter 8, where the overall assessment of RPTs will be considered in some detail. However, before turning our attention to those theories which do yield the Kolmogorov spectrum as a solution, we shall make a rather limited analysis of certain unsatisfactory features of the older theories.

Our reason for doing this at the present stage, is that it provides a perspective into which we can fit the development of the newer theories. Inevitably this means that the discussion is conducted in terms of failures to achieve the Kolmogorov form of the energy spectrum. Nevertheless, this should not be taken to mean that we are asserting that the Kolmogorov distribution provides a crucial test for analytic turbulence theories. That, in itself, is a potentially controversial topic and will receive separate treatment in Chapter 8.

7.1 The low-wavenumber catastrophe

Our terminology in this section is borrowed from quantum field theory, where unphysical divergences at low frequency or at high frequency are referred to (respectively) as 'infra-red' and 'ultraviolet' divergences. Historically, the first example of this kind of effect was probably the well-known failure (at the end of the nineteenth century) of classical physics to predict the spectral distribution of cavity radiation. The Rayleigh–Jeans distribution predicted a divergence in the energy as the wavelength tended to zero—an unphysical result which is usually known as the 'ultraviolet catastrophe'. However, we are concerned here with the other end of the spectrum, so to speak. We shall in fact consider the behaviour of certain RPTs at low frequencies, or (what is the same thing) at low wavenumbers.

7.1.1 *The infra-red divergence*

Let us return to the Edwards's 'thought experiment', which we discussed in Section 6.2.7. We concluded that the EFP energy equation (6.101) was compatible with the Kolmogorov inertial-range energy spectrum, which, in the limit of infinite Reynolds number, must apply at all wavenumbers (it should be noted that this conclusion also applies—by implication—to the DIA and SCF forms as well).

However, we pointed out that a second equation was needed in order to determine values for the two constants of proportionality α and β, in (6.120) and (6.121) for the Kolmogorov forms for the spectrum and the modal decay rate. It is here that the problem arises.

The obvious second equation in the EFP theory is eqn (6.107), which is the analogue of the response equation in DIA. We rewrite this by interchanging dummy variables $\mathbf{j}$ and $\mathbf{k} - \mathbf{j}$ to obtain

$$\omega(k) = \int \frac{d^3 j\, L(\mathbf{k}, \mathbf{k} - \mathbf{j}) q(j)}{\omega(k) + \omega(j) + \omega(|\mathbf{k} - \mathbf{j}|)}. \tag{7.1}$$

The coefficient $L(\mathbf{k}, \mathbf{k} - \mathbf{j})$ is derived in Appendix E. Its analytical form has been given previously in eqn (6.45), and for convenience we repeat this here,

$$L(\mathbf{k}, \mathbf{k} - \mathbf{j}) = (k^4 + 2k^3 j\mu - kj^3\mu)(1 - \mu^2)\frac{1}{k^2 - 2kj\mu + j^2}.$$

Now we can obtain the form of (7.1) which is appropriate for the limit of infinite Reynolds number merely by substituting (6.120) and (6.121) for $q(k)$ and $\omega(k)$, just as we did in the case of the energy equation (6.101). The result is

$$\beta\varepsilon^{1/3}k^{2/3} = \frac{\alpha\varepsilon^{1/3}}{4\pi\beta}\int d^3 j\, L(\mathbf{k}, \mathbf{k} - \mathbf{j}) j^{-11/3}\frac{1}{k^{2/3} + j^{2/3} + |\mathbf{k} - \mathbf{j}|^{2/3}}. \tag{7.2}$$

Then we cancel $\varepsilon^{1/3}$ across, divide both sides by β, and make the change of variable

$$\mathbf{j} = |\mathbf{k}|\,\mathbf{J}$$

to obtain (7.2) as

$$k^{2/3} = k^{2/3}\left(\frac{\alpha}{4\pi\beta^2}\right)\int d^3 J\, L(\mathbf{1}, \mathbf{1} - \mathbf{J}) J^{-11/3}\frac{1}{1 + J^{2/3} + |\mathbf{1} - \mathbf{J}|^{2/3}}, \tag{7.3}$$

where $\mathbf{1}$ is a vector of unit magnitude in the direction of $\mathbf{k}$. If we further cancel $k^{2/3}$ on both sides and rearrange, we have the required second equation:

$$\frac{\alpha D}{\beta^2} = 1, \tag{7.4}$$

where D is given by the integral

$$D = \frac{1}{4\pi}\int d^3 J\, L(\mathbf{1}, |\mathbf{1} - \mathbf{J}|) J^{-11/3}\frac{1}{1 + J^{2/3} + |\mathbf{1} - \mathbf{J}|^{2/3}}. \tag{7.5}$$

Equation (7.4) can be solved simultaneously with (6.125) for the two inertial-range constants. The result for the Kolmogorov spectral constant is easily found to be

$$\alpha = \frac{D^{1/3}}{C^{2/3}} = \frac{D^{1/3}}{(0.19)^{2/3}}, \qquad (7.6)$$

where we have previously noted in Section 6.2.7 that the numerical value of C is known to be 0.19.

If we now consider the problem of evaluating the constant D, it is easily seen from (7.5) that the integral is well behaved as J tends to infinity, and in fact the contribution from the upper limit is zero. However, the situation in the neighbourhood of $J = 0$ is quite different. Let us perform the integration from $J = Z$, where Z is a small quantity which we ultimately set equal to zero, up to infinity. At the lower limit the leading contribution is readily obtained by expanding in powers of J:

$$D \propto \int_Z^\infty J^{-5/3} \, dJ \times \text{integration over angular factors}$$

$$\propto Z^{-2/3} \times \text{integration over angular factors}. \qquad (7.7)$$

Clearly this diverges as $Z \to 0$, and therefore it follows from (7.6) that the EFP prediction is that the Kolmogorov constant is infinite.

7.1.2 *Spurious convection effects*

It is clear from the similarity between the EFP response equation (6.107) and the time-independent DIA equivalent, in the form of (6.113), that the infra-red divergence affects the latter theory too. In fact this really only amounts to independent confirmation of the failure of the DIA to give the Kolmogorov spectrum as its solution at large Reynolds number. As we have already seen in eqn (6.51), the DIA prediction of the inertial-range spectrum is quite different from the Kolmogorov law. However, the experimental results of Grant, Stewart, and Moilliet (1962) were sufficiently clear cut to rule out the DIA solution, as given by eqn (6.51), and to offer considerable support to the Kolmogorov spectrum. As a result, Kraichnan (1964) was motivated to analyse the reasons for the differences between the Kolmogorov and DIA theories. We shall follow the arguments of this paper closely in this and subsequent sections. We begin with a brief summary statement of the difference between the two theories and then go on to a more quantitative analysis based on an idealized convection problem.

The Kolmogorov analysis proceeds from assumptions which are equivalent to a postulate that energy transfers in wavenumber are local: in other words, there is an energy cascade which takes place in a large number of small steps. This means, for instance, that the energy transfer process at large wavenumbers is not affected by details of the energy-containing range of wavenumbers.

In contrast, DIA—although apparently yielding a local cascade of energy (Kraichnan 1959)—predicts that the rate of energy transfer in the inertial range depends on the amount of energy contained in the low wavenumbers through the r.m.s. velocity v_0. This is because the processes which relax response and correlation functions (i.e. non-linear mixing) are—on the DIA picture—dominated by the time-scale of the energy-containing range of wavenumbers (see eqn (6.50)).

Following Kraichnan, our procedure now is to set up an idealized convection problem which is simple enough for us to be able to deduce the form of correlation functions (and even the infinitesimal response function), and against which we can assess the performance of the DIA.

Let us consider a velocity field

$$\mathbf{U}(\mathbf{x}, t) = \mathbf{u}(\mathbf{x}, t) + \mathbf{v}, \tag{7.8}$$

where $\mathbf{v}$ is constant in space and time in any one realization, but fluctuates from one realization to another in a way which is governed by a Gaussian distribution. We shall take $\mathbf{u}$ to be very small compared with the uniform convection velocity. Thus viscous-range time-scales will be too long to be dynamically relevant and we therefore neglect viscous effects. Hence, from the Navier–Stokes equation (1.4)—now written in vector form with the viscous term neglected—we have, upon substitution of (7.8),

$$\frac{\partial \mathbf{u}}{\partial t} + (\mathbf{u} + \mathbf{v}) \cdot \nabla \mathbf{u} = -\nabla P. \tag{7.9}$$

Now, for $\mathbf{u} \ll \mathbf{v}$, we can neglect terms which are quadratic in $\mathbf{u}$. Also, from eqn (2.8), we should note that this means neglecting the pressure term as well, and (7.9) takes a linearized form where the only convective effect is due to the constant $\mathbf{v}$; thus

$$\frac{\partial \mathbf{u}(\mathbf{x}, t)}{\partial t} = -(\mathbf{v} \cdot \nabla)\mathbf{u}(\mathbf{x}, t). \tag{7.10}$$

Then we transform back into wavenumber space, using eqn (2.71), to obtain

$$\frac{\partial \mathbf{u}(\mathbf{k}, t)}{\partial t} = -i\mathbf{k} \cdot \mathbf{v}\mathbf{u}(\mathbf{k}, t), \tag{7.11}$$

and, for prescribed $u(\mathbf{k}, 0)$, this is readily solved to give the time evolution of $\mathbf{u}(\mathbf{k}, t)$ due to uniform convection by $\mathbf{v}$. The result is

$$\mathbf{u}(\mathbf{k}, t) = \mathbf{u}(\mathbf{k}, 0) \exp\{-i(\mathbf{k} \cdot \mathbf{v})t\}. \tag{7.12}$$

Now, in order to obtain correlation functions we must first consider the averaging processes. We shall also take $\mathbf{u}$ to have a Gaussian probability distribution, but only at $t = 0$, when we impose the requirement that the $\mathbf{u}$

and **v** fields are statistically independent of each other—something which is not true at subsequent times (Kraichnan 1964).

We shall denote the average over the distribution of the **u** field by $\langle\ \rangle$ and that over the ensemble specified by the distribution of the **v** field by $\langle\ \rangle_v$. Joint averages can be represented by $\langle\langle\ \rangle\rangle_v$. Then, substituting from (7.12), we find the covariance of the **u** field as

$$\langle\langle\mathbf{u}(\mathbf{k},t)\mathbf{u}(-\mathbf{k},t')\rangle\rangle_v = \langle\exp\{-i(\mathbf{k}\cdot\mathbf{v})(t-t')\}\rangle_v\langle\mathbf{u}(\mathbf{k},0)\mathbf{u}(-\mathbf{k},0)\rangle, \quad (7.13)$$

where we have invoked the initial condition of statistical independence of **u** and **v**. This result allows us to evaluate the correlation function $R(k;t,t')$ defined by

$$R(k;t,t') = \langle\langle\mathbf{u}(\mathbf{k},t)\mathbf{u}(-\mathbf{k},t')\rangle\rangle_v\frac{1}{\{\langle|\mathbf{u}(\mathbf{k},t)|^2\rangle\langle|\mathbf{u}(-\mathbf{k},t')|^2\rangle\}^{1/2}}, \quad (7.14)$$

which is readily shown to be

$$R(k;t,t') = \langle\exp\{-i(\mathbf{k}\cdot\mathbf{v})(t-t')\}\rangle_v$$
$$= \exp\left\{\frac{-v_0^2 k^2(t-t')^2}{2}\right\}, \quad (7.15)$$

where the second line follows upon expanding out the exponential, averaging term by term (recalling the rules for factoring out Gaussian even-order moments of any order in terms of second-order moments), and re-summing. The r.m.s. value of v is defined by

$$v_0^2 = \langle|\mathbf{v}|^2\rangle_v. \quad (7.16)$$

Thus the two-point two-time correlation dies away over a correlation time determined by the uniform convection **v** and of order $(v_0 k)^{-1}$.

We also need the third-order moment, as this characterizes the non-linear energy transfer. Unfortunately, the existence of a third-order moment at $t=0$ is incompatible with our assumption of an initially Gaussian distribution for the **u** field. Kraichnan gets round this difficulty by means of an ingenious artifice. He introduces a fluctuation in mode **k** at $t=0$ which is due to the coupling of two other modes; thus

$$\Delta u_\alpha(\mathbf{k},0) = ik_\sigma D_{\alpha\varepsilon}(\mathbf{k})u_\varepsilon(-\mathbf{j},0)u_\sigma(-\mathbf{l},0), \quad (7.17)$$

and the wavevectors **j** and **l** satisfy the relation

$$\mathbf{k}+\mathbf{j}+\mathbf{l}=0.$$

The inclusion of $D_{\alpha\varepsilon}(\mathbf{k})$ ensures that the fluctuation satisfies the incompressibility condition, in the form

$$k_\alpha D_{\alpha\varepsilon}(\mathbf{k})=0.$$

The triple moment can then be set up in terms of the fluctuation in mode **k**, and is defined as

$$S(\mathbf{k}, \mathbf{j}, \mathbf{l}; t, t', t'') = \langle\langle k_\alpha u_\alpha(\mathbf{j}, t') u_\beta(\mathbf{l}, t'') \Delta u_\beta(\mathbf{k}, t)\rangle\rangle_v \times$$

$$\times \frac{1}{\langle k_\alpha u_\alpha(\mathbf{j}, 0) u_\beta(\mathbf{l}, 0) \Delta u_\beta(\mathbf{k}, 0)\rangle}$$

$$= \langle \exp\{-i(\mathbf{k}\cdot\mathbf{v}t + \mathbf{j}\cdot\mathbf{v}t' + \mathbf{l}\cdot\mathbf{v}t'')\}\rangle_v$$

$$= \exp\left\{-\frac{v_0^2|\mathbf{k}t + \mathbf{j}t' + \mathbf{l}t''|^2}{2}\right\}, \tag{7.18}$$

where the last two lines follow from the same steps used in the derivation of eqn (7.15).

The important thing to note for later reference is that if we are on the time diagonal $t = t' = t''$, then the time dependence of S vanishes owing to the condition $\mathbf{k} + \mathbf{j} + \mathbf{l} = 0$. Clearly, away from the time diagonal, there is a time dependence induced by uniform convection.

It is possible to generalize the model to the case of a convection velocity $\mathbf{v}$ which is not uniform in space, thus increasing the resemblance of the idealized problem to that of turbulence. If $\mathbf{v}$ is only slowly varying in space, then eqn (7.11) becomes

$$\frac{\partial \mathbf{u}(\mathbf{k}, t)}{\partial t} = -\sum_{\mathbf{k}'} \mathbf{k}\cdot\mathbf{v}(\mathbf{k}')\mathbf{u}(\mathbf{k} - \mathbf{k}'; t). \tag{7.19}$$

That is, the $\mathbf{v}$ field and the $\mathbf{u}$ field are now coupled together through the non-linear term.

The results given in eqns (7.15) and (7.18) for R and S still apply (Kraichnan 1964) when (7.19) replaced (7.11) as equation of motion, provided that the following conditions hold:

(1) $\langle|\mathbf{u}(\mathbf{k}, 0)|^2\rangle$ is a smooth function of k.
(2) Equation (7.17) for $\Delta u_\alpha(\mathbf{k}, 0)$ is replaced by

$$\Delta u_\alpha(\mathbf{k}, 0) = ik_\sigma \sum_{\mathbf{j}}' \sum_{\mathbf{l}}' D_{\alpha\varepsilon}(\mathbf{k})u_\varepsilon(-\mathbf{j}, 0)u_\sigma(-\mathbf{l}, 0), \tag{7.20}$$

where $\mathbf{k}$ is a wavevector within some finite volume $V(\mathbf{k})$ of wavevector space and the primes on the double summation indicate that $\mathbf{l}$ and $\mathbf{j}$ are confined to finite volumes $V(\mathbf{j})$ and $V(\mathbf{l})$ respectively, where there is no intersection between any pair of the three volumes $V(\mathbf{k})$, $V(\mathbf{j})$, and $V(\mathbf{l})$.
(3) Wavevectors excited in the $\mathbf{v}$ field must be very small compared with $\mathbf{k}$, $\mathbf{j}$, or $\mathbf{l}$, or compared with the width of any of the volumes $V(\mathbf{k})$, $V(\mathbf{j})$, and $V(\mathbf{l})$.
(4) Times larger than $(v_0 k)^{-1}$ are not considered.

Condition (1) is a technicality, in that we require smooth initial conditions in order to apply an ergodic theorem. Condition (2) excludes the non-linear interactions which would couple distinct wavenumbers, and this exclusion allows us to impose condition (3) which ensures that the shear associated with the $\mathbf{v}$ field produces negligible distortion of the $\mathbf{u}$ field in $\mathbf{x}$ space. Condition (4) provides an upper bound on the time over which the system can evolve and conditions (1)–(3) still hold.

Now let us return to the original idealized convection problem, with eqn (7.11) as the equation of motion. We wish to test the DIA theory and we begin by adding a stirring force $\mathbf{f}(\mathbf{k}, t)$ to the r.h.s. of (7.11):

$$\frac{\partial \mathbf{u}(\mathbf{k}, t)}{\partial t} + i(\mathbf{k} \cdot \mathbf{v})\mathbf{u}(\mathbf{k}, t) = \mathbf{f}(\mathbf{k}, t). \tag{7.21}$$

Then, following the same derivation as in the full turbulence problem, which leads from eqn (6.1) to eqn (6.6), we obtain

$$\frac{\partial \hat{G}_{\alpha\beta}(k; t, t')}{\partial t} + i(\mathbf{k} \cdot \mathbf{v})\hat{G}_{\alpha\beta}(k; t, t') = \delta(t - t') \tag{7.22}$$

where $\hat{G}$ is as defined by eqn (6.3). The ensemble-averaged infinitesimal response tensor is then obtained by

$$\begin{aligned} G_{\alpha\beta}(k; t, t') &= \langle \hat{G}_{\alpha\beta}(k; t, t') \rangle \\ &= \langle \exp\{-i(\mathbf{k} \cdot \mathbf{v})(t - t')\} \rangle_v \\ &= R(k; t, t') \qquad \text{for } t > t', \end{aligned} \tag{7.23}$$

where $R(k; t, t')$ is given by eqn (7.15).

The application of DIA to the idealized convection problem results in an infinitesimal response function of the form

$$G(k; t, t') = J_1\left[\frac{2v_0 k(t - t')}{v_0 k(t - t')}\right]. \tag{7.24}$$

Note that this is just eqn (6.50), with the kinematic viscosity set equal to zero. Under these circumstances, we also have the DIA prediction that the correlation function is equal to the response function.

Comparison with eqns (7.15) and (7.23) shows that the DIA solution is qualitatively incorrect, as the Bessel function of (7.24) has some damped oscillations, unlike the monotonically damped behaviour of the Gaussian curve of (7.15). On the other hand, near the central peak of the Bessel function, the qualitative agreement between the DIA form and the 'exact solution' is quite good, and both results scale on the convective time-scale $(v_0 k)^{-1}$.

The trouble really starts when we examine the DIA prediction for the triple moment. This is (Kraichnan 1964)

$$S(k, j, l; t, t', t'') = G(k, t)R(j, t')R(l, t''). \tag{7.25}$$

Our next step is clear. We substitute (7.24) for G and each of the two R. It therefore follows that the DIA prediction for the triple moment possesses a serious fault. Even on the time diagonal $t = t' = t''$, eqn (7.25) indicates that the triple moment is time dependent. This is in complete contrast with the exact result as given by eqn (7.18). As we saw, this is independent of time when $t = t' = t''$.

This erroneous behaviour can be illustrated by substituting the exact forms of G and R into the DIA triple moment, rather than the Bessel function of eqn (7.24). The reason for this particular manoeuvre is that the squared exponentials are easier to manipulate than the Bessel functions. Accordingly, substituting (7.15) for both G and R, we can write eqn (7.25) as

$$S(k, j, l; t, t', t'') = \exp\left\{-\frac{v_0^2(k^2t^2 + j^2t'^2 + l^2t''^2)}{2}\right\}, \tag{7.26}$$

which can be compared with eqn (7.18) for the exact result. As indicated above, the important thing to note about our (approximate) DIA prediction is that it is still time dependent, even when on the time diagonal and with $\mathbf{k} + \mathbf{j} + \mathbf{l} = 0$. It was this behaviour which Kraichnan concluded was at the root of the failure of DIA to yield inertial-range dynamics which were independent of the energy-containing range of wavenumbers.

7.1.3   *Postulate of random Galilean invariance*

When we added a uniform convective velocity to the general turbulent velocity field (see eqn (7.8)), we were, in effect, making a Galilean transformation of the velocity field. We shall now consider the work of the preceding section from this new point of view (see also Appendix C). Let us refer to our usual frame of reference as S. Then we wish to consider some other frame of reference S', which is moving with velocity $-\mathbf{v}$ relative to S. It is a cardinal principle of physics that the description of any dynamical process should be the same in both systems. This is known as 'Galilean invariance' and is the classical form (i.e. when $|\mathbf{v}| \ll$ speed of light) of Lorentz invariance. A transformation from one system to the other is known as a 'Galilean transformation', and is represented mathematically by

$$\mathbf{x}' = \mathbf{x} + \mathbf{v}t$$

$$\mathbf{u}'(\mathbf{x}', t) = \mathbf{u}(\mathbf{x}, t) + \mathbf{v}, \tag{7.27}$$

where primed quantities refer to variables measured in S'; this form of notation is a purely temporary measure, which we shall use in this section only. Also, without loss of generality, we assume that S and S' were in coincidence at $t = 0$.

We obtain the corresponding transformation for the Fourier components of the velocity field by invoking the general formula, as given by eqn (D.22), and substituting from (7.27):

$$\mathbf{u}'(\mathbf{k}, t) = \left(\frac{1}{2\pi}\right)^3 \int d^3x' \, \mathbf{u}'(\mathbf{x}', t) \exp(-i\mathbf{k} \cdot \mathbf{x}')$$

$$= \mathbf{v}\delta(\mathbf{k}) + \mathbf{u}(\mathbf{k}, t) \exp\{-i(\mathbf{k} \cdot \mathbf{v})t\}, \qquad (7.28)$$

where the delta function follows from the generalization of eqn (D.11) to three dimensions. It should be noted that, as far as the Fourier components are concerned, the main effect of the Galilean transformation is a phase change identical with the one obtained previously in eqn (7.12).

The implication of this result for single-time moments of any order follows quite readily. We have

$$\langle \mathbf{u}'(\mathbf{k}, t)\mathbf{u}'(\mathbf{j}, t)\mathbf{u}'(\mathbf{l}, t)\ldots \rangle = \langle \mathbf{u}(\mathbf{k}, t)\mathbf{u}(\mathbf{j}, t)\mathbf{u}(\mathbf{l}, t)\ldots \rangle, \qquad (7.29)$$

provided only that $\mathbf{k} + \mathbf{j} + \mathbf{l} + \cdots = 0$, as required by spatial homogeneity; the phase factors cancel out. Thus the single-time moments, of any order, are unaffected by the Galilean transformation. That is, they are Galilean invariant.

However, this is not the case for the non-simultaneous moments; as we saw in connection with eqn (7.15) (first line on the r.h.s.), such moments exhibit a time dependence which depends on $\mathbf{v}$.

The physical interpretation of this latter effect in $\mathbf{x}$-space is quite straightforward. In the reference frame S, we have, let us say, a two-point two-time correlation

$$Q(\mathbf{r}, T) = \langle \mathbf{u}(\mathbf{x}, t)\mathbf{u}(\mathbf{x} + \mathbf{r}, t + T) \rangle,$$

which depends only on $\mathbf{r}$ and T, for homogeneous stationary fields. Then in S' the distance between the measuring points would be increased from $\mathbf{r}$ to $\mathbf{r} + \mathbf{v}T$, and hence, for (as is usual) Q a monotonically declining function of the separation distance between the measuring points, we have $Q' < Q$.

If we now take $\mathbf{v}$ to be a random variable, with statistics as specified in Section 7.1.2, it is clear that single-time moments will still be invariant under each individual transformation making up the ensemble. Hence single-time moments must therefore be unaffected by the entire ensemble of random Galilean transformations. In Kraichnan's phrase, single-time moments of any order exhibit 'random Galilean invariance'.

However, let us consider the case of $Q(\mathbf{r}, T)$ as an example of a non-simultaneous correlation. Then the situation under random Galilean transformations is quite different. Evidently Q' is now to be seen as a random variable, with a different value for each realization of $\mathbf{v}$. Sometimes we will have $Q' > Q$ and sometimes we will have $Q' < Q$, with the ensemble-averaged value of Q' depending on the difference time T. Thus the effect of repeated Galilean transformations will be to tend to smear out the correlation. Qualitatively, therefore, the effect of averaging two-time correlations with respect to the distribution of $\mathbf{v}$, as given by eqn (7.15), can be seen to be quite reasonable, although that particular quantitative result is an approximation.

Now we can sum up the analysis of the difficulties with DIA, as treated in Section 7.1.2, from our new point of view. Single-time correlations should be invariant under Galilean transformation, both random and deterministic. Thus the equation for $R(k; t, t)$, which contains a triple moment evaluated at $t = t' = t''$, should itself be invariant under Galilean transformation. But in the DIA formulation the triple moment is expressed in terms of two-time correlation and response functions, which transform (at least to an approximation) according to eqn (7.12). Hence the DIA triple moment violates the requirement of random Galilean invariance and exhibits a spurious time decay, even on the time diagonal $t = t' = t''$.

7.1.4   *Response integrals with an arbitrary cut-off in wavenumber*

In the preceding two sections, we have discussed the failure of DIA in terms of spurious convection effects of low wavenumbers on high wavenumbers. Having made this diagnosis, Kraichnan also provided further support for it by demonstrating that the removal of such convection effects from the Navier–Stokes equation in turn removed the difference between the DIA and Kolmogorov theories. We shall complete this section by giving a brief account of this work.

Let us consider the Navier–Stokes equation in $\mathbf{x}$-space. From eqn (2.2) we have

$$\frac{\partial u_\alpha}{\partial t} + \frac{\partial(u_\alpha u_\beta)}{\partial x_\beta} = -\frac{\partial p}{\partial x_\alpha} + \nu \nabla^2 u_\alpha.$$

The straightforward interpretation of the non-linear term is that it represents the convection of u_α by $u_\beta \partial/\partial x_\beta$, where the latter term is the convective derivative.

If we now go into $\mathbf{k}$-space, then (7.30) becomes

$$\frac{\partial u_\alpha(\mathbf{k})}{\partial t} + i k_\beta \sum_{\mathbf{j}} u_\alpha(\mathbf{k} - \mathbf{j}) u_\beta(\mathbf{j}) = -i k_\alpha p(\mathbf{k}) - \nu k^2 u_\alpha(\mathbf{k}) \qquad (7.30)$$

(e.g. see eqn (D.32)), and we have not shown the time dependence explicitly. From a comparison of (7.30) and (2.2), we see that the convective term is $k_\beta u_\beta(\mathbf{j})$. Evidently, the equivalent of eliminating the convection of small eddies by large eddies is to eliminate values of $u_\beta(\mathbf{j})$ for which $j < k$ or $j < |\mathbf{k} - \mathbf{j}|$. In fact Kraichnan (1964) imposes the conditions

$$\frac{|\mathbf{k} - \mathbf{j}|}{n} \leqslant j < \infty$$

$$\frac{k}{n} \leqslant j < \infty \qquad (7.31)$$

$$n \geqslant 2,$$

and shows that these result in the separate conservation of energy by wave-number triads which are excluded, as well as by those which are included in the convolution sum in eqn (7.30).

Numerical computation of the DIA, when applied to eqns (7.30) and (7.31), was found (Kraichnan 1964) to yield the Kolmogorov spectrum. This improvement was attributed to a change in the behaviour of the response and correlation functions. The elimination of low-wavenumber convection effects results in $R(k; t, t')$ and $G(k; t, t')$ having decay times which depend on the local (in wavenumber) intensity level, rather than on the energy range value v_0.

As we saw in Section 7.1.1, the energy equation has internal cancellations which make it well behaved at the infra-red divergence. Accordingly, it might be expected that the energy equation would be fairly insensitive to the removal of low wavenumber modes from the Navier–Stokes equation, in which case the above procedure would be equivalent to cutting off the wavenumber integral in the response equation. In order to mimic the cancellation behaviour, the low wavenumber cut-off must be proportional to k. That is, the EFP response equation (6.107) or the time-independent DIA response equation (6.113) should be integrated over the range $mk < j < \infty$, where m is an arbitrary cut-off ratio. Such procedures have been tried (e.g. Nakano 1972; Leslie 1973, p. 106), and certainly—as was the case with Kraichnan's similar procedure on the Navier–Stokes equation—plausible choices of the cut-off ratio lead to reasonable values of the Kolmogorov constant. Nevertheless, as one can see from eqns (7.6) and (7.7), the value obtained in this way for the Kolmogorov constant must in turn depend on the value chosen for the cut-off ratio. This latter choice is, of course, arbitrary. In the rest of this chapter we shall be concerned with less arbitrary ways of obtaining improved renormalized perturbation theories of turbulence.

7.2 Lagrangian-history direct-interaction theories

Up until now we have relied on the Eulerian or field description of fluid motion. That is, our principle dependent variable has been the velocity field $\mathbf{u}(\mathbf{x}, t)$, which tells us the value of the fluid velocity at any space–time point $(\mathbf{x}, t)$. However, we can also describe the motion of a fluid in Lagrangian coordinates, in which we follow the motion of a particular point which moves as part of the fluid. In fact we shall refer to such a point as a 'fluid particle', but it should always be remembered that this term stands for a mathematical abstraction.

The basic Lagrangian coordinates can be set up as follows. Imagine that we have tagged some particular fluid particle. We draw a vector $\mathbf{X}(t)$ from the origin of (Eulerian) coordinates to the particle. Then the information that the fluid particle is at point $\mathbf{x}$ at time t can be expressed as

$$\mathbf{X}(t) = \mathbf{x}, \tag{7.32}$$

while its velocity is given by

$$\mathbf{V}(t) = \frac{\partial \mathbf{X}(t)}{\partial t}. \tag{7.33}$$

We can identify the specific particle tagged by stating that the velocity is that of the particle which was at, say, the point x_0 at the time t_0. That is, we can write the Lagrangian velocity as $\mathbf{V}(x_0, t)$.

The physical connection between the two systems can be established quite readily. At any fixed point $\mathbf{x}$ in space, $\mathbf{u}(\mathbf{x}, t)$ represents a time history of the individual fluid particles passing through $\mathbf{x}$ at successive instants and making up the continuous variation of fluid properties at $\mathbf{x}$ with time. Clearly, when any particular particle is known to be at a specific space–time point, its Lagrangian velocity must be equal to the Eulerian field value at that space–time point. It follows that the connection between the Eulerian and Lagrangian velocities is given by

$$\mathbf{V}(t) = \mathbf{u}[\mathbf{X}(t), t], \tag{7.34}$$

where the assignment $\mathbf{x} = \mathbf{X}(t)$ samples the Eulerian velocity field at all points along the particle trajectory.

It should perhaps be emphasized that, in order to use (7.34) to transform from one system to another, we need to obtain a form for $\mathbf{X}(t)$. This is, in fact, an unsolved problem. We shall discuss this particular topic further in Chapters 12 and 13, where we shall consider the diffusive properties of turbulence.

In diagnosing the difficulties with DIA, Kraichnan (1964) noted that the elimination of spurious convection effects by wavenumber cut-offs—see the preceding section—was equivalent to representing the Navier–Stokes equation in quasi-Lagrangian coordinates. He went on to propose a mixed Lagrangian-Eulerian coordinate system (Kraichnan 1964), and was later able to rework the DIA in the new coordinates, thus producing a theory which was in agreement with the Kolmogorov picture (Kraichnan 1965).

7.2.1   *The Lagrangian-history formulation*

The key step is the introduction of the generalized velocity $\mathbf{u}(\mathbf{x}, t|s)$, which is defined as the velocity measured at time s of a fluid particle which was at the position $\mathbf{x}$ at time t. The two distinct times are known as

$$t = \text{labelling time} \quad \text{and} \quad s = \text{measuring time.}$$

It follows at once from this definition that the generalized velocity satisfies two limiting conditions:

$$\mathbf{u}(\mathbf{x}, t|t) = \mathbf{u}(\mathbf{x}, t) \tag{7.35}$$

$$\mathbf{u}(\mathbf{x}_0, t_0|s) = \mathbf{V}(\mathbf{x}_0, s). \tag{7.36}$$

In other words, the dependence on t is a Eulerian characteristic, whereas the dependence on s is a Lagrangian characteristic.

When $t = s$, the generalized velocity is just the Eulerian field and hence satisfies the Navier–Stokes equation. However, when t is not equal to s, we need to derive a special equation. This is quite easily done. Consider a particle at point $\mathbf{x}$ at time t. At time $t + dt$ the particle will have moved to $\mathbf{x} + d\mathbf{x}$. Thus we can write

$$\mathbf{u}(\mathbf{x} + d\mathbf{x}, t + dt|s) = \mathbf{u}(\mathbf{x}, t|s). \tag{7.37}$$

Note that both sides of this equation give the velocity of the particle at time s, and hence they have the same value. Now expand the l.h.s. out in Taylor series in dt and $d\mathbf{x}$, and subtract the r.h.s. from it. In the limit of $dt \to 0$, we then have

$$\left[\frac{\partial}{\partial t} + \mathbf{u}(\mathbf{x}, t) \cdot \mathbf{V}\right] \mathbf{u}(\mathbf{x}, t|s) = 0 \tag{7.38}$$

as the required equation of motion for the generalized velocity. It should be noted that—once we have (7.37)—the derivation of (7.38) is just the standard Eulerian form and will be found in any elementary textbook on fluid dynamics.

Equation (7.34) relates the Lagrangian velocity to the Eulerian field at any time. We can write down an analogous relationship for the generalized velocity by introducing a displacement function $\mathbf{X}(\mathbf{x}, t|s)$, which is defined to be the displacement of the fluid particle during the time interval $s - t$. The relationship between the generalized velocity and the Eulerian field is readily seen to be

$$\mathbf{u}(\mathbf{x}, t|s) = \mathbf{u}[\mathbf{x} + \mathbf{X}(\mathbf{x}, t|s), s]. \tag{7.39}$$

It should be appreciated that the generalized velocity does not have to satisfy the incompressibility condition. That is, unless $t = s$, when of course it is just the same as the Eulerian velocity field and must satisfy eqn (1.1). This can raise problems with the definition of the infinitesimal response tensor, and we shall anticipate our discussion of these quantities in the next section by following Kraichnan and introducing a division of the generalized velocity into a solenoidal part $\mathbf{u}^s$ and a curl-free part $\mathbf{u}^c$. It is a standard theorem that any arbitrary vector can be written as the sum of a solenoidal vector and a curl-free vector; thus the generalized velocity becomes

$$\mathbf{u}(\mathbf{x}, t|s) = \mathbf{u}^s(\mathbf{x}, t|s) + \mathbf{u}^c(\mathbf{x}, t|s), \tag{7.40}$$

where the solenoidal (or divergenceless) part satisfies

$$\mathbf{V} \cdot \mathbf{u}^s(\mathbf{x}, t|s) = 0 \tag{7.41}$$

and the curl-free part satisfies (naturally) the equation

$$\mathbf{V} \times \mathbf{u}^c(\mathbf{x}, t|s) = 0. \tag{7.42}$$

We have previously seen in Section 2.1 how the equations of motion can be put in solenoidal form, with the introduction of the projection operator $D_{\alpha\beta}(\mathbf{V})$, as defined by eqn (2.11). We can use this approach to relate $\mathbf{u}^s(\mathbf{x}, t|s)$ to the generalized velocity:

$$u_\alpha^s(\mathbf{x}, t|s) = D_{\alpha\beta}(\mathbf{V}) u_\beta(\mathbf{x}, t|s). \tag{7.43}$$

It is easily verified that a corresponding result for the curl-free part is

$$u_\alpha^c(\mathbf{x}, t|s) = \Pi_{\alpha\beta}(\mathbf{V}) u_\beta(\mathbf{x}, t|s), \tag{7.44}$$

where the operator $\Pi_{\alpha\beta}(\mathbf{V})$ is introduced by writing the projection operator in the form

$$D_{\alpha\beta}(\mathbf{V}) = \delta_{\alpha\beta} - \Pi_{\alpha\beta}(\mathbf{V}) \tag{7.45}$$

and comparing this equation with (2.11).

We now wish to consider how the equations of motion should be modified to take account of the decomposition of the generalized velocity according to (7.40). For the case where $s \neq t$, this is easily done. We simply rewrite eqn (7.38) as

$$\left[\frac{\partial}{\partial t} + \mathbf{u}^s(\mathbf{x}, t|t) \cdot \mathbf{V}\right] \mathbf{u}(\mathbf{x}, t|s) = 0, \tag{7.46}$$

where we have just replaced the Eulerian field in the convective derivative by the solenoidal part of the generalized velocity at $t = s$.

For the case where $t = s$, the generalized velocity just becomes the Eulerian velocity field and hence satisfies the Navier–Stokes equation. This is given in solenoidal form by eqn (2.15). As we are restricting our attention to isotropic turbulence occupying either a finite box with cyclic boundary conditions or an infinite system with all variables vanishing sufficiently rapidly at infinity, we neglect the surface term and the external pressure gradient alike, and write (2.15) as

$$\left[\frac{\partial}{\partial t} - \nu \nabla^2\right] u_\alpha(\mathbf{x}, t) = M_{\alpha\beta\gamma}(\mathbf{V}) [u_\beta(\mathbf{x}, t) u_\gamma(\mathbf{x}, t)] \tag{7.47}$$

where the inertial transfer operator $M_{\alpha\beta\gamma}(\mathbf{V})$ is given by eqn (2.13).

Now we substitute from eqn (7.40), with $t = s$, for $\mathbf{u}(\mathbf{x}, t)$ in (7.47). We note that, as (7.47) is in solenoidal form, it follows at once that $u_\alpha^s(\mathbf{x}, t|t)$ satisfies it as it stands; thus

$$\left[\frac{\partial}{\partial t} - \nu\nabla^2\right] u_\alpha^s(\mathbf{x}, t|t) = M_{\alpha\beta\gamma}(\mathbf{V})\left[u_\beta^s(\mathbf{x}, t|t)u_\gamma^s(\mathbf{x}, t|t)\right], \tag{7.48}$$

and hence $\mathbf{u}^c$ satisfies

$$\left[\frac{\partial}{\partial t} - \nu\nabla^2\right] u_\alpha^c(\mathbf{x}, t|t) = 0. \tag{7.49}$$

Lastly, it should be noted that $\mathbf{u}^c$ is a fictitious velocity which does not affect the dynamics of $\mathbf{u}^s(\mathbf{x}, t|t)$. It is clear from (7.49) that if $\mathbf{u}^c$ is zero initially, then it remains zero.

7.2.2 The statistical formulation

The statistical formulation in terms of the Lagrangian-history coordinates of displacement $\mathbf{X}(\mathbf{x}, t|s)$ and velocity $\mathbf{u}(\mathbf{x}, t|s)$ is a straightforward generalization of the Eulerian version which we have been using hitherto. We begin by defining the generalized covariance

$$Q_{\alpha\beta}(\mathbf{x}, t|s; \mathbf{x}', t'|s') = \langle u_\alpha(\mathbf{x}, t|s)u_\beta(\mathbf{x}', t'|s')\rangle, \tag{7.50}$$

which is an obvious extension of the Eulerian form (2.24). Clearly higher-order moments—as in (2.25)—can be extended to Lagrangian-history coordinates in an equally obvious way.

A new feature of the present formulation arises owing to the decomposition of the generalized velocity, as given by (7.40). That is, we can define

$$Q_{\alpha\beta}^s = \langle u_\alpha^s u_\beta \rangle; \, Q_{\alpha\beta}^s = \langle u_\alpha u_\beta^s \rangle$$

and so on, where the superscript refers to the tensor index immediately beneath it. In all, $Q_{\alpha\beta}$ can be decomposed into the following tensors:

$$Q_{\alpha\beta}^s, \, Q_{\alpha\beta}^s, \, Q_{\alpha\beta}^c, \, Q_{\alpha\beta}^c, \, Q_{\alpha\beta}^{sc}, \, Q_{\alpha\beta}^{cs}, \, Q_{\alpha\beta}^{ss}, \, Q_{\alpha\beta}^{cc}.$$

Note that any one of the above can be related to $Q_{\alpha\beta}$ by substituting from equations (7.43) and (7.44) as appropriate. For instance,

$$\begin{aligned} Q_{\alpha\beta}^s(\mathbf{x}, t|s; \mathbf{x}', t'|s') &= \langle u_\alpha^s(\mathbf{x}, t|s)u_\beta(\mathbf{x}', t'|s')\rangle \\ &= D_{\alpha\gamma}(\mathbf{V})\langle u_\gamma(\mathbf{x}, t|s)u_\beta(\mathbf{x}', t'|s')\rangle \\ &= D_{\alpha\gamma}(\mathbf{V})Q_{\gamma\beta}(\mathbf{x}, t|s; \mathbf{x}', t'|s'), \end{aligned} \tag{7.51}$$

and all the others can be obtained in a similar fashion.

For the case of homogeneous turbulence, the two-point correlations depend only on $\mathbf{x} - \mathbf{x}'$. So, just as in the Eulerian case, we shall find it convenient to resort to Fourier transformation into $\mathbf{k}$-space. With the addition of reflectional symmetry—a requirement for isotropy—the covariances must be invariant under the interchange of $\mathbf{x}$ and $\mathbf{x}'$. This means that $Q_{\alpha\beta}^{sc} = Q_{\alpha\beta}^{cs} = 0$ (see Appendix C for examples of this kind of argument). It is not difficult to show that, in homogeneous isotropic turbulence, the decomposition of $Q(\mathbf{k}; t|s; t', s')$ is fully specified by

$$Q_{\alpha\beta}^{s}(\mathbf{k}; t|s; t'|s') = D_{\alpha\gamma}(\mathbf{k})Q_{\gamma\beta}(\mathbf{k}; t|s; t'|s') \tag{7.52}$$

$$Q_{\alpha\beta}^{c}(\mathbf{k}; t|s; t'|s') = \Pi_{\alpha\gamma}(\mathbf{k})Q_{\gamma\beta}(\mathbf{k}; t|s; t'|s'), \tag{7.53}$$

where

$$\Pi_{\alpha\beta}(\mathbf{k}) = \delta_{\alpha\beta} - D_{\alpha\beta}(\mathbf{k}) = k_\alpha k_\beta |\mathbf{k}|^{-2} \tag{7.54}$$

and $D_{\alpha\beta}(\mathbf{k})$ is defined by eqn (2.78). It should be noted that the relevant Fourier transformations will be found in Appendix D.

7.2.3 DIA adapted to Lagrangian-history coordinates

In Section 6.1 we saw that the starting point for the derivation of DIA was the introduction of the infinitesimal response tensor through eqn (6.3). In **x**-space this can be written

$$\hat{G}_{\alpha\beta}(\mathbf{x}, t; \mathbf{x}', t') = \frac{\delta u_\alpha(\mathbf{x}, t)}{\delta f_\beta(\mathbf{x}', t')} \tag{7.55}$$

$$\hat{G}_{\alpha\beta}(\mathbf{x}, t; \mathbf{x}', t') = 0 \qquad t < t'$$

where the latter condition indicates (as usual) that the effect cannot precede the cause.

The straightforward generalization of (7.55) to the case of Lagrangian-history coordinates is then just

$$\hat{G}_{\alpha\beta}(\mathbf{x}, t|s; \mathbf{x}', t'|s') = \frac{\delta u_\alpha(\mathbf{x}, t|s)}{\delta f_\beta(\mathbf{x}', t'|s')} \tag{7.56}$$

$$\hat{G}_{\alpha\beta}(\mathbf{x}, t|s; \mathbf{x}', t'|s') = 0 \qquad s < s'.$$

However, there are other aspects which we must consider. In the Eulerian case, the incompressibility condition requires the velocity field to be solenoidal. We can ensure that this is so by simply choosing the stirring force to be always solenoidal, which in turn ensures that any induced fluctuations in the velocity must also be solenoidal. In this way, eqn (1.1) is satisfied at all times.

Of course this constraint still applies to the present formulation, when we set $t = s$ in the generalized velocity. The problem arises in the general case, where $t \neq s$ and there is no requirement on the velocity $\mathbf{u}(\mathbf{x}, t|s)$ to be solenoidal. Hence an arbitrary stirring force may induce unphysical disturbances in the generalized velocity. Clearly, we need to ensure that any fluctuations in $\mathbf{u}(\mathbf{x}, t|s)$ which propagate to the time diagonal ($t = s$) are solenoidal. In practice, the necessary discrimination can be achieved by the decomposition of

$\mathbf{u}(\mathbf{x}, t|s)$ into $\mathbf{u}^s$ and $\mathbf{u}^c$, as in eqn (7.40), and indeed this was Kraichnan's original reason for introducing this step.

The governing equations for the response tensors can be obtained by adding a force term to the r.h.s. of (7.46), if $t \neq s$, and to the r.h.s. of eqns (7.48) and (7.49) combined, if $t = s$. Then we follow the procedures given in Section 6.1, in order to carry out the functional differentiations. The results are

$$\frac{\partial \hat{G}_{\alpha\beta}(\mathbf{x}, t|s; \mathbf{x}', t'|s')}{\partial t} = -u_\gamma^s(\mathbf{x}, t|s) \frac{\partial \hat{G}_{\alpha\beta}(\mathbf{x}, t|s; \mathbf{x}', t'|s')}{\partial x_\gamma} -$$

$$- \hat{G}_{\alpha\beta}^s(\mathbf{x}, t|t; \mathbf{x}', t'|s') \frac{\partial u_\alpha(\mathbf{x}, t|s)}{\partial x_\gamma} \tag{7.57}$$

and

$$\left[\frac{\partial}{\partial t} - v\nabla^2 \right] \hat{G}_{\alpha\beta}(\mathbf{x}, t|t; \mathbf{x}', t'|s')$$

$$= 2M_{\alpha\gamma\sigma}(\nabla)[u_\sigma^s(\mathbf{x}, t|t)\hat{G}_{\alpha\beta}^s(\mathbf{x}, t|t; \mathbf{x}', t'|s')], \tag{7.58}$$

where the solenoidal Green tensor is given by

$$\hat{G}_{\alpha\beta}^s(\mathbf{x}, t|s; \mathbf{x}', t'|s') = D_{\alpha\gamma}(\nabla)\hat{G}_{\gamma\beta}(\mathbf{x}, t|s; \mathbf{x}', t'|s'). \tag{7.59}$$

It should be noted that the operator ∇ acts on $\mathbf{x}$ and not on $\mathbf{x}'$. Kraichnan's method also requires equations in which derivatives are taken with respect to $\mathbf{x}'$ and t'. Details of these complementary equations of motion can be found in the original paper (Kraichnan 1965, Appendix B) or in the book by Leslie (1973, Section 9.2).

We can complete the basic formulation by introducing the ensemble-averaged response tensor

$$G_{\alpha\beta}(\mathbf{x}, t|s; \mathbf{x}', t'|s') = \langle \hat{G}_{\alpha\beta}(\mathbf{x}, t|s; \mathbf{x}', t'|s') \rangle, \tag{7.60}$$

which is an obvious extension of eqn (6.7) in the Eulerian case. However, in the Lagrangian case, $G_{\alpha\beta}$ (or $\hat{G}_{\alpha\beta}$) can take a superscript s or c, but only above the first tensor index. Apart from that minor restriction, all the general conditions on $Q_{\alpha\beta}$ in the previous section also apply to $G_{\alpha\beta}$.

Our objective at this point is to derive closed equations for the covariance tensor $Q_{\alpha\beta}(\mathbf{k}; t|s; t'|s')$ and the response tensor $G_{\alpha\beta}(\mathbf{k}; t|s; t'|s')$, where the latter quantity is the Fourier transform of the l.h.s. of eqn (7.60). Kraichnan's approach to this problem involved two stages. First, he extended the re-normalized perturbation theory to the case of Lagrangian-history coordinates and checked that the DIA still possessed the same conservation and invariance properties in the new representation. Second, he altered the new DIA equations in order to produce invariance under random Galilean transformations. The resulting equations constitute the Lagrangian-history direct interaction (LHDI) theory. We shall only give a very brief outline of their derivation here, and, in the interests of conciseness, this will be based on our account of the derivation of Eulerian DIA.

In Section 6.1, we showed how the perturbation expansions (6.8) and (6.9) for the Eulerian velocity field and response tensor were applied to the Navier–Stokes equation (5.56) and to eqn (6.6) for the infinitesimal response tensor. In the present case, we make analogous expansions of the generalized velocity and the response tensor in order to obtain iterative solutions of (7.46)–(7.48)— these equations replace the Navier–Stokes equation in the Eulerian problem— and to eqns (7.56)—(7.58), which replace eqn (6.6) in the Eulerian problem.

In the Eulerian case, we obtain (6.21) for the response tensor and (6.26) for the covariance tensor. In both these equations, renormalization is achieved by setting the bookkeeping parameter equal to unity, replacing the zero-order quantities G^0, Q^0 by exact quantities G and Q, and truncating at second order.

The equivalent statistical equations for the Lagrangian-history case are, in exact form, as follows:

$$\left(\frac{\partial}{\partial t} + vk^2\right) Q_{\alpha\beta}(\mathbf{k}; t|t; t'|s') = S_{\alpha\beta}(\mathbf{k}; t|t; t'|s') \tag{7.61}$$

$$\left(\frac{\partial}{\partial t} + vk^2\right) G_{\alpha\beta}(\mathbf{k}; t|t; t'|s') = H_{\alpha\beta}(\mathbf{k}; t|t; t'|s') \qquad t \geqslant s' \tag{7.62a}$$

$$\left(\frac{\partial}{\partial t'} - vk^2\right) G_{\alpha\beta}(\mathbf{k}; t|s; t'|t') = H_{\alpha\beta}^{+}(\mathbf{k}; t|s; t'|t') \qquad s \geqslant t' \tag{7.62b}$$

$$\frac{\partial Q_{\alpha\beta}(\mathbf{k}; t|s; t'|s')}{\partial t} = M_{\alpha\beta}(\mathbf{k}; t|s; t'|s') \tag{7.63}$$

$$\frac{\partial G_{\alpha\beta}(\mathbf{k}; t|s; t'|s')}{\partial t} = N_{\alpha\beta}(\mathbf{k}; t|s; t'|s') \qquad s \geqslant s' \tag{7.64a}$$

$$\frac{\partial G_{\alpha\beta}(\mathbf{k}; t|s; t'|s')}{\partial t'} = N_{\alpha\beta}^{+}(\mathbf{k}; t|s; t'|s') \qquad s \geqslant s'. \tag{7.64b}$$

The terms on the right-hand sides are given by the Fourier transforms, with respect to $\mathbf{x} - \mathbf{x}'$, of the following quantities:

$$2S_{\alpha\beta}(\mathbf{x}, t|t; \mathbf{x}', t'|s') = M_{\alpha\gamma\sigma}(\nabla)\langle u_{\sigma}^{s}(\mathbf{x}, t|t)u_{\gamma}^{s}(\mathbf{x}, t|t)u_{\beta}(\mathbf{x}', t'|s')\rangle \tag{7.65}$$

$$H_{\alpha\beta}(\mathbf{x}, t|t; \mathbf{x}', t'|s') = 2M_{\alpha\gamma\sigma}(\nabla)\langle u_{\sigma}^{s}(\mathbf{x}, t|t)\hat{G}_{\gamma\beta}^{s}(\mathbf{x}, t|t; \mathbf{x}', t'|s')\rangle \tag{7.66}$$

$$2M_{\alpha\beta}(\mathbf{x}, t|s; \mathbf{x}', t'|s') = -\left\langle u_{\gamma}^{s}(\mathbf{x}, t|t)u_{\beta}(\mathbf{x}', t'|s')\frac{\partial u_{\alpha}(\mathbf{x}, t|s)}{\partial x_{\gamma}}\right\rangle, \tag{7.67}$$

$$N_{\alpha\beta}(\mathbf{x}, t|s; \mathbf{x}', t'|s') = -\left\langle u_{\gamma}^{s}(\mathbf{x}, t|t)\frac{\partial \hat{G}_{\alpha\beta}(\mathbf{x}, t|s; \mathbf{x}', t'|s')}{\partial x_{\gamma}}\right\rangle -$$

$$\qquad - \left\langle \hat{G}_{\gamma\beta}^{s}(\mathbf{x}, t|t; \mathbf{x}', t'|s')\frac{\partial u_{\alpha}(\mathbf{x}, t|s)}{\partial x_{\gamma}}\right\rangle. \tag{7.68}$$

The right-hand sides of (7.62b) and (7.64b)—$H_{\alpha\beta}^+$ and $N_{\alpha\beta}^+$—are the adjoint forms of $H_{\alpha\beta}$ and $N_{\alpha\beta}$, and can be obtained by interchanging $(\mathbf{x}, t)$ with $(\mathbf{x}', t')$ in (7.66) and (7.68).

The DIAs for the terms given by eqns (7.65)–(7.68) are constructed in exactly the same way as for (6.21) and (6.26) in the pure Eulerian case. Just as in that case, the resulting equations contain space–time integrals which take account of the memory and relaxation effects of the turbulence. But these integrals turn out to have the form of purely Eulerian time histories.

Kraichnan coped with this by altering these integrals to take on a purely Lagrangian character, while at the same time preserving the conservation and invariance properties of the purely Eulerian approximation. Apparently the only casualty of this process was the loss of the model representation which was held to guarantee the realizability of DIA. In the altered approximation, the space–time integrals are changed to be over Lagrangian histories or fluid-point space–time trajectories, with the desirable result that the final equations are invariant under random Galilean transformations.

We shall not go into great detail about these derivations, but for completeness we state the alteration principle as follows: change each intermediate (dummy) labelling time to t if it arises from the expansion of a factor with labelling time t. An identical process is carried out for the expansion of those factors with labelling time t'.

We shall not state the final equations here. This is partly because they are very complicated and partly because we shall be considering an abridged form of the theory in the next section. The interested reader will find the LHDI approximations for $S_{\alpha\beta}$, $H_{\alpha\beta}$, $M_{\alpha\beta}$, and $N_{\alpha\beta}$ elsewhere (Kraichnan 1965, eqns (8.9)–(8.12)). We shall confine ourselves here to the related topics of random Galilean invariance and the Kolmogorov distribution.

The demonstration of random Galilean invariance relies on the ensemble discussed in Sections 7.1.2 and 7.1.3, where the transforming velocity $\mathbf{v}$ is a random variable, which is constant in space and time, and which has the r.m.s. value v_0. Following Kraichnan, we use the notation $[\;\;]_v$ to denote a mean quantity which takes account of the distribution of $\mathbf{v}$.

The Galilean transformations are given by equation (7.27), and the corresponding effect on the Fourier components of the velocity field by (7.28). If we accept eqn (7.15)—and it must be borne in mind that this result is hedged about with many restrictions upon its validity—then we can use this, along with eqns (7.27) and (7.28), to express the effect of random Galilean transformation upon the velocity covariance and response tensor by the following forms:

$$[Q_{\alpha\beta}^s(\mathbf{k}; t|s; s'|s')]_v$$

$$= \delta_{\alpha\beta}\delta(k)v_0^2 + \exp\left\{-\frac{v_0^2 k^2 (t - t')^2}{2}\right\} Q_{\alpha\beta}^s(\mathbf{k}; t|s; t'|s') \qquad (7.69)$$

$$[G_{\alpha\beta}(\mathbf{k}; t|s; t'|s')]_v$$

$$= \exp\left\{-\frac{v_0^2 k^2 (t-t')^2}{2}\right\} G_{\alpha\beta}(\mathbf{k}; t|s; t'|s'). \tag{7.70}$$

As we saw in Section 7.1.3, we should require the theory to exhibit Galilean invariance on the time diagonal when $t = t'$. Clearly the approximate forms given by eqns (7.69) and (7.70) satisfy this requirement, as the exponential factor becomes unity when $t = t'$. Kraichnan (1965) has shown that, with the substitution of (7.69) and (7.70), the LHDI approximations for $S_{\alpha\beta}$, $H_{\alpha\beta}$, $M_{\alpha\beta}$, and $N_{\alpha\beta}$ are all invariant under random Galilean transformation for the case $t = t'$. It has also been shown (Kraichnan 1965) that these LHDI equations are compatible with the Kolmogorov inertial-range spectrum.

7.2.4 Abridged LHDI theory

We now consider the LHDI equations in what appears to be their simplest form. First, we make our usual simplification, in which we take the turbulence to be isotropic. Then we can eliminate tensors in favour of scalar functions, and the appropriate reduction for Lagrangian-history coordinates is

$$Q_{\alpha\beta}(\mathbf{k}; t|s; t'|s') = D_{\alpha\beta}(\mathbf{k})Q^s(k; t|s; t'|s') + \Pi_{\alpha\beta}(\mathbf{k})Q^c(k; t|s; t'|s') \tag{7.71}$$

$$G_{\alpha\beta}(\mathbf{k}; t|s; t'|s') = D_{\alpha\beta}(\mathbf{k})G^s(k; t|s; t'|s') + \Pi_{\alpha\beta}(\mathbf{k})G^c(k; t|s; t'|s'), \tag{7.72}$$

which is just a straightforward generalization of the usual Eulerian forms with $\Pi_{\alpha\beta}(\mathbf{k})$ given by (7.54) and $D_{\alpha\beta}(\mathbf{k})$ by (2.78).

This reduction to isotropic forms would still leave us with a very complicated set of equations. However, Kraichnan (1965) has made an abridgement of the theory which leads to a set of equations which is very little more complicated than Eulerian DIA. The procedure is conveniently explained in two steps.

(1) Generate a subset of the LHDI equations (7.61)–(7.64) which is purely Lagrangian in character. This is done by setting $t = t'$, and results in the turbulence being described in terms of the functions $Q_{\alpha\beta}(\mathbf{k}; t|t; t|s)$ and $G_{\alpha\beta}(\mathbf{k}; t|t; t|s)$.

(2) An examination of the LHDI approximations for $S_{\alpha\beta}$ etc. shows convolution integrals involving time arguments of the form $(t|s'; t|s'')$ (see Kraichnan 1965, eqns (8.9)–(8.12)). The abridgement is completed by approximating all such quantities as follows:

$$Q_{\alpha\beta}(\mathbf{k}; t|s'; t|s'') \to Q_{\alpha\beta}(\mathbf{k}; s'|s'; s'|s'') \qquad s' > s''$$

$$\to Q_{\alpha\beta}(-\mathbf{k}; s''|s''; s''|s') \qquad s' < s'' \tag{7.73}$$

$$G_{\alpha\beta}(\mathbf{k}; t|s'; t|s'') \to G_{\alpha\beta}(\mathbf{k}; s'|s'; s'|s''). \tag{7.74}$$

This step is exact for the particular case where $\mathbf{k} = 0$, but Kraichnan has offered no justification for its extension to all values of $\mathbf{k}$.

In addition to reducing the number of variables, the abridgement brings a bonus: the curl-free or compressive part of the generalized velocity does not have to be included in the equations. This follows from (7.49), (7.35), and the requirement that the Eulerian velocity field should be solenoidal. Thus we have $\mathbf{u}^c(k; t|t) = 0$, and hence $Q^c(k; t|t; t|r) = 0$.

With all these points in mind, we can introduce a simpler notation:

$$Q(k; t|s) = Q^s(k; t|t; t|s) \tag{7.75}$$

$$G(k; t|s) = G^s(k; t|t; t|s). \tag{7.76}$$

Then the abridged forms of equations (7.61)–(7.64) can be written for the case of isotropic turbulence as

$$\left(\frac{\partial}{\partial t} + 2vk^2\right) Q(k; t|t)$$

$$= 2 \int d^3j \, L(\mathbf{k}, \mathbf{j}) \int_0^t Q(|\mathbf{k} - \mathbf{j}|; t|r) \times$$

$$\times [G(k; t|r)Q(j; t|r) - G(j; t|r)Q(k; t|r)] \, dr \tag{7.77}$$

$$\left(\frac{\partial}{\partial t} + vk^2\right) Q(k; t|s)$$

$$= -\left(\frac{2k^2}{3}\right) Q(k; t|s) \int d^3j \int_0^t Q(|\mathbf{k} - \mathbf{j}|; t|r) \, dr +$$

$$+ \int d^3j \, N(\mathbf{k}, \mathbf{j}) Q(j; t|s) \int_s^t Q(|\mathbf{k} - \mathbf{j}|; t|r) \, dr +$$

$$+ \int d^3j \int_0^s [L(\mathbf{k}, \mathbf{j})G(k; s|r)Q(j; t|r) -$$

$$- N(\mathbf{j}, \mathbf{k})G(j; s|r)Q(k; t|r)]Q(|\mathbf{k} - \mathbf{j}|; t|r) \, dr -$$

$$- \int d^3j \int_0^t [L(\mathbf{k}, \mathbf{j})G(j; t|r)Q(k; t|r) -$$

$$- N(\mathbf{k}, \mathbf{j})G(k; t|r)Q(j; s|r)]Q(|\mathbf{k} - \mathbf{j}|; t|r) \, dr -$$

$$- \int d^3j \int_0^t [L(\mathbf{k}, \mathbf{j}) - N(\mathbf{j}, \mathbf{k})] \times$$

$$\times G(j; t|r)Q(k; t|r)Q(|\mathbf{k} - \mathbf{j}|; s|r) \, dr \tag{7.78}$$

$$\left(\frac{\partial}{\partial t} + vk^2\right) G(k; t|s)$$

$$= -\left(\frac{2k^2}{3}\right) G(k; t|s) \int d^3j \int_s^t Q(|\mathbf{k} - \mathbf{j}|; t|r) \, dr +$$

$$+ \int d^3j \, N(\mathbf{k}, \mathbf{j}) G(j; t|s) \int_s^t Q(|\mathbf{k} - \mathbf{j}|; t|r) \, dr +$$

$$+ \int d^3j [N(\mathbf{k}, \mathbf{j}) - L(\mathbf{k}, \mathbf{j})] \times$$

$$\times G(j; t|s) Q(|\mathbf{k} - \mathbf{j}|; t|s) \int_s^t G(k; r|s) \, dr -$$

$$- \int d^3j \int_s^t [L(\mathbf{k}, \mathbf{j}) G(j; t|r) G(k; r|s) -$$

$$- N(\mathbf{k}, \mathbf{j}) G(k; t|r) G(j; r|s)] Q(|\mathbf{k} - \mathbf{j}|; t|r) \, dr \qquad (7.79)$$

$$G(k; s|s) = 1. \qquad (7.80)$$

The coefficient $L(\mathbf{k}, \mathbf{j})$ is given by eqn (2.162), while $N(\mathbf{k}, \mathbf{j})$ can be written as

$$N(\mathbf{k}, \mathbf{j}) = L(\mathbf{k}, \mathbf{j}) + A(\mathbf{k}, \mathbf{j}), \qquad (7.81)$$

where $A(\mathbf{k}, \mathbf{j})$ can be expressed in terms of μ, which, as before, is the cosine of the angle between the vectors $\mathbf{k}$ and $\mathbf{j}$, and η, which is the cosine of the angle between the vectors $\mathbf{k}$ and $\mathbf{k} - \mathbf{j}$:

$$A(\mathbf{k}, \mathbf{j}) = \frac{k^2}{2} (\mu^2 - \eta^2). \qquad (7.82)$$

It is of interest to compare these abridged LHDI (ALHDI) equations with those of the Eulerian DIA as derived in Chapter 6. If first we consider the covariance equation on the time diagonal, then the Eulerian result is given by eqn (6.35). Note that the arbitrary input term should be set equal to zero in order to facilitate the comparison. Evidently this is identical with the ALHDI equivalent, as given by eqn (7.77), provided that we simply replace time arguments like (t, t) by $(t|t)$!

However, eqns (7.78) and (7.79) clearly involve a few more terms than their Eulerian counterparts—eqns (6.29) and (6.31)—although it is fair to say that their intrinsic level of complication is no greater.

7.2.5   Other Lagrangian theories

In the preceding sections, we have discussed the original derivation (Kraichnan 1965) of the ALHDI equations. To recapitulate, there were essentially three stages.

(1) Modify the derivation of DIA by replacing the Eulerian coordinates by the generalized Lagrangian-history coordinates.

(2) Alter the time arguments to generate a purely Lagrangian formulation which is invariant under random Galilean transformations. The result is the LHDI theory.

(3) Alter the time arguments to produce a further abridged set of equations. This is an imponderable approximation and results in the ALHDI equations.

Kraichnan (1977) later used the method of reversion of power series (see Section 6.1.7 for a brief discussion) to develop systematic renormalized perturbation expansions, in which each order is invariant under random Galilean transformations. When the new expansions are truncated at the lowest order, the LHDI and ALHDI approximations are recovered.

From our present point of view, the most important practical consequence of this work was the development of a new Lagrangian-history theory (Kraichnan and Herring 1978). While we do not want to become involved in the quantitative aspects here—that topic will be the subject of the next chapter—we should just mention that the motivation for the new theory was an awareness that the quantitative performance of ALHDI was, despite its qualitative success, less than satisfactory. In particular, we can pick on the tendency of ALHDI to overestimate the rate of inertial transfer of energy to higher wavenumbers. This is of interest because the opposite tendency has been associated with the failure of Eulerian DIA.

The general approach is based on the use of the straining field (or rate-of-strain tensor) $b_{\alpha\beta}(\mathbf{x}, t)$, instead of the velocity field, as the basic variable of the primitive perturbation series. The straining field is defined in terms of the Eulerian velocity as

$$b_{\alpha\beta}(\mathbf{x}, t) = \frac{\partial u_\alpha(\mathbf{x}, t)}{\partial x_\beta} + \frac{\partial u_\beta(\mathbf{x}, t)}{\partial x_\alpha}. \tag{7.83}$$

By analogy with the introduction of the generalized velocity $\mathbf{u}(\mathbf{x}, t|s)$, with properties as defined by (7.35) and (7.36), a generalized straining field is introduced and defined by

$$\frac{\partial b_{\alpha\beta}(\mathbf{x}, t|s)}{\partial t} + \mathbf{u}(\mathbf{x}, t) \cdot \nabla b_{\alpha\beta}(\mathbf{x}, t|s) = 0$$

$$b_{\alpha\beta}(\mathbf{x}, s|s) = b_{\alpha\beta}(\mathbf{x}, s), \tag{7.84}$$

which can be compared with eqn (7.38) for the generalized velocity. It is important to note that the definition of $b_{\alpha\beta}(\mathbf{x}, t)$, as given by (7.83), cannot be generalized by replacing $\mathbf{u}(\mathbf{x}, t)$ by $\mathbf{u}(\mathbf{x}, t|s)$, except for the trivial case where $t = s$.

Kraichnan and Herring (1978) have shown that the reversion procedures used previously with the velocity field also work with the straining field, and are able to derive SBLHDI and SBALHDI approximations, where SB stands for 'strain-based'. We shall return to the subject of the quantitative performance of these theories in the next chapter.

Lastly, for completeness, we should mention the non-Eulerian renormalized expansion method of Horner and Lipowsky (1979), in which the formalism of Martin, Siggia, and Rose (1973) (see Chapter 5) is used to construct Galilean-invariant expansions, and the Lagrangian method of Kaneda (1981), who produced a variant of Kraichnan's Lagrangian-history formulation by working with measuring-time derivatives rather than labelling-time derivatives.

7.3 Modified EFP theories

In Section 6.2.5, we discussed the EFP 'second equation' in the form of (6.107) for $\omega(k)$, and from a comparison with DIA in Section 6.2.6, its interpretation as a response equation. In Sections 6.2.7 and 7.1.1, we saw that the energy-balance equation (6.101) is compatible with the Kolmogorov distribution, and that the failure of the EFP theory can be traced quite specifically to an infra-red divergence in eqn (6.107). Hence, in view of both the arbitrariness and the incorrectness of eqn (6.107), the need for an appropriate second equation—the energy equation (6.101) is, of course, the 'first equation'—is manifest.

In this section we shall present two contrasting approaches to this problem. In the first (Edwards and McComb 1969) it is argued that one needs an additional principle in order to derive an equation for the response of the system. In the second, it is claimed (McComb 1974, 1976) that the response equation can be determined on purely physical grounds from the unmodified EFP theory, although it is not suggested that such a prescription is unique.

7.3.1 Maximal entropy principle

We have argued previously that, although the methods of near-equilibrium statistical mechanics are inapplicable to turbulence, we can to some extent be guided by analogies with the classical methods and may even borrow some of the general strategies. In this section we shall discuss a particularly bold example of such borrowing.

It is well known that the progress of an isolated system towards equilibrium is encapsulated in Boltzmann's H-theorem. That is, the entropy (H, in Boltzmann's terminology) of the system increases until it reaches a maximum, at which time all parts of the phase space are occupied with equal weight.

Clearly this is not the case for turbulence, which is always far from equilibrium. Yet it is nevertheless possible that some modified version of the H-theorem may be applicable. For instance, Kraichnan has put forward the view that turbulence is in a state of maximal randomness. This picture of turbulence—that it is in the most chaotic state permitted by the (deterministic) equation of motion—has considerable appeal, and raises the possibility that any specific measure of the randomness (such as the entropy!) might also be described as maximal.

It is in this spirit that Edwards and McComb (1969) argued that entropy, when interpreted as (negative) information, is available as a concept for any system, without reference to thermal equilibrium. Thus, if the general form (Shannon and Weaver 1949) is given by

$$S = -\kappa \int P \ln P, \qquad (7.85)$$

where κ is the Boltzmann constant, and the integration is over all the variables of the system, then the straightforward generalization to turbulence is just

$$S = -\int P[\mathbf{u(k)}] \ln P[\mathbf{u(k)}] \prod_{\mathbf{k}} d\mathbf{u(k)}, \qquad (7.86)$$

where we have put $\kappa = 1$, as the value of κ does not enter into the subsequent calculation.

If we now maximize S, as given by (7.86) with (6.73) for P, it is arguable that this is equivalent to choosing P such that it contains the least information. Of course, the resultant extremum will not be an absolute, such as one would have in thermal equilibrium, but will depend on our choice of variables. However, against this one might argue that it has been adopted as a principle of wide application—see Shore and Johnson (1980) for a recent paper which contains many other references—that one should choose as the most probable of many system states the one with the largest entropy. In this sense, even for a dissipative system like turbulence, it seems not too unreasonable to choose the probability distribution such that the system has the largest entropy possible. In other words, one is presumably ennunciating a principle of maximal entropy.

Therefore, taking this pragmatic view, we can argue that, through eqns (7.86) and (6.73), we have the relationship $S = S[q(k), \omega(k)]$, and that we already have an equation for $q(k)$ in the form of (6.101), so that we should obtain an equation for $\omega(k)$ through the condition

$$\frac{\delta S}{\delta \omega(k)} = 0. \qquad (7.87)$$

However, S is a function of both $q(k)$ and $\omega(k)$, two quantities which in turn are connected through the energy equation (6.101). Thus the variation in (7.87) is not free, but is in fact subject to the constraint of the energy balance. Hence we should write (7.87) as

$$\frac{\delta S}{\delta \omega(k)} + \sum_{\mathbf{j}} \left[\frac{\delta S}{\delta q(j)} \right] \frac{\delta q(j)}{\delta \omega(k)} = 0, \qquad (7.88)$$

where the coefficient in the second term on the r.h.s. can be obtained from eqn (6.84):

$$\frac{\delta q(j)}{\delta \omega(k)} = -\frac{d(k)}{2\omega^2(k)\delta(k-j)} + \frac{1}{2\omega(j)}\frac{\delta d(j)}{\delta \omega(k)}. \tag{7.89}$$

A problem now arises, in that the second term on the r.h.s. of (7.89) must be calculated iteratively by using eqn (6.84) to put $d(k) = 2q(k)\omega(k)$ in the non-linear term of the energy equation. Edwards and McComb (1969) simply drop this term altogether. The computation needed to keep it would be tremendous, and its neglect does not affect the question of whether or not a solution exists. Therefore, with this approximation, eqn (7.88) becomes

$$\frac{\delta S}{\delta \omega(k)} - \frac{d(k)}{2\omega^2(k)}\frac{\delta S}{\delta q(k)} = 0, \tag{7.90}$$

where the summation over $\mathbf{j}$ has been eliminated along with the delta function.

Now, from eqn (7.86), and with the probability distribution P given by the expansion (6.73), we can calculate the turbulent entropy to second order in the non-linearity as

$$S = -\int (P_0 + \lambda P_1 + \lambda^2 P_2)\ln(P_0 + \lambda P_1 + \lambda^2 P_2)\delta\mathbf{u}$$

$$= -\int (P_0 + \lambda P_1 + \lambda^2 P_2)\ln[P_0(1 + \lambda P_0^{-1}P_1 + \lambda^2 P_0^{-1}P_2)]\delta\mathbf{u}$$

$$= -\int P_0 \ln P_0 \delta\mathbf{u} - \lambda^2 \int P_1 P_0^{-1}P_1 \delta\mathbf{u} + o(\lambda^4), \tag{7.91}$$

where terms of order λ and λ^3 vanish as they involve the integration of an odd function of $\mathbf{u}$ over all values of $\mathbf{u}$, and terms of order λ^2 involving P_2 vanish because of the conditions imposed on the expansion for $P[\mathbf{u}(k)]$ through eqns (6.84), (6.85), and (6.94).

Then, upon substitution of (6.83) for P_0 and (6.92) for P_1, it can be shown that the entropy takes the form

$$S = \left(\frac{2\pi}{L}\right)^3 \sum_k -\frac{1}{2}\sum_k \ln q(k) -$$

$$-\left(\frac{2\pi}{L}\right)^9 \sum_k \sum_j \sum_l [L(\mathbf{k},\mathbf{j},\mathbf{l})q(j)q(l) - L(\mathbf{l},\mathbf{j},\mathbf{k})q(k)q(j)] \times$$

$$\times \frac{1}{q(k)[\omega(k) + \omega(j) + \omega(l)]^2}, \tag{7.92}$$

where the bookkeeping parameter λ has now been put equal to unity.

A full derivation of this result has been given elsewhere, and so we shall only make a few points here. First, we note that the first term on the r.h.s. is essentially just the number of degrees of freedom of the system. As it is a constant, it does not contribute when we perform the differentiation according to eqn (7.90). The second term arises from the normalization in (6.83) which is

$$N = \prod_{\mathbf{k}} \{q(k)\}^{-1/2}.$$

The third term involves a manipulation of the kind which should be familiar by now (e.g. see Appendix E). Note that $L(\mathbf{k}, \mathbf{j}, \mathbf{l})$ is a generalized form of the familiar $L(\mathbf{k}, \mathbf{j})$—which is given by eqn (2.163)—and which contains a delta function such that

$$\int d^3l\, L(\mathbf{k}, \mathbf{j}, \mathbf{l}) f(\mathbf{k}, \mathbf{j}, \mathbf{l}) = L(\mathbf{k}, \mathbf{j}) f(\mathbf{k}, \mathbf{j}, \mathbf{k} - \mathbf{j}), \qquad (7.93)$$

where $f(\mathbf{k}, \mathbf{j}, \mathbf{l})$ stands for any kernel which occurs in turbulence theory.

Finally, if S given by eqn (7.92) is varied with respect to some particular $\omega(A)$, then it can be shown (McComb 1967) that (7.90) becomes

$$\frac{1}{\omega(A)} - \int\int d^3l\, d^3k [L(\mathbf{k}, \mathbf{A}, \mathbf{l}) q(A) q(l) -$$

$$- L(\mathbf{l}, \mathbf{k}, \mathbf{A}) q(k) q(A)] \frac{1}{q(k)\omega(A)[\omega(k) + \omega(A) + \omega(l)]^2} -$$

$$- \int\int d^3j\, d^3k\, L(\mathbf{k}, \mathbf{j}, \mathbf{A}) q(A) q(j) \frac{1}{q(k)\omega(A)[\omega(k) + \omega(j) + \omega(A)]^2} +$$

$$+ \int\int d^3j\, d^3l\, L(\mathbf{A}, \mathbf{j}, \mathbf{l}) q(j) q(l) \frac{1}{\omega(A)q(A)[\omega(A) + \omega(j) + \omega(l)]^2} -$$

$$- 4\left\{ \int\int d^3k\, d^3l [L(\mathbf{k}, \mathbf{A}, \mathbf{l}) q(A) q(l) - L(\mathbf{l}, \mathbf{A}, \mathbf{k}) q(A) q(k)] \times \right.$$

$$\times \frac{1}{q(k)[\omega(k) + \omega(A) + \omega(l)]^3} +$$

$$+ 2 \int\int d^3j\, d^3l [L(\mathbf{A}, \mathbf{j}, \mathbf{l}) q(j) q(l) - L(\mathbf{l}, \mathbf{A}, \mathbf{j}) q(A) q(j)] \times$$

$$\left. \times \frac{1}{q(A)[\omega(A) + \omega(j) + \omega(l)]^3} \right\}$$

$$= 0, \qquad (7.94)$$

where we have taken the limit of infinite system volume.

This looks rather formidable, but in the limit of large Reynolds numbers we can follow the procedures discussed in Sections 6.2.7 and 7.1.1, with $q(k)$ and $\omega(k)$ being given by the Kolmogorov forms (6.120) and (6.121) respectively. Equation (7.94) can be used to replace the original EFP response equation (6.107), and, following the procedures used in deriving equation (7.4), we can show (McComb 1967) that (7.94) reduces to

$$\frac{\alpha D'}{\beta^2} = 1. \tag{7.95}$$

The important point about this result is that the integrals which determine the value of D' all exist owing to cancellations at the various singularities (McComb 1967). Thus the maximal entropy method does eliminate the infra-red divergence which was found with (6.107).

Equation (7.95) can be solved simultaneously with (6.125), so that the Kolmogorov spectral constant becomes (compare eqn (7.6))

$$\alpha = \frac{(D')^{1/3}}{(0.19)^{2/3}}. \tag{7.96}$$

Edwards and McComb (1969) report a value $D' = 2.0$, and hence $\alpha = 3.8$, which is roughly twice as large as the accepted experimental value.

Overall, the maximal entropy method is qualitatively successful in eliminating the infra-red divergence, but in view of the neglect of the second term on the r.h.s. of eqn (7.89), no conclusions can be drawn about its quantitative performance.

Lastly, we should note that an alternative variational method has been proposed by Quian (1983), but this treats the variation as if $\omega(k)$ and $q(k)$ are independent. In effect, this is the same as if the second term on the r.h.s. of our eqn (7.88) had been neglected. Hence, it follows that Quian's method is mathematically incorrect.

7.3.2   The response function determined by a local energy balance

In the original EFP theory, we saw that the response equation for the system is identified through the decomposition of the energy equation (6.101) into the form (6.106). This particular decomposition was guided by a physical model, in which the non-linear term is interpreted as a random force which is uncorrelated with the velocity field (see Section 6.2.5 and the arguments leading to eqn (6.105) for $s(k)$).

Once made, this particular division of the non-linearity into input and output terms looks physically very reasonable, with $2r(k)q(k)$ representing the loss of energy from mode **k** owing to inertial transfer. Note also the relevant discussion of the DIA energy equation, as given in Section 6.1.5.

However, the decomposition of (6.101) into (6.106) is arbitrary, and McComb (1974, 1976) has proposed a variant on this procedure. This later method of dividing up the energy equation leads to some interesting results.

Let us begin by multiplying each term in (6.101) by $4\pi k^2$ and rewriting the equation as

$$-\int_0^\infty dj\, B(k,j) = 4\pi k^2 W(k) - 2vk^2 E(k),\tag{7.97}$$

where the energy spectrum $E(k)$ is defined by eqn (2.101) and $B(k,j)$ is given by

$$B(k,j) = 16\pi^2 k^2 j^2 \int_{-1}^{1} d\mu\, L(\mathbf{k},\mathbf{j})q(|\mathbf{k}-\mathbf{j}|)[q(j) - q(k)] \times$$

$$\times \frac{1}{\omega(k) + \omega(j) + \omega(|\mathbf{k}-\mathbf{j}|)}.\tag{7.98}$$

Here μ is the cosine of the angle between the vectors $\mathbf{k}$ and $\mathbf{j}$, and $L(\mathbf{k},\mathbf{j})$ depends on μ through the relationship (2.162). It should be noted that $B(k,j)$ can be expressed in terms of the transfer spectrum $T(k)$, as defined by eqn (2.118), through

$$T(k) = \int_0^\infty dj\, B(k,j).\tag{7.99}$$

It should also be noted that one cannot obtain (7.99) from the DIA form merely by putting $t = 0$ in (6.37) for $T(k,t)$. The correct procedure is to be found in Section 6.2.6.

It is a simple matter to show that this form for $T(k)$ satisfies the fundamental requirement of (2.126):

$$\int_0^\infty T(k)\, dk = \int_0^\infty dk \int_0^\infty dj\, B(k,j) = 0,\tag{7.100}$$

or in other words the inertial term conserves energy. As before, this result relies on the antisymmetry of $B(k,j)$ under interchange of k and j,

$$B(k,j) = -B(j,k).\tag{7.101}$$

Now let us replace the original EFP postulates, as given by eqns (6.79) and (6.80), by

$$d(k) = W(k) + H(k)\tag{7.102}$$

$$\omega(k) = vk^2 + v(k)k^2,\tag{7.103}$$

where $v(k)$ is the effective turbulent viscosity. The idea that the molecular viscosity is augmented (or renormalized) by non-linear interactions has been implicit in all the theories discussed so far. Now we are simply making this hypothesis explicit.

The implication of these steps is that we are replacing the EFP decomposition of (6.101), which implies that the inertial transfer term can be written

$$T(k) = 4\pi k^2 [s(k) - 2r(k)q(k)], \tag{7.104}$$

by the new form

$$T(k) = 4\pi k^2 [H(k) - 2v(k)k^2 q(k)]. \tag{7.105}$$

But, of course, one has to find new ways of defining $H(k)$ and $v(k)$ in order to give substance to this.

Let us now consider the steady state energy balance, as discussed in general terms in Section 4.3.2 and in the context of the inertial range in Section 6.1.5. For a wavevector **k** in the inertial range, we have from Kraichnan's definition of the transport power $\Pi(k)$ the general condition

$$\Pi(k) = \varepsilon, \tag{6.38}$$

which can in itself be regarded as defining the inertial range. It follows from eqns (7.100) and (7.101) that we can obtain this condition in one of two ways. Either we can integrate both sides of (7.97) from zero up to k, or we can integrate from k up to infinity. This will be true irrespective of the value of k, provided only that it is in the inertial range of wavenumbers. In both cases we assume that eqn (4.90) can be approximated by

$$\int_0^k 4\pi k'^2 W(k') \, dk' = \int_k^\infty 2vk'^2 E(k') \, dk' = \varepsilon. \tag{7.106}$$

Hence, from the integration of (7.97), the two forms of (6.38) can be written as

$$\int_0^k \left\{ 4\pi k'^2 W(k') + \int_k^\infty dj \, B(k',j) \right\} dk' = 0 \tag{7.107}$$

$$\int_k^\infty \left\{ \int_0^k dj \, B(k',j) - 2vk'^2 E(k') \right\} dk' = 0, \tag{7.108}$$

where, in both cases, the limits on the integration with respect to j follow from the antisymmetry of $B(k,j)$ under interchange of k and j (see eqn (7.101)).

Equations (7.107) and (7.108) together represent the steady state energy balance. From the definitions implied by (7.102) and (7.103), it seems natural to rewrite the steady state energy balances as

$$\int_0^k \left\{ 4\pi k'^2 W(k') - 2v(k')k'^2 E(k') \right\} dk'$$

$$= \int_0^k \left\{ W(k') - 2v(k')k'^2 q(k') \right\} 4\pi k'^2 \, dk'$$

$$= 0 \tag{7.109}$$

$$\int_k^\infty \{4\pi k'^2 H(k') - 2vk'^2 E(k')\} \, dk'$$

$$= \int_k^\infty \{H(k') - 2vk'^2 q(k')\} 4\pi k'^2 \, dk'$$

$$= 0 \tag{7.110}$$

respectively. Hence we can make the identifications

$$v(k) = -\frac{\{\int_k^\infty k^{-2} \, dj \, B(k,j)\}}{2q(k)}$$

$$= k^{-2} \int_{j \geqslant k} d^3j \, L(\mathbf{k},\mathbf{j}) q(|\mathbf{k}-\mathbf{j}|)[q(k) - q(j)] \times$$

$$\times \frac{1}{q(k)[\omega(k) + \omega(j) + \omega(|\mathbf{k}-\mathbf{j}|)]} \tag{7.111}$$

$$H(k) = \left(\frac{1}{4\pi k^2}\right) \int_0^k dj \, B(k,j)$$

$$= 2 \int d^3j \, L(\mathbf{k},\mathbf{j}) q(|\mathbf{k}-\mathbf{j}|)[q(j) - q(k)] \times$$

$$\times \frac{1}{\omega(k) + \omega(j) + \omega(|\mathbf{k}-\mathbf{j}|)}. \tag{7.112}$$

We can treat eqn (7.111) in the same way as we did previously with (7.1) or (7.94). For the case of infinite Reynolds numbers, we have

$$\omega(k) = k^2 v(k)$$

$$= \int_{j \geqslant k} d^3j \, L(\mathbf{k},\mathbf{j}) q(|\mathbf{k}-\mathbf{j}|)[q(k) - q(j)] \times$$

$$\times \frac{1}{q(k)[\omega(k) + \omega(j) + \omega(|\mathbf{k}-\mathbf{j}|)]}. \tag{7.113}$$

Substituting the Kolmogorov forms (6.120) and (6.121) for $q(k)$ and $\omega(k)$, we can reduce (7.113) to

$$\frac{\alpha D''}{\beta^2} = 1, \tag{7.114}$$

which can be solved simultaneously with (6.125) for the Kolmogorov spectral constant—this can be compared with the similar procedures which lead to eqns (7.95) and (7.96). McComb (1976) has reported the value $D'' = 0.573$ and hence $\alpha = 2.5$, which is only marginally outside the most probable experimental range.[1]

Equation (7.113) is very much simpler than (7.94), and so in this case it is easy to see that the integral for D'' exists and that there is no infra-red divergence. The essential point is that a potential divergence at $|\mathbf{k} - \mathbf{j}| = 0$ is eliminated by a cancellation in the integrand, whereas the potential divergence at $j = 0$ is excluded by the lower limit at $j = k$.

7.3.3   Local energy-transfer equations

The Fourier transformed Navier–Stokes equations indicate that turbulence is, in principle, a non-local phenomenon in wavenumber space. All modes— however far apart in $\mathbf{k}$-space—are coupled together. However, there is a widespread belief that, in practice, turbulent inertial transfers are dominated by (as it were) nearest-neighbour interactions between modes in $\mathbf{k}$-space, or, in other words, these energy transfers are predominantly local. Such ideas are, after all, fundamental to the Kolmogorov hypotheses, and indeed to the very cascade picture of turbulence.

Yet, despite this prevalent belief in the localness of energy transfers, there have been surprisingly few attempts to describe spectral processes by means of differential equations. We regard this as surprising because the use of Taylor series to produce governing equations in differential form is a basic tool of other subjects ranging from statistical mechanics through continuum mechanics to engineering models of turbulence. Of the few attempts known to us, McComb (1969) and Edwards and McComb (1971, 1972) have expanded the kernel of the energy equation (6.101) about an assumed Kolmogorov distribution in order to produce simpler formulations which can then be applied to more complex problems such as shear flows. Nakano (1972) made expansions in wavenumber of the non-linearity in the unaveraged Navier–Stokes equation, and used the resulting equations as the basis of his RPT, whereas McComb (1974) took a renormalized perturbation theory as a starting point and made similar expansions of eqns (7.112) and (7.113) in order to obtain differential equations for the energy spectrum and modal lifetime. We shall give a very brief account of only the last reference here.

We begin by rewriting eqns (7.112) and (7.113) with $\mathbf{k} - \mathbf{j}$ replacing $\mathbf{j}$. The resulting equations are readily obtained in the form

$$H(k) = 2 \int_{j \leqslant k} d^3 j \, L(\mathbf{k}, \mathbf{k} - \mathbf{j}) q(j) [q(|\mathbf{k} - \mathbf{j}|) - q(k)] \times$$

$$\times \frac{1}{\omega(k) + \omega(j) + \omega(|\mathbf{k} - \mathbf{j}|)}, \tag{7.115}$$

$$k^2 v(k) = \int_{j \geqslant k} d^3 j \, L(\mathbf{k}, \mathbf{k} - \mathbf{j}) q(j) [q(k) - q(|\mathbf{k} - \mathbf{j}|)] \times$$

$$\times \frac{1}{q(k) [\omega(k) + \omega(j) + \omega(|\mathbf{k} - \mathbf{j}|)]}, \tag{7.116}$$

where $L(\mathbf{k}, \mathbf{k} - \mathbf{j})$ is given by eqn (6.46). The reader may find it helpful to recall that we did this to the DIA response equation previously when we replaced (6.29) by (6.44). It should also be recalled that it is permissible to interchange $\mathbf{j}$ and $\mathbf{k} - \mathbf{j}$ in this way because they are both dummy variables.

Our next step is to set up the problem formally. We do this by using eqns (7.97), (7.99), (7.102), (7.103), and (7.105) to write the energy equation as

$$4\pi k^2 H(k) + 4\pi k^2 W(k) - 2\omega(k)k^2 E(k) = 0. \tag{7.117}$$

Then the problem is to obtain approximate forms for $H(k)$ and $\omega(k)$ which are local in wavenumber space.

We start with the former quantity. From eqn (7.115) we see that we always have $j/k \leqslant 1$, and so we can expand $L(\mathbf{k}, \mathbf{k} - \mathbf{j})$, $q(|\mathbf{k} - \mathbf{j}|)$, and $\omega(|\mathbf{k} - \mathbf{j}|)$, all in powers of j/k. After some detailed algebra (McComb 1974), we find

$$H(k) = \left(\frac{\partial}{\partial k} + \frac{2}{k}\right)\left\{A_2(k)\frac{\partial q(k)}{\partial k}\right\} - A_1(k)\frac{\partial q(k)}{\partial k}, \tag{7.118}$$

where terms of order $(j/k)^3$ have been discarded, and the coefficients $A_1(k)$ and $A_2(k)$ are given by

$$A_1(k) = \frac{k^2}{15\omega(k)}\int_0^\infty \delta(k-j)j^2 E(j)\,dj \tag{7.119}$$

$$A_2(k) = \frac{k^2}{15\omega(k)}\int_0^\infty j^2 E(j)\,dj. \tag{7.120}$$

Then, multiplying $H(k)$, as given by (7.118), by $4\pi k^2$ and substituting into (7.117) leads us to the required form for the energy equation:

$$\frac{\partial[k^2 A_2(k)\partial\{E(k)/k^2\}/\partial k]}{\partial k} - k^2 A_1(k)\frac{\partial\{E(k)/k^2\}}{\partial k} +$$

$$+ 4\pi k^2 W(k) - 2\omega(k)k^2 E(k)$$

$$= 0. \tag{7.121}$$

We now need a corresponding equation for $\omega(k)$ and we start with (7.116) for the effective viscosity. This time we expand in powers of k/j and again retain second-order terms. The result is

$$\omega(k) = \{v + v(k)\}k^2$$

$$= vk^2 + \frac{2k^2}{15}\int_k^\infty \frac{dj\, E(j)[1 + \{j/4\omega(j)\}\partial\omega(j)/\partial j]}{\omega(j)}. \tag{7.122}$$

Equations (7.121) and (7.122) have been solved simultaneously, in the inertial range of wavenumbers, and the result is a value for the Kolmogorov constant which is $\alpha = 1.5$ (McComb 1974). This actually compares quite well with experimental results, but the real attraction of this work lies in the relative simplicity of the final equations for $E(k)$ and $\omega(k)$. Quite similar results have been obtained, using different methods, by Nakano (1972), but unfortunately his equations contain a term which cannot be expressed in terms of the main statistical variables. A detailed comparison of the two sets of equations has been given elsewhere (McComb 1974).

7.4   Local energy-transfer theory of non-stationary turbulence (LET)

The idea that the turbulent response function can be determined from an energy balance, which is local in wavenumber space, has been extended to the more general case of time-dependent turbulence (McComb 1978). The resulting time-dependent local energy-transfer (LET) theory is the subject of this section.

LET belongs to that class of theories (such as DIA or Wyld's analysis) which relies on a direct renormalization of the primitive expansion of the turbulent velocity field. This method has been discussed in a general way in Sections 3.5.2 and 5.5, and, with particular reference to DIA, in Section 6.1. Naturally this means that there are similarities between LET and (Eulerian) DIA. However, at this stage, our principal interest lies in how LET differs from DIA, and this really boils down to the choice of ansatz (or basic hypothesis).

In the case of DIA, this is the introduction of the infinitesimal response function which relates fluctuations in the stirring forces to the resulting fluctuations in the velocity field through eqn (6.3). The corresponding step in the LET theory is a hypothesis that the turbulent response can be represented by the introduction of an exact propagator which connects the velocity field, associated with mode $\mathbf{k}$, to itself at a later time.

7.4.1   The velocity-field propagator

We have previously met the zero-order propagator in Section 5.5.1, where we formulated the general perturbative approach to the Navier–Stokes equation. We saw that it could be regarded as the (zero-order) response to a forcing term, as in eqn (5.61), or as the relationship between $\mathbf{u}(\mathbf{k}, t)$ and $\mathbf{u}(\mathbf{k}, t')$, as in eqn (5.62). It is the latter interpretation which is of interest to us here. We begin by summarizing some of the properties of the zero-order propagator.

First, we rewrite (5.57) as

$$\left[\frac{\partial}{\partial t} + \nu k^2 \right] H_{\alpha\beta}^{(0)}(\mathbf{k}; t, t') = D_{\alpha\beta}(\mathbf{k})\delta(t - t'), \qquad (7.123)$$

where $G^{(0)}$ has been renamed $H^{(0)}$ for later convenience. As always, in this (and the preceding) chapter, we restrict our attention to isotropic turbulence; hence it follows that the zero-order propagator takes the form

$$H_{\alpha\beta}^{(0)}(\mathbf{k};t,t') = \begin{cases} D_{\alpha\beta}(\mathbf{k})\exp\{-vk^2(t-t')\} & t > t', \\ 0 & t < t'. \end{cases} \tag{7.124}$$

As usual, our starting point is the Navier–Stokes equation. This time we take this to be in the form given by (5.56), with a bookkeeping parameter λ in front of the non-linear term, but with the stirring forces set equal to zero:

$$\left[\frac{\partial}{\partial t} + vk^2\right]u_\alpha(\mathbf{k},t) = \lambda M_{\alpha\beta\gamma}(\mathbf{k}) \sum_{\mathbf{j}} u_\beta(\mathbf{j},t)u_\gamma(\mathbf{k}-\mathbf{j},t). \tag{7.125}$$

Evidently eqn (7.125) is just the inhomogeneous version of (7.123). Hence $H^{(0)}$ is the appropriate Green function, and the solution of (7.125), for the velocity field at any time, is given by

$$u_\alpha(\mathbf{k},t) = H_{\alpha\sigma}^{(0)}(\mathbf{k};t,t_0)u_\sigma(\mathbf{k},t_0) +$$

$$+ \lambda \int_{t_0}^t ds\, H_{\alpha\sigma}^{(0)}(\mathbf{k};t,s) \sum_{\mathbf{j}} M_{\sigma\beta\gamma}(\mathbf{k})u_\beta(\mathbf{j},s)u_\gamma(\mathbf{k}-\mathbf{j},s) \tag{7.126}$$

where the velocity field is prescribed at $t = t_0$.

Thus the role of $H^{(0)}$ as a zero-order propagator is clear: it has the property

$$u_\alpha^{(0)}(\mathbf{k},t) = H_{\alpha\sigma}^{(0)}(\mathbf{k};t,s)u_\sigma^{(0)}(\mathbf{k},s), \tag{7.127}$$

which is the solution of (7.125) when we set $\lambda = 0$. It is also easily seen that it satisfies the conditions

$$H_{\alpha\sigma}^{(0)}(\mathbf{k};t,s)H_{\sigma\beta}^{(0)}(\mathbf{k};s,t') = H_{\alpha\beta}^{(0)}(\mathbf{k};t,t')$$

$$H_{\alpha\sigma}^{(0)}(k;t,t) = 1. \tag{7.128}$$

Let us now consider eqn (7.126) for the velocity field at any time. For simplicity, we assume the initial condition to be that $\mathbf{u}(\mathbf{k},t) = 0$ at $t = 0$ and we also put the bookkeeping parameter $\lambda = 1$, and (7.126) reduces to

$$u_\alpha(\mathbf{k},t) = \int_0^\infty ds\, H_{\alpha\sigma}^{(0)}(\mathbf{k};t,s) \sum_{\mathbf{j}} M_{\sigma\beta\gamma}(\mathbf{k})u_\beta(\mathbf{j},s)u_\gamma(\mathbf{k}-\mathbf{j},s). \tag{7.129}$$

This is, of course, the exact solution of the Navier–Stokes equation, but in its present form it is of very little use. The basic assumption of the LET theory is that we can postulate the existence of a rernormalized version of (7.127), and hence that the exact solution (7.129) can be approximated by

$$u_\alpha(\mathbf{k},t) = H_{\alpha\sigma}(\mathbf{k};t,s)u_\sigma(\mathbf{k},s), \tag{7.130}$$

where $H_{\alpha\sigma}(\mathbf{k};t,s)$ is the renormalized velocity-field propagator and, like the zero-order form, satisfies the conditions

$$H_{\alpha\sigma}(\mathbf{k};t,s)H_{\sigma\beta}(\mathbf{k};s,t') = H_{\alpha\beta}(\mathbf{k};t,t')$$

$$H_{\alpha\beta}(\mathbf{k};t,t) = 1. \tag{7.131}$$

It should perhaps be emphasized that Wyld's analysis (see Section 5.5) tells us that a renormalized propagator exists in the form of an infinite series. Specific turbulence theories then correspond to specific choices about ways of truncating, and otherwise approximating, this series. If we consider the EFP theory, for example, then what we are saying (in effect) is that the accuracy of the energy-balance equation (6.101) depends on our choice of $\omega(k)$.

This can be understood in the following way. If we were to follow simple perturbation theory, then the denominator on the r.h.s. of (6.101) would contain only modal decay parameters of the form νk^2. That is, the response times would be governed by the kinematic viscosity of the fluid. Renormalization replaces the bare viscosity by the renormalized form, $\nu k^2 \to \{\nu + \nu(k)\} k^2 = \omega(k)$, and the accuracy of our calculation of $q(k)$ is vastly improved (we shall return to this specific point when we discuss almost Markovian models in Section 7.5). It is then arguable that there are infinitely many ways of choosing $\omega(k)$, and in general the accuracy of eqn (6.101) for $q(k)$ will depend on our choice. Thus the original choice of Edwards (see Section 6.2.5) and the later choice of McComb (see Section 7.3.2) are both quite arbitrary. Indeed, the very fact that they are both so reasonable in physical terms rather underlines their arbitrariness. It is in this context that Edwards and McComb (1969) argued that a most probable (but not unique) choice of $\omega(k)$ could be made by maximizing the entropy.

It can be seen that, as a corollary of the above discussion, the concept of a renormalized propagator is necessarily statistical. That is, the renormalized propagator can only depend on the velocity field through its moments. As far as any one realization is concerned, the renormalized propagator must be statistically independent of the velocity field and satisfy

$$\langle H_{\alpha\sigma}(\mathbf{k}; t, t') \rangle = H_{\alpha\sigma}(\mathbf{k}; t, t'). \tag{7.132}$$

It also follows that the propagator should be determined by some statistical principle. We introduce the statistical form of the LET theory's basic hypothesis as follows. Multiplying both sides of (7.129) by $u_\varepsilon(-\mathbf{k}, t')$ and averaging, we obtain

$$Q_{\alpha\varepsilon}(\mathbf{k}; t, t') = \int_0^t ds\, H_{\alpha\sigma}^{(0)}(\mathbf{k}; t, s) \left(\frac{L}{2\pi}\right)^3 \sum_{\mathbf{j}} M_{\alpha\beta\gamma}(\mathbf{k}) \times$$

$$\times \langle u_\beta(\mathbf{j}, t) u_\gamma(\mathbf{k} - \mathbf{j}, t) u_\varepsilon(-\mathbf{k}, t') \rangle, \tag{7.133}$$

where eqn (2.88) has been used to introduce the covariance tensor $Q_{\alpha\varepsilon}(\mathbf{k}; t, t')$. Exactly the same procedures applied to eqn (7.130) then yield

$$Q_{\alpha\varepsilon}(\mathbf{k}; t, t') = H_{\alpha\sigma}(\mathbf{k}; t, s) Q_{\sigma\varepsilon}(\mathbf{k}; s, t'), \tag{7.134}$$

where we have invoked (7.132) in order to carry out the averaging.

The LET theory then requires the equivalence of eqns (7.133) and (7.134). This postulate is the basis for the calculation of $H_{\alpha\sigma}(\mathbf{k}; t, t')$, and is the statistical version of the previous requirement that the propagator be defined through the postulated equivalence of eqns (7.129) and (7.130).

7.4.2   The generalized covariance equation

The renormalized perturbation theory is implemented as before. We take the zero-order velocity field to have a Gaussian distribution with zero mean. The perturbation series in the velocity is just eqn (6.8), which we repeat here for convenience:

$$u_\alpha(\mathbf{k}, t) = u_\alpha^{(0)}(\mathbf{k}, t) + \lambda u_\alpha^{(1)}(\mathbf{k}, t) + O(\lambda^2).$$

In practice we shall only work to second order in λ, and this means that the only higher coefficient we shall need in the expansion for $\mathbf{u}(\mathbf{k}, t)$ will be $\mathbf{u}^{(1)}(\mathbf{k}, t)$. This is given by (6.12) and we rewrite it here in terms of $H^{(0)}$ rather than $G^{(0)}$ as follows:

$$u_\alpha^{(1)}(\mathbf{k}, t) = \int_0^t ds\, H_{\alpha\sigma}^{(0)}(\mathbf{k}; t, s) M_{\sigma\beta\gamma}(\mathbf{k}) \sum_{\mathbf{j}} u_\beta^{(0)}(\mathbf{j}, s) u_\gamma^{(0)}(\mathbf{k} - \mathbf{j}, s). \quad (7.135)$$

In addition, we expand the propagator out as

$$H_{\alpha\beta}(\mathbf{k}; t, s) = H_{\alpha\beta}^{(0)}(\mathbf{k}; t, s) + \lambda^2 H_{\alpha\beta}^{(2)}(\mathbf{k}; t, s) + O(\lambda^4). \quad (7.136)$$

It should be noted that this is (effectively) an expansion in terms of the moments of the zero-order velocity field, and hence only even-order terms occur.

The LET theory is based on the derivation of a generalized equation for the covariance $Q_{\alpha\beta}(\mathbf{k}; t, t')$. We begin by using eqn (7.130) for the exact propagator, in order to rewrite (7.125) in the form

$$\left(\frac{\partial}{\partial t} + vk^2\right) H_{\alpha\sigma}(\mathbf{k}; t, s) u_\sigma(\mathbf{k}, s) = \lambda M_{\alpha\beta\gamma}(\mathbf{k}) \sum_{\mathbf{j}} u_\beta(\mathbf{j}, t) u_\gamma(\mathbf{k} - \mathbf{j}, t). \quad (7.137)$$

The generalized covariance equation is obtained if we first multiply both sides of (7.137) by $u_\varepsilon(-\mathbf{k}, t')$ and then average:

$$\left(\frac{\partial}{\partial t} + vk^2\right) H_{\alpha\sigma}(\mathbf{k}; t, s) \langle u_\sigma(\mathbf{k}, s) u_\varepsilon(-\mathbf{k}, t') \rangle$$
$$= \lambda \sum_{\mathbf{j}} M_{\alpha\beta\gamma}(\mathbf{k}) \langle u_\beta(\mathbf{j}, t) u_\gamma(\mathbf{k} - \mathbf{j}, t) u_\varepsilon(-\mathbf{k}, t') \rangle, \quad (7.138)$$

where we note eqn (7.132) when carrying out the averaging.

The r.h.s. of (7.138) can be expanded out using (6.8) for the velocity field to yield

$$\left(\frac{\partial}{\partial t} + vk^2\right) H_{\alpha\sigma}(\mathbf{k}; t, s) \langle u_\sigma(\mathbf{k}, s) u_\varepsilon(-\mathbf{k}, t')\rangle$$

$$= \lambda \sum_{\mathbf{j}} M_{\alpha\beta\gamma}(\mathbf{k}) [\langle u_\beta^{(0)}(\mathbf{j}, t) u_\gamma^{(0)}(\mathbf{k} - \mathbf{j}, t) u_\varepsilon^{(0)}(-\mathbf{k}, t')\rangle +$$

$$+ \lambda \langle u_\beta^{(0)}(\mathbf{j}, t) u_\gamma^{(0)}(\mathbf{k} - \mathbf{j}, t) u_\varepsilon^{(1)}(-\mathbf{k}, t')\rangle +$$

$$+ 2\lambda \langle u_\beta^{(1)}(\mathbf{j}, t) u_\gamma^{(0)}(\mathbf{k} - \mathbf{j}, t) u_\varepsilon^{(0)}(-\mathbf{k}, t')\rangle + O(\lambda^2)]. \qquad (7.139)$$

It should be noted that the factor of 2 in the last term on the r.h.s. arises because we have combined two terms, with an appropriate renaming of dummy variables. We have previously taken just this step in going from (6.24) to (6.25), while in the process of deriving the DIA equations. If we then substitute for $\mathbf{u}^{(1)}$ from (7.135), and note that odd-order moments of the $\mathbf{u}^{(0)}$ field vanish, we obtain the generalized covariance equation in the form

$$\left(\frac{\partial}{\partial t} + vk^2\right) H_{\alpha\sigma}(\mathbf{k}; t, s) Q_{\sigma\varepsilon}(\mathbf{k}; s, t')$$

$$= \lambda^2 \sum_{\mathbf{j}} \sum_{\mathbf{p}} M_{\alpha\beta\gamma}(\mathbf{k}) \left\{ M_{\rho\eta\delta}(-\mathbf{k}) \int_0^{t'} dt'' \, H_{\varepsilon\rho}(\mathbf{k}; t', t'') \times \right.$$

$$\times \langle u_\beta^{(0)}(\mathbf{j}, t) u_\gamma^{(0)}(\mathbf{k} - \mathbf{j}, t) u_\eta^{(0)}(\mathbf{p}, t'') u_\delta^{(0)}(\mathbf{k} - \mathbf{p}, t'')\rangle +$$

$$+ 2M_{\rho\eta\delta}(\mathbf{j}) \int_0^t dt'' \, H_{\beta\rho}(\mathbf{j}; t, t'') \times$$

$$\times \langle u_\eta^{(0)}(\mathbf{p}, t'') u_\delta^{(0)}(\mathbf{j} - \mathbf{p}, t'') u_\gamma^{(0)}(\mathbf{k} - \mathbf{j}, t) u_\varepsilon^{(0)}(-\mathbf{k}, t')\rangle +$$

$$\left. + O(\lambda^4)\right\}. \qquad (7.140)$$

Now we follow much the same general procedure which was used to derive the DIA. In particular, we shall take the same steps as we did in going from eqn (6.25) to eqn (6.31). Of course, our present starting point is eqn (7.139) and we are seeking the equivalent of (6.131) for the LET theory. We can summarize the various steps as follows.

(1) Evaluate the moments of the zero-order velocity field in terms of $Q^{(0)}$.
(2) Truncate the expansion at second order, put the bookkeeping parameter λ equal to unity, and make the replacements $Q^{(0)} \to Q$, and $H^{(0)} \to H$.
(3) Specialize to the case of isotropic turbulence, replacing $Q_{\alpha\beta}(\mathbf{k}; t, t')$ by the correlation function, according to eqn (2.97), and introducing the propagator function $H(k; t, t')$ in an analogous way:

$$Q_{\alpha\beta}(\mathbf{k}; t, t') = D_{\alpha\beta}(\mathbf{k}) Q(k; t, t')$$

$$H_{\alpha\beta}(\mathbf{k}; t, t') = D_{\alpha\beta}(\mathbf{k}) H(k; t, t'). \qquad (7.141)$$

When all these steps are taken and tensor indices are summed over $\alpha = \varepsilon$, eqn (7.140) becomes

$$\left(\frac{\partial}{\partial t} + vk^2\right) H(k;t,s)Q(k;s,t')$$

$$= \int d^3j \, L(\mathbf{k},\mathbf{j})\left\{\int_0^{t'} dt'' \, H(k;t',t'')Q(j;t,t'')Q(|\mathbf{k}-\mathbf{j}|;t,t'') - \right.$$

$$\left. - \int_0^t dt'' \, H(j;t,t'')Q(k;t'',t')Q(|\mathbf{k}-\mathbf{j}|;t,t'')\right\} \qquad (7.142)$$

which is the required generalized covariance equation for isotropic turbulence. It is quite easily shown that the coefficient $L(\mathbf{k},\mathbf{j})$ is given by eqn (2.162).

7.4.3 *Equations for the correlation and propagator functions*

The equation for the correlation function $Q(k;t,t')$ is readily obtained from (7.142). We first reduce (7.134) to isotropic form by invoking equation (7.141), so that

$$Q(k;t,t') = H(k;t,s)Q(k;s,t') \qquad t > s > t', \qquad (7.143)$$

and use the resulting relationship to contract time arguments on the l.h.s. of (7.142) to obtain

$$\left(\frac{\partial}{\partial t} + vk^2\right) Q(k;t,t')$$

$$= \int d^3j \, L(\mathbf{k},\mathbf{j})\left\{\int_0^{t'} dt'' \, H(k;t',t'')Q(j;t,t'')Q(|\mathbf{k}-\mathbf{j}|;t,t'') - \right.$$

$$\left. - \int_0^t dt'' \, H(j;t,t'')Q(k;t'',t')Q(|\mathbf{k}-\mathbf{j}|;t,t'')\right\}. \qquad (7.144)$$

The equation for the propagator function is a little more tricky. Essentially we are carrying out the same approach as for the identification of the effective viscosity in Section 7.3.2. There we interpreted the stationary energy-balance equation (7.107) in terms of the renormalized viscosity, through eqn (7.109), thus leading to (7.111) for $v(k)$.

In the present case the problem is complicated by the equations being non-stationary and by the plethora of time arguments involved. Nevertheless, our procedure is really very much the same: we wish to equate coefficients of $Q(k;t,t)$ in order to obtain the response equation. When doing this, we should bear in mind that the labelling wavenumber $\mathbf{k}$ is what matters here, and not the time arguments.

Thus, setting $s = t'$ on the l.h.s. of eqn (7.142), we can write the general covariance equation as

$$\left(\frac{\partial}{\partial t} + vk^2\right) H(k; t, t') Q(k; t', t')$$

$$= \int d^3j\, L(\mathbf{k}, \mathbf{j}) \left\{ \int_0^{t'} dt''\, H(k; t', t'') Q(j; t, t'') Q(|\mathbf{k} - \mathbf{j}|; t, t'') - \right.$$

$$\left. - \int_0^t dt''\, Q(k, t'', t') H(j; t, t'') Q(|\mathbf{k} - \mathbf{j}|; t, t'') \right\}, \qquad (7.145)$$

and this is the LET response equation.

7.4.4   Comparison with DIA

The DIA equation for the correlation function is given by (6.31). If we set the arbitrary input term equal to zero, and take the non-linear term to the r.h.s., then it is clear that (6.31) is identical with the LET equation (7.144) for the correlation function.

This result is hardly surprising—we have seen that second-order RPTs all lead to the same energy equation. In fact the difference between the LET and DIA theories lies in the differing forms of their response equations. We can see this as follows. Divide the integral over $0 < t'' < t$ on the r.h.s. of eqn (7.145) into two ranges $0 < t'' < t'$ and $t' < t'' < t$, use (7.143) to write

$$Q(k; t', t'') = Q(k; t'', t') = H(k; t'', t') Q(k; t', t'),$$

and rearrange to obtain

$$\left[\frac{\partial}{\partial t} + vk^2\right] H(k; t, t') +$$

$$+ \int d^3j\, L(\mathbf{k}, \mathbf{j}) \int_{t'}^t dt''\, H(k; t'', t') H(j; t', t'') Q(|\mathbf{k} - \mathbf{j}|; t, t'')$$

$$= \frac{1}{Q(k; t', t')} \int d^3j\, L(\mathbf{k}, \mathbf{j}) \int_0^{t'} dt''\, Q(|\mathbf{k} - \mathbf{j}|; t, t'') \times$$

$$\times \{H(k; t', t'') Q(j; t, t'') - H(j; t, t'') Q(k; t', t'')\}. \qquad (7.146)$$

The comparable equation in DIA is eqn (6.29) for the response function $G(k; t, t')$, which is analogous to $H(k; t, t')$ in the LET theory. Comparison shows that the two left-hand sides are the same, but that eqn (7.146) has additional non-linear terms on the r.h.s.

As we saw in Section 7.1.1, the DIA response integral is divergent when the limit of infinite Reynolds number is taken. (In fact this was actually shown for the EFP response equation, but in this context there is no significant difference between the two.) The form of the integrand studied in Section 7.1.1 was divergent at $\mathbf{j} = 0$, whereas the integrand of (6.29) would diverge at $\mathbf{k} - \mathbf{j} = 0$. It is quite easily shown that the additional term on the r.h.s. of the LET response equation (7.146) will cancel this divergence as we take the limit $\mathbf{k} \to \mathbf{j}$.

We conclude by noting that the LET theory was originally derived for the stationary case (McComb 1978) and later generalized in an *ad hoc* fashion in order to permit numerical calculation of freely decaying (i.e. non-stationary) turbulence; we discuss such calculations in the next chapter. However, the more general derivation given here is much simpler than the original, and also eliminates some minor errors. It is also of interest that Nakano (1988) has apparently arrived at the LET response equation by considering the DIA in a wave-packet representation in which $\mathbf{k}$-space is divided up into shells of finite thickness.

7.5 Near-Markovian model closures

In this section we shall briefly discuss a class of turbulence models which has proved rather popular in recent years. As before, we use the term 'model' to mean a theory which relies on some quite specific assumption, resulting in a free parameter which is not determined by the theory and is normally adjusted to make the theoretical predictions agree with experiment.

The models to be discussed rely on an analogy being drawn between turbulence processes and Markov processes, such as Brownian motion for example. We have previously discussed Markov processes in Section 4.1.2 in connection with the derivation of the Chapman–Kolmogorov equation. The basic idea is one of a random walk in which the current step depends only on the preceding step, but not on any step before that. Thus turbulence is not strictly Markovian, but, as we shall see, in certain circumstances it is natural to argue that it is almost Markovian, when turbulent time-scales are compared with those determined solely by the kinematic viscosity of the fluid.

7.5.1 *Quasi-normal Markovian approximations*

We have discussed the quasi-normality hypothesis and the failure of the resulting spectral equation (2.163) in Section 2.8.2. The reasons for this failure have been analysed in some detail by Orszag (1970), who also suggested a remedy. It is the latter aspect which will interest us here.

Orszag pointed out that the memory integral in the quasi-normal spectral equation—see the r.h.s. of eqn (2.163)—was ultimately limited by time-scales determined by the kinematic viscosity of the fluid. That is, the exponential factor in the integrand cuts off the integration on time-scales $T(k)$ such that

$$T(k) \simeq (\omega(k))^{-1},$$

where the modal decay rate $\omega(k)$ is purely viscous; thus

$$\omega(k) = vk^2.$$

As the effect of turbulence is to destroy correlations, it is arguable that (2.163) should be replaced by

$$\left(\frac{d}{dt} + 2vk^2\right) Q(k, t) = \int d^3j\, L(\mathbf{k}, \mathbf{j}) \times$$

$$\times \int_0^t ds \exp[\{\omega(k) + \omega(j) + \omega(|\mathbf{k} - \mathbf{j}|)\}(t - s)] \times$$

$$\times Q(|\mathbf{k} - \mathbf{j}|, s)\{Q(j, s) - Q(k, s)\}, \tag{7.147}$$

where the modal decay rate should take some account of the 'de-correlating' effects of the turbulent dynamics.

If we are specifically interested in the inertial range, then we have, on dimensional grounds, the result for the modal decay rate already given as eqn (6.120):

$$\omega(k) = \beta \varepsilon^{1/3} k^{2/3}. \tag{6.120}$$

It follows from the discussion of the EFP theory in Section 6.2.7, that (7.147) and (6.121) together lead to the Kolmogorov inertial-range spectrum.

Some justification for replacing (2.163) by eqn (7.147) can be found in a consideration of the DIA energy equation. The eddy-damped quasi-normal spectral equation—that is, eqn (7.147)—can be obtained from (6.35) by making the approximations

$$G(k; t, t') = \exp\{-\omega(k)(t - t')\} \qquad t > t'$$

$$Q(k; t, t') = Q(k, t') \exp\{-\omega(k)(t - t')\}. \tag{7.148}$$

In a sense, this way of 'fixing up' the quasi-normality hypothesis rather underlines what was wrong with it in the first place. In renormalized perturbation theories, the primitive velocity field has Gaussian statistics, and so the factoring of fourth-order (or higher) moments into products of second-order moments is rigorously correct. The next step—the replacement of $Q^{(0)}$ by Q—is always accompanied by a procedure which changes $G^{(0)}$ to G—the renormalization process! In quasi-normality, only the step $Q^{(0)} \to Q$ has been taken, and so Orszag's proposal really amounts to making the change $G^{(0)} \to G$ afterwards.

Orszag also proposed that the theory should be modified to take on a Markovian character, with memory effects being eliminated. This was achieved by updating the time integrals on the r.h.s. of eqn (7.147) to time t:

$$\left(\frac{d}{dt} + 2\nu k^2\right)Q(k,t) = \int d^3j\, L(\mathbf{k},\mathbf{j})D(k,j,|\mathbf{k}-\mathbf{j}|;t) \times$$

$$\times\, Q(|\mathbf{k}-\mathbf{j}|,t)\{Q(j,t) - Q(k,t)\}, \qquad (7.149)$$

where $D(k,j,|\mathbf{k}-\mathbf{j}|;t)$ is a memory time which depends on the modal decay (or decorrelation) rate $\omega(k,t)$. In an almost-Markovian approximation, the memory time satisfies

$$\frac{dD}{dt} = 1 - \{\omega(k,t) + \omega(j,t) + \omega(|\mathbf{k}-\mathbf{j}|,t)\}D \qquad (7.150)$$

with initial condition

$$D(k,j,|\mathbf{k}-\mathbf{j}|;0) = 0. \qquad (7.151)$$

If the turbulence is stationary, then $\omega(k,t) = \omega(k)$, and (7.150) becomes

$$D(k,j,|\mathbf{k}-\mathbf{j}|) = \frac{1}{\omega(k) + \omega(j) + \omega(|\mathbf{k}-\mathbf{j}|)}. \qquad (7.152)$$

We note that eqns (7.149) (with $t = 0$) and (7.152) together are identical with the original form of the EFP theory (see eqns (6.101) and (6.107)). This has prompted the suggestion (Kraichnan 1971) that eqns (7.149) and (7.152) constitute an extension of the EFP theory to nonstationary turbulence.

The precise form of the Markovian modification may seem less intuitively appealing than the introduction of the eddy viscosity. Nevertheless, despite these strictures, eqns (7.149) and (7.150) form an easy set of equations for numerical computation. Thus it is not really surprising that variants of this model have proved helpful in tackling problems ranging from the generation of magnetic fields in conducting fluids to sub-grid-scale modelling. We shall refer again to some of these aspects when we consider the numerical simulation of turbulence in Chapter 10.

7.5.2 The test-field model

The above work led to further formulations of the DIA, based on a generalized Langevin equation which provided a new model representation of the velocity field (Kraichnan 1970; Leith 1971), a development which, as we mentioned in Section 6.3.2, was also influenced by the work of Phythian (Kraichnan 1970).

Kraichnan (1971) further extended this approach to (a) almost-Markovian approximations and (b) approximations which were modified to be invariant under random Galilean transformations. This latter method is also known as the test-field model.

The modification of the almost-Markovian model to give random Galilean invariance involves the introduction of a test field which can be decomposed into solenoidal and compressive parts (just as in the Lagrangian-history case; see Section 7.2). We shall not go into further detail here; the interested reader should consult Kraichnan (1971), or there is a good summary of the test-field model in the paper by Leith and Kraichnan (1972).

Note

1. This value of D'' may be too large. In a different context, Kraichnan (see Section 10.3.2) has reported a result equivalent to $D'' = 0.44$. Substitution of the smaller value of D'' into eqn (7.114) would lead to a theoretical prediction of a Kolmogorov constant $\alpha = 2.30$, in much better agreement with experiment.

References

EDWARDS, S. F. and McCOMB, W. D. (1969). *J. Phys. A* **2**, 157.
—— and —— (1971). *Proc. R. Soc. A* **325**, 313.
—— and —— (1972). *Proc. R. Soc. A* **330**, 495.
GRANT, H. L., STEWART, R. W., and MOILLIET, A. (1962). *J. fluid Mech.* **12**, 241.
HORNER, H. and LIPOWSKY, R. (1979). *Z. Phys. B* **33**, 223.
KANEDA, Y. (1981). *J. fluid Mech.* **107**, 131.
KRAICHNAN, R. H. (1959). *J. fluid Mech.* **5**, 497.
—— (1964). *Phys. Fluids* **7**, 1723.
—— (1965). *Phys. Fluids* **8**, 575.
—— (1970). *J. fluid Mech.* **41**, 189.
—— (1971). *J. fluid Mech.* **47**, 513.
—— (1977). *J. fluid Mech.* **83**, 349.
—— and HERRING, J. R. (1978). *J. fluid Mech.* **88**, 355.
LEITH, C. E. (1971). *J. atmos. Sci.* **28**, 145.
—— and KRAICHNAN, R. H. (1972). *J. atmos. Sci.* **29**, 1041.
LESLIE, D. C. (1973). *Developments in the theory of turbulence.* Clarendon Press, Oxford.
McCOMB, W. D. (1967). An entropy definition of the total viscosity in isotropic turbulence. *M.Sc. Thesis*, Victoria University of Manchester.
—— (1969). Turbulent shear flow. *Ph.D. Thesis*, Victoria University of Manchester.
—— (1974). *J. Phys. A* **7**, 632.
—— (1976). *J. Phys. A* **9**, 179.
—— (1978). *J. Phys. A* **11**, 613.
MARTIN, P. C., SIGGIA, E. D., and ROSE, H. A. (1973). *Phys. Rev. A* **8**, 423.
NAKANO, T. (1972). *Ann. Phys. (NY)* **73**, 326.
—— (1988). *Phys. Fluids* **31**, 1420.
ORSZAG, S. A. (1970). *J. fluid Mech.* **41**, 363.
QUIAN, J. (1983). *Phys. Fluids* **26**, 2098.
SHANNON, C. E., and WEAVER, W. (1949). *The mathematical theory of communication.* University of Illinois Press, Urbana, Il.
SHORE, J. E., and JOHNSON, R. W. (1980). *IEEE Trans. Inform. Theory*, **26**, 26.

8

AN ASSESSMENT OF
RENORMALIZED PERTURBATION
THEORIES

Our principal source of information on the performance of RPT, is a handful
of numerical investigations into the application of RPT to the problem of
freely decaying isotropic turbulence. Accordingly, this will be our main topic
in the present chapter, although we shall also revive the question of whether
such theories should be compatible with the Kolmogorov spectrum at high
Reynolds numbers. We shall also examine the problems involved in applying
RPTs to inhomogeneous flows—along with a glance in passing at some
other approaches to shear flows—before attempting to come to some overall
conclusion about the various theories.

8.1   Free decay of isotropic turbulence as a test problem

Our account of this subject is based on the following investigations: Kraichnan
(1964a, 1965), Herring and Kraichnan (1972, 1979), and McComb and Shanmugasundaram (1984). For reasons of convenience, most of the figures in this
chapter have been taken from the last reference.

An introduction to the problem of free turbulent decay has already been
given in Section 4.3.1, and so accordingly we proceed directly to a consideration of initial spectra. In all cases these were of the form

$$E(k, 0) = c_1 k^{c_2} \exp\{-c_3 k^{c_4}\}, \tag{8.1}$$

with an appropriate choice of constants c_1–c_4, subject to the constraint

$$\int_0^\infty E(k, 0)\, dk = 3/2. \tag{8.2}$$

That is, the mean-square level of the turbulence at $t = 0$ is taken to be unity.

We shall classify the calculations to be described into one of two categories,
according to the value of the Reynolds number. That is, calculations will be
categorized as 'low Reynolds number' or 'high Reynolds number', according
to whether or not the Reynolds number is large enough for there to be an
inertial range of wavenumbers in the energy spectrum. The two cases will be
considered separately in Sections 8.2 and 8.3, but first we shall discuss the
choice of trial spectra here, along with a summary of useful formulae plus some
remarks on the nature of free decay as a test problem.

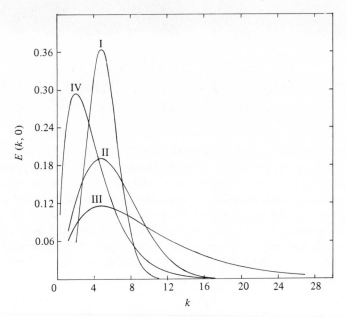

Fig. 8.1. Initial wavenumber spectra.

TABLE 8.1   *Values of the constants in eqn* (8.1)

Spectrum number	c_1	c_2	c_3	c_4
I	0.0052	4	0.0884	2
II	0.0663	1	0.0221	2
III	0.0663	1	0.2102	1
IV	0.4000	1	0.5000	1

We shall be interested in four particular trial spectra for calculations at low Reynolds numbers. These are shown in Fig. 8.1, and the corresponding values of the constants c_1–c_4 are given in Table 8.1.

Of these initial spectra. spectrum I was introduced by Ogura (1963) as part of his numerical study of the quasi-normality approximation. Spectra II and III are just Kraichnan's spectra C and D. Noteworthy points are that spectrum II is self-preserving under purely viscous decay, while spectrum III rapidly becomes self-preserving under the combined actions of inertial transfer and viscous decay (Kraichnan 1964a). All three spectra peak at the same wavenumber which is

$$k_{max} = 4 \times 2^{1/4} = 4.75683 \text{ arbitrary units.}$$

Spectrum IV is a modified form of spectrum III. It was chosen as a test spectrum by McComb and Shanmugasundaram (1984) in the light of some comments made by Van Atta and Chen (1969), who found that their measured dissipation spectra peaked at lower wavenumbers than did calculated spectra (Kraichnan 1964a). Accordingly spectrum IV has been chosen to have a much lower peak wavenumber at $k_{max} = 2$ arbitrary units.

For calculations at high wavenumbers, the constants c_1–c_4 have usually been set to give $E(k, 0)$ as a simple power law. We shall discuss this further in Section 8.3.

At this point it is convenient to collect together relationships defining the various parameters which are to be calculated for each theory. To begin with, from the energy spectrum we can obtain the energy $E(t)$ per unit mass of fluid, the r.m.s. value of any velocity component $u(t)$, and the rate of dissipation of energy $\varepsilon(t)$ per unit mass of fluid as follows:

$$E(t) = \int_0^\infty E(k,t)\,dk = (3/2)\{u(t)\}^2 \tag{8.3}$$

$$\varepsilon(t) = 2v \int_0^\infty k^2 E(k,t)\,dk. \tag{8.4}$$

The transfer spectrum $T(k, t)$, as defined by (2.119), has been discussed in connection with the DIA, and written as eqn (6.37) in terms of $P(k; t, t)$. For the case of the DIA, $P(k; t, t')$ is given by (6.34), and for completeness we repeat (6.37) here

$$T(k,t) = 8\pi k^2 P(k; t, t)$$

along with (6.42) which defines the transport power

$$\Pi(k,t) = \int_k^\infty T(k',t)\,dk' = -\int_0^k T(k',t)\,dk'.$$

The dissipation spectrum is $2vk^2 E(k,t)$, and is sometimes given its own symbol and referred to as $D(k,t)$. Also, comparison with experiment is helped by the introduction of the one-dimensional energy spectrum. We rewrite (2.109) as

$$\phi_1(k,t) = \int_k^\infty \left\{ \left(1 - \frac{k^2}{j^2}\right) \frac{E(j,t)}{2j} \right\} dj, \tag{8.5}$$

with $\phi_1(k,t)$ replacing $E_{11}(k,t)$. Either notation is acceptable, but the former seems to be more commonly employed by experimentalists.

The integral scale $L(t)$ and the Taylor microscale $\lambda(t)$ can be expressed in terms of the energy spectrum by (Batchelor 1971)

$$L(t) = \frac{\frac{3\pi}{4} \int_0^\infty k^{-1} E(k, t)\, dk}{E(t)} \qquad (8.6)$$

$$\lambda(t) = \left\{ \frac{5E(t)}{\int_0^\infty k^2 E(k, t)\, dk} \right\}^{1/2}, \qquad (8.7)$$

along with their associated Reynolds numbers

$$R_L(t) = \frac{L(t)u(t)}{\nu} \qquad (8.8)$$

$$R_\lambda(t) = \frac{\lambda(t)u(t)}{\nu}. \qquad (8.9)$$

As we shall see, the skewness of the longitudinal derivative (or, more simply, the skewness factor) $S(t)$ may be the most sensitive indicator of differences between the various theories. It can be written in the form (Batchelor 1971)

$$S(t) = \frac{2}{35} \left\{ \frac{\lambda(t)}{u(t)} \right\}^3 \int_0^\infty k^2 T(k, t)\, dk. \qquad (8.10)$$

Also, following Kraichnan (1964a), we introduce the modal time-correlation

$$R(k; t, t') = \frac{Q(k; t, t')}{\{Q(k; t, t)Q(k; t', t')\}^{1/2}}, \qquad (8.11)$$

along with characteristic velocity and wavenumber scales

$$v_S = \left(\frac{R_\lambda}{15^{1/2}} \right)^{-1/3} u(t) \qquad (8.12)$$

$$k_S = (15R_\lambda)^{1/3} \lambda^{-1}. \qquad (8.13)$$

Kraichnan found that, for low R_λ, scaling with v_S and k_S produced a better collapse of data than did the Kolmogorov scales v and k_d (as defined by our equations (2.132) and (2.133)). It should be noted that our present use of subscripts on the wavenumber scales is exactly opposite to Kraichnan's usage, but conforms better to current conventions.

Our last general point, before turning to specific investigations, concerns the relevance of the free decay calculation to stationary turbulence at large Reynolds numbers. The Kolmogorov hypotheses essentially demand not only local isotropy, but also quasi-stationarity in the inertial range of wavenumbers. In order to see how this can be achieved in decay calculations, let us consider the definition of the inertial range as embodied in eqn (6.43). We require the rate at which energy is transferred locally through wavenumber k to be equal to the global dissipation rate. Wavenumbers k, for which this condition is satisfied, belong to the inertial range.

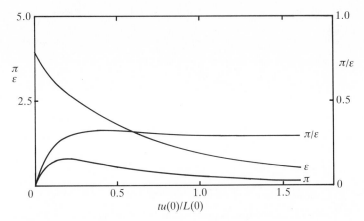

Fig. 8.2. Variation of the dissipation rate ε and the maximum value of the transport power $\Pi(t)$ with time as calculated by the LET theory for $R_\lambda(t_f) = 16.8$. The condition for the existence of an inertial range ($\Pi(t) = \varepsilon$) is not satisfied at this low value of the Reynolds number.

Suppose that we now consider the case where the above condition is satisfied for only one wavenumber (i.e. this case would also constitute a lower bound on values of the Reynolds number for which an inertial range could occur). Then the critical wavenumber would correspond to a maximum value of the transport power. We shall denote this maximum value by

$$\text{maximum value of } \Pi(k, t) = \Pi(t). \tag{8.14}$$

Clearly an extended inertial range would imply that the maximum value of the transport power would take the form of a plateau, in which the condition (8.14) held for the entire inertial range of wavenumbers. In addition, eqn (6.43) would also have to be satisfied by $\Pi(t)$.

In Figs. 8.2 and 8.3 we plot $\Pi(t)$—as calculated by the LET theory— against time for two different values of the Reynolds number. In both cases, it can be seen that the dissipation rate falls off from the value prescribed initially as the decay proceeds. In contrast, $\Pi(t)$—initially zero, as triple correlations are zero by prescription at $t = 0$—first increases and then starts to decay away. However, for $R_\lambda = 16.8$, the ratio $\Pi(t)/\varepsilon$ in never greater than 0.3, whereas for $R_\lambda = 392$ the ratio becomes unity—indicating the presence of an inertial range. In this latter case, once the ratio of transport power to dissipation rate becomes unity, it should be noted that each of the two quantities becomes slowly varying with time, thus demonstrating the requisite quasi-stationarity.

It should also be noted that in both cases the time variable has been scaled by the initial large eddy turnover time $L(0)/u(0)$.

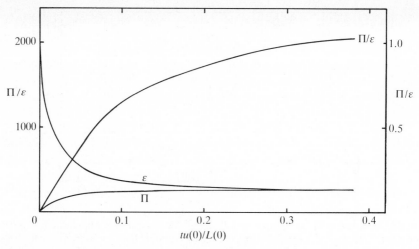

Fig. 8.3. Variation of the dissipation rate ε and the maximum value of the transport power $\Pi(t)$ with time, as calculated by the LET theory for $R_\lambda(t_f) = 392$. Evidently the criterion for the existence of an inertial range is satisfied in this case, and this is confirmed by the energy spectrum which shows a region of 5/3 power law.

8.2   Calculations of decaying turbulence at low Reynolds numbers

In this section, we shall discuss the basic numerical methods as part of our account of Kraichnan's (1964a) investigation of the DIA in 8.2.1. Figures illustrating Kraichnan's results are deferred to Section 8.2.3, where they are taken from McComb and Shanmugasundaram (1984), who used virtually the same methods in calculating the LET theory and comparing its results to those of DIA.

8.2.1   *The direct-interaction approximation (DIA)*

Kraichnan (1964a) investigated the Eulerian form of the DIA using four different initial spectra. His spectrum A was a rectangular pulse, while his spectra B, C, and D were the same as our Spectra I, II, and III. Initial values of the Taylor–Reynolds number up to $R_\lambda(0) = 42$ were considered.

Equations (6.29) and (6.31) for the response function $G(k;t,t')$ and the correlation function were discretized in all variables and numerically integrated forward in time. We shall only give very brief details of this procedure here, and the interested reader should consult the original reference.

As treated by Kraichnan, the calculation of $Q(k;t,t')$ involves the double sum over the wavevector magnitudes j and l (where $l = |\mathbf{k} - \mathbf{j}|$, in our notation). All elements in the summation must satisfy the condition that k, j, l can form a triangle. The three wavenumbers were then truncated to the finite range

(k_{bot}, k_{top}) and divided up into (in general, non-uniform) intervals. A feature of this scheme is the necessity to apply a correcting factor in order to eliminate additional errors which occur when logarithmic steps are used and conditions like $k \gg j$ are encountered.

Time integrations were carried out for the interval $(0, t_f)$ using an implicit integration scheme which was based on a first-order predictor–corrector method.

The main dynamical feature of all the calculations involving finite values of fluid viscosity, was found to be a transfer of energy from low to high wavenumbers, along with a consequent moving apart of the maxima of the energy and dissipation spectra. An interesting feature was the early transfer of energy in the reverse direction to lower wavenumbers which were presumably insufficiently excited by the arbitrary initial spectrum—thus echoing the trend to equipartition previously found with inviscid equilibrium ensembles.

Most of the integral parameters were found to decay with time, but the skewness $S(t)$ and the transport power $\Pi(t)$ increased initially from zero, and passed through a maximum. Thereafter, $\Pi(t)$ behaved like other parameters and decayed with time, but $S(t)$ showed a tendency to take on an asymptotic constant value. An overshoot of $S(t)$ at short times above its asymptotic value turned out to be more pronounced when the initial spectrum was more concentrated about the peak wavenumber.

The evolved DIA spectra showed some degree of universal behaviour. At high wavenumbers there was a tendency to independence of the initial spectrum shape. Also, there was strong tendency to self-preservation, especially at the higher wavenumbers, although a comparison with experimental results obtained in grid turbulence was quite favourable at low wavenumbers as well.

All in all, the general behaviour of DIA, when integrated forward in time at low Reynolds numbers, was found to be quite good in physical terms. The only jarring note was an unphysical oscillation with time in the case of the response function.

We shall return to the subject of DIA in Section 8.2.3, where we present the results of similar calculations by McComb and Shanmugasundaram (1984).

8.2.2 *Comparison of various theories: Herring and Kraichnan*

Kraichnan's (1964a) calculations were later extended by Herring and Kraichnan (1972) to make a comparison of five different theories. These were as follows: DIA (see Section 6.1), SCF (see Section 6.3), a non-stationary form of EFP (see Section 6.2 and eqns (7.149) and (7.152); this theory was referred to as EDW), and two variants of the test-field model (TFM) (see Section 7.5.2).

The two variants of TFM arose from the fact that this theory contains an adjustable scaling parameter g. Originally the value $g = 1.064$ was chosen so that TFM would give the same results as DIA for the relaxation of small departures from absolute equilibrium. This value of $g = 1.064$ resulted in a Kolmogorov constant of $\alpha = 1.40$. Herring and Kraichnan also considered a TFM with $g = 1.5$, because this led to a Kolmogorov constant of $\alpha = 1.76$, which is virtually the same as the value of $\alpha = 1.77$ predicted by ALHDI (see Section 7.2). This variant of TFM was denoted by TFM'.

The various theories were compared with each other, with laboratory and field experiments, and—a significant step forward—with the results of a numerical simulation (i.e. a computer experiment) which had the same pre-scribed initial conditions (Orszag and Patterson 1972). The overall qualitative conclusion was that the various theories performed in a rather similar way, and gave a physically reasonable picture of the energy transfer processes involved in the free decay of isotropic turbulence.

However, there were significant quantitative differences between theories, and these are illustrated by the evolved dissipation spectra shown in Fig. 8.4. In these calculations the initial conditions were determined by spectrum I. It can be seen that the results for SCF and DIA are quite close to each other. In fact, for other choices of initial spectrum, these two theories gave virtually identical results. Herring and Kraichnan seem inclined to put this down to (in effect) coincidence.

In Fig. 8.5 the corresponding theoretical predictions for the skewness factor are plotted as a function of time, and compared with the results of the

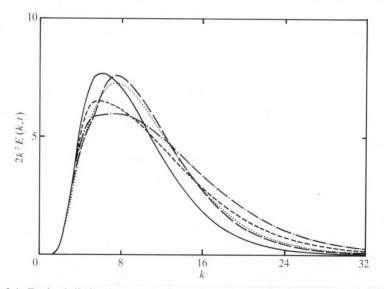

Fig. 8.4. Evolved dissipation spectra for various renormalized perturbation theories (after Herring and Kraichnan 1972). Spectrum I: —— EDW; — — SCF; — · — TFM'; — — — TFM; · · · · · DIA.

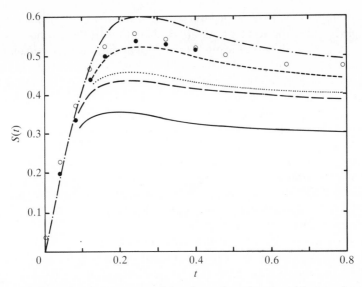

Fig. 8.5. Comparison of the skewness factor for various theories with some data points from two computer simulations (after Herring and Kraichnan 1972). Spectrum I: ———, EDW; – – –, SCF; – · –, TFM′; ----, TFM; · · · ·, DIA.

computer simulation by Orszag and Patterson (1972). It is noticeable that only the theories which are compatible with the Kolmogorov spectrum at high Reynolds numbers (i.e. TFM and TFM′) agree at all well with the computer simulation at low Reynolds numbers.

The skewness factor can also be seen as a measure of the efficiency of the various theories at transferring energy through wavenumber space. Calculations of transfer spectra and transport power revealed that EDW transferred energy at the lowest rate, DIA and SCF were intermediate, and TFM and TFM′ were the most efficient. This ranking shows up quite clearly in the asymptotic values of $S(t)$ in Fig. 8.5.

Lastly, it is interesting to see that all the theories exhibit an overshoot in the skewness at short evolution times, and that this also happens with the computer simulation. Presumably this is a consequence of the particular (arbitrary) choice of initial spectrum.

8.2.3 The LET theory

The application of the LET theory to the decay of isotropic turbulence was studied by McComb and Shanmugasundaram (1984). The methods used closely followed those of Kraichnan (1964a), with essentially only two minor variations. First, there was the introduction of spectrum IV. Second, wavenumber integrations were carried out with k, j, and μ, where μ is the cosine of the angle between the vectors $\mathbf{k}$ and $\mathbf{j}$. Mathematically, there is no difference between this procedure and using (as Kraichnan did) the scalar magnitudes k, j, and l. In the one case, we take $l = |\mathbf{k} - \mathbf{j}|$, while in the other, k, j, and l must always be able to form a triangle. However, in practice, the rectangular field of integration, as defined by $k_{\text{bot}} \leqslant j \leqslant k_{\text{top}}$ and $-1 \leqslant \mu \leqslant 1$, presents fewer numerical problems than the case of the (j, l) integration.

Some of the results of this investigation are summarized in Figs. 8.6–8.12. In most cases, results are presented for both LET and DIA. The calculations of DIA were found to agree quite well with those of Kraichnan (1964a). To take a specific example, the skewness factor—which seems to be particularly sensitive to differences between theories or calculations—the value calculated by McComb and Shanmugasundaram was less than 5 per cent larger than Kraichnan's equivalent value (see Fig. 8.12).

The variation of integral parameters, as presented in Fig. 8.6, shows that LET behaved very much like DIA, and indeed very much like all the theories discussed in Section 8.2.2. However, in quantitative terms there are some significant differences between LET and DIA. For instance, the LET value for $S(t)$ was some 16 per cent larger than the corresponding DIA value. This is interesting, in view of the surmise by Herring and Kraichnan (1972), that the underestimation of $S(t)$ by DIA was a real physical effect, associated with lack of random Galilean invariance.

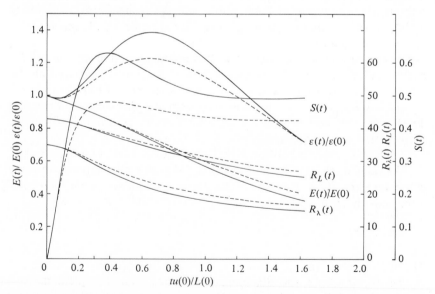

Fig. 8.6. Variation of integral parameters for LET and DIA theories (after McComb and Shanmugasundaram 1984). Spectrum I: ——— LET; – – – DIA.

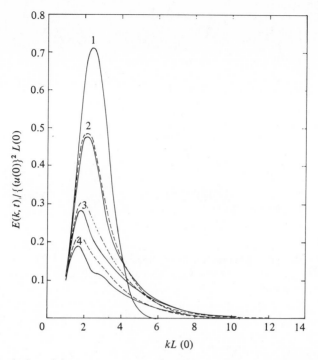

Fig. 8.7. Evolution of the energy spectrum for LET and DIA theories (after McComb and Shanmugasundaram 1984). Spectrum I: ——— LET; – – – DIA. Curve 1, $tu(0)/L(0) = 0$; curve 2, $tu(0)/L(0) = 0.5$; curve 3, $tu(0)/L(0) = 1.0$; curve 4, $tu(0)/L(0) = 1.6$.

Results for energy spectra, dissipation spectra, and transfer spectra are shown in Figs 8.7–8.9 respectively. An obvious qualitative difference between the two approximations is the development of kinks in the evolving LET spectrum but not in the corresponding DIA spectrum. This behaviour did not occur with either spectrum II or spectrum III, and is presumably attributable to the more efficient energy transfer mechanism of LET attempting to cope with an initial spectrum (i.e. spectrum I) which was highly peaked. This feature may well not be an artefact of the LET theory, for similar kinks have been found experimentally by Stewart and Townsend (1951).

Like DIA, the LET theory gave spectra which tended to become independent of the initial spectrum shape at high wavenumbers, and also which rapidly became self-preserving. The former behaviour is illustrated for LET only in Fig. 8.10, where the dissipation spectra corresponding to spectra I, II and III are plotted.

Dissipation spectra are also shown—this time in one-dimensional form—in Fig. 8.11, where LET and DIA results are compared with some representative experimental values. Clearly both approximations agree quite well with

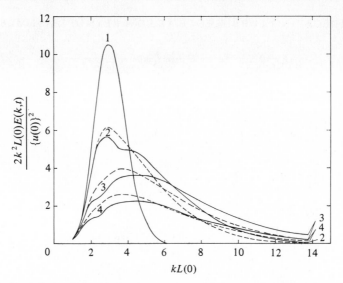

Fig. 8.8. Evolution of the dissipation spectrum for LET and DIA theories (after McComb and Shanmugasundaram 1984). Spectrum I: ——— LET; ––– DIA. Curve 1, $tu(0)/L(0) = 0$; curve 2, $tu(0)/L(0) = 0.5$; curve 3, $tu(0)/L(0) = 1.0$; curve 4, $tu(0)/L(0) = 1.6$.

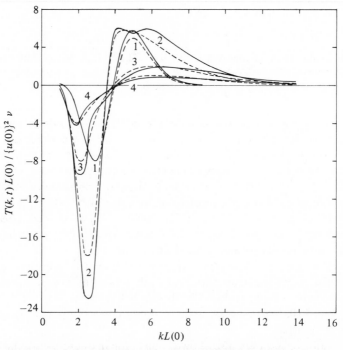

Fig. 8.9. Evolution of the transfer spectrum for the LET and DIA theories (after McComb and Shanmugasundaram 1984). Spectrum I: ——— LET; ––– DIA. Curve 1, $tu(0)/L(0) = 0$; curve 2, $tu(0)/L(0) = 0.5$; curve 3, $tu(0)/L(0) = 1.0$; curve 4, $tu(0)/L(0) = 1.6$.

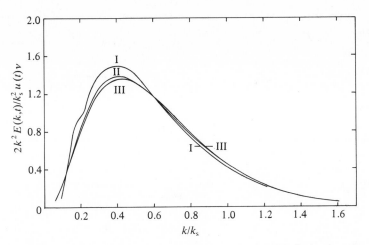

Fig. 8.10. Effect of initial spectrum shape on the evolved LET dissipation spectrum (after McComb and Shanmugasundaram 1984).

Spectrum	$tu(0)/L(0)$	R_λ
I	1.2	16.3
II	1.6	16.6
III	1.9	16.6

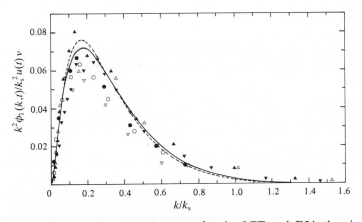

Fig. 8.11. Evolved one-dimensional spectra for the LET and DIA theories after McComb and Shanmugasundaram 1984). Spectrum IV: ——— LET; – – – DIA. Experimental results: $\triangledown$, $R_\lambda = 39.4$ (Stewart and Townsend 1951); $\circ$, $R_\lambda = 49.0$ and $\bullet$, $R_\lambda = 35.0$ (Chen 1968); $\triangle$, $R_\lambda = 38.1$ and $\blacktriangle$, $R_\lambda = 36.6$ (Comte-Bellot and Corrsin 1971); $\blacktriangledown$, $R_\lambda = 45.2$ (Frenkiel and Klebanoff 1971).

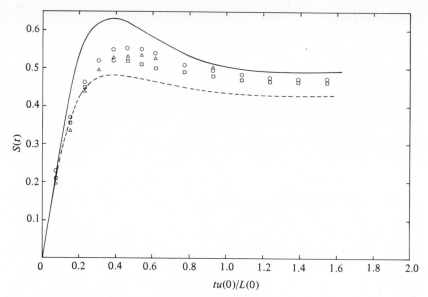

Fig. 8.12. Evolution of the skewness factor for the LET and DIA theories (after McComb and Shanmugasundaram 1984). Spectrum I: ——— LET; – – – DIA; ○, □, △, direct numerical simulation (Orszag and Patterson 1972).

experiment. But it should be noted that the initial spectrum in this case was spectrum IV, which was specifically chosen to improve the basis of this sort of comparison.

In Fig. 8.12, both LET and DIA predictions of the skewness are plotted against time, along with values obtained from the numerical simulation of Orszag and Patterson (1972). It is of some interest to compare this figure with Fig. 8.5, but at the same time it should be borne in mind that the calculation of the DIA value of $S(t)$ by McComb and Shanmugasundaram is about 4 per cent higher than the corresponding value obtained by Herring and Kraichnan (1972).

In Section 8.2.1, we mentioned the unphysical oscillations in time of the DIA response function. This behaviour violates the realizability requirement that $G(k; t, t')$ should be non-negative. Similar behaviour was found by McComb and Shanmugasundaram for both LET and DIA. In addition, the LET propagator exhibited a small overshoot at short evolution times, and hence violated the other realizability condition $G(k; t, t') < 1$.[1]

8.3   Calculations of decaying turbulence at high Reynolds numbers

In this section we shall be concerned with isotropic turbulence where the Reynolds numbers are large enough to ensure the existence of an inertial range. Also, the various theories to be discussed all have some *a priori* reason for us to expect that they should have the Kolmogorov 5/3 power law as their inertial-range solution. That is to say, the Lagrangian-history theories were constructed to be invariant under random Galilean transformations in the hope of remedying the deficiencies of DIA, whereas in the case of the LET theory the response equation was known to be free of the infra-red divergence.

We begin with the work of Kraichnan (1966), who investigated the ALHDI approximation for the free decay of isotropic turbulence by using the same methods as for the Eulerian case (Kraichnan 1964a). This time the initial spectrum was taken as

$$E(k, 0) = 2\pi k^{-5/3}, \tag{8.15}$$

the kinematic viscosity as $v = 0.008$, and the evolved value of the Taylor–Reynolds number was $R_\lambda(t_f) = 440$.

When the ALHDI approximation—summarized here in eqns (7.77)–(7.80)—had been integrated forward in time, the evolved spectra were compared with the experimental results of Grant, Stewart, and Moilliet (1962). Comparison with the experimental spectra, as taken from tidal-channel data at $R_\lambda = 2000$, indicated a satisfactory degree of agreement between the ALHDI theory and the experimental energy and dissipation spectra.

In view of this, it may seem surprising that the ALHDI prediction of the Kolmogorov constant $\alpha = 1.77$ does not agree with the stated experimental value of $\alpha = 1.44$ of Grant *et al*. However, this discrepancy probably reflects the difficulties inherent in making a determination of α from the experimental data (see the discussion of this point in Section 2.9.2).

An interesting result of this investigation was the elimination of the spurious oscillations in the response function. Such oscillations with time arose in the Eulerian case—see Section 8.2.1—and it is a definite point in favour of ALHDI that the change of coordinate system suppressed this particular unphysical behaviour.

Naturally the success of the ALHDI approximation at high Reynolds numbers raises the question of how it will perform at low Reynolds numbers. To some extent this question was answered by Herring and Kraichnan (1972) who made a calculation of free decay with ALHDI at an initial Taylor–Reynolds number of $R_\lambda(0) = 19$. The resulting prediction of the dissipation spectrum was not significantly different from theories like DIA (the Eulerian version) or EDW. But the calculation of the skewness factor resulted in an asymptotic value of around $S(t) = 0.63$, which is very much larger than the value from the numerical simulations (i.e. $S = 0.48$) and indicates a poorer agreement with experiment than the other theories considered by Herring and Kraichnan (1972).

Deficiencies of this kind naturally provided a motivation to develop a better Lagrangian-history theory and, as we have noted in Section 7.2.5, Kraichnan and Herring later produced the strain-based ALHDI or SBALHDI. This was followed (Herring and Kraichnan 1979) by a numerical comparison of the Eulerian DIA, ALHDI, and SBALHDI with initial conditions $R_\lambda = 42$ (spectrum I) and $R_\lambda = 19.7$ (spectrum III).

The first of these runs was compared with the numerical simulation of Orszag and Patterson (1972), with the clear result that ALHDI greatly overestimates the spectral transfer of energy. The opposite failing has, of course, been associated with the lack of random Galilean invariance of the Eulerian DIA.

In contrast, SBALHDI was found to behave very much like DIA, but with a slightly higher level of energy transfer—as indicated by the evolved skewness

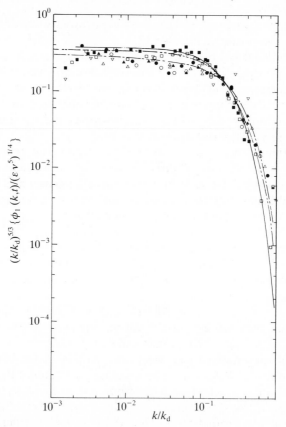

Fig. 8.13. Evolved one-dimensional spectrum at high Reynolds number: comparison of theory with experiment. Theory: ——— LET (McComb and Shanmugasundaram 1984); ——— ALHDI; ———— SBALHDI (Herring and Kraichnan 1979). Experiment: o, ●, △, ▲, $R_\lambda = 2000$ (Grant et al. 1962); ■, $R_\lambda = 538$ (Kistler and Vrebalovich 1966); ▽, $R_\lambda = 308$ (Uberoi and Freymuth 1969); □, $R_\lambda = 850$ (Coantic and Favre 1974).

factor, which agrees rather well with the value obtained from the numerical simulation.

At high Reynolds numbers, the calculation was limited to the Lagrangian-history theories, with initial conditions as given by eqn (8.15). In this case, a repetition of Kraichnan's earlier (1966) comparison of ALHDI with the data of Grant *et al.* (1962) indicated that both theories were in good agreement with the experimental results, although SBALHDI gave a Kolmogorov constant of about $\alpha = 2.0$ compared with the value $\alpha = 1.77$ of ALHDI.

More recently, McComb and Shanmugasundarm (1984) investigated the LET theory (as summarized here by eqns (7.144) and (7.145)) using similar initial conditions to those discussed above. Again, good agreement was found with experimental data, and the overall conclusion was that the LET theory behaved very much like the Lagrangian-history theories at these high Reynolds numbers. This is illustrated in Fig. 8.13, where we show the one-dimensional spectrum, plotted as $k^{5/3}\phi_1(k)$, so that the Kolmogorov 5/3 inertial range appears as a plateau. Theoretical results are presented for LET, ALHDI, and SBALHDI, along with data from several experimental investigations. Clearly the scatter in experimental data is such that one can only conclude that the three theories agree equally well with experiment, although it is marginally the case that LET and SBALHDI seem to be closer together than any other possible pairing.

This behaviour seems to hold for low Reynolds numbers as well. In Fig. 8.14 we show the three-dimensional dissipation spectrum, with LET, DIA, ALHDI, and SBALHDI all computed from Spectrum III, with $R_\lambda = 16.7$.

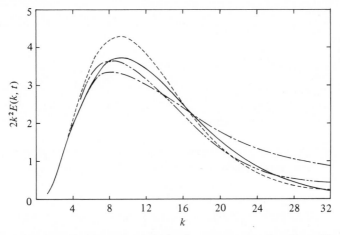

Fig. 8.14. Evolved three-dimensional dissipation spectrum at low Reynolds number. Comparison of various theoretical predictions, with initial spectrum given by Spectrum III: ——— LET (McComb and Shanmugasundaram 1984); – – – DIA; —–—– ALHDI; — – – — SBALHDI (Herring and Kraichnan 1979).

Again, marginally perhaps, LET and SBALHDI seem to be closer together than any other possible pair of theories.

8.4   The Kolmogorov spectrum as a test problem

In Section 3.2.2 we discussed the possible effects of fine-structure intermittency on the inertial-range form of the structure functions of arbitrary order. Here we shall be specifically concerned with the second-order structure function and its Fourier transform: the energy spectrum.

We begin by reminding ourselves that structure functions are based on two space points, $\mathbf{x} + \mathbf{r}$ and $\mathbf{x}$, and that we restrict our attention to components of the velocity field u' (at $\mathbf{x} + \mathbf{r}$) and u (at $\mathbf{x}$) in the direction of the vector $\mathbf{r}$. Then from the general Kolmogorov theory of the structure functions, as embodied in eqn (3.14), we have for the case $n = 2$

$$\langle (u' - u)^2 \rangle = C_2 \varepsilon^{2/3} r^{2/3}, \tag{8.16}$$

which Fourier transforms to the well-known spectral form

$$E(k) = \alpha \varepsilon^{2/3} k^{-5/3}. \tag{8.17}$$

Then we saw that, on the basis of a particular hypothesis ('log-normality' in this case, but there are others), the effects of intermittency would lead to eqn (3.14) being replaced by eqn (3.20), which for the particular case $n = 2$ reduces to

$$\langle (u' - u)^2 \rangle = \tilde{C}_2 \varepsilon^{2/3} r^{2/3} \left(\frac{L}{r} \right)^{\mu/9}, \tag{8.18}$$

and Fourier transformation of this result leads to the intermittency-corrected spectrum

$$E(k) = \alpha' \varepsilon^{2/3} k^{-5/3} (kL)^{\mu/9}, \tag{8.19}$$

which is supposed to replace the well-known 5/3 law of eqn (8.17). Here L is an external length scale, which is a measure of the size of the largest eddies, and μ is a constant in the assumed (log-normal) distribution of the dissipation $\varepsilon(\mathbf{x}, t)$.

The rapid growth of attention to the question of exponent modification in the light of intermittency effects led during the 1970s to a widespread feeling that the Kolmogorov 5/3 inertial-range spectrum could no longer be regarded as a crucial test of an analytical turbulence theory. It is therefore surprising that this aspect of the topic has not received any significant attention in the literature. Indeed, to the present writer's knowledge, there are only two cases of this view actually appearing in print (Martin and De Dominicis 1978; Frisch, Lesieur, and Schertzer 1980). Nevertheless, while there is any uncertainty about whether (8.17) or (8.19) applies, there is an obvious need for caution and clearly it is a very proper view to take.

In the remainder of this section we shall take a properly cautious approach to this tricky but extremely important subject. We shall do this in two parts. First, we examine the question of whether intermittency corrections are large enough to affect our perception of the various RPTs. Second, we consider some of the most recent developments which tend to suggest that the pendulum may already be swinging back to the point where we are inclined to regard the 5/3 law as something which a good RPT should predict.

8.4.1   *Do intermittency corrections have any bearing on our assessment of RPTs?*

Our aim now is to consider how large a difference there is between the 5/3 and the $5/3 + \mu$ laws, solely in terms of the implications for renormalized perturbation theories. This imposes a narrow perspective on the subject, but it is one which is by no means without interest.

We begin with the investigation by Champagne (1978) of the validity of the Kolmogorov (1941) similarity hypotheses. This was based on a consideration of the experimental evidence from many different flows. Among the interesting features of this work, were the conclusions that the Kolmogorov normalized spectral shapes were universal for flows with the same value of Taylor–Reynolds number, and that corrections for deviations from Taylor's hypothesis of frozen convection could have a non-trivial effect on the high-frequency spectrum.

However, from our present point of view, the particularly interesting conclusion was to the effect that it would be difficult, if not impossible, to distinguish between the two forms of spectrum, as given here by eqns (8.17) and (8.19). In arriving at this conclusion, Champagne took a value of the intermittency constant $\mu = 0.5$ which nowadays would be regarded as rather large.

In Section 3.2.2 we mentioned that an analysis of earlier data by Van Atta and Chen (1967) implied $\mu = 0.25$, if the data were to agree with the predictions of the scale-similarity theories. More recently, Antonia, Phan-Thien, and Satyaprakash (1981), Antonia, Satyaprakash, and Hussain (1982), and Anselmet, Gagne, Hopfinger, and Antonia (1984) all suggested $\mu = 0.2$. It should perhaps be mentioned in passing that these values of μ were determined from measured higher-order structure functions (e.g. $n = 6$ in the last case). Thus the difficulty in distinguishing between exponents of $-5/3$ and $-5/3 + \mu$ would be even greater than Champagne supposed.

The difference between these two possible exponents can be put into context by comparing it with the difference between the Eulerian DIA 3/2 law and the 5/3 law. Taking the intermittency correction to be $\mu = 0.2$, we can write each of the three exponents to two places of decimals as follows:

$$-5/3 + 0.2/9 = -1.65$$
$$-5/3 = -1.67$$
$$-3/2 = -1.50.$$

Clearly, therefore, the existence of an intermittency correction of this magnitude would not modify the conclusion that Eulerian DIA does not agree with experiment at high Reynolds numbers.

Equally clearly, the change from $-5/3$ to $-5/3 + \mu$ would not cure the infra-red divergence discussed in Section 7.1.1. From eqn (7.7) it may be seen that this would require $\mu/9 > 2/3$ or $\mu > 6$!

8.4.2 Is the Kolmogorov 5/3 law correct after all?

We begin by reminding ourselves that intermittency corrections of the kind we have just been discussing are essentially rather arbitrary. That is, models like 'log-normality' and the β-model are based on hypotheses which may or may not be true. For instance, if other—equally valid—hypotheses are made about the way in which the dissipation is distributed, then it is possible to arrive at the conclusion that the exponent in the spectral law is unaffected, although the constant of proportionality and the extent of the inertial range of wavenumbers may both be changed by intermittency (Grant et al. 1962; Wyngaard and Tennekes 1970).

This point of view has recently received some theoretical attention. Mjolsness (1980) has proposed that the Kolmogorov similarity hypotheses can be replaced by two weaker forms, which essentially do not require the condition of local isotropy. One consequence of this approach is that the Kolmogorov constant is replaced by a function which is only constant for a particular kind of flow, and can be expected to vary from one flow to another.

Of course the Kolmogorov theory itself is purely phenomenological and—apart from its greater intuitive appeal—has no better fundamental status than its successors. For this reason the recent work of Foias, Manley, and Temam (1987) is particularly interesting. These authors claim to have derived the 5/3 law rigorously direct from the Navier–Stokes equation, something which has not previously been done. We shall not go into details here, but note only that the theory relies on the dimensional homogeneity of the statistical solutions of the Navier–Stokes equation (as obtained from the properties of the Hopf equation). The theory may still be regarded as somewhat heuristic, as the authors assume (and cannot prove) the existence of a family of homogeneous statistical solutions of the Navier–Stokes equation, and then demonstrate that invariance under a time-dependent two-parameter group of transformations implies that the second-order velocity correlation takes the Kolmogorov form (i.e as in eqn (8.16)).

The question must then arise: if the second-order structure function has the K41 form, how do we reconcile this with the very definite effects of intermittency on the higher-order structure functions? After all, it is the existence of the latter effect which permits the experimentalist to measure the constant μ. In fact two recent pieces of work can be seen as (in effect) addressing that point. Nakano (1986) has produced a new model of the scale-similar type, which is different from both the log-normal model and β-model. This model is claimed to exhibit a crossover from intermittent scaling to Kolmogorov scaling. That is, higher-order structure functions show intermittent scaling, whereas low-order structure functions (such as $n = 2$ or $n = 3$, although a critical value does not appear to have been determined) show K41 scaling.

On the other hand, Chorin (1988) considers only the second-order moment and argues that the K41 result is compatible with the existence of intermittency. Unfortunately only brief details of his argument are given and these do not seem altogether convincing.[2]

8.5 Application to non-isotropic turbulence

The renormalized perturbation theories which we have been discussing are not limited to the special case of isotropic (or indeed homogeneous) turbulence. By this we mean that those aspects of the theory which are imponderable— essentially the second-order truncations of line-renormalized perturbation expansions—are no more or less probable when applied to general inhomogeneous turbulence. Yet, despite this intrinsic generality, there have been (to our knowledge) only two attempts to provide RPT formulations of turbulent shear flow. These were an extension of the EFP theory (Allen 1963) and an application of DIA (Kraichnan 1964b). We shall discuss the second of these theories. Not only is it simpler, but the original source is more readily available.

8.5.1 *Application of DIA to inhomogeneous turbulence*

The general statistical formulation for inhomogeneous velocity fields is discussed in Sections 2.1 and 2.2. In the first of these sections we derive the solenoidal form of the Navier–Stokes equation, and in the second we introduce the moment hierarchy. Working now in terms of the kinematic pressure (i.e. we shall set the density equal to unity) and making some minor rearrangements, we shall restate the principal equations here for completeness. Thus eqn (2.28) for the mean velocity becomes

$$\left[\frac{\partial}{\partial t} - \nu\nabla^2\right]\bar{U}_\alpha(\mathbf{x}, t) = -\frac{\partial P_{\text{ext}}}{\partial x_\alpha} - L_{\alpha\beta}(\mathbf{V})\bar{U}_\beta(\mathbf{x}, t) +$$

$$+ M_{\alpha\beta\gamma}(\mathbf{V})[\bar{U}_\beta(\mathbf{x}, t)\bar{U}_\gamma(\mathbf{x}, t) + Q_{\beta\gamma}(\mathbf{x}, \mathbf{x}; t, t)], \quad (8.20)$$

where the operators $L_{\alpha\beta}(\mathbf{V})$ and $M_{\alpha\beta\gamma}(\mathbf{V})$ are given by eqns (2.12) and (2.13) respectively.

We note that this equation for the mean velocity contains the unknown single-point moment $Q_{\beta\gamma}(\mathbf{x}, \mathbf{x}; t, t)$. This can be obtained in principle from the solution of eqn (2.30) for the general two-point two-time correlation $Q_{\beta\gamma}(\mathbf{x}, \mathbf{x}'; t, t')$. With some slight rearrangements, this equation can be written as

$$\left[\frac{\partial}{\partial t} - \nu\nabla^2\right]Q_{\alpha\sigma}(\mathbf{x}, \mathbf{x}'; t, t') = -L_{\alpha\beta}(\mathbf{V})Q_{\beta\sigma}(\mathbf{x}, \mathbf{x}'; t, t') +$$

$$+ 2M_{\alpha\beta\gamma}(\mathbf{V})[\bar{U}_\beta(\mathbf{x}, t)Q_{\gamma\sigma}(\mathbf{x}, \mathbf{x}'; t, t')] +$$

$$+ M_{\alpha\beta\gamma}(\mathbf{V})Q_{\beta\gamma\sigma}(\mathbf{x}, \mathbf{x}, \mathbf{x}'; t, t, t'), \quad (8.21)$$

where the moments $Q_{\alpha\sigma}(\mathbf{x}, \mathbf{x}'; t, t')$ and $Q_{\beta\gamma\sigma}(\mathbf{x}, \mathbf{x}', \mathbf{x}''; t, t', t'')$ are defined by eqns (2.24) and (2.25).

A closure approximation is now required to express the triple moment $Q_{\beta\gamma\sigma}$ in terms of the other dependent variables, namely $Q_{\alpha\sigma}$ and $\bar{U}_\alpha$. The DIA can be used for this purpose (cf. Section 6.1.1) by introducing the infinitesimal response tensor $G_{\alpha\beta}(\mathbf{x}, \mathbf{x}'; t, t')$:

$$G_{\alpha\beta}(\mathbf{x}, \mathbf{x}'; t, t') = \begin{cases} \left\langle \dfrac{\partial u_\alpha(\mathbf{x}, t)}{\partial f_\beta(\mathbf{x}', t')} \right\rangle & t > t' \\ 0 & t < t' \end{cases} \quad (8.22)$$

where, as usual, $\langle \ \rangle$ denotes ensemble average and $\mathbf{f}(\mathbf{x}, t)$ is an arbitrary stirring force. The equation of motion for $G_{\alpha\beta}(\mathbf{x}, \mathbf{x}'; t, t')$ can be obtained by the functional differentiation of the solenoidal Navier–Stokes equation (i.e. eqn (2.15)) after the term $\mathbf{f}(\mathbf{x}, t)$ has been added to the r.h.s. The result is

$$\left[\frac{\partial}{\partial t} - \nu\nabla^2\right]G_{\alpha\sigma}(\mathbf{x}, \mathbf{x}'; t, t') + L_{\alpha\beta}(\mathbf{V})G_{\beta\sigma}(\mathbf{x}, \mathbf{x}'; t, t') -$$

$$- 2M_{\alpha\beta\gamma}(\mathbf{V})[\bar{U}_\gamma(\mathbf{x}, t)G_{\beta\sigma}(\mathbf{x}, \mathbf{x}'; t, t')] -$$

$$- M_{\alpha\beta\gamma}(\mathbf{V})H_{\beta\gamma\sigma}(\mathbf{x}, \mathbf{x}, \mathbf{x}'; t, t, t')$$

$$= \delta_{\alpha\sigma}\delta(\mathbf{x} - \mathbf{x}')\delta(t - t'), \quad (8.23)$$

where

$$H_{\beta\gamma\sigma}(\mathbf{x}, \mathbf{x}', \mathbf{x}''; t, t', t'') = \left\langle u_\gamma(\mathbf{x}', t')\frac{\partial u_\beta(\mathbf{x}, t)}{\partial f_\sigma(\mathbf{x}'', t'')} \right\rangle. \quad (8.24)$$

The solution of (8.23) and (8.24) for $G_{\alpha\sigma}(\mathbf{x}, \mathbf{x}'; t, t')$ is subject to the conditions implied by eqn (8.22), and must also satisfy the boundary conditions on the velocity field.

The general renormalized perturbation theory, as discussed in Section 6.1 for $\mathbf{k}$-space, also goes through for $\mathbf{x}$-space, and Kraichnan (1964b) has given the DIA expressions for $Q_{\beta\gamma\sigma}$ and $H_{\beta\gamma\sigma}$ as

$$Q_{\beta\gamma\sigma}(\mathbf{x}, \mathbf{x}', \mathbf{x}''; t, t', t'')$$

$$= 2 \int d^3y \left[\int_0^t G_{\beta\rho}(\mathbf{x}, \mathbf{y}; t, s) M_{\rho\delta\varepsilon}(\mathbf{V}) \times \right.$$

$$\times \{Q_{\gamma\delta}(\mathbf{x}', \mathbf{y}; t', s) Q_{\sigma\varepsilon}(\mathbf{x}'', \mathbf{y}; t'', s)\} \, ds +$$

$$+ \int_0^{t'} G_{\gamma\rho}(\mathbf{x}', \mathbf{y}; t', s) M_{\rho\delta\varepsilon}(\mathbf{V}) \times$$

$$\times \{Q_{\beta\delta}(\mathbf{x}, \mathbf{y}; t, s) Q_{\sigma\varepsilon}(\mathbf{x}'', \mathbf{y}; t'', s)\} \, ds +$$

$$+ \int_0^{t''} G_{\sigma\rho}(\mathbf{x}'', \mathbf{y}; t'', s) M_{\rho\delta\varepsilon}(\mathbf{V}) \times$$

$$\left. \times \{Q_{\beta\delta}(\mathbf{x}, \mathbf{y}; t, s) Q_{\gamma\varepsilon}(\mathbf{x}', \mathbf{y}; t', s)\} \, ds \right] \tag{8.25}$$

and

$$H_{\beta\gamma\sigma}(\mathbf{x}, \mathbf{x}', \mathbf{x}''; t, t', t'') = 2 \int d^3y \int_{t''}^t G_{\beta\rho}(\mathbf{x}, \mathbf{y}; t, s) \times$$

$$\times M_{\rho\delta\varepsilon}(\mathbf{V}) \{Q_{\gamma\delta}(\mathbf{x}', \mathbf{y}; t', s) G_{\varepsilon\sigma}(\mathbf{y}, \mathbf{x}''; s, t'')\}. \tag{8.26}$$

It may be helpful if we mention that eqns (8.25) and (8.26) can be reduced to their isotropic homogeneous equivalents, as discussed in Section 6.1.4. The appropriate procedure is to set the surface terms (i.e those involving $L_{\alpha\beta}$), terms involving the mean velocity, and the external pressure gradient all equal to zero. Then, taking statistical quantities to be homogeneous in the space variables, Fourier transformation into $\mathbf{k}$-space, followed by the use of the isotropic forms of Section 2.6.4, leads to the required result.

8.5.2 *The computational problems*

Equation (8.20) is essentially just the most general form of the Reynolds equation for the mean velocity in inhomogeneous turbulent flow. In order to solve it, we need the Reynolds stress tensor, and this can be obtained by putting $\mathbf{x} = \mathbf{x}'$ in the general covariance tensor $Q_{\alpha\beta}(\mathbf{x}, \mathbf{x}'; t, t')$, which in turn requires the solution of eqns (8.21)–(8.26). Thus, to second order in renormalized perturbation theory, eqns (8.20)–(8.26) offer a complete prescription for calculating the statistical parameters of interest for any general (but incompressible) turbulent flow.

In principle, the solution of these equations by numerical means is no more difficult than for the case of homogeneous isotropic turbulence, as discussed in Sections 8.2 and 8.3. However, in practice, the difficulty we face is one of size. Even for the simplest non-trivial example of a shear flow, an enormous number of arithmetic operations will be required in order to solve the above set of equations.

We can enlarge on this point by considering the specific case of two-dimensional (mean) flow through a plane channel. The general specification of this flow configuration has already been given in Section 1.4.5, and the application of the general statistical formulation to channel flow has been discussed in Section 2.3. From our present point of view, the main result is contained in (2.44) for the two-point covariance tensor. We repeat that result here for the sake of convenience:

$$\langle u_\alpha(\mathbf{x},t)u_\beta(\mathbf{x}',t')\rangle = Q_{\alpha\beta}(\mathbf{r}, R_2; t, t'),$$

where $\mathbf{r} = \mathbf{x} - \mathbf{x}'$ is the relative coordinate and $\mathbf{R}$ is the centroid (absolute) coordinate.

Let us now make a comparison with the calculation of freely decaying turbulence. In this case homogeneity eliminates the dependence of $Q_{\alpha\beta}$ on $\mathbf{R}$. Then isotropy further reduces the spatial dependence from three scalar variables (r_1, r_2, r_3) to one $(|\mathbf{r}|)$, and the nine scalar components of $Q_{\alpha\beta}$ can be replaced by a single scalar function. Thus, after Fourier transformation with respect to $\mathbf{r}$, we have $q(k; t, t')$ to calculate, along with a similarly reduced response function.

In contrast, for plane channel flow, we have four components of $Q_{\alpha\beta}$ (i.e. the diagonal elements and $Q_{12} = Q_{21}$), each of which is a function of six scalar variables: r_1, r_2, r_3, R_2, t, and t'. In addition, we have (8.20) for the mean velocity as an additional equation which must be solved simultaneously with the equations for the covariance and the response tensor. Also, everything that we have said about $Q_{\alpha\beta}$ also applies to $G_{\alpha\beta}$. Hence it is clear that the simplest shear flow presents a very much larger computational problem for the DIA than that of freely decaying turbulence.

When the theory was originally published (1964), its computational needs were too great for the existing computers. This is probably still the case today. As far as the present writer knows, DIA has not been applied directly to inhomogeneous turbulence. The most demanding application to date seems to be the next problem after isotropic turbulence in the hierarchy of difficulty: axisymmetric homogeneous turbulence (see Batchelor 1946). Herring (1974) used DIA to study the return of initially axisymmetric turbulence to isotropy. Later, Schumann and Herring (1976) compared the DIA calculation with the results of a direct numerical simulation.

Further approximations are made in the process of carrying out the above calculations so that the results (although physically reasonable) are not really conclusive. However, two points made by Schumann and Herring seem particularly relevant to our discussion here. First, they note that the numerical solution of the DIA equations is not trivial and that the computing time is actually comparable with that required for a single realization of the numerical simulation of the Navier–Stokes equation.

Second, they argue that it would be necessary to improve the numerical analysis of the DIA before practical flow problems, such as plane channel flow, could be tackled.

They also speculate that simplified turbulence theories—and particularly the TFM (see Kraichnan (1972) for the generalization of TFM to the case of inhomogeneous turbulence)—would be needed for high Reynolds numbers. However, almost-Markovian models do not offer the full generality of the renormalized perturbation theories. And, despite the fact that they are much easier to integrate numerically than the RPTs, the most demanding application which appears to have been tackled so far is non-isotropic turbulence (Cambon, Jeandel, and Mathieu 1981).

In fact, it is difficult to imagine that the physics of turbulence can be satisfactorily captured by any theory which is very much simpler than the DIA. As we have seen in preceding chapters, all RPTs seem to lead to similar forms and virtually the same energy equation. If one accepts this view, then there seem to be two possibilities. First, as computers advance in power and speed, it may be possible to develop hybrid methods in which the RPT can be applied in a limited way to only part of the problem, while the direct numerical simulation takes care of the rest. This topic is discussed further in Chapter 10, where we consider numerical simulation of turbulence.

Secondly, there is the possibility that RPTs can be simplified by means of physical and mathematical approximations which are based on our knowledge of turbulence. The ultimate aim of this approach might be a set of differential equations, comparable in computational difficulty with, say, the engineering models which we discussed in Sections 3.3.3 and 3.3.4.

Some efforts have been made in this direction by Leslie (1970, 1973) and by Edwards and McComb (1971, 1972). Both investigations are based on a consideration of plane channel flow, and have a number of assumptions in common. For instance, both approaches involve the introduction of a spectral density for channel flow by Fourier transforming eqn (2.44) with respect to $\mathbf{r}$ to obtain

$$Q_{\alpha\beta}(\mathbf{k}, y; t, t') = \left(\frac{1}{2\pi}\right)^3 \int d^3r \, Q_{\alpha\beta}(\mathbf{r}, \mathbf{y}; t, t') \exp\{-i\mathbf{k}\cdot\mathbf{r}\}, \qquad (8.27)$$

where we have invoked eqn (2.90) and we have also put $R_2 = y$ in order to simplify the notation. Then, for y measured from the centre of the channel, symmetry implies that the turbulence at $y = 0$ is isotropic. This means that we can expand the inhomogeneous spectral tensor (and indeed all other statistical quantities) in powers of y about $y = 0$, where the general formulation can be reduced to the isotropic case.

As an illustration of the algebra involved, we cite an example from Edwards and McComb (1972) who expanded the inhomogeneous spectral tensor as

$$Q_{\alpha\beta}(\mathbf{k}, y) = \left[D_{\alpha\sigma}(\mathbf{k})D_{\rho\beta}(\mathbf{k}) + \frac{1}{2k^2} D_{\alpha\sigma}(\mathbf{k})D_{\rho\beta}(\mathbf{k}) \frac{d^2}{dy^2} - \right.$$

$$- \frac{1}{4k^2}\{(\delta_{\alpha1}\delta_{\sigma1} + \delta_{\alpha3}\delta_{\sigma3})D_{\rho\beta}(\mathbf{k}) +$$

$$+ (\delta_{\beta1}\delta_{\rho1} + \delta_{\beta3}\delta_{\rho3})D_{\alpha\sigma}(\mathbf{k})\} \frac{d^2}{dy^2} +$$

$$\left. + O\left(\frac{d^4}{dy^4}\right) \right] q_{\sigma\rho}(\mathbf{k}, y), \tag{8.28}$$

where time dependences are not shown explicitly. This expansion is accompanied by an approximation in which $q(\mathbf{k}, y)$—a kinetic energy function equal to one-third of the sum of the diagonal elements of $Q_{\alpha\beta}(\mathbf{k}, y)$—is introduced. The off-diagonal elements ($Q_{12} = Q_{21}$) are represented by a function $q_R(\mathbf{k}, y)$. Then the modified spectral tensor $q_{\sigma\rho}(\mathbf{k}, y)$ can be defined jointly by eqn (8.28) and

$$q_{\sigma\rho}(\mathbf{k}, y) = \delta_{\sigma\rho}q(\mathbf{k}, y) + (\delta_{\sigma1}\delta_{\rho2} + \delta_{\sigma2}\delta_{\rho1})q_R(\mathbf{k}, y). \tag{8.29}$$

Other functions (and operators) can be treated in this way, and angular dependences on the vector $\mathbf{k}$ can be integrated out.

At this stage, the two approaches diverge. Leslie adopted a kind of similarity solution in which he assumed that the spectral part of (8.27) has a 'constant shape'. He also expanded the anisotropy out in spherical harmonics—a technique which subsequently proved helpful in the axisymmetric problem which we discussed earlier—and ended up with a single-point formulation which bears some resemblances to the Reynolds stress closure of Hanjalic and Launder (1972).

In contrast, Edwards and McComb followed a more classical route in statistical physics, deriving local (differential) transport equations from the integral equations of the EFP (or DIA) closure. These equations were solved for the mean and r.m.s. velocity profiles, which agreed quite well with the experimental shapes, at least for the core region, where the expansion in the centroid coordinate would be valid. It is perhaps worth observing that these results—whatever their shortcomings—can claim to be the first calculations of such turbulence parameters entirely from a general renormalized perturbation theory.

Since the above research was carried out, this type of approach seems to have been very much neglected—although there has been some recent work by Ortiz and Ruiz de Elvira (1984). In a field noted both for its difficulty and for the paucity of original ideas, this neglect of a potentially very fruitful way of tackling real turbulent flows seems rather surprising.

8.5.3 *Other applications*

In considering the application of RPTs to other problems, we have to bear in mind the shaky status of these theories. Thus any such application must be seen as being as much a check on the theory as a way of studying another problem.

For the sake of completeness, we therefore list the following applications of RPTs without further comment; DIA has been used to study the problem of a non-linear oscillator with random forcing (Morton and Corrsin 1970); TFM was used to study atmosphere predictability (Leith 1971); EFP has been used to investigate qualitative changes in energy spectra due to non-Newtonian effects (McComb 1974, 1976); the eddy-damped quasi-normal Markovian model has been applied to magnetohydrodynamic turbulence (Pouquet, Frisch, and Leorat 1976) and the DIA applied to one-dimensional plasma turbulence (Pesme and duBois 1982); lastly, the LET theory has been used to calculate the total energy decay in three-dimensional turbulence (Hosokawa and Yamamoto 1986).

This list does not attempt to be exhaustive, and some specific applications (e.g. passive scalar transport, subgrid modelling) will be dealt with in other chapters.

8.6 Appraisal of the theories

In this section we shall try to come to some tentative conclusions about the current status of renormalized perturbation theories, as applied to turbulence. We should, however, stress that none of the theories discussed so far can claim to be a rational approximation to the Navier–Stokes equation. Accordingly, it would be quite inappropriate to try to establish which of the various theories is 'best', or to attempt to rank them into some sort of pecking order, as it were. Our point of view is that our ignorance about the correct way to formulate turbulence theories remains profound, and that one of the few possible ways of remedying this is to study the existing theories in an impartial and open-minded fashion. That does not mean that we should suspend the critical process, but merely that we should not allow ourselves to be hypnotized by the engineer's goal of a successful turbulence theory. That would indeed be premature. It should be borne in mind that physics as a subject has had many successes by not being unduly concerned by the ultimate practical applications!

8.6.1 Critique of DIA: the wider justification of RPT approaches

In his original formulation of DIA, Kraichnan (1959) put forward three postulates as the foundations of his theory. These were (a) maximal randomness condition, (b) weak dependence principle, and (c) direct-interaction approximation, and they were held to justify various detailed steps in the derivation of closed equations for the correlation and response functions.

Subsequently this formulation of Kraichnan's theory was subject to very detailed criticism by Proudman (1962) and later by Saffman (1968). The interested reader may wish to consult these references, as we do not intend to go into further detail here. Our purpose in raising the matter is to make a more general point. We wish to suggest that this kind of detailed criticism and the detailed justification which inspired it in the first place are both equally irrelevant.

If we consider the above hypotheses, then the first of them does not really touch on the closure problem at all but is relevant to the evaluation of the correlation between the velocity field and the stirring forces (see eqn (6.32), along with eqns (6.23) and (6.27c)). That leaves us with (b) and (c), and we shall discuss each of these in turn.

The hypothesis of weak dependence is to the effect that the coupling between any few Fourier modes (i.e. distinct wavenumbers) is very weak when the system size is large. That is, as the system size increases, the modes are more densely packed and a finite interaction strength has to be shared out among more of them. Hence any specific triad of wavenumbers must have an individually small effect. Ignoring the likelihood (or otherwise) of such an effect's being dynamically significant, we would wish to make just one point. There is no way of knowing whether the hypothesis survives the subsequent global renormalization, which is a complicated mathematical operation of unknown properties.

The term 'direct interaction' refers to the interaction between the three modes $(\mathbf{k}, \mathbf{j}, \mathbf{l})$ of the Navier–Stokes equation, as opposed to higher-order terms in the perturbation series in which the interaction is mediated indirectly by other modes. Again we would argue that the question of whether this is correct or not for the primitive perturbation series is irrelevant, as once renormalization is carried out the coupling strengths in the new expansion are affected by the replacement of G_0 by G. Hence, the DIA is effectively just a decision to truncate the renormalized perturbation series at second order—much like all other renormalized perturbation theories!

Our general view therefore is that renormalized perturbation theory—as introduced in the present book by the discussion in Chapter 5—is a general method of approach to problems involving collective strong interactions and has had its successes in other fields of physics. In the case of Kraichnan's DIA theory, the feature which gives it its own identity (i.e. which distinguishes it from all other RPTs) is surely the introduction of the infinitesimal response function. Thus the Wyld diagrammatic formulation (as discussed in Section 5.5) tells us that once we identify the need to renormalize the viscous response to the arbitrary stirring forces, then Kraichnan's DIA follows from purely formal (indeed, topological) procedures. We would argue that that is sufficient *a priori* justification, and that it is pointless to be unduly concerned about physical arguments which, if they are not imponderable to begin with, certainly are after renormalization.

If we adopt this point of view (and basically the only virtue we would claim for it would be its pragmatism) then we are forced to consider the theories solely in terms of the properties of the renormalized expansions. That is to say, we accept that the primitive perturbation expansion is essentially a power series in the interaction strength (i.e. the Reynolds number) and is therefore widely divergent in all cases of practical interest. Then, also accepting that the global renormalization is too complicated a procedure to be other than imponderable in nature, we must consider the renormalized perturbation series on its own merits. Are its properties better than those of the primitive expansion?

Kraichnan (1975) has given a rather pessimistic answer to this question. He points out that the removal of reducible diagrams by renormalization (i.e partial summation; see Section 5.5) does not stop the rapid growth of terms with order of expansion. Thus, judged by the number of terms in each order, the renormalized series is nearly as bad as the primitive series.

But this does not seem to the present writer to be a valid criterion for divergence or otherwise. What we are surely concerned with is the total quantitative effect of each order. There would seem to be at least three possible ways in which the renormalized series could be better than the primitive one.

(a) The RP series might be convergent.
(b) There might be cancellations at higher orders, which would justify the retention of second-order terms only.
(c) The RP series might be asymptotic, that is, divergent, but with the lowest non-trivial order approximately equal to the sum of the series.

In fact very little is known about the analytical properties of the renormalized perturbation series based on the Navier–Stokes equation. The only attempt to treat this question at all quantitatively appears to be a remark by Edwards (1965) that a derivation of the energy equation by primitive perturbation theory would result in purely viscous time-scales appearing in the denominator of the non-linear term. As it is, the renormalized form is the EFP energy equation (see eqn (6.101)), with the viscous time-scale $\{vk^2 + vj^2 + vl^2\}^{-1}$

replaced by the renormalized time scale $\{\omega(k) + \omega(j) + \omega(l)\}^{-1}$. Bearing in mind that this non-linear term is only the second-order term of an expansion which is a power series in (roughly speaking; see Edwards (1964) for the precise result) $q(k)/\omega(k)$, then the possible relevance to the convergence of the underlying RP expansion is clear.

However, the bedrock difficulty in the way of any quantitative assessment of this important point is the numerical problem of calculating even the second order of the renormalized perturbation expansions. We saw this in Sections 8.2 and 8.3, where we were in fact concerned with the calculation of just this order. Thus relatively little is known about the convergence properties of these series, and we are thrown back on the need to compute the theories and judge them by results.

8.6.2 *Some comments on random Galilean invariance*

It may seem surprising that the straightforward application of renormalized perturbation theory can give good results at low Reynolds numbers, but fail (if only slightly) at high Reynolds numbers. Given the generality of the approach, it is by no means clear why there should be this apparent dependence on Reynolds number.

As a result of his analysis of the failure of DIA to give the Kolmogorov spectrum, Kraichnan has been led to the rather extreme conclusion that there is nothing intrinsically wrong with DIA, but that the Eulerian coordinate system is unsuited to distinguishing between the two possible effects which large eddies can have on smaller eddies. These are uniform convection (which does not affect energy transfer) and distortion (which does affect energy transfer). This failure is associated with a failure to maintain Galilean invariance when simultaneous moments are calculated in terms of non-simultaneous moments.

To put it another way, Kraichnan seems to be asserting that theories involving non-simultaneous moments cannot successfully describe turbulence in the laboratory coordinate frame. When put like that, this seems every bit as surprising as the failure of DIA, EFP, and SCF at large values of the Reynolds number. Although we have given a detailed account of that work in Sections 7.1.2–7.1.4, we think it may be interesting if we take a slightly more sceptical look at it here. In particular, it may be helpful if we first make some succinct statements about the requirement that a theory sould possess the property of random Galilean invariance (RGI).

Let us list the requirements as follows:

(1) The Navier–Stokes equation must be invariant under Galilean transformation by constant velocity **v**.

(1a) The Navier–Stokes equation must be invariant under every Galilean transformation which makes up the ensemble $\{v\}$, and hence must be invariant under random Galilean transformation.

(2) Correlations of velocities at two or more space points are invariant under Galilean transformation by constant velocity v, provided that the correlations are simultaneous.

(2a) Correlations of velocities at two or more space points are invariant under every Galilean transformation which makes up the set $\{v\}$, and hence must be invariant under random Galilean transformation, provided that the correlations are simultaneous.

(3) Non-simultaneous correlations at two or more space points are not invariant under Galilean transformation by constant velocity v because the distance between the measuring points changes according to v and the difference between the measuring times.

(3a) Non-simultaneous correlations at two or more space points are not invariant under any member of the set $\{v\}$ of random Galilean transformations, and hence are not invariant under random Galilean transformation.

If follows, therefore, that RGI is indeed a rigorous property of both the Navier–Stokes equation and the single-time multipoint correlations (or moments). It is equally clear that, if any RPT constructs the triple moment from two-time moments, there is at least a possibility that the single-time form of the resulting triple moment will not possess the requisite RGI. And, of course, failure in this must be seen as failure of the theory.

In all of this we agree with Kraichnan. Where we would wish to exercise a little more caution is in the way that these ideas can be used in rigorously testing theories for RGI. Essentially, Kraichnan bases the possibility of doing this on eqn (7.15), with the formal extension of that result being (7.69) and (7.70), as the test for RGI. But (7.15) is an approximation and is hedged about with many restrictions. This in itself rules it out as a rigorous test for RPTs.

We should also note that the diagnosis of the failure of DIA would not be the same if it were tested for deterministic Galilean invariance. Referring back to the model problem in Section 7.1.2, we see that if we took the correlation function in (7.15) for one v (i.e. we do not average) then the DIA triple moment—as given by eqn (7.26)—would have the same phase factor as the exact form prior to averaging, as given by the penultimate line of eqn (7.18). Thus this interpretation of the failure of the DIA depends to at least some extent on the nature of the approximation for the turbulent correlation function when further averaged over the random Galilean ensemble.

This is not a severe criticism of the work which ultimately led to the Lagrangian-history theories. Rather, it is just an attempt to sound a note of caution about the general applicability of these ideas to other theories. The nature of the test for RGI is such that it depends heavily on the actual correlation and response or propagator functions behaving as if they scale on large-eddy convection times $(uk)^{-1}$, where u is the r.m.s. velocity of the turbulence. In theories other than DIA, that may not be the case, and hence our argument is that there are other possible diagnoses for the failure of RPTs and other possible tests to satisfy than a test for RGI which is based on an approximation. In short, we think that a conclusion that RPTs cannot work in an Eulerian coordinate frame may be unduly pessimistic (see also the recent results discussed in Section 8.8).

8.7   General remarks

We have argued that the strength of renormalized perturbation theories lies in their generality and in the absence of *ad hoc* assumptions or disposable constants. Yet one would not need to be unduly cynical to argue that RPTs (without exception) are cut off from true fundamental status, on the one hand, by their inability to predict their own errors, and from engineering utility, on the other, by their enormous complexity when formulated for inhomogeneous turbulence.

There seems to be a growing belief that the answer to the first of these problems lies with the renormalization group. An introduction to this topic is to be found in Section 3.5.3, and a fuller discussion occupies Chapter 9. However, as we saw in Sections 8.2 and 8.3, the RPTs have had their successes and should not be underestimated.

The second problem—the analytical complexity of RPTs—is really, as we saw in Section 8.5, a matter of dimensionality. Without the simplifications of homogeneity and isotropy, the calculations described in preceding sections would be much too large for present-day computers. What is needed is an attack on the problem of analytical reduction of the second-order equations for shear flows, with the aim of reducing their complexity to the level of, say, the equations computed for the case of freely decaying isotropic turbulence. In view of the fact that most RPTs give good results at low Reynolds numbers, there is some reason to feel optimistic about their potential application to other problems with low Reynolds numbers. As we pointed out in Section 8.5.2, it is surprising that such a potentially fruitful research topic has been so comprehensively neglected.

8.8   Postscript: some current work

The calculations described in Sections 8.2.3 and 8.3 have been carried on and extended to Taylor–Reynolds numbers over the range $0.5 < R_\lambda(t_f) < 1009$, where t_f is the final time of computation (McComb, Shanmugasundaram, and Hutchinson 1989). It was found that, for $R_\lambda(t_f) \leqslant 5$, the calculations for LET and DIA were almost indistinguishable, with any difference between the two theories tending to disappear as the Taylor–Reynolds number tended to zero. Somewhat more surprisingly, the two theories were again found to behave in a very similar way as the Reynolds number became large, and on the basis of these calculations it seemed clear that the calculated energy spectrum was closer to the Kolmogorov 5/3 law than to the analytical prediction of a 3/2 law (see Section 6.1.6), even for DIA.

An even more surprising (and certainly more controversial) finding was that two-time correlation and propagator (or response) functions were found to scale on the Kolmogorov time-scale rather than on the convective sweeping time-scale, which is associated with the energy-containing range of wave-numbers. It has been argued (McComb *et al.* 1989) that this result raises questions about the diagnosis of the failure of DIA (see Sections 7.1.2 and 7.1.3), and the relevance of the postulate of random Galilean invariance has been questioned.

More recently, McComb and Shanmugasundaram (1989) have carried out similar calculations, but this time using the more general (and corrected) form of LET theory, as represented by eqns (7.144) and (7.145), for $Q(k;t,t')$ and $H(k;t,t')$. The results are virtually the same as before (i.e. using an *ad hoc* generalization of the stationary form as given by McComb (1978)), but the computations are now much faster and the unphysical overshoots no longer occur in the calculated propagator.

Notes

1. These remarks are based on an *ad hoc* generalization of the original LET theory to non-stationary turbulence. When the correct non-stationary form of LET theory (as derived in Section 7.4) is used, these unphysical overshoots do not occur (see Section 8.8).
2. This note should be read in conjunction with Section 3.2.1 and, in particular with the derivation of eqn (3.29). Chorin (1988) argues that a paradox arises if one attempts to derive (3.29) by a different method. This is summarized as follows.

   We begin by making the same assumption that all the energy on scales l_n is transferred to scales l_{n+1}. Hence, Chorin argues, the average energy in eddies of size l_n is $E_0 t_n/T$, where T is the characteristic decay time of the vortical structures (*sic*), $t_n = l_n/v_n$ and E_0 is the total available energy.

   This result seems to imply a much more extreme assumption than that which underlies the shell model leading to eqn (3.29). Although the present writer is unclear about the definition of the time-scale T and the physical significance of the ratio t_n/T, it seems as if Chorin is saying:

   amount of energy on scales l_n = total amount of energy put into the system in unit time E_0 × (time taken to cascade to l_n)/(time taken for E_0 to be fully dissipated as heat).

   If this interpretation is correct, then Chorin's argument rests on the assumption that the decay process is linear and that t_n is the time taken for the energy to reach scales l_n.

Chorin's second assumption is that $v_n = E^{1/2}$, where E is a constant but is otherwise undefined. Hence, with this substitution, $E_n \sim l_n$, and there is a disagreement with (3.29) which can only be resolved by invoking intermittency. However, this also seems unconvincing. In fact we really have $v_n = E_n^{1/2}$, and substitution of this leads back to $v_n^2 \approx l_n^2$, in agreement with (3.29).

Evidently, Chorin's arguments need some clarification.

References

ALLEN, J. R. (1963). Formulation of the theory of turbulent shear flow *Ph.D. Thesis*, Victoria University of Manchester.

ANSELMET, F., GAGNE, Y., HOPFINGER, E. J. and ANTONIA, R. A. (1984). *J. fluid Mech.* **140**, 63.

ANTONIA, R. A., PHAN-THIEN, N., and SATYAPRAKASH, B. R. (1981). *Phys. Fluids* **24**, 554.

—— SATYAPRAKASH, B. R., and HUSSAIN, A. K. M. F. (1982). *J. fluid Mech.* **119**, 55.

BATCHELOR, G. K. (1946). *Proc. R. Soc. A* **186**, 480.

—— (1971). *The theory of homogeneous turbulence*, 2nd edn. Cambridge University Press, Cambridge.

CAMBON, C., JEANDEL, D., and MATHIEU, J. (1981). *J. fluid Mech.* **104**, 247.

CHAMPAGNE, F. H. (1978). J. fluid Mech. **86**, 67.

CHEN, W. Y. (1968). Spectral energy transfer and higher-order correlations in grid turbulence. *Ph.D. Thesis*, University of California, San Diego, CA.

CHORIN, A. J. (1988). *Commun. math. Phys.* **114**, 167.

COANTIC, M., and FAVRE, A. (1974). *Adv. Geophys.* **18A**, 391.

COMTE-BELLOT, G., and CORRSIN, S. (1971). *J. fluid Mech.* **48**, 273.

EDWARDS, S. F. (1964). *J. fluid Mech.* **18**, 239.

—— (1965). *Proc. Int. Conf. on Plasma Physics, Trieste*, p. 595. IAEA, Vienna.

—— and McCOMB, W. D. (1971). *Proc. Roy. Soc. A* **325**, 313.

—— and —— (1972). *Proc. Roy. Soc. A* **330**, 495.

FOIAS, C., MANLEY, O. P. and TEMAM, R. (1987). *Phys. Fluids* **30**, 2007.

FRENKIEL, N. F. and KLEBANOFF, P. S. (1971). *J. fluid Mech.* **48**, 183.

FRISCH, U., LESIEUR, M. and SCHERTZER, D. (1980). *J. fluid Mech.* **97**, 181.

GRANT, H. L., STEWART, R. W. and MOILLIET, A. (1962). *J. fluid Mech.* **12**, 241.

HANJALIC, K. and LAUNDER, B. E. (1972). *J. fluid Mech.* **52**, 609.

HERRING, J. R. (1974). *Phys. Fluids* **17**, 859.

—— and KRAICHNAN, R. H. (1972). Comparison of some approximations for isotropic turbulence. In *Statistical models and turbulence* (eds M. ROSENBLATT and C. VAN ATTA). *Lecture Notes in Physics*, vol. **12**, 148. Springer, Berlin.

—— and —— (1979). *J. fluid Mech.* **91**, 581.

HOSOKAWA, I. and YAMAMOTO, K. (1986). *Phys. Fluids* **29**, 2013.

KISTLER, A. L. and VREBALOVICH, T. (1966). *J. fluid Mech.* **26**, 37.

KOLMOGOROV, A. N. (1941). *C. R. Acad. Sci. URSS* **30**, 301; **32**, 16.

KRAICHNAN, R. H. (1959). *J. fluid Mech.* **5**, 497.

—— (1964a). *Phys. Fluids* **7**, 1030.

—— (1964b). *Phys. Fluids* **7**, 1048.

—— (1966). *Phys. Fluids* **9**, 1728.

—— (1972). *J. fluid Mech.* **56**, 287.

—— (1975). *Adv. Math.* **16**, 305.

LEITH, C. E. (1971). *J. atmos. Sci.* **28**, 145.

LESLIE, D. C. (1970). *J. Phys. A* **3**, L16.

—— (1973). *Developments in the theory of turbulence*. Clarendon Press, Oxford.

McComb, W. D. (1974). Proc. *R. Soc. Edin. A* **72**, 18.

—— (1976). *Int. J. eng. Sci.* **14**, 239.

—— and Shanmugasundaram, V. (1984). *J. fluid Mech.* **143**, 95.

—— and —— (1990). Submitted to *J. fluid Mech.*

——, —— and Hutchinson, P. (1989). *J. fluid Mech.* **208**, 91.

Martin, P. C. and De Dominicis, C. (1978). *Supp. Prog. theor. Phys.* **64**, 108.

Mjolsness, R. C. (1980). In *Turbulent shear flows 2*, (eds L. J. S. Bradbury, F. Durst, B. E. Launder, F. W. Schmidt, and J. H. Whitelaw). Springer, Berlin.

Morton, J. B. and Corrsin, S. (1970). *J. stat. Phys.* **2**, 153.

Nakano, T. (1986). *Prog. theor. Phys.* **75**, 1295.

Ogura, Y. (1963). *J. fluid Mech.* **16**, 33.

Orszag, S. A. and Patterson, G. S. (1972). *Phys. Rev. Lett.* **28**, 76.

Ortiz, M. J. and Ruiz De Elvira, A. (1984). *J. Phys. A* **17**, L555.

Pesme, D. and Dubois, D. (1982). In *Nonlinear problems: present and future* (eds A. R. Bishop, D. K. Campbell, and B. Nicolaenko) North-Holland, Amsterdam.

Pouquet, A., Frisch, U. and Leorat, J. (1976). *J. fluid Mech.* **77**, 321.

Proudman, I. (1962). *Mecanique de la turbulence, Colloq. Int. du CNRS à Marseille*, p. 107. CNRS, Paris.

Saffman, P. G. (1968). Lectures on homogeneous turbulence. In *Topics in nonlinear physics* (ed. N. J. Zabusky). Springer, New York.

Schumann, U. and Herring, J. R. (1976). *J. fluid Mech.* **76**, 755.

Stewart, R. W. and Townsend, A. A. (1951). *Phil. Trans. R. Soc. Lond. A* **243**, 359.

Uberoi, M. S. and Freymuth, P. (1969). *Phys. Fluids* **12**, 1359.

Van Atta, C. W. and Chen, W. Y. (1969). *J. fluid Mech.* **38**, 743.

Wyngaard, J. C. and Tennekes, H. (1970). *Phys. Fluids* **13**, 1962.

9

RENORMALIZATION GROUP THEORIES

In Section 3.5.3 we introduced the basic idea of renormalization group (RG) methods, and illustrated the general method by discussing how it might be applied in principle to fluid turbulence. The key phrase here is 'in principle', for in practice there are quite formidable problems in the way. Some of the attempts to overcome these problems will form the subject of this chapter.

However, before turning to this—the main topic—a few remarks about the background to the subject may not be out of place. To begin with, RG was a procedure for eliminating divergences in field theory, and later became famous when it was successfully applied to critical phenomena in the early 1970s (e.g. see Wilson and Kogut 1974; Wilson 1975). More recently there has been an interplay between field-theoretical and renormalization group methods (Amit (1984) emphasizes this aspect), and this tends to be reflected in papers describing attempts to apply RG to turbulence.

One unfortunate consequence of this is that papers which are of interest to us here are often couched in esoteric jargon, with apparently little attempt being made to communicate the basic ideas to turbulence researchers. Instead the authors often seem content to conduct a dialogue between aficionados, in which there is much allusive reference back and forward between various schools of field theory and statistical mechanics. In this activity, macroscopic Navier–Stokes turbulence appears only to be an excuse to relabel the variables in a quest for novelty, if not originality!

The overall effect to the outsider is off-putting, and even bizarre. This is a pity, for the basic ideas are really very simple and pragmatic.

In an attempt to counteract this impression, we shall preface our discussion of RG applied to turbulence, with an outline account of its implementation in the case of the Ising model of a ferromagnet, along with a very brief explanation of the meaning of a perturbation method based on the use of non-integral space dimension.

9.1 Background: RG applied to critical phenomena

The subject of critical phenomena deals with matter in the neighbourhood of a phase transition. Familiar examples are a liquid–gas system near the critical point (the upper limit on temperature and pressure at which a liquid and its vapour can coexist) or a ferromagnet at the Curie point (the highest temperature at which there can be a finite overall magnetism, assuming that there is no externally imposed magnetic field). We shall discuss the ferromagnet as an example of the application of RG.

9.1.1 Ferromagnetism and the Ising model

Magnetism arises because spins at lattice sites become aligned with each other. The tendency for spins to align is opposed by thermal effects, which of course tend to make the spin vectors take up all possible orientations at random. Thus alignments occur as random fluctuations, with length scales ranging from the lattice spacing (L_0, say) up to some correlation length ξ. The correlation length is a function of temperature and, as the temperature is reduced to Curie point, $\xi \to \infty$. Thus, at the critical point, fluctuations occur on all wavelengths from L_0 (about 1 $\mathring{A}$) up to infinity, and therefore a net overall magnetism can appear.

Fluctuations of this kind are characteristic of certain phase transitions, and are not just a feature of magnetic systems. For instance, there is the phenomenon of critical opalescence in liquids. This is due to density fluctuations scattering light, and indicates the presence of correlations on scales of the order of microns, as compared with the lower limit of the order of angstroms.

The theoretical objective is to calculate the Hamiltonian (and hence—via the partition function—the thermodynamic properties) of the system; see the relevant discussions in Sections 3.5.1, 4.1.1, and 5.3. But this simple aim is not easily achieved. The Hamiltonian H is dominated by a collective term, which is the sum (over configurations) of all spin interactions. This leads to formidable problems, and in practice the exact form has to be modelled, the classic approach being 'mean field theories', in which any one spin is supposed to experience a mean field due to the collective effect of all the others. In passing we should note that mean field theories do not agree with experiment, but that they are not wildly out either.

Another simplifying concept which has proved fruitful is the Ising model. This imposes a Boolean characteristic on the lattice, spins being either positive or negative with intermediate states not permitted. The concept is perhaps easiest to envisage in one dimension (i.e. space dimension $d = 1$), where we can imagine a long horizontal line of spins, each of which is either up or down. The model can be constructed for any arbitrary (integral) number of dimensions, and the case $d = 2$ was solved exactly by Onsager (1944). In view of our comments above, it is worth noting that the mean field theory for the two-dimensional Ising model differs quite appreciably from the exact solution.

9.1.2 Block spins and RG

The application of RG to magnetism is often interpreted as giving a quantitative meaning to the concept of block spins. This latter idea was due to

Kadanoff (1966) and was proposed in order to explain the observed self-similarity of certain thermodynamic relationships under scaling transformations. Essentially Kadanoff suggested that a group of aligned spins (the 'block') would behave like a single large spin if viewed from some larger scale still. Then a number of these 'blocks' could be amalgamated to form an even larger block, and so on, leading to a self-similar structure.

The corresponding RG method is to start with an interaction Hamiltonian H_0, which is associated with two spins separated by a distance L_0 (i.e. the lattice spacing). Then one calculates an effective Hamiltonian H_1, associated with regions of size $2L_0$ (the factor of 2 is arbitrary), which means averaging out the effects of scales L_0. Next one calculates H_2, associated with a region of size $4L_0$, with the effects of scales less than or equal to $2L_0$ averaged out. Thereafter, the general operation can be denoted by the calculation of the Hamiltonian H_n, with associated region of size $2^n L_0$, and the elimination of scales less than or equal to $2^{n-1} L_0$.

The above process can be expressed as a transformation T which is applied repeatedly

$$T(H_0) = H_1, \qquad T(H_1) = H_2, \qquad T(H_2) = H_3 \ldots . \tag{9.1}$$

At each stage, the length scales are changed

$$L_0 \to 2L_0, \qquad 2L_0 \to 4L_0 \ldots,$$

and, in order to compensate, the spin variables are also scaled in an appropriate fashion such that the Hamiltonian always looks the same in scaled coordinates. It is this rescaling which leads to renormalization and the transformations (9.1) define a simple group.

If iterating the transformation leads to the result

$$H_{n+1} = H_n, \tag{9.2}$$

where $H_{n+1} = T(H_n)$, then $H_n = H_N$, say, is a fixed point which corresponds to the critical point. Intuitively this can be understood in terms of the fact that the fluctuations of infinite wavelength (which occur at the critical point) will be invariant under scaling transformations.

If we think of the physical system being represented by a point in a multidimensional space, the coordinates of which are the interaction forces, then scaling moves the representative point. Thus the action of RG is to move the system along a trajectory, with the sequence of scaling operations playing the part of time. The resulting fixed point is determined by the solution of the equation

$$T(H_N) = H_H, \tag{9.3}$$

and is a property of the transformation T rather than the initial interaction H_0. This is associated with the idea of universality of critical behaviour. In the case of turbulence, the corresponding property would be that the renormalized effective viscosity v would not depend on the molecular viscosity v_0.

9.1.3 Space dimension and the epsilon expansion

If we repeat the above discussion in terms of wavenumber, the Fourier transform of the spin variables (strictly of the spin field) is introduced and this operation brings with it the space dimension d as an explicit part of the analysis.

We now talk about forming H_1 from H_0 by integrating out modes $k \geqslant 2\pi/L_0$. Thereafter, modes are eliminated in bands $\pi/L_0 \leqslant k \leqslant 2\pi/L_0$, $\pi/2L_0 \leqslant k \leqslant \pi/L_0 \ldots$, as we form H_2, H_3, $\ldots$. It is found that there two fixed points. For $d > 4$, the fixed point corresponds to classical mean field theory. For $d < 4$ there are non-trivial corrections to the classical results.

In the language of the subject, one refers to the upper critical dimension d_c (for technical reasons one should differentiate between this and the 'lower critical dimension', but this need not concern us here). Thus the Ising model of a ferromagnet has $d_c = 4$.

Put in very simple terms, what we are saying is that RG works rather well for the Ising model for a lattice with $d > d_c$, with the results being equivalent to those of mean field theory. However, what we would really like is a theory for the case of $d = 3$, but here it seems that correction terms arise which are difficult to evaluate.

The answer to this problem appears to be to perform a perturbation expansion in $\epsilon = d_c - d$. In the present case this means $\epsilon = 4 - d$, where the perturbation expansion is justified in terms of small ϵ. Then ϵ is set equal to unity and excellent agreement with experiment is obtained for $d = 3$.

The idea that dimension may be treated as a variable is a commonplace to the practitioners of field theory[1], where a theory constructed for one value of d can be extended to another by the process of analytical continuation. Also, those readers who are not familiar with field theory may well have encountered the notion that the dimension may be non-integral as part of the study of the phenomenology of fractal curves (although it should be emphasized that the fractal dimension applies to the curve, not the space). However, the concept of non-integral dimension in the epsilon expansion can be seen as purely pragmatic. One is faced with the situation that RG is exact in $d > 4$, and subject to only small logarithmic corrections in $d = 4$. The expansion about $d = 4$ turns out to be asymptotic, so that the first term is approximately equal to the sum of the series irrespective of the value of the expansion parameter ϵ.

Physically this amounts to an expansion about mean field theory in order to include the effect of fluctuations. In this case, one is relying on the equivalence of RG and mean field theory for $d > 4$. This is a property of a particular class of critical phenomena, of which the ferromagnetic transition on the Ising lattice is one. But the RG method is (as stated in Section 3.5.3) more general than this, and the epsilon expansion should really be seen as only one method of implementing RG for one particular class of problems.

9.2   Application of RG to turbulence

For our present purposes, we can classify attempts to apply RG to turbulence into three broad groups, according to the nature of the problem studied. These are as follows: (a) the laminar–turbulent transition; (b) calculation of scaling laws for asymptotic turbulent energy spectra, with specified noise inputs; (c) reduction of the number of degrees of freedom of well-developed turbulence, in conjuction with numerical simulation of the motion on a computer.

The first of these topics involves a transition from quasi-periodic behaviour to chaos, under the influence of external noise inputs. This is beyond our present scope, and so we shall only consider the other two topics here.

9.2.1   *Determination of scaling laws*

We begin by giving a slightly more formal definition of RG than that of Section 3.5.3. To do this, we shall adapt the form given by Ma and Mazenko (1975) for dynamical critical phenomena (i.e. where the conserved variables of a system are allowed to vary slowly with time), to the case of the Navier–Stokes equation for the velocity field $u_\alpha(\mathbf{k}, t)$. The system will be taken to be subject to random stirring forces $f_\alpha(\mathbf{k}, t)$ which, as before, have prescribed statistics and noise spectrum.

The RG procedure involves two stages:

1. The Fourier decomposition of the velocity is taken to be cut off for $k < \Lambda$ (often referred to as an ultraviolet cut off). Divide the velocity field up as follows:

$$u_\alpha(\mathbf{k}, t) = \begin{cases} u_\alpha^<(\mathbf{k}, t) & \text{for } 0 < k < b\Lambda \\ u_\alpha^>(\mathbf{k}, t) & \text{for } b\Lambda < k < \Lambda, \end{cases} \qquad (9.4)$$

where the scaling parameter b satisfies $0 < b < 1$. Now eliminate the high-$\mathbf{k}$ modes by solving the equation of motion for $u_\alpha^>(\mathbf{k}, t)$ and substituting the solution into the equation for $u_\alpha^<(\mathbf{k}, t)$. Note, because the non-linear mixing term induces a sum over modes, that the solution for $\mathbf{u}^>$ will contain $\mathbf{u}^<$. Average over $f^>(\mathbf{k}, t)$.
2. Rescale $\mathbf{k}, t, \mathbf{u}^<$, and $\mathbf{f}^<$ so that the new equation looks like the original Navier–Stokes equation. This last step involves the introduction of re-normalized transport coefficients.

The above scheme has been implemented by low-order perturbation theory. The pioneers were Forster, Nelson, and Stephen (1976, 1977; to be referred to as FNS), who chose the ultraviolet cut off Λ to be low enough to exclude cascade effects. With this limitation, they posed a rather artificial problem which is amenable to Wilson-type theory, complete with upper critical dimension and epsilon expansion.

The exclusion of the cascade means that the FNS theory does not strictly deal with turbulence at all. Nevertheless, their theory is an impressively rigorous approach to statistical hydrodynamics, with renormalization of the viscosity, the stirring forces, and the coupling constants. Under RG iteration, the Navier–Stokes equation is reduced to a (linear) Langevin equation, valid in the limit $k \to 0$.

FNS theory has been followed by other asymptotic theories for scaling behaviour in the infra-red. For example, Fournier and Frisch (1978), who investigated the relationship between FNS and closures of the eddy-damped quasi-normality type, and DeDominicis and Martin (1979), who used field-theoretical methods. Both these investigations concluded that the FNS theory might be valid away from the crossover dimension, and this is a point to which we shall return in Section 9.3, where we discuss FNS theory in detail.

Scaling behaviour has also been studied in the ultraviolet, where $k \to \infty$ (e.g. Grossman and Schnedler 1977; Levich 1980; Yakhot 1981; Levich and Tsinober 1984). A particular motivation was to overcome a specific problem encountered in the FNS approach, which is the restriction to low wavenumbers as the effective coupling constant increases with increasing wavenumber. In these studies, the usual RG procedure is reversed, and it is the low wavenumbers which are progressively eliminated. In all cases the theories yield corrections to the Kolmogorov spectrum. An interesting side issue is that Levich and Tsinober (1984) stesss that their non-integral dimension is the fractal dimension D (Mandelbrot 1977), and not the non-integer dimension used for analytical continuation in Wilson-type theory (i.e. the epsilon expansion).

It is difficult to know how to assess these theories. On the one hand, they seem to be very limited in their abilities. As Kraichnan (1982) has pointed out, they only lead to scaling exponents (at best) and do not fix the constants of proportionality. Moreover, they are only asymptotically valid for situations remote from the local-in-wavenumber energy tranfer which is such a characteristic feature of turbulence. Yet, on the other hand, they bring a welcome rigour into turbulence theory.

We shall return to these points in Section 9.3, where we examine the FNS theory in some detail.

9.2.2 Subgrid-scale modelling

It is well known that turbulent flows of any practical significance lie far beyond the scope of full numerical simulation. The number of degrees of freedom is simply too large for present-day computers. For this reason, much attention has recently been given to the idea of large-eddy simulation (LES). In general terms this technique presents two problems. First there is what one might call the 'software aspect', i.e. the problem of developing the numerical and computational methods needed to simulate the large eddies on a grid. This topic will not concern us here. Second, there is the problem of modelling the 'subgrid drain' or transfer of energy from the explicit scales to the unresolved subgrid scales.

This general topic is discussed in Sections 3.3.2 and 10.2. We shall confine ourselves here to noting that the very idea of LES sounds like a crude version of the RG. Therefore, it would seem that the subgrid modelling problem is an ideal candidate for an RG approach.

This was first recognized by Rose (1977), who applied RG methods to the subgrid modelling of passive scalar convection. He was able to obtain a renormalized (eddy) diffusivity, which had a weak dependence on the explicit wavenumbers and represented the mean effect of the velocity-field subgrid scales on the scalar-field explict scales. He also found two additional terms which (a) represented the noise injected from the subgrid scales and (b) represented the coupling between the large eddies and the eddies just below the resolution of the grid. We shall refer again to these results presently.

The more general approach of iterative averaging was later introduced by McComb (1982) as a way of reformulating the statistical equations for inhomogenous turbulence. Initially this work was motivated by a feeling that Reynolds averaging was too inflexible to allow consideration of intermittent effects (such as the bursting process) or even slow external time variations. Statistical equations (analogous to the Reynolds equation) were derived by progressively averaging the Navier–Stokes equation over a series of increasing time periods.

Averaging over the shortest period smooths out that part of the field which corresponds to the highest-frequency fluctuations. The mean effect of these fluctuations was calculated from the time-averaged equation of motion and so eliminated from the equation for the rest of the velocity field (that is, the unaveraged part). This procedure was repeated for a longer time period and in this way an iteration led to equations for the mean and covariance of the fluctuating field.

The properties of this general statistical reformulation have not been explored to any great extent. But a simpler situation can be obtained if we restrict our attention to fluid motions which are remote from solid boundaries, so that we can assume homogeneity and isotropy. Connection with RG can then be made (McComb 1982) by (1) Fourier transforming with into $(\mathbf{k}, \omega)$-space and (2) involing the Taylor hypothesis of frozen convection. Then, iterative averaging is found to lead to an effective viscosity which is a fixed point of the Navier–Stokes equations.

With these changes, iterative averaging becomes much closer to Rose's method. The two theories have been compared (McComb and Shanmugasundaram 1983) by applying Rose's method to the velocity field rather than the scalar field. It was found to differ from iterative averaging by the appearance of a triple moment involving only explicit scales. It was later argued (McComb 1985) that this term is an artefact of Rose's procedure, and that it is the spurious presence of this triple moment which leads the occurrence of terms (a) and (b), as mentioned in the discussion above.

The procedure of iterative averaging has been subject to various developments over the last few years. In particular, there was the change to conditional ensemble averaging in **k**-space (McComb and Shanmugasundaram 1984). In the most recent account (McComb 1986), an outline of a more general treatment has been given and the prospect of a subgrid model based on rational approximations held out. This will be the basis of our account of iterative averaging in Section 9.4.

9.3   The Forster–Nelson–Stephen (FNS) theory

In Section 5.5 we introduced the primitive perturbation expansion of the Navier–Stokes equation and showed how it could be renormalized by summing terms to all orders. Subsequently, in Chapters 6 and 7, we have seen that renormalized perturbation theories (which consist of truncations of the line-renormalized perturbation series at second order) give very good results when applied to the prediction of isotropic turbulence, despite the fact that the mathematical properties of the renormalized expansions are very largely unknown.

In this section, we shall consider a different method of renormalizing the perturbation series, where the reverse situation obtains. That is, the mathematical rigour of the methods is good but the resulting restrictions rule out even the simplest of turbulence calculations! The essential trick is to perform the perturbation theory in a narrow band of wavenumbers, which allows us to retain only low orders in the primitive expansion. Then the low-order expansion is renormalized by iteration to successively lower wavenumber bands.

Our first task is to generalize and adapt the definitions and results of Section 5.5 to suit our present purposes. It will also be noted that we conform to the modern practice (in this particular topic) of using the Fourier integral notation directly for the velocity field, without resorting to either limiting procedures or the Fourier–Stieltjes measure. This is the only section of the book where we do this, but there is no practical reason for not doing it everywhere else. It is probably very largely a matter of personal prejudice.

9.3.1   *Formulation of the problem*

We begin by imposing a maximum wavenumber Λ, where Λ (the ultraviolet cut-off) is very much smaller than the dissipation wavenumber. This means that the Fourier decomposition of the velocity field in the wavenumber–frequency domain, as given by eqn (5.63), should not be written as

$$u_\alpha(\mathbf{x}, t) = \left(\frac{1}{2\pi}\right)^{d+1} \int_{k \leqslant \Lambda} d^d k \int d\omega \, u_\alpha(\mathbf{k}, \omega) \exp(i\mathbf{k} \cdot \mathbf{x} + i\omega t), \qquad (9.5)$$

where d is the number of space dimensions. We shall only be interested in the case $d = 3$.

Similarly, the equation of motion, as given by eqn (5.64), should be modified to

$$(i\omega + v_0 k^2) u_\alpha(\mathbf{k}, \omega) = D_{\alpha\beta}(\mathbf{k}) f_\beta(\mathbf{k}, \omega) +$$

$$+ \lambda_0 M_{\alpha\beta\gamma}(\mathbf{k}) \int_{j \leqslant \Lambda} d^3 j \int d\Omega \, u_\beta(\mathbf{j}, \Omega) u_\gamma(\mathbf{k} - \mathbf{j}, \omega - \Omega),$$
$$(9.6)$$

where v_0 is the unrenormalized viscosity and $\lambda_0(=1)$ is the unrenormalized expansion parameter.

As usual, the stirring forces are specified by their autocorrelation. With the appropriate restrictions to homogeneity, isotropy, and stationarity, we modify eqn (5.67) to

$$\langle f_\alpha(\mathbf{k}, \omega) f_\beta(\mathbf{k}', \omega') \rangle = 2W(k)(2\pi)^{d+1} D_{\alpha\beta}(\mathbf{k}) \delta(\mathbf{k} + \mathbf{k}') \delta(\omega + \omega'), \qquad (9.7)$$

where $W(k)$ is a measure of the rate at which the stirring force does work on the fluid.

If we assume a power law form for $W(k)$,

$$W(k) = W_0 k^{-y}, \qquad (9.8)$$

then $y = -2$ gives us Model A of FNS, corresponding to thermal equilibrium, and $y = 0$ gives us their Model B, which corresponds to the case of macroscopic stirring of the fluid. It is the latter case which will interest us here, but we shall leave y unspecified for the present, as this will allow us to examine certain later conjectures about the FNS theory.

We complete our specification of the RG approach by dividing up the velocities and forces into low-frequency and high-frequency parts as follows:

$$u_\alpha(\mathbf{k}, \omega) = \begin{cases} u_\alpha^<(\mathbf{k}, \omega) & 0 < k < \Lambda \exp(-l) \\ u_\alpha^>(\mathbf{k}, \omega) & \Lambda \exp(-l) < k < \Lambda, \end{cases} \qquad (9.9)$$

$$f_\alpha(\mathbf{k}, \omega) = \begin{cases} f_\alpha^<(\mathbf{k}, \omega) & 0 < k < \Lambda \exp(-l) \\ f_\alpha^>(\mathbf{k}, \omega) & \Lambda \exp(-l) < k < \Lambda, \end{cases} \qquad (9.10)$$

where l is chosen such that $0 < \exp(-l) < 1$.

The corresponding decomposition of the Navier–Stokes equation can be obtained by substituting (9.9) and (9.10) into (9.6):

$$(i\omega + \nu_0 k^2)u_\alpha^<(\mathbf{k}, \omega)$$

$$= f_\alpha^<(\mathbf{k}, \omega) + \lambda_0 M_{\alpha\beta\gamma}^<(\mathbf{k}) \int_{j \leqslant \Lambda} d^3j \int d\Omega \{u_\beta^<(\mathbf{j}, \Omega)u_\gamma^<(\mathbf{k} - \mathbf{j}, \omega - \Omega) +$$

$$+ 2u_\beta^<(\mathbf{j}, \Omega)u_\gamma^>(\mathbf{k} - \mathbf{j}, \omega - \Omega) + u_\beta^>(\mathbf{j}, \Omega)u_\gamma^>(\mathbf{k} - \mathbf{j}, \omega - \Omega)\}$$

$$(9.11)$$

$$(i\omega + \nu_0 k^2)u_\alpha^>(\mathbf{k}, \omega)$$

$$= f_\alpha^>(\mathbf{k}, \omega) + \lambda_0 M_{\alpha\beta\gamma}^>(\mathbf{k}) \int_{j \leqslant \Lambda} d^3j \int d\Omega \{u_\beta^<(\mathbf{j}, \Omega)u_\gamma^<(\mathbf{k} - \mathbf{j}, \omega - \Omega) +$$

$$+ 2u_\beta^<(\mathbf{j}, \Omega)u_\gamma^>(\mathbf{k} - \mathbf{j}, \omega - \Omega) + u_\beta^>(\mathbf{j}, \Omega)u_\gamma^>(\mathbf{k} - \mathbf{j}, \omega - \Omega)\},$$

$$(9.12)$$

where the superscript on $M_{\alpha\beta\gamma}(\mathbf{k})$ has the obvious interpretation.

The aim now is to eliminate the $\mathbf{u}^>$ from (9.11) by solving equation (9.12) for $\mathbf{u}^>$ in terms of $\mathbf{u}^<$. It is fairly obvious that this can only be done approximately, and in the next section we consider the use of perturbation theory for this purpose.

9.3.2 The perturbation series

Formally, the perturbation theory is developed, as in Section 5.5, about eqn (5.68) as the zero-order solution. At this stage we shall temporarily adopt a compact notation in which wavevector, frequency, and tensor index are all combined into one symbol. With these changes, a subscript is now reserved to indicate the order of a term in either the perturbation expansion or the RG iteration. Thus eqn (5.68) can be translated into our new notation as

$$u_0^>(\hat{k}) = G_0(\hat{k})f^>(\hat{k}), \tag{9.13}$$

and the associated perturbation expansion can be written as

$$u^>(\hat{k}) = u_0^>(\hat{k}) + \lambda_0 u_1^>(\hat{k}) + \lambda_0^2 u_2^>(\hat{k}) \cdots + \lambda_0^n u_n(\hat{k}) \cdots. \tag{9.14}$$

Note that while either of these expressions can be applied to both the low-frequency and the high-frequency parts of the velocity field, we are only interested in a perturbation solution for $\mathbf{u}^>$ in order to eliminate the effect of the small scales from the equation for $\mathbf{u}^<$.

The coefficients $u_1^>, u_2^>, \ldots$, can all be expressed in terms of $u_0^>$ in the usual way. That is, we substitute (9.14) for $\mathbf{u}^>$ into both sides of (9.12) and equate coefficients at each order in the expansion parameter λ_0. The result, up to second order, is

$$n = 0: \quad u_0^>(\hat{k}) = G_0(\hat{k})f^>(\hat{k}) \tag{9.15a}$$

$$n = 1: \quad u_1^>(\hat{k}) = G_0(\hat{k})M^>(\hat{k}) \sum_j \{u^<(\hat{j})u^<(\hat{k} - \hat{j}) +$$

$$+ 2_u^<(\hat{j})u_0^>(\hat{k} - \hat{j}) + u_0^>(\hat{j})u_0^>(\hat{k} - \hat{j})\} \tag{9.15b}$$

$$n = 2: \qquad u_2^>(\hat{k}) = G_0(\hat{k})M^>(\hat{k})\sum_j \{2u^<(\hat{j})u_1^>(\hat{k} - \hat{j}) +$$

$$+ 2u_0^>(\hat{j})u_1^>(\hat{k} - \hat{j})\}. \qquad (9.15c)$$

In addition, we can substitute (9.15b) for $u_1^>$ into eqn (9.15c), so that $u_2^>$ can be expressed entirely in terms of $u_0^>$, but we shall not pursue that here.

Our next step is to replace $\mathbf{u}^>$, whenever it occurs on the r.h.s. of eqn (9.11) for $\mathbf{u}^<$, by inserting the perturbation series given by eqn (9.14):

$$u^<(\hat{k}) = G_0(\hat{k})f^<(\hat{k}) + \lambda_0 G_0(\hat{k})M^<(\hat{k})\sum_j u^<(\hat{j})u^<(\hat{k} - \hat{j}) +$$

$$+ \lambda_0 G_0(\hat{k})M^<(\hat{k})\sum_j \{2u^<(\hat{j})u^>(\hat{k} - \hat{j}) +$$

$$+ 2\lambda_0 u^<(\hat{j})u_1^>(\hat{k} - \hat{j}) + 2\lambda_0^2 u^<(\hat{j})u_2^>(\hat{k} - \hat{j})\} +$$

$$+ \lambda_0 G_0(\hat{k})M^<(\hat{k})\sum_j \{u_0^>(\hat{j})u_0^>(\hat{k} - \hat{j}) + 2\lambda_0 u_0^>(\hat{j})u_1^>(\hat{k} - \hat{j}) +$$

$$+ \lambda_0^2 u_1^>(\hat{j})u_1^>(\hat{k} - \hat{j}) + 2\lambda_0^2 u_2^>(\hat{j})u_0^>(\hat{k} - \hat{j})\} + O(\lambda_0^4). \qquad (9.16)$$

Then we average out the effect of the high frequencies according to the following rules.

(1) The low-frequency components are statistically independent of the high-frequency components and are invariant under the averaging process: $\langle f^<\rangle = f^<$ and $\langle u^<\rangle = u^<$.

(2) Averages involving $u_0^>$ can be evaluated using (9.13) and the statistics of the $f^>$, as G_0 is statistically sharp.

(3) The stirring forces are statistically homogeneous (see eqn (9.7)), and thus $M^<(\hat{k})\langle u_0^>(\hat{j})u_0^>(\hat{k} - \hat{j})\rangle = 0$, as $M^<(0) = 0$.

(4) The stirring forces have zero mean; hence $\langle u_0^>\rangle = 0$, as $\langle f^>\rangle = 0$.

(5) The probability distribution of the stirring forces is Gaussian. Therefore $\langle f^> f^> f^>\rangle = 0$, and hence it follows that $\langle u^> u^> u^>\rangle = 0$.

With all these points in mind, eqn (9.16) may be written as

$$u^<(\hat{k}) = G_0(\hat{k})f^<(\hat{k}) + \lambda_0 G_0(\hat{k})M^<(\hat{k})\sum_j u^<(\hat{j})u^<(\hat{k} - \hat{j}) +$$

$$+ 2\lambda_0^2 G_0(\hat{k})M^<(\hat{k})\sum_j \sum_p G_0(\hat{k} - \hat{j})M^>(\hat{k} - \hat{j})u^<(\hat{j})u^<(\hat{p})u^<(\hat{k} - \hat{j} - \hat{p}) +$$

$$+ \left\{8\lambda_0^2 G_0(\hat{k})M^<(\hat{k})\sum_j G_0(\hat{k} - \hat{j})M^>(\hat{k} - \hat{j})|G_0(\hat{j})|^2 \times \right.$$

$$\left. \times D(\hat{j})(2\pi)^{d+1}W(\hat{j})\right\}u^<(\hat{k}) + O(\lambda_0^3), \qquad (9.17)$$

where dummy variables $\hat{j}$ and $\hat{p}$ have been renamed as appropriate. Note also, that in evaluating the second moment we invoked eqn (9.7) and eliminated the sum over $\hat{p}$ along with the delta function $\delta(\hat{k} - \hat{p})$.

9.3.3 *The effective viscosity*

At this point, eqn (9.17) may seem to be very little improvement on (9.11). Nevertheless, only two steps are needed to show that it is. First, we multiply across by the factor $(i\omega + v_0 k^2)$, in order to eliminate $G_0(\hat{k})$ from the r.h.s. Then, with some rearrangement, (9.17) can be written in terms of an increased viscosity $v_0 + \Delta v_0(k)$:

$$\{i\omega + v_0 k^2 + \Delta v_0(k)k^2\}u_\alpha^<(\mathbf{k}.\,\omega)$$

$$= f_\alpha^<(\mathbf{k}, \omega) + \lambda_0 M_{\alpha\beta\gamma}^<(\mathbf{k}) \int_{j\leqslant\Lambda} d^3j \int d\Omega\, u_\beta^<(\mathbf{j}, \Omega)u_\gamma^<(\mathbf{k} - \mathbf{j}, \omega - \Omega) +$$

$$+ 2\lambda_0^2 M_{\alpha\beta\gamma}^<(\mathbf{k}) \int d^3j \int d^3\hat{p} M_{\beta\rho\sigma}^>(\hat{k} - \hat{j})G_0(\hat{k} - \hat{j}) \times$$

$$\times u_\gamma^<(\hat{k} - \hat{j})u_\rho^<(\hat{p})u_\sigma^<(\hat{k} - \hat{j} - \hat{p}). \tag{9.18}$$

Note that the full notation has been restored in all terms except the last one on the r.h.s. The reason for this will quickly become apparent. Note also that a specific form of $\Delta v_0(k)$—the increment to the viscosity—can be established from a comparison of eqns (9.18) and (9.17), and we shall return to this shortly. But, before doing that, we shall consider our second main simplification.

This is to the effect that it can be shown (Forster, Nelson, and Stephen) that the last term on the r.h.s. of (9.18) is an irrelevant variable. In other words, as the iteration proceeds to the fixed point, this term vanishes. The proof of this is subject to certain restrictions, and we shall consider this aspect in more detail later on. But, for the moment, we should note that the principal restriction amounts to the requirement that $k \to 0$ for us to be able to neglect the triple product of the $\mathbf{u}^<$. It is the existence of this restriction which means that we cannot regard the analysis which follows as a theory of turbulence. Strictly speaking, it is (as FNS claimed) a theory of the long-wavelength properties of a randomly stirred fluid.

However, if we drop the last term on the r.h.s. of (9.18), it follows that we have the basis of an iteration. That is, if we replace $\mathbf{u}^<$ by $\mathbf{u}$ and $\mathbf{f}^<$ by $\mathbf{f}$, the resulting equation will look like the original Navier–Stokes equation, but now defined on the interval $0 < k < \Lambda \exp(-1)$, and with an increased viscosity $v_1 = v_0 + \Delta v_0(k)$.

Formally, therefore, we now drop the term in $u^< u^< u^<$, and write eqn (9.18) as the Navier–Stokes equation in its intermediate form:

$$\{i\omega + v_0 k^2 + \Delta v_0(k)k^2\}u_\alpha^<(\mathbf{k}, \omega)$$

$$= f_\alpha^<(\mathbf{k}, \omega) + \lambda_0 M_{\alpha\beta\gamma}^<(\mathbf{k}) \int_{j \leqslant \Lambda} d^3 j \int d\Omega\, u_\beta^<(\mathbf{j}, \Omega)u_\gamma^<(\mathbf{k} - \mathbf{j}, \omega - \Omega), \quad (9.19)$$

where comparison of eqn (9.18) with eqn (9.17) allows us to deduce the equation for the effective viscosity as

$$\Delta v_0(k) = 8\lambda_0^2 k^{-2} M_{\rho\beta\gamma}^<(\mathbf{k}) \int_> d^3 j \int d\Omega\, G_0(|\mathbf{k} - \mathbf{j}|, \omega - \Omega) \times$$

$$\times |G_0(j, \Omega)|^2 M_{\gamma\sigma\rho}^<(\mathbf{k} - \mathbf{j})D_{\beta\sigma}(\mathbf{j})W(j), \quad (9.20)$$

where $\int_>$ denotes an integration over the band of wavenumbers being eliminated and we have made the replacement

$$M_{\alpha\beta\gamma}(\mathbf{k})u_\rho^<(\mathbf{k}, \omega) = M_{\rho\beta\gamma}(\mathbf{k})u_\alpha^<(\mathbf{k}, \omega).$$

This step just anticipates the result that the angular integration associated with the vector $\mathbf{j}$ leads to a Kronecker delta which picks out the α component of the velocity $u_\rho^<$. Note that, although we are only really interested in the case $d = 3$, we shall continue to keep the treatment general for the present, in order to see the significance of arbitrary space dimension d for the renormalized viscosity.

An explicit form can be obtained for the viscosity increment by substituting from (9.8) for the spectrum of the stirring forces $W(k)$ into (9.20). Then $\Delta v_0(k)$ can be evaluated as follows:

(a) Substitute for each $G_0(k, \omega)$ using (5.65) rewritten as

$$G_0(k, \omega) = \frac{1}{i\omega + v_0 k^2}$$

and perform the convolution integral over frequency.
(b) Make the change of variable in the wavenumber integral $\mathbf{j} \to \mathbf{j} + \mathbf{k}/2$, and take the limit $\omega \to 0$, $k \to 0$.
(c) Perform the integral over angles in wavenumber space, using the standard identities[2]

$$\int j_\alpha j_\beta\, d^d j = \frac{S_d}{d} \delta_{\alpha\beta} \int j^{d+1}\, dj \quad (9.21)$$

$$\int j_\alpha j_\beta j_\gamma j_\delta\, d^d j = \frac{S_d}{d(d+2)}(\delta_{\alpha\beta}\delta_{\gamma\delta} + \delta_{\alpha\gamma}\delta_{\beta\delta} + \delta_{\alpha\delta}\delta_{\gamma\beta}) \int j^{d+3}\, dj, \quad (9.22)$$

where S_d is the area of the unit sphere in d dimensions and is given by

$$S_d = \frac{2\pi^{d/2}}{\Gamma(d/2)}, \quad (9.23)$$

with Γ, as usual, being the gamma function.

The overall result of this procedure is readily found to be

$$\Delta v_0(0) = \frac{K(d)\lambda_0^2 W_0}{v_0^2 \Lambda^\epsilon} \frac{\exp\{\epsilon l\} - 1}{\epsilon},$$

(9.24)

where

$$\epsilon = 4 + y - d$$

(9.25)

and

$$K(d) = \frac{A(d)S_d}{(2\pi)^d}$$

(9.26)

$$A(d) = \frac{d^2 - d - \epsilon}{2d(d + 2)}.$$

(9.27)

Thus the total viscosity v_1, after the elimination of the modes in the band $\Lambda \exp(-1) < k < \Lambda$, is given by

$$v_1 = v_0 + \Delta v_0(0)$$

$$= v_0 \left\{ 1 + \frac{K(d)\lambda_0^2 W_0}{v_0^2 \Lambda^\epsilon} \frac{\exp\{\epsilon l\} - 1)}{\epsilon} \right\}$$

$$= v_0 \left\{ 1 + K(d)\bar{\lambda}_0^2 \frac{\exp\{\epsilon l\} - 1}{\epsilon} \right\},$$

(9.28)

where the modified strength parameter $\bar{\lambda}_0$ is given by

$$\bar{\lambda}_0 = \frac{\lambda_0 W_0^{1/2}}{v_0^{3/2} \Lambda^{\epsilon/2}}.$$

(9.29)

This last step highlights the fact that this is a different way of performing perturbation theory. In renormalized perturbation theory—as discussed in Chapter 5—λ merely plays the part of a bookkeeping parameter. The procedure there was to sum certain classes of terms to all orders and then put $\lambda = 1$ at the end of the calculation. Here, however, the strength parameter becomes renormalized as a consequence of the viscosity's being renormalized, and the process leading to this is the combination of iteration with rescaling.

Lastly, for completeness, we note that the propagator becomes modified to

$$G_1(k, \omega) = \frac{1}{i\omega + v_1 k^2}.$$

(9.30)

9.3.4   Recursion relations

Our aim now is to make (9.19) look as much as possible like the original Navier–Stokes equation. We do this by scaling both the independent and the dependent variables of the problem. For instance, in (9.19) the wavenumber

k is now defined on the interval $0 < k < \Lambda \exp(-l)$. If we divide k by the scaling factor $\exp(-1)$, then we obtain a new variable which is defined on the original interval. Thus we introduce

$$\tilde{k} = k \exp(l), \qquad (9.31)$$

where $\tilde{k}$ is defined on $0 < \tilde{k} < \Lambda$.

Also, we shall obviously want the coefficient of the term $i\omega u_\alpha^<(\mathbf{k}, \omega)$ to remain equal to unity. Accordingly, we introduce the more general scaling

$$\hat{\omega} = \omega \exp\{a(l)\}$$
$$\tilde{u}_\alpha(\tilde{k}, \tilde{\omega}) = u_\alpha^<(\mathbf{k}, \omega) \exp\{-c(l)\} \qquad (9.32)$$

where $a(l)$ and $c(l)$ are to be determined.

With these considerations in mind, eqn (9.19), with the replacement $v_1 = v_0 + \Delta v_0$, can now be written in terms of the new variables. Substituting from (9.31) and (9.32), we have

$$\{i\tilde{\omega} + v(l)\tilde{k}^2\}\tilde{u}_\alpha(\tilde{k}, \tilde{\omega}) = \tilde{f}_\alpha(\tilde{k}, \tilde{\omega}) + \lambda(l)M_{\alpha\beta\gamma}(\tilde{k}) \times$$

$$\times \int_{j \leqslant \Lambda} d^3j \int d\Omega \, \tilde{u}_\beta(\tilde{j}, \tilde{\Omega})\tilde{u}_\gamma(\tilde{k} - \tilde{j}, \tilde{\omega} - \tilde{\Omega}). \qquad (9.33)$$

The scaled stirring force, viscosity, and strength parameter must then satisfy, respectively, the following relationships:

$$\tilde{f}_\alpha(\tilde{k}, \tilde{\omega}) = f_\alpha^<(\mathbf{k}, \omega) \exp(a - c) \qquad (9.34)$$

$$v(l) = v_1 \exp(a - 2l) \qquad (9.35)$$

$$\lambda(l) = \lambda_0 \exp\{c - (d + 1)l\}. \qquad (9.36)$$

Note that we temporarily put the general dimensionality d (instead of the specific value $d = 3$) in eqn (9.36). Also note that eqns (9.34)–(9.36) stem purely from the homogeneity requirement that the factor $\exp(c - a)$, which arises from the scaling of the term $i\omega u_\alpha^<(\mathbf{k}, \omega)$ in eqn (9.19), should appear in each of the other terms of (9.33), and hence can be eliminated by cancellation.

Although the stirring forces are rescaled, the rate at which they do work on the system must be unaffected by this procedure. This requirement gives us a constraint on the scaling factors, which can be used to relate $c(l)$ and $a(l)$. We do this in the following way. From eqns (9.7) and (9.8) we have the correlation of the stirring forces as

$$\langle f_\alpha(\mathbf{k}, \omega)f_\beta(\mathbf{k}', \omega')\rangle = 2(2\pi)^{d+1} W_0 D_{\alpha\beta}(\mathbf{k})k^{-y} \times$$

$$\times \delta(\mathbf{k} + \mathbf{k}')\delta(\omega + \omega'). \qquad (9.37)$$

Now apply the transformation (9.34) to the l.h.s., and the transformations (9.31) and (9.32) to the r.h.s., with the result

$$\langle \tilde{f}_\alpha(\tilde{\mathbf{k}}, \tilde{\omega}) \tilde{f}_\beta(\tilde{\mathbf{k}}', \tilde{\omega}') \rangle = 2(2\pi)^{d+1} W_0 D_{\alpha\beta}(\tilde{\mathbf{k}}) \tilde{k}^{-y} \times$$
$$\times \delta(\tilde{\mathbf{k}} + \tilde{k}') \delta(\tilde{\omega} + \tilde{\omega}'), \tag{9.38}$$

provided that the homogeneity requirement

$$2c = 3a + (y + d)l \tag{9.39}$$

is satisfied.

Forster *et al.* (1977) carried out their iteration using infinitesimal wave-number bands at each step. This means that the recursion relations can be turned into differential equations with l as a continuous independent variable. The general procedure is discussed by Reichl (1980), and here we shall only quote the final results. From eqns (9.28), (9.29), (9.35), and (9.36) we obtain

$$\frac{dv}{dl} = v(l)\{z - 2 + K(d)\bar{\lambda}^2\} \tag{9.40}$$

$$\frac{dW_0}{dl} = 0 \tag{9.41}$$

$$\frac{d\lambda}{dl} = \lambda(l)\left(\frac{3z}{2} - 1 - \frac{d - y}{2}\right), \tag{9.42}$$

where z is defined by

$$z = \frac{da}{dl}. \tag{9.43}$$

Also, by an obvious analogy with eqn (9.29), the modified strength parameter $\bar{\lambda}$ is defined by

$$\bar{\lambda}^2 = \frac{\lambda^2 W_0}{v^3 \Lambda^\epsilon}. \tag{9.44}$$

The recursion relation for $\bar{\lambda}$ can be obtained from eqns (9.40)–(9.42), and takes the form

$$\frac{d\bar{\lambda}}{dl} = \frac{\bar{\lambda}}{2}\{\epsilon - 3K(d)\bar{\lambda}^2\}. \tag{9.45}$$

9.3.5 *Behaviour near the fixed point*

In earlier chapters, we have often had to keep in mind the fact that turbulence is a problem of strong interactions. From this point of view, eqn (9.45) is perhaps the single most interesting feature of the preceding analysis. Here one is actually controlling the effective interaction strength (insofar as it affects perturbation theory) by deciding systematically just how the coupling parameter $\bar{\lambda}(l)$ will be renormalized. Of course, as we remarked previously, the resulting theory has only limited validity in the limit $k \to 0$ and hence does not qualify as a theory of turbulence. Nevertheless, the technique is of considerable academic interest, and is worth pursuing here.

Clearly the nature of solutions to eqn (9.45) depends on the value of ϵ, with a crossover from one form of behaviour to another at $\epsilon = 0$. Thus we can distinguish three cases as follows:

$\epsilon < 0$: $\bar{\lambda}(l)$ tends exponentially to zero as $l \to \infty$.

$\epsilon = 0$: $\bar{\lambda}(l)$ tends to zero as $1/l$, and there are logarithmic corrections to $v(l)$ (see FNS, eqn (3.78)).

$\epsilon > 0$: $\bar{\lambda}$ tends to the fixed point $\bar{\lambda}^*$.

The latter case is the interesting one, and the solution for $\bar{\lambda}$ at the fixed point is given by

$$\bar{\lambda}^* = \left\{ \frac{\epsilon}{3A(d)} \right\}^{1/2} \tag{9.46}$$

as $l \to \infty$. It is readily seen from the substitution of this result into (9.40) that the renormalized viscosity becomes independent of l at the fixed point, provided only that

$$z = 2 - \epsilon/3, \tag{9.47}$$

which fixes our one remaining free parameter $a(l)$ through eqn (9.43).

The above dependence on a crossover value of ϵ recalls the role of the 'upper critical dimension' in the theory of phase transitions. However, in the present case the arbitrary choice of the correlation of stirring forces plays a major part. Suppose, for instance, that we concentrate on the case of practical interest and put $d = 3$. Then the above criterion for the existence of a non-trivial fixed point becomes

$$y > -1, \tag{9.48}$$

which is a lower bound on the arbitrarily chosen exponent in eqn (9.8).

We can derive an upper bound on y by considering the condition for the neglected triple product $u^< u^< u^<$ to be an irrelevant variable. This term was simply dropped in going from (9.17) to (9.18). However, suppose that we associated with it a coupling parameter g_0 and retained it in our derivation of the rescaled Navier–Stokes equation, then it would appear in (9.33) with the modified coupling parameter

$$g(l) = g_0 \exp\{-(d - y)l\}, \tag{9.49}$$

a result which follows from the scaling arguments in the same way as did eqns (9.34)–(9.36).

It is clear, therefore, that contributions from the triple product vanish exponentially as $l \to \infty$ provided that

$$y < d. \tag{9.50}$$

Hence the power-law exponent in (9.8) is bounded by

$$-1 < y < 3 \tag{9.51}$$

in space dimension $d = 3$.

Lastly, the scaling behaviour of the spectrum can be deduced from the fact that, for $k < \Lambda \exp(-l)$, the correlation function of the fluctuating velocities can be computed from both the original and the reduced set of equations. FNS (see their eqn 3.9)) show that this leads to a homogeneity relationship, which in turn yields the spectrum

$$E(k) \simeq k^{-5/3 + 2(d-y)/3}, \tag{9.52}$$

as rewritten in terms of our present notation.

If we put $y = 0$, then we recover their model B, with

$$E(k) \simeq k^{1/3} \tag{9.53}$$

in space dimension $d = 3$.

9.3.6 Some later conjectures about FNS theory

The FNS theory was later generalized by DeDominicis and Martin (1979, hereafter referred to as DDM), who applied field-theoretical methods to the problem. Their overall conclusion was that the FNS results were valid to all orders in the anomalous dimension (ϵ), although this conclusion depended on an assumption (said to be 'plausible') that neglected operators remained irrelevant. Even so, for this to be true, one must have $\epsilon < 4$, corresponding to the upper bound given in eqn (9.51) for the particular case of $d = 3$.

An interesting feature of this work was the remark that a choice of forcing exponent $y = d$ causes the energy spectrum to take the Kolmogorov form. This is easily seen if we substitute $y = d$ in the r.h.s. of equation (9.52).

It is also easily seen, from eqn (9.49), that this choice of y corresponds to the boundary separating regions where variables treated as irrelevant decay from regions where such variables grow as $l \to \infty$. This particular point was also made by DDM.

In assessing the practical significance of this result, we must first consider the applicability of the FNS theory to macroscopic hydrodynamic turbulence. After doing that, we shall then make some pertinent remarks about DDM's proposal.

We have already seen that the FNS procedure is only valid in the limit $k \to 0$. The converse statement is that the renormalized expansion parameter increases as k increases, thus invalidating low-order perturbation theory. It was for this reason that FNS only claimed to have studied the small-k properties of the correlations of a randomly stirred fluid.

However, even this modest claim may not be justified. FNS truncate the Navier–Stokes equation on the interval $0 < k < \Lambda$, where Λ is chosen to be much smaller than any dissipation wavenumber. Yet, if a real fluid is stirred in this wavenumber band, it is an inevitable consequence of non-linearity that energy must be transferred to wavenumbers greater than Λ. The resulting dissipation wavenumber—as defined by eqn (2.133)—is then determined only by that rate of energy transfer and the fluid viscosity. It does not automatically follow, therefore, that the condition $\Lambda \ll k_d$ will be satisfied in practice. And, moreover, given the non-local (in wavenumber) structure of the inertial terms, it would be unsafe to assume that the shape of the spectrum at small k would be unaffected by energy transfers to $k > \Lambda$.

These comments do not affect the FNS theory as such, for it can be assessed purely as a model on its own terms. However, the significance of the Kolmogorov distribution is that it appears to be a property of real turbulence and accordingly we must at least consider the applicability of the background theory as a preliminary to assessing the proposal made by DDM.

Now let us turn to the idea that a force correlation of k^{-3} (in $d = 3$) leads to an energy spectrum $E(k) \simeq k^{-5/3}$. Our first point is the general observation that the Kolmogorov theory is based on the assumption of universal behaviour, in which the cascade at high enough k becomes independent of conditions at small k, including stirring forces! Thus $W(k) \simeq k^{-3}$ seems less meritorious than the assumption of $W(k) \simeq \delta(k)$ in the Edwards 'thought experiment' (see Section 6.2.7), which has the same dimensions but produces the Kolmogorov spectrum without forcing the result at high wavenumbers.

Our second point is a corollary of our first. The choice of forcing spectrum $W(k) = W_0 k^{-3}$ does not lead to a well-posed problem. For a steady state, we must have the dissipation rate equal to the rate of doing work on the system. Or,

$$
\varepsilon = 4\pi \int_{k_{\min}}^{k_{\max}} W_0 k^{-3} k^2 \, \mathrm{d}k
$$

$$
= 4\pi W_0 \ln k |_{k_{\min}}^{k_{\max}}, \tag{9.54}
$$

which is log divergent as $k_{\max} \to \infty$. Thus the DDM 5/3 spectrum cannot be extended beyond the ultraviolet cut-off Λ, which in turn must satisfy the condition $\Lambda \ll k_d$.

Overall, the DDM proposal is not without its intrinsic interest, but in the present state of knowledge, the above considerations would seem to rule out any practical application of the idea (although this has not prevented some extravagant claims being made, in just this connection[3]).

9.4 Application of RG by iterative averaging

In this section we are interested in the potential of RG as a method of assisting the numerical computation of turbulence by reducing the number of degrees of freedom which have to be computed. As usual, we interpret the degrees of freedom as being the independently excited Fourier modes. Then we envisage a situation where modes $\{\mathbf{k}\}$ such that $k < k_c$, are to be simulated numerically, while the non-linear transfer of energy to modes $k > k_c$ is to be represented analytically by an effective viscosity $v(k|k_c)$. In the present approach, $v(k|k_c)$ is to be calculated from the Navier–Stokes equations by progressively eliminating the mean effects of modes at high wavenumbers. It should be noted that, from our point of view, the cut-off wavenumber k_c is to be chosen arbitrarily, but in an actual calculation it is fairly obvious that any such choice would be influenced by quite practical considerations to do with the size and speed of the computer available.

9.4.1 *General formulation*

Consider the velocity field to be represented by the Fourier series

$$u_\alpha(\mathbf{x}, t) = \sum_{|k| \leqslant k_0} u_\alpha(\mathbf{k}, t) \exp\{i\mathbf{k} \cdot \mathbf{x}\} \tag{9.55}$$

where k_0 is, in some sense, the largest wavenumber present. Evidently this just involves a straightforward modification to our original Fourier series representation, as given by eqn (2.71), and the correspondingly trivial modification to the Navier–Stokes equation in the form eqn (2.76) is

$$\left(\frac{\partial}{\partial t} + v_0 k^2\right) u_\alpha(\mathbf{k}, t) = M_{\alpha\beta\gamma}(\mathbf{k}) \sum_{j \leqslant k_0} u_\beta(\mathbf{j}, t) u_\gamma(\mathbf{k} - \mathbf{j}, t), \tag{9.56}$$

and clearly the inertial transfer and projection operators remain the same as defined by eqns (2.77) and (2.78).

As any real flow system is of finite physical extent, we shall regard the Fourier components of the velocity field as a denumerable finite set. Then a general statement of the problem is to the effect that we wish to solve eqn (9.56) for each member of this set, subject to the boundary conditions

$$u_\alpha(\mathbf{k}, t) = 0, \quad \text{for } \mathbf{k} = 0$$

$$u_\alpha(\mathbf{k}, t) \text{ to be prescribed for } t = 0. \tag{9.57}$$

Of course, when we take averages, the resulting functions will be assumed to be smoothly varying functions of the continuous variable k, and all the results of Section 2.6 will be taken to apply just as if we had rigorously taken the limit $L \to \infty$.

The maximum wavenumber k_0 is defined approximately through the dissipation integral:

$$\varepsilon = \int_0^\infty 2v_0 k^2 E(k) \, dk \simeq \int_0^{k_0} 2v_0 k^2 E(k) \, dk. \tag{9.58}$$

That is, in general terms k_0 is such that, say, 99.9 per cent of dissipation takes place at wavenumbers $k < k_0$. Clearly k_0 should be of the same order of magnitude as the Kolmogorov dissipation wavenumber k_d.

At this stage it may be helpful if we summarize the general approach by adapting our earlier general outline of the RG procedure (see Section 9.2.1) to the present case.

We choose a cut-off wavenumber k_1 such that $k_1 < k_0$, but where $k_1 \simeq k_0 \simeq k_d$. The two stages of the RG procedure then become the following.

> (a) Solve eqn (9.56) on $k_1 < k < k_0$. Substitute the solution for the mean effect of the high-k modes into eqn (9.56) now on the interval $0 < k < k_1$. This results in an increment to the viscosity: $v_0 \to v_1 = v_0 + \Delta v_0$.
> (b) Rescale the basic variables so that the Navier–Stokes equation on $0 < k < k_1$ looks like the original Navier–Stokes equation on the original interval $0 < k < k_0$.

This procedure is then repeated for $k_2 < k < k_1$, and so on.

The aim is to find a fixed point at which the molecular viscosity can be replaced by a renormalized viscosity of universal form. The physics underlying this process can be understood in the following way. In the viscous range of wavenumbers, the dominant physical process is (by definition) the viscous dissipation of turbulent kinetic energy. Indeed, it is reasonable to think of modes in this range as being critically damped. That is, any mode in the band $k_1 < k < k_0$ is driven by energy transfer from modes $k < k_1$, and this energy is dissipated locally by being turned into heat.

In principle, therefore, one can expect to solve the Navier–Stokes equation for $u_\alpha(\mathbf{k}, t)$ in the band $k_1 < k < k_0$ in terms of the bilinear energy transfer from $k < k_1$ into the band, while the quadratic nonlinearity—which involves transfers through the band—is neglected (although, as we shall see later, this term can be treated systematically to all orders).

As a result of all this, the new Navier–Stokes equation has an increased effective viscosity v_1 in the range of wavenumbers $0 < k < k_1$. Let us therefore define an effective dissipation wavenumber for the new Navier–Stokes equation:

$$k_d^{(1)} = \left(\frac{\varepsilon}{v_1^3} \right)^{1/4}. \tag{9.59}$$

Then, for $v_1 > v_0$, it follows that

$$k_d^{(1)} < k_d^{(0)} = \left(\frac{\varepsilon}{v_0^3} \right)^{1/4}. \tag{9.60}$$

Thus the above arguments may all go through again for some $k_2 \simeq k_d^{(1)}$, and (if so) one can again solve the (suitably scaled) Navier–Stokes equation in the band $k_2 < k < k_1$.

As this procedure is carried on, the existence of a fixed point can be associated with the onset of scaling behaviour in the inertial range of wavenumbers. This will be seen later when we consider the full mathematical treatment.

9.4.2 Partial averaging of the small scales

Our first step is to decompose the velocity field into two sets of Fourier components, i.e. those which are in the band $k_1 < k < k_0$, and those which are not. Or

$$u_\alpha(\mathbf{k}, t) = \begin{cases} u_\alpha^-(\mathbf{k}, t) & \text{for } 0 < k < k_1, \\ u_\alpha^+(\mathbf{k}, t) & \text{for } k_1 < k < k_0. \end{cases} \tag{9.61}$$

Using this decomposition, we can introduce the operation of partial averaging over fluctuations in the band of wavenumbers $k_1 < k < k_0$. We denote this operation by $\langle \ \rangle_0$, where the subscript serves to distinguish it from the general (global) average, denoted by $\langle \ \rangle$. We define it through its properties, which we list as follows:

$$\langle u_\alpha(\mathbf{k}, t) \rangle_0 = \begin{cases} u_\alpha^-(\mathbf{k}, t) & \text{for } 0 < k < k_1, \\ 0 & \text{for } k_1 < k < k_0, \end{cases}$$

$$\langle u_\alpha^-(\mathbf{k}, t) \rangle_0 = u_\alpha^-(\mathbf{k}, t)$$

$$\langle u_\alpha^+(\mathbf{k}, t) \rangle_0 = 0 \tag{9.62}$$

and

$$M_{\alpha\beta\gamma}(\mathbf{k})\langle u_\beta^+(\mathbf{j}, t)u_\gamma^-(\mathbf{k} - \mathbf{j}, t)\rangle_0 = 0. \tag{9.63}$$

The first set of properties, as listed in eqn (9.62), would be satisfied by a simple filter. That is, the operation of partial averaging need only consist of multiplying the Fourier components of the velocity field by a unit step function which is zero for $k > k_1$.

However, the property stated in eqn (9.63) would only be true to some level of approximation for a filtering operation, and clearly requires some method of taking the phase into account. Therefore we must think of partial averaging as either a combined filter and ensemble average, or as a conditional average in which $\mathbf{u}^-$ is held constant while $\mathbf{u}^+$ is averaged.

This procedure can be seen as a generalization of Reynolds averaging, as discussed in Section 1.3. If we temporarily return to x-space, then the decomposition given in eqn (9.61) becomes

$$u_\alpha(\mathbf{x}, t) = u_\alpha^-(\mathbf{x}, t) + u_\alpha^+(\mathbf{x}, t), \tag{9.64}$$

where the low-frequency and high-frequency parts of the velocity field are defined by

$$u_\alpha^-(\mathbf{x}, t) = \sum_k u_\alpha^-(\mathbf{k}, t) \exp(i\mathbf{k} \cdot \mathbf{x})$$

$$= \sum_{|k| \geqslant k_1} u_\alpha(\mathbf{k}, t) \exp(i\mathbf{k} \cdot \mathbf{x}), \qquad (9.65)$$

$$u_\alpha^+(\mathbf{x}, t) = \sum_k u_\alpha^+(\mathbf{k}, t) \exp(i\mathbf{k} \cdot \mathbf{x})$$

$$= \sum_{k_1 \leqslant k \leqslant k_0} u_\alpha(\mathbf{k}, t) \exp(i\mathbf{k} \cdot \mathbf{x}), \qquad (9.66)$$

Clearly, if we let $k_1 \to 0$, then $u_\alpha^-(\mathbf{x}, t)$ tends to the mean value $\bar{u}_\alpha(x, t) = \langle u_\alpha(\mathbf{x}, t) \rangle$, and $u_\alpha^+(\mathbf{x}, t)$ tends to the fluctuating velocity $u_\alpha(\mathbf{x}, t) - \langle u_\alpha(\mathbf{x}, t) \rangle_0$.

An important point to note is that $u_\alpha^-(\mathbf{x}, t)$ and $u_\alpha^+(\mathbf{x}, t)$ are uncorrelated parts of the velocity field (see Monin and Yaglom 1975, pp. 18–19). In this connection, we should also note that the Navier–Stokes equation induces coupling—but not correlation—between different Fourier modes.

It follows therefore that, under global averaging, eqn (9.63) would just be a particular example of the general result

$$M_{\alpha\beta\gamma}(\mathbf{k}) \langle u_\beta(\mathbf{j}, t) u_\gamma(\mathbf{k} - \mathbf{j}, t) \rangle = M_{\alpha\beta\gamma}(\mathbf{k}) \delta(\mathbf{k}) Q_{\beta\gamma}(\mathbf{k} - \mathbf{j}; t, t)$$

$$= 0, \qquad (9.67)$$

where we have invoked eqn (2.93) along with the linear dependence of $M_{\alpha\beta\gamma}(\mathbf{k})$ on $\mathbf{k}$. Nevertheless, we should emphasize that eqn (9.63) applies only for partial averaging and indeed should be regarded as part of the definition of that process.

We can complete this part of the work by pursuing the analogy with Reynolds averaging and deriving equations for $\mathbf{u}^-$ and $\mathbf{u}^+$ in much the same way as we previously obtained eqns (1.12) and (1.14) for the mean and fluctuating velocities. Thus, substituting from (9.61) into (9.56) and partially averaging both sides according to (9.62), we obtain an equation for $u_\alpha^-(\mathbf{k}, t)$ in the form

$$\left(\frac{\partial}{\partial t} + \nu_0 k^2 \right) u_\alpha^-(\mathbf{k}, t) = M_{\alpha\beta\gamma}(\mathbf{k}) \sum_{j \leqslant k_0} \langle u_\beta(\mathbf{j}, t) u_\gamma(\mathbf{k} - \mathbf{j}, t) \rangle_0. \qquad (9.68)$$

Note from eqn (9.67) that the r.h.s. of (9.68) would vanish if this were a global averaging. Also note that each term in (9.68) fluctuates on the interval $0 < k < k_1$.

The equation for fluctuations on the interval $k_1 < k < k_0$ is obtained by subtracting (9.68) from (9.56):

$$\left(\frac{\partial}{\partial t} + \nu_0 k^2 \right) u_\alpha^+(\mathbf{k}, t) = M_{\alpha\beta\gamma}(\mathbf{k}) \sum_{j \leqslant k_0} \{ u_\beta(\mathbf{j}, t) u_\gamma(\mathbf{k} - \mathbf{j}, t) -$$

$$- \langle u_\beta(\mathbf{j}, t) u_\gamma(\mathbf{k} - \mathbf{j}, t) \rangle_0 \}. \qquad (9.69)$$

These two equations form the basis of a modified version of the usual turbulent moment hierarchy, in which we only need consider moments of velocities fluctuating in the band $k_1 < k < k_0$.

9.4.3   *The statistical equations of motion*

We now wish to derive explicit forms for the right-hand sides of eqns (9.68) and (9.69) in terms of the u^- and u^+ variables. We begin with (9.68) for the explicit scales. Substituting (9.61) for each of the velocity vectors in the non-linear term and partially averaging according to eqns (9.62) and (9.63), we obtain

$$\left(\frac{\partial}{\partial t} + v_0 k^2\right) u_\alpha^-(\mathbf{k}, t) - M_{\alpha\beta\gamma}(\mathbf{k}) \sum_{\mathbf{j}} \langle u_\beta^+(\mathbf{j}, t) u_\gamma^+(\mathbf{k} - \mathbf{j}, t) \rangle_0$$

$$= M_{\alpha\beta\gamma}(\mathbf{k}) \sum_{\mathbf{j}} u_\beta^-(\mathbf{j}, t) u_\gamma^-(\mathbf{k} - \mathbf{j}, t). \tag{9.70}$$

We also need to obtain a governing equation for $\langle u_\beta^+(\mathbf{j}, t) u_\gamma^+(\mathbf{k} - \mathbf{j}, t) \rangle_0$, which represents the explicit coupling in (9.70) to the wavenumber band $k_1 < k < k_0$. In order to do this, we first derive an explicit equation for $u_\alpha^+(\mathbf{k}, t)$ by substituting (9.61) for velocities on the r.h.s., and partially averaging according to (9.62) and (9.63). Thus

$$\left(\frac{\partial}{\partial t} + v_0 k^2\right) u_\alpha^+(\mathbf{k}, t) = M_{\alpha\beta\gamma}(\mathbf{k}) \sum_{\mathbf{j}} 2u_\beta^-(\mathbf{j}, t) u_\gamma^+(\mathbf{k} - \mathbf{j}, t) +$$

$$+ M_{\alpha\beta\gamma}(\mathbf{k}) \sum_{\mathbf{j}} \{u_\beta^+(\mathbf{j}, t) u_\gamma^+(\mathbf{k} - \mathbf{j}, t) -$$

$$- \langle u_\beta^+(\mathbf{j}, t) u_\gamma^+(\mathbf{k} - \mathbf{j}, t) \rangle_0\}. \tag{9.71}$$

It should be noted that the term $\langle u_\beta^+(\mathbf{j}, t) u_\gamma^+(\mathbf{k} - \mathbf{j}, t) \rangle_0$ gives zero if we multiply through by $u_\sigma^+(\mathbf{k}', t)$ and carry out the partial average. Then, with this in mind, we can form an equation for $\langle u_\beta^+(\mathbf{j}, t) u_\gamma^+(\mathbf{k} - \mathbf{j}, t) \rangle_0$ by taking the following steps.

(a) Generalize (9.71) to form an equation for $u_\beta^+(\mathbf{j}, t)$. This involves re-labelling $\mathbf{k} \to \mathbf{j}$, $\alpha \to \beta$, the dummy wavevector $\mathbf{j}$ is renamed $\mathbf{p}$, and so on. Then multiply through by $u_\gamma^+(\mathbf{k} - \mathbf{j}, t)$ and average $\langle \ \rangle_0$.
(b) Generalize (9.71) to form an equation for $u_\gamma^+(\mathbf{k} - \mathbf{j}, t)$, multiply through by $u_\beta^+(\mathbf{j}, t)$, and average $\langle \ \rangle_0$.
(c) Add the two resulting equations together and rename dummy variables as appropriate to obtain

$$\left(\frac{\partial}{\partial t} + v_0 j^2 + v_0 |\mathbf{k} - \mathbf{j}|^2\right)\langle u_\beta^+(\mathbf{j}, t)u_\gamma^+(\mathbf{k} - \mathbf{j}, t)\rangle_0$$

$$= 2M_{\beta\sigma\rho}(\mathbf{j})\sum_{\mathbf{p}}\{2\langle u_\sigma^-(\mathbf{p}, t)u_\rho^+(\mathbf{j} - \mathbf{p}, t)u_\gamma^+(\mathbf{k} - \mathbf{j}, t)\rangle_0 +$$

$$+ \langle u_\sigma^+(\mathbf{p}, t)u_\rho^+(\mathbf{j} - \mathbf{p}, t)u_\gamma^+(\mathbf{k} - \mathbf{j}, t)\rangle_0\}. \qquad (9.72)$$

Formally, this equation can be solved in terms of the zero-order (or viscous) Green function. But this leads us into a new version of the turbulent moment hierarchy and hence a new form of the closure problem, which we shall discuss in the following section.

9.4.4   Moment hierarchy from partial averaging

The hierarchy which results from partial averaging is more complicated than that associated with global averaging, if only because there are now two different types of moment. This means that the algebra can become rather complicated. However, it turns out that the only information needed explicitly is the following:

(1) the number of $\mathbf{u}$ factors in a product;
(2) whether any given $\mathbf{u}$ is $\mathbf{u}^-$ or $\mathbf{u}^+$;
(3) the nature of the average: $\langle\;\rangle_0$ or $\langle\;\rangle$.

Accordingly, we can temporarily drop all wavevector and coordinate labels, provided only that we retain an indication of how many wavevector arguments are associated with any particular Green function:

$$G_{01} = \frac{1}{i\omega + v_0 k^2}, \qquad G_{02} = \frac{1}{i\omega + v_0 k^2 + v_0 j^2},$$

and so on. Then the solution to (9.72) can be written schematically as

$$\langle u^+ u^+\rangle_0 = 2G_{02}\sum 2M\langle u^- u^+ u^+\rangle_0 + G_{02}\sum M\langle u^+ u^+ u^+\rangle_0. \quad (9.73)$$

Now we can see the complication, as mentioned above, when we compare the present procedure with the usual turbulence hierarchy. Evidently we can distinguish moments which contain one factor $\mathbf{u}^-$ from those which do not.

From our earlier discussions, we can guess that the term $\langle u^- u^+ u^+\rangle_0$ is the important one, as it represents energy transfer into the band $k_1 < k < k_0$, whereas the term $\langle u^+ u^+ u^+\rangle_0$ must represent energy transfers through the band. (The underlying argument was based on the wavenumber band's being in the viscous range, but it may still go through elsewhere if the band is, in some sense, thin.)

Hence, we can guess that $\langle u^+ u^+ u^+\rangle_0$ can be treated as a small correction term, which we seek to calculate by forming an equation for it, just as we did for $\langle u^+ u^+\rangle_0$. This is readily done, and we show the result schematically, as follows:

$$\langle u^+ u^+ u^+ \rangle_0 = G_{03} \sum 2M \langle u^- u^+ u^+ u^+ \rangle_0 + G_{03} \sum M \langle u^+ u^+ u^+ u^+ \rangle_0. \quad (9.74)$$

Then we can go through the same arguments all over again in order to derive an equation for $\langle u^+ u^+ u^+ u^+ \rangle_0$ and so on, to all orders.

This process can be accompanied by repeated substitution back into (9.73), thus yielding the moment expansion

$$\langle u^+ u^+ \rangle_0 = 2G_{02} \sum 2M \langle u^- u^+ u^+ \rangle_0 +$$
$$+ 2G_{02} \sum M G_{03} \sum 2M \langle u^- u^+ u^+ u^+ \rangle_0 +$$
$$+ 2G_{02} \sum M G_{03} \sum M G_{04} \sum 2M \langle u^- u^+ u^+ u^+ u^+ \rangle_0 +$$
$$+ 2G_{02} \sum M G_{03} \sum M G_{04} \sum M \langle u^+ u^+ u^+ u^+ u^+ \rangle_0. \quad (9.75)$$

and so on. Thus the term $\langle u_\beta^+ (\mathbf{j}, t) u_\gamma^+ (\mathbf{k} - \mathbf{j}, t) \rangle_0$ can be written as the sum of a moment expansion, where (to all orders) every moment contains a single factor $\mathbf{u}^-$.

It follows that we can conclude that $\langle u_\beta^+ (\mathbf{j}, t) u_\gamma^+ (\mathbf{k} - \mathbf{j}, t) \rangle_0$ is a linear functional of $\mathbf{u}^-$ and, in principle therefore, eqn (9.70)—when taken in conjuction with (9.75)—can be regarded as a closed equation for $\mathbf{u}^-$.

9.4.5 A mean field approximation

Let us now turn our attention to the relationship between the operation of partial averaging, on the one hand, and conventional (or global) averaging, on the other. If, for instance, we average the product $u_\beta(\mathbf{j}, t) u_\gamma(\mathbf{k} - \mathbf{j}, t)$ in the usual way, then for stationary homogeneous turbulence we obtain

$$\langle u_\beta(\mathbf{j}, t) u_\gamma(\mathbf{l}, t) \rangle = Q_{\beta\gamma}(\mathbf{j}) \delta(\mathbf{j} + \mathbf{l}), \quad (9.76)$$

where $Q_{\alpha\beta}(\mathbf{j})$ is constant with respect to time. Further, if we consider the product $u_\beta^+ (\mathbf{j}, t) u_\gamma^+ (\mathbf{l}, t)$, then the obvious extension of (9.76) is

$$\langle u_\beta^+ (\mathbf{j}, t) u_\gamma^+ (\mathbf{l}, t) \rangle = Q_{\beta\gamma}^+(\mathbf{j}) \delta(\mathbf{j} + \mathbf{l}), \quad (9.77)$$

and here Q^+ is related to Q by a simple filtering operation:

$$Q_{\beta\gamma}^+(\mathbf{j}) = Q_{\beta\gamma}(\mathbf{j}) \qquad \text{for } k_1 < j < k_0. \quad (9.78)$$

However, if we take the partial average of the last product, then we are forced to write something like

$$\langle u_\beta^+ (\mathbf{j}, t) u_\gamma^+ (\mathbf{l}, t) \rangle_0 = \begin{cases} P_{\beta\gamma}^{++} (\mathbf{j} + \mathbf{l}, t) & \text{for } 0 < |\mathbf{j} + \mathbf{l}| < k_1 \\ 0 & \text{for } k_1 < |\mathbf{j} + \mathbf{l}| < k_0, \end{cases} \quad (9.79)$$

where $P_{\beta\gamma}^{++}$ fluctuates randomly for $0 < |\mathbf{j} + \mathbf{l}| < k_1$, whereas the superscripts indicate that $k_1 < j, l < k_0$. Naturally, in these circumstances, neither homogeneity nor stationarity can come to our aid.

The generalization to moments of any order is readily made. For global averages, we have the usual result

$$\langle u_\beta(\mathbf{j}, t) u_\gamma(\mathbf{l}, t) \ldots u_\alpha(\mathbf{p}, t) \rangle$$

$$= Q_{\beta\gamma\ldots\sigma}(\mathbf{j}, \mathbf{l}, \ldots, -\mathbf{j} - \mathbf{l} \ldots) \qquad \text{for } \mathbf{j} + \mathbf{l} + \cdots + \mathbf{p} = 0,$$

$$= 0 \qquad \text{for } \mathbf{j} + \mathbf{l} + \cdots + \mathbf{p} \neq 0, \tag{9.80}$$

while for partial averages the corresponding result is easily seen to be

$$\langle u_\beta(\mathbf{j}, t) u_\gamma(\mathbf{l}, t) \ldots u_\alpha(\mathbf{p}, t) \rangle_0$$

$$= P_{\beta\gamma\ldots\sigma}(\mathbf{j}, \mathbf{l} + \cdots + \mathbf{p}, t) \qquad \text{for } 0 \leqslant |\mathbf{j} + \mathbf{l} + \cdots + \mathbf{p}| \leqslant k_1,$$

$$= 0 \qquad \text{for } k_1 \leqslant |\mathbf{j} + \mathbf{l} + \cdots + \mathbf{p}| < k_0. \tag{9.81}$$

Note that a plus or minus superscript can be added to P, above the appropriate tensor subscript, as required.

Now from the moment expansion on the r.h.s. of eqn (9.75), it is clear that we need only consider moments which contain a single factor $\mathbf{u}^-$. Our aim must be to factor out the low-k dependence, and to do this we make the following proposal:

$$\langle u_\beta^-(\mathbf{j}, t) u_\gamma^+(\mathbf{l}, t) u_\rho^+(\mathbf{r}, t) \ldots u_\sigma^+(\mathbf{p}, t) \rangle_0$$

$$= u_\beta^-(\mathbf{j}, t) \langle u_\gamma^+(\mathbf{l}, t) u_\rho^+(\mathbf{r}, t) \ldots u_\sigma^+(\mathbf{p}, t) \rangle$$

$$= u_\beta^-(\mathbf{j}, t) Q_{\gamma\rho\ldots\sigma}(\mathbf{l}, \mathbf{r}, \ldots, -\mathbf{l} - \mathbf{r} \ldots)$$

$$\text{if } \mathbf{l} + \mathbf{r} + \cdots + \mathbf{p} = 0$$

$$= 0 \qquad \text{otherwise.} \tag{9.82}$$

Clearly this is a plausible way of factoring out the low-k dependence, in the sense that the two conditions on the constituent wavenumbers ensure that $\mathbf{j}$ must satisfy the condition $0 < j < k_1$. Thus the fluctuation in the explicit scales is being represented by a term of the form: constant $\times \mathbf{u}^-$. Nevertheless, apart from its obvious utility, this step is rather less than transparent.

However, it can be argued that (9.82) can be interpreted as a mean field approximation. This interpretation can be understood most easily in the context of a specific case. Consider the first term on the r.h.s. of (9.75). This will turn out to be our leading approximation, and so it is worth treating in detail. Its explicit form can be deduced from eqn (9.72):

$$4M_{\beta\sigma\rho}(\mathbf{j}) \sum_\mathbf{p} \langle u_\sigma^-(\mathbf{p}, t) u_\rho^+(\mathbf{j} - \mathbf{p}, t) u_\gamma^+(\mathbf{k} - \mathbf{j}, t) \rangle_0$$

$$= 4M_{\beta\sigma\rho}(\mathbf{j}) \sum_\mathbf{p} u_\sigma^-(\mathbf{p}, t) \langle u_\rho^+(\mathbf{j} - \mathbf{p}, t) u_\gamma^+(\mathbf{k} - \mathbf{j}, t) \rangle$$

$$= 4M_{\beta\sigma\rho}(\mathbf{j}) \sum_\mathbf{p} u_\sigma^-(\mathbf{p}, t) \delta(\mathbf{k} - \mathbf{p}) Q_{\rho\gamma}^+(|\mathbf{k} - \mathbf{j}|)$$

$$= 4M_{\beta\sigma\rho}(\mathbf{j}) D_{\rho\gamma}(\mathbf{k} - \mathbf{j}) Q^+(|\mathbf{k} - \mathbf{j}|) u_\alpha^-(\mathbf{k}, t), \tag{9.83}$$

where the summation over $\mathbf{p}$ and the delta function have been eliminated together. We have also made the assumption (for later convenience) that the fluctuations in the band $k_1 < k < k_0$ are isotropic.

Reverting temporarily to the more symbolic notation, we can interpret (9.83) in the following way. Write the product $u^+ u^+$ in terms of its mean value:

$$u^+ u^+ = \langle u^+ u^+ \rangle + [u^+ u^+ - \langle u^+ u^+ \rangle]$$
$$= \langle u^+ u^+ \rangle + C^+.$$

Now let us neglect the correction term C^+ for the moment, i.e. we are approximating the $u^+ u^+$ field by its mean value. This allows us to evaluate the mixed triple moment as

$$\langle u^- u^+ u^+ \rangle_0 \simeq \langle u^- \langle u^+ u^+ \rangle\rangle_0$$
$$= \langle u^- \rangle_0 \langle u^+ u^+ \rangle$$
$$= u^- \langle u^+ u^+ \rangle$$

because $\langle u^+ u^+ \rangle$ is a constant and is therefore invariant under partial (or any other) averaging.

Of course this is an approximation, but there are various ways in which the effects of the correction term C^+ could be taken into account. At the time of writing this has not yet been done.

Lastly, it should be noted that (9.63)—which is one of the properties defining the partial average—is consistent with (9.82). This is easily verified and will be left as an exercise for the reader.

9.4.6 *The RG equations*

In this section we shall give a brief general discussion of the RG method as a preliminary to the detailed calculation to be given in the following section. As before, we resort to a highly symbolic notation, the better to let the general ideas stand out. We begin by recapitulating the overall position.

In general terms, we wish to solve eqn (9.70), which, for the sake of convenience, we shall rewrite here in symbolic notation as

$$\left(\frac{\partial}{\partial t} + v_0 k^2\right) u^-(\mathbf{k}) - \sum_{\mathbf{j}} M(\mathbf{k}) \langle u^+(\mathbf{j}) u^+(\mathbf{k} - \mathbf{j}) \rangle_0$$
$$= \sum_{\mathbf{j}} M(\mathbf{k}) u^-(\mathbf{j}) u^-(\mathbf{k} - \mathbf{j}), \tag{9.84}$$

where the coupling to the band $k_1 < k < k_0$ is given by the non-linear term on the l.h.s. This term can be expressed as a linear functional of $\mathbf{u}^-$ by the moment expansion in (9.75). Further, the 'partially averaged' moments can be evaluated in terms of the conventional moments, using (9.82) as a general (mean field) approximation and eqn (9.83) as a specific guide. We note that the selection rules for wavenumbers on the r.h.s. of (9.75) are always such that $\mathbf{u}^-$ appears with the argument $\mathbf{k}$. It can therefore be shown that the non-linear coupling in (9.84) takes the general form

$$\sum_j M(\mathbf{k}) \langle u^+(\mathbf{j}) u^+(\mathbf{k} - \mathbf{j}) \rangle_0$$

$$= 2MG_{02} \sum 2Q_2^+ u^-(\mathbf{k}) + 2MG_{02} \sum M \sum G_{03} \sum 2MQ_3^+ u^-(\mathbf{k}) +$$
$$+ 2MG_{02} \sum M \sum G_{03} \sum MG_{04} \sum 2M Q_4^+ u^-(\mathbf{k}) + \cdots$$
$$= \delta v_0(k) k^2 u^-(\mathbf{k}), \tag{9.85}$$

where $Q_2^+ = \langle u^+ u^+ \rangle$, $Q_3^+ = \langle u^+ u^+ u^+ \rangle$, etc.

We note that this expansion is linear in $u^-(\mathbf{k})$ at every order, and hence can be written as an increment to the viscosity. Making the appropriate substitution back into eqn (9.84), we then have

$$\left[\frac{\partial}{\partial t} + \{ v_0 + \delta v_0(k) \} k^2 \right] u^-(\mathbf{k}) = \sum_j M(\mathbf{k}) u^-(\mathbf{k} - j) \tag{9.86}$$

as the 'new' Navier–Stokes equation, defined on the interval $0 < k < k_1$. Of course the fluid viscosity has apparently increased to v_1 where

$$v_1(k) = v_0 + \delta v_0(k),$$

and correspondingly the Green function is changed:

$$G_0(\mathbf{k}, \omega) = G_1(\mathbf{k}, \omega).$$

Then we put $u^-(\mathbf{k}) = u(\mathbf{k})$, in eqn (9.86), and carry out the whole procedure again for $k_2 < k < k_1$.

As there are no stirring forces and the interaction strength is not renormalized, it is sufficient to scale the viscosity from one iteration cycle to another. Carrying this on for successive bands

$$k_n < k_{n-1} < \cdots k_2 < k_1 < k_0$$

results in a fixed point, where

$$G_{n+1}(\mathbf{k}, \omega) = G_n(\mathbf{k}, \omega) = G^*(\mathbf{k}, \omega)$$

when $G_{n+1}(\mathbf{k}, \omega)$ is scaled to the same wavenumber interval as $G_n(\mathbf{k}, \omega)$.

9.4.7 Second-order calculation of the effective viscosity

In order to make a specific calculation, we shall truncate the moment expansion on the r.h.s. of eqn (9.85), retaining only the first term (which is second order in u^+). The detailed form of this term is readily obtained from eqn (9.72), with the second term on the r.h.s. neglected. We have previously justified this approximation on physical grounds and we shall later consider, more quantitatively, the conditions under which it can be expected to be valid.

The required solution is found by two steps. First, we make what is often called a 'Markovian' approximation. That is, we assume that the u^+ evolve much more rapidly than the u^-, and that we can therefore neglect the differentiation with respect to time on the l.h.s. of eqn (9.71). The Green function then takes a particularly simple form. Second, we evaluate the 'mixed' partial average by means of the general mean field approximation (see eqn (9.82)) which has been carried out explicitly for this particular case in eqn (9.83). The result is

$$\langle u_\beta^+ (\mathbf{j}, t) u_\gamma^+ (\mathbf{k} - \mathbf{j}, t) \rangle_0 = \left\{ 4 \left(\frac{2\pi}{L} \right)^3 M_{\beta\sigma\rho}(\mathbf{j}) D_{\rho\gamma}(\mathbf{k} - \mathbf{j}) \times \right.$$

$$\left. \times \frac{Q^+(|\mathbf{k} - \mathbf{j}|)}{(v_0 j^2 + v_0 |k - j|^2)} \right\} u_\sigma^- (\mathbf{k}, t). \qquad (9.87)$$

Then the non-linear coupling term on the l.h.s. of eqn (9.70) can be written as

$$- M_{\alpha\beta\gamma}(\mathbf{k}) \sum_{\mathbf{j}} \langle u_\beta^+ (\mathbf{j}, t) u_\gamma^+ (\mathbf{k} - \mathbf{j}, t) \rangle_0$$

$$= \left\{ 2 \int \mathrm{d}^3 j \frac{L(\mathbf{k}, \mathbf{j}) Q^+(|\mathbf{k} - \mathbf{j}|)}{(v_0 j^2 + v_0 |\mathbf{k} - \mathbf{j}|^2)} \right\} u_\alpha^- (\mathbf{k}, t)$$

$$= \delta v_0(k) k^2 u_\alpha^- (\mathbf{k}, t), \qquad (9.88)$$

where we have taken the limit $L \to \infty$ in converting the sum over $\mathbf{j}$ into an integral. The coefficient $L(\mathbf{k}, \mathbf{j})$ is the usual form met previously in second-order turbulence theories (see eqn (6.30), or Appendix E). Note that the second line on the r.h.s. of (9.88) is equivalent to (9.85), to second order in u^+.

It follows from eqn (9.88) that we should write the increment to the viscosity—which represents the non-linear transfer of energy to the eliminated modes—as

$$\delta v_0(k) = \frac{2}{k^2} \int \mathrm{d}^3 j \frac{L(\mathbf{k}, \mathbf{j}) Q^+(|\mathbf{k} - \mathbf{j}|)}{v_0 j^2 + v_0 |\mathbf{k} - \mathbf{j}|^2}$$

$$\text{for } 0 < k < k_1; k_1 < j, |\mathbf{k} - \mathbf{j}| < k_0. \qquad (9.89)$$

It should perhaps be noted that the increment to the viscosity is positive because the inertial transfer operators are pure imaginary and $M^2 = -|M^2|$.

In order to repeat this process, we choose wavenumber bands by putting

$$k_n = h^n k_0 \qquad 0 < h < 1 \qquad (9.90)$$

where the scaling parameter h is chosen arbitrarily, within the limits given. Then, following the procedure outlined in the preceding section, we obtain the recursion relation

$$v_{n+1}(k) = v_n(k) + \delta v_n(k), \qquad (9.91)$$

where

$$\delta v_n(k) = \frac{2}{k^2} \int d^3 j\, L(\mathbf{k}, \mathbf{j}) Q^+(|\mathbf{k} - \mathbf{j}|) \times$$

$$\times \frac{1}{v_n(j)j^2 + v_n(|\mathbf{k} - \mathbf{j}|)|\mathbf{k} - \mathbf{j}|^2}$$

$$0 < k < k_{n+1};\; k_{n+1} < j, |\mathbf{k} - \mathbf{j}| < k_n. \quad (9.92)$$

Also, eqn (9.70) for the explicit scales can now be written as

$$\left(\frac{\partial}{\partial t} + v_{n+1}(k)k^2 \right) u_\alpha^-(\mathbf{k}, t)$$

$$= M_{\alpha\beta\gamma}(\mathbf{k}) \sum_{\mathbf{j}} u_\beta^-(\mathbf{j}, t) u_\gamma^-(\mathbf{k} - \mathbf{j}, t)$$

$$\text{for } 0 < k, j, |\mathbf{k} - \mathbf{j}| < k_{n+1}, \quad (9.93)$$

where $v_{n+1}(k)$ is given by eqns (9.91) and (9.92).

We can form an energy equation for the explicit scales by multiplying each term in (9.93) by $u_\alpha^-(-\mathbf{k}, t)$ and averaging. If we then integrate both sides of the resulting equation with respect to $\mathbf{k}$ over the range $0 < k < k_{n+1}$, the result is the renormalized dissipation equation

$$\int_0^{k_{n+1}} 2v_{n+1}(k)k^2 E(k)\, dk = \varepsilon, \qquad (9.94)$$

which can be compared with eqn (9.58), which is the conventional dissipation integral.

If we now assume a power-law form for the energy spectrum

$$E(k) = \alpha \varepsilon^r k^s \qquad (9.95)$$

and make the scaling transformation

$$k = k_n k', \qquad (9.96)$$

it follows from eqns (9.91) and (9.92) that the effective viscosity can be written as

$$v_n(k_n k') = \alpha^{1/2} \varepsilon^{r/2} k^{(s-1)/2} \tilde{v}_n(k') \qquad (9.97)$$

where α is the constant of proportionality in the assumed spectrum.

Substitution of (9.97) into (9.94) then fixes the exponents as

$$r = 2/3 \qquad s = -5/3,$$

the well-known Kolmogorov spectrum. With these results, eqns (9.97), (9.91), and (9.92) become

$$v_n(k_n k') = \alpha^{1/2} \varepsilon^{1/3} k_n^{-4/3} \tilde{v}_n(k') \tag{9.98}$$

$$v_{n+1}(k') = h^{4/3} \{ \tilde{v}_n(hk') + \delta \tilde{v}_n(hk') \} \tag{9.99}$$

$$\delta v_n(k') = (2\pi k'^2)^{-1} \int d^3 j \, L(\mathbf{k}', \mathbf{j}') |\mathbf{k}' - \mathbf{j}'|^{-11/3} \times$$

$$\times \frac{1}{(\tilde{v}_n(j') j'^2 + \tilde{v}_n(|\mathbf{k}' - \mathbf{j}'|) |\mathbf{k}' - \mathbf{j}'|^2}$$

$$0 < k' < 1; \, 1 < j', |\mathbf{k}' - \mathbf{j}'| < h^{-1}. \tag{9.100}$$

Iteration of eqns (9.99) and (9.100) reaches a fixed point, with $\tilde{v}_{n+1} = \tilde{v}_n = v^*$ (see McComb 1986; earlier references are based on a slightly different form of the viscosity increment and this leads to some difference in earlier numerical calculations). Some quantitative results of this work are given in Chapter 10 in connection with the problem of subgrid modelling.

9.4.8 *The effect of higher-order moments*

The one approximation which is amenable to systematic treatment is the truncation of the moment expansion on the r.h.s. of eqn (9.85). The various moments of the u^+ may be related to each other, as if u^+ had a Gaussian distribution. This means that all odd-order moments vanish, while all even-order moments can be factored into products of second-order moments.

It should perhaps be emphasized that such a procedure is not equivalent to the quasi-normality hypothesis, as discussed in Section 2.8.2. The reason is that the $u_\alpha^+(\mathbf{k}, t)$ are band-filtered projections of the non-Gaussian velocity field $u_\alpha(\mathbf{k}, t)$. It seems to be the case that such band-filtered variables tend to a Gaussian distribution, as the filter bandwidth becomes small (Lumley and Takeuchi 1976).

It has been conjectured (McComb 1986) that the moment expansion in (9.85) is bounded from above by a power series in $\delta v_n/v_n$. At the fixed point it is found that $\delta v^*/v^* = 1$ at $h = 0.55$, and that $\delta v^*/v^*$ tends rapidly to zero as h tends to unity (i.e. corresponding to the bandwidth shrinking to zero). Accordingly, for h somewhere in the range $0.55 < h < 1$, a second-order calculation can be established as a rational approximation.

9.5 Concluding remarks

At the present stage, we seem to be faced with the choice of rigour (but not applicability), on the one hand, or applicability (but not rigour), on the other. In view of the obvious intuitive appeal of RG as a method of tackling turbulence, it is difficult to believe that this situation will not improve.

Perhaps the most interesting comparison to make is with the successful RPT theories, as discussed in Chapters 6–8. Such theories (as we have seen) lead to very comprehensive predictions about virtually any statistical parameter characterizing turbulence. In contrast, the RG method only seems to yield power laws for scaling behaviour, although the iterative averaging method can also yield the excitation level (see the discussion in Chapter 10).

However, although it can be argued that both types of theory begin by chopping up the dynamics, in order to make the non-linearity tractable (Kraichnan 1982), they really have very little more in common than that. RPTs formulate the problem in a perturbation series which is wildly divergent. They then renormalize this series in a global operation of unknown properties, thus producing a renormalized perturbation series, also with unknown properties.

In contrast, RG begins by tackling a tractable situation (in iterative averaging this is the viscous range, which is characterized by a low Reynolds number). Then the renormalization is carried out systematically in a series of steps which allows one to keep everything under control. The very existence of a fixed point, therefore, is some evidence that one has succeeded in this aim. Certainly there is no comparable claim to be made for RPTs and it seems that in this respect the advantage lies with RG.

Notes

1. This idea may also be familiar to those who are acquainted with perturbation theory in classical aerodynamics. Garabedian (1956) employed an expansion in $\epsilon/(2 + \epsilon)$, where $2 + \epsilon$ is the number of space dimensions, to solve for axially symmetric flows with free boundaries. The case $\epsilon = 1$ is the physically interesting situation where the flow is genuinely axisymmetric in three-dimensional space.
2. Readers who are uncomfortable with general manipulations in d dimensions may find it helpful to begin by considering the case $d = 3$. For instance, the basic volume integral in wavenumber space is just (in spherical polar coordinates) given by

$$\int d^3 j = \int_0^{2\pi} d\phi \int_0^\pi \sin\theta \, d\theta \int j^2 \, dj$$

$$= S_d \int j^{d-1} \, dj \qquad \text{for } d = 3.$$

Clearly $S = 4\pi$ is the area of a sphere of unit radius. The concept can be generalized to d dimensions, with

$$S_d = \int d\phi \int \sin\theta_1 \, d\theta_1 \int \sin^2\theta_2 \, d\theta_2 \dots \int \sin^{d-2}\theta_{d-2} \, d\theta_{d-2},$$

and it can be shown (e.g. Ramond 1981, pp. 387–8) that this takes the form given by eqn (9.23).

Then, eqn (9.21) follows quite straightforwardly. The angular integration vanishes by symmetry, unless $\alpha = \beta$. Hence we write the integrand as

$$j_\alpha j_\beta = \frac{j^2 \delta_{\alpha\beta}}{\operatorname{tr} \delta_{\alpha\beta}} = \frac{j^2 \delta_{\alpha\beta}}{d},$$

and, using the result given above for the volume integral, we obtain eqn (9.21). Similar methods lead to eqn (9.22), and this derivation is left as an exercise for the reader.

3. Yakhot and Orszag (1986; henceforth referred to as YO) have applied the FNS theory to practical turbulence problems, simply by deciding to take the ultra-violet cut-off Λ to be larger than the dissipation wavenumber, irrespective of whether or not this gives a valid extension of the FNS theory. They argue that, in some sense, the numerics of strongly stirred turbulence and inertially cascading turbulence will be equivalent, provided that the stirring forces are chosen correctly. Their ultimate justification for this assertion is that it leads to some good theoretical values for representative numbers, ranging from the Kolmogorov spectral constant to the von Karman constant for shear flows.

This work has been the subject of some comment at a fairly technical level (Ronis 1987; Bhattacharjee 1988), but here we shall be concerned only with rather practical considerations. We have two points to make.

First, YO follow DDM in choosing the stirring forces such that $y = d$ (we use the notation of Section 9.3 here), thus encountering the problem that this energy input is divergent (see eqn (9.54)). They circumvent this in an ingenious way by adopting a relationship from RPT. They argue that their results for the effective viscosity and the energy spectrum are equivalent to $v(k) = \beta \varepsilon^{1/3} k^{-4/3}$ and $E(k) = \alpha \varepsilon^{2/3} k^{-5/3}$, where (from RPT) the constants of proportionality must satisfy (see eqn (6.125)) $\alpha^2 C/\beta = 1$ where numerical integration (Edwards 1965) gives $C = 0.19$. By incorporating this relationship into their theory, YO are able to relate W_0 to a finite dissipation rate; thus $W_0 = 11.12\varepsilon$, which implies (via conservation of energy) that their stirring forces are restricted to a band such that $k_{max}/k_{min} = 1.007$! This in itself would seem to pose a serious problem for their 'principle of equivalence', but (at the very least) the dependence on another class of theory (with unknown convergence properties) raises doubts about whether the work of YO can be properly described as an RG theory.

Our second point concerns an arithmetical inconsistency. The authors evaluate the Kolmogorov constant using $\epsilon = 0$ and $\epsilon = 4$ at different points in the same calculation. Their result is $\alpha = 1.617$. However, if they were to use $\epsilon = 4$ consistently, their result would become $\alpha = 1.113$, which is outside the normally accepted range of $1.2 < \alpha < 2.2$. The claims made for the numerical success of this theory would appear to be unfounded.

References

AMIT, D. J. (1984). *Field theory, the renormalisation group, and critical phenomena* (2nd edn). World Scientific.

BHATTACHARJEE, J. K. (1988) *J. Phys. A* **21**, L551.

DEDOMINICIS, C. and MARTIN, P. C. (1979). *Phys. Rev. A* **19**, 419.

EDWARDS, S. F. (1965). *Proc. Int. Conf. on Plasma Physics*, Trieste, p. 595. IAEA, Vienna.

FORSTER, D., NELSON, D. R., and STEPHEN, M. J. (1976). *Phys. Rev. Lett.* **36**, 867.

——, ——, and —— (1977). *Phys. Rev. A* **16**, 732.

FOURNIER, J.-D. and FRISCH, U. (1978). *Phys. Rev. A* **17**, 747.

GARABEDIAN, P. R. (1956). *Pacific J. Math.* **6**, 611.

GROSSMANN, S. and SCHNEDLER, E. (1977). *Z. Phys. B* **26**, 307.

KADANOFF, L. P. (1966). *Physics* **2**, 263.

KRAICHNAN, R. H. (1982). *Phys. Rev. A* **25**, 3281.

LEVICH, E. (1980). *Phys. Lett.* **79A**, 171.

—— and Tsinober, A. (1984). *Phys. Lett.* **101A**, 265.

Lumley, J. L. and Takeuchi, K. (1976). *J. fluid Mech.* **74**, 433.

Ma, S. K. and Mazenko, G. F. (1975). *Phys. Rev. B* **11**, 4077.

McComb, W. D. (1982). *Phys. Rev. A* **26**, 1078.

—— (1985). In *Theoretical approaches to turbulence* (eds D. L. Dwoyer, M. Y. Hussaini, and R. G. Voigt). *Applied mathematical sciences*, Vol. 58. Springer, Berlin.

—— (1986). In *Direct and large eddy simulation of turbulence* (eds U. Schumann and R. Friedrich). *Notes on numerical fluid mechanics*, Vol. 15. Vieweg, Braunschweig).

—— and Shanmugasundaram, V. (1983). *Phys. Rev. A* **28**, 2588.

—— and —— (1984). *J. Phys. A* **18**, 2191.

Mandelbrot, B. (1977). *The fractal geometry of nature.* Freeman, San Francisco, CA.

Monin, A. S. and Yaglom, A. M. (1975). *Statistical fluid mechanics*, Vol. 2, *Mechanics of turbulence.* MIT Press, Cambridge, Ma.

Onsager, L. (1944). *Phys. Rev.* **65**, 117.

Ramond, P. (1981). *Field theory: a modern primer.* Benjamin/Cummings.

Reichl, L. E. (1980). *A modern course in statistical physics.* Edward Arnold, London.

Ronis, D. (1987). *Phys. Rev. A* **36**, 3322.

Rose, H. A. (1977). *J. fluid Mech.* **81**, 719.

Wilson, K. G. (1975). *Adv. Math.* **16**, 170.

—— and Kogut, J. (1974). *Phys. Rep.* **12C**, 75.

Yakhot, V. (1981). *Phys. Rev. A* **23**, 1486.

—— and Orszag, S. A. (1986). *Phys. Rev. Lett.* **57**, 1722.

10

NUMERICAL SIMULATION OF TURBULENCE

In recent times, the rapid growth in both size and speed of digital computers has tended to change our perception of the turbulence problem. Twenty or so years ago, the clear aim—shared by engineers and theorists alike—was to derive transport equations by averaging the Navier–Stokes equation.

Nowadays, the possibility of computers becoming powerful enough to simulate some flows of engineering interest leads us to consider a simplified form of the turbulence problem: in broad terms, can we use fundamental approaches to reduce the number of degrees of freedom? That is, can we reduce the size of the task for numerical simulation?

It is because we must now consider questions of this kind that we include here a short treatment of numerical simulation (in itself, a large and important subject) as a coda to our treatment of the statistical theory of turbulence. In doing this, we shall have three broad purposes in mind.

The first of these purposes is to establish just what can actually be achieved, by way of numerical simulation of turbulence, at the present time. In doing this, we shall consider both full simulation (all scales resolved) and large-eddy simulation (LES) (only the large scales are resolved). As a supplement to the latter topic, we shall also consider current practice and achievements in subgrid modelling. It should be noted that Section 3.3 can be regarded as an introduction to this part of the current chapter.

Then we turn our attention to our second purpose, which is to consider how renormalization methods can be applied to the problem. Specifically, we formulate the LES equations for isotropic turbulence, and discuss the application of renormalization methods to the subgrid modelling problem (in the context of spectra).

Our third, and last, purpose is to give a brief account of 'other' methods of simulating turbulence on a computer. By this, we mean methods which either do not depend directly on the Navier–Stokes equation, or which have some other novel or original feature.

10.1 Full simulations

In Section 3.3.1, we have already considered how to estimate the size of the numerical problem involved in performing a full turbulent simulation. To do this, we interpreted the Fourier modes as the degrees of freedom of the system, and related the number N of such degrees of freedom to the value of the Reynolds number R for the case of isotropic turbulence.

Now, in practice, what we would really like to know is how large a value of the Reynolds number we could achieve in a turbulence simulation on a given computer. If we take N to be the number of mesh points (in k-space), then this is simply related to the storage capacity of the machine. Thus we can obtain the desired estimate by rearranging eqn (3.42), and, combining this with eqn (3.57) in order to express the result in terms of the Taylor–Reynolds number, we find

$$R_\lambda \sim N^{7/36}. \tag{10.1}$$

Equivalently, if we wish to increase the Taylor–Reynolds number of a turbulence simulation by a factor of 2, then we must increase the available computer storage by a factor of about 64!

This depressing conclusion has been borne out in practice. In the decade following the pioneering simulation of Orszag and Patterson (a decade, incidentally, which included the development of parallel processors) the Taylor–Reynolds number achieved in isotropic simulations was only increased by a factor of 2 (Kerr 1985)!

The interest to the theorist of such isotropic simulations lies in their utility as computer experiments in which the initial conditions can be very precisely controlled. In particular, we wish to obtain very high Reynolds numbers in order to obtain universal behaviour. This provides a motivation to exploit available computer power in order to obtain the largest possible Reynolds number in an isotropic simulation. However, the impetus towards more realistic situations (for engineering applications) suggests an equally valid strategy: that one should accept a modest Reynolds number and use increments in computer power to explore the difficulties inherent in treating inhomogeneity and anisotropy. This latter strategy also has considerable fundamental interest, and so we shall discuss the topic from both these points of view.

10.1.1 *Isotropic turbulence*

The spectral methods pioneered by Orszag and Patterson (1972) have been discussed in Section 3.3.1, and so we shall concentrate on their results here. Their calculation of decaying isotropic turbulence was carried out with a grid of 32^3 points in k-space (a '32-cubed grid', in the jargon of the subject). The highest Taylor–Reynolds number for which all scales would be satisfactorily resolved was $R_\lambda = 45$, but the authors presented data taken from runs with $R_\lambda = 35$. Evidently, these particular results could be regarded as being especially reliable (although note the comments made later in connection with Kerr's (1985) comparison of 32-cubed and 64-cubed simulations).

Results were presented for dissipation and transfer spectra, along with the time evolution of the dissipation rate and the skewness, and were compared with comparable results obtained from a calculation of the direct-interaction approximation (DIA) made with the same initial conditions. For the first three quantities, agreement between the simulation and DIA is very close indeed, underlying alike the value of this type of computer experiment to the theorist, and the remarkable success of DIA as a 'first-principles' turbulence theory at low Reynolds numbers. However, the simulation gives evolved skewness factors which are significantly larger than the value calculated from DIA. Orszag and Patterson put this down to the theory's underestimating the magnitude of vortex stretching by turbulence and hence the transfer of energy to higher wavenumbers. This feature is of course associated with the failure of DIA to give the Kolmogorov spectrum at higher Reynolds numbers (see Section 7.1).

A last point worth noting about these results is that the authors concluded that the small-scale structure, which determines the skewness factor, is independent of the Reynolds number, although their evolved skewness was about $S = 0.47$, which is quite a bit larger than typical values from wind-tunnel data. The authors attribute part of this numerical discrepancy to the very peaked nature of their initial spectrum. We shall return to the Reynolds number independence of the skewness factor when we discuss the corresponding part of Kerr's work later in this section.

Various other applications of these spectral methods have been reported. A particularly interesting example is the use of the Orszag–Patterson Super-box code to make a numerical study of pressure fluctuations in isotropic turbulence (Schumann and Patterson 1978a). It is very difficult to measure pressure fluctuations in laboratory experiments, with the result that such measurements are rare. Evidently computer experiments have a potentially important role in providing experimental values for quantities (such as pressure or vorticity) which are not easily measured. For completeness, we mention that these authors have used the same technique to study pressure fluctuations when the turbulence is initially subject to an axisymmetric distortion (Schumann and Patterson 1978b).

The Superbox code was also used by Siggia and Patterson (1978) to investigate fine-structure intermittency. They attempted to get round the limitations of a 32-cubed simulation by employing a combination of forcing at low wavenumbers and enhanced viscosity at high wavenumbers. Associated free parameters were adjusted to produce a stationary field with a Kolmogorov $-5/3$ spectrum over the restricted range of wavenumbers resolved.

This is undoubtedly an ingenious way to obtain a Kolmogorov spectrum, but (in the jargon of the field) it is not an 'honest' simulation. However, we would be inclined to take the view that it is the enhanced viscosity to which one should take exception. In effect, this step turns the method into an LES. In contrast, the forcing of the low wavenumbers seems an entirely reasonable and physical way of obtaining a stationary field, although one would wish to make two provisos: (1) the fluid viscosity should be used in the simulation; (2) the existence of universal behaviour, independent of the nature of the forcing must be properly and convincingly demonstrated. This work has been carried on (Siggia 1981), with the code being adapted to run a 64-cubed simulation on a vector processor.

The modified Superbox code has also been run on a vector processor by Herring and Kerr (1982), who simulated passive scalar convection, and by Kerr (1985), who studied derivative correlations and their relationship to small-scale structure. We shall not pursue the topic of passive scalar convection in the present chapter, but the simulation by Herring and Kerr also obtained some results for the velocity field, which were then compared with the predictions of the DIA and test-field model (TFM) theories. In a 32-cubed simulation of decaying isotropic turbulence, they obtained values for the skewness factor over a range of Taylor–Reynolds numbers up to about 30. A comparison with the theoretical curves for S against R_λ showed that DIA agreed well with the simulation at low R_λ, but diverged from it as R_λ increased above about 8. This, of course, was consistent with the results of Orszag and Patterson. However, the results from TFM did not agree well with either the simulation or DIA for any appreciable range of wavenumbers.

This investigation was later extended by Kerr (1985), who used a method of forcing the lowest wavenumber shell in order to obtain stationary turbulence on 32-cubed, 64-cubed, and 128-cubed grids wavenumber space. Three-dimensional energy spectra were obtained for R_λ ranging from 9 to 83. When these spectra were scaled using the Kolmogorov variables, all (except for the lowest value of the Reynolds number) collapsed to universal form in the dissipation region. At the highest Reynolds number, the simulated spectrum had a short range of $-5/3$ spectrum at the lowest wavenumbers.

Although these results are open to some criticism—for instance, the predicted Kolmogorov constant is rather high at $\alpha = 2.45$, and turn-ups in the spectra at high wavenumbers indicate that the smallest scales are not fully resolved—this investigation seems to be a significant step forward in the numerical simulation of turbulence.

Two other features of Kerr's investigation are worthy of special mention. First, he has assessed the effect of aliasing errors by comparing simulations (initially identical) carried out on 32-cubed and 64-cubed grids. On the basis of the flatness factors, he concluded that $R_\lambda = 29$ was slightly too high for a 32-cubed grid. This is perhaps rather a sensitive criterion, but clearly such a result must affect our confidence in the results presented by Orszag and Patterson for $R_\lambda = 35$ and higher.

However, the second feature of interest, from our present point of view, supports the conclusion of Orszag and Patterson that the skewness is independent of Reynolds number. This result would appear to have considerable implications for the study of fine-scale intermittency (see Section 3.2).

10.1.2 *Shear flows*

If we try to get away from isotropic homogeneous fields towards more realistic flows, the next step up in difficulty is what is often referred to as 'slab geometry'. This is where there is a mean rate of flow in one particular direction, say x_1, mean quantities vary in the x_2 direction owing to boundary conditions at $x_2 = 0$ and $x_2 = 2a$, and the fluid extends to infinity in the x_3 direction. Thus the velocity field is statistically homogeneous in two directions, but not in the third (x_2) direction. In practice the only flow which rigorously satisfies these conditions is well-developed flow through a plane channel, as described in Section 1.4.5. But the plane wake, the plane jet, the mixing layer, and the boundary layer on a flat plate can all be approximated by a slab geometry, providing only that in each case mean quantities vary slowly in the streamwise direction.

The spectral methods (and in this we include pseudo-spectral methods; see Section 3.3.1) discussed in the previous section are usually extended to slab geometry by using the conventional Fourier series expansion in the two homogeneous directions $(x_1$ and $x_3)$ and invoking periodic boundary conditions. In the direction (x_2) in which mean quantities vary, some other expansion technique is needed for use in conjunction with no-slip boundary conditions. The usual method is to expand in Chebyshev polynomials (see Orszag 1971), which has the advantage of giving (in effect) a non-uniform grid in x-space, resulting in good resolution near boundaries. This is, of course, likely to be of particular benefit near solid boundaries, where mean quantities tend to be rapidly varying (compare the experimental results given for channel flow and jet flow in Chapter 1).

The isotropic 32-cubed simulation of Orszag and Patterson (1972) was first extended to turbulent shear flows by Orszag and Pao (1974), who simulated the momentumless wake (due to a self-propelled body) on a grid of 16 × 16 × 32 points in k-space. Their results for radial and axial variation of turbulent intensities were in fair agreement with laboratory measurements, but the authors emphasize the difficulty in taking spatial averages when compared with the earlier simulation of isotropic turbulence. They note the need to use either time or ensemble averages to improve their statistics (although it is not clear whether or not they did this).

However, the point is worth making more generally. When one considers the problems involved in numerically simulating the Navier–Stokes equations, there is really no essential difference between isotropic turbulence and (for example) pipe flow. These are, by their very nature, statistical concepts, and as far as the raw velocity field is concerned, the computational box is the same in both cases. The difference arises when we want to form averages. In the isotropic homogeneous case, symmetries in mean quantities can be exploited in order to allow the use of shell averages. If we consider slab geometry, then for well-developed flow, the best we can hope for is to be able to average over grid points in planes parallel to the $(x_1 x_3)$ plane. (It may be noted that we have ignored the effect of differences in boundary conditions on the raw simulation, but such differences have their main effect on the statistical quantities.)

Further work on these lines has been carried out by Metcalfe and Riley (1981), who made simulations of (1) the axisymmetric wake of a towed body and (2) a plane mixing layer. They used both the pseudo-spectral 32-cubed code of Orszag and Pao, and a spectral 64-cubed code of Rogallo (cited by Orszag and Pao 1974). The main problem encountered was the setting up of initial conditions, but the various mean values obtained seem to agree fairly well with laboratory data and to collapse quite well onto self-similar forms. Particular features of the work include the development of intermittency at the free edge of the wake and some indication of Brown–Roshko roll vortices in the mixing layer.

Well-developed turbulent flow in a plane channel has recently been simulated by Kim, Moin, and Moser (1987) at a Reynolds number (based on the mean velocity at the centre line and the semiwidth of the channel) of 3300. Their simulation involved $192 \times 129 \times 160$ grid points in x_1, x_2, x_3, and is claimed to resolve all scales. Results for correlations and spectra bear out this claim. Mean velocities and turbulent intensities were found to agree well with laboratory results, once the latter had been rescaled to eliminate inconsistencies between sets of results taken on different occasions with the same apparatus. Also, the ability of the simulation to provide statistical correlations not available from laboratory experiments was demonstrated. However, detailed comparisons in the wall region indicated some systematic discrepancies between the numerical and laboratory experiments. The authors discuss these at some length, and essentially attribute them to difficulties in using hot-film probes near the wall and to the failure of the actual laboratory experiment to achieve the idealization of two-dimensional mean flow. In some ways a significant result of this simulation is to underline the unsatisfactory state of laboratory data for simple classic flows.

Lastly, we note the large simulation ($438 \times 80 \times 320$ grid points in x_1, x_2, x_3) by Spalart (1988) of a turbulent boundary layer on a flat plate. The Reynolds number quoted (1410) is apparently based on the momentum thickness and the free-stream velocity. Unfortunately no value is given for the Taylor–Reynolds number, so that we cannot say just how comparable the results should be with those of Kerr (1985) (see the previous section). Another imponderable in such a comparison arises because Spalart used an approximate analytic method to take account of slow streamwise variations, which allowed him to use periodic boundary conditions in the x_1 direction.

Nevertheless, the most interesting result of Spalart's simulation was the finding of a distinct inertial range in the one-dimensional streamwise energy spectrum. This extended over half a decade in wavenumber, and was sufficient to allow the theoretical determination of the Kolmogorov constant as $\alpha = 1.68$, in good agreement with experiment. Certainly the spectra presented would indicate that all scales were satisfactorily resolved, with only the spanwise spectrum showing a slight trace of a turn-up at high wavenumbers.

10.2 Large-eddy simulations

In Section 3.3.2, we discussed the general formulation of large-eddy simulations (LESs), and mentioned some of the pioneering simulations which were carried out during the 1960s and 1970s. In recent years there has been a steady growth of interest in this subject, with isotropic LES becoming of great interest to fundamental theorists. We shall defer this topic to the next section, where we will consider it in the context of renormalization methods. Here we shall briefly review work done on other flow configurations.

Progress in this field has taken place in various directions. On the one hand, the pioneering work of Schumann, on finite-difference simulations of flow through channels and annuli, has been extended (Schumann, Grotzbach, and Kleiser 1980) by improvements to the subgrid model, in which the Smagorinsky form has been supplemented by a procedure to solve partial differential equations for the subgrid correlation terms. On the other hand, the method can be extended to new problems. For example, Eidson (1985) has used finite-difference methods to carry out an LES of turbulence which is generated by thermal convection (i.e. Rayleigh–Benard convection). In this work the Smagorinsky model was supplemented by the addition of buoyancy terms.

However, just as in the case of full simulations, slab geometry offers the most rewarding class of fairly realistic problems. Fourier series can be used in the two homogeneous directions, in conjunction with periodic boundary conditions. But in LESs finite-difference methods are favoured for the cross-stream direction. The LES of plane channel flow by Moin and Kim (1982), as briefly discussed in Section 3.3.2, is an example of this approach, as is the application to a shear-free boundary layer (Biringen and Reynolds 1981) and to a planetary boundary layer (Moeng 1984). The first two relied on the Smagorinsky model, whereas Moeng followed the example of Deardorff (cited by Moeng 1984) and used a form of turbulence model to predict the subgrid stresses (also see the reference to Schumann *et al.* (1980)).

The successful LES of a classic flow by Moin and Kim is undoubtedly a major achievement. They used a grid of $64 \times 63 \times 128$ points in the (x_1, x_2, x_3) directions, with the points in the cross-stream (x_2) direction being distributed non-uniformly (see their eqn (4.1) for details). They also used a variable filter width (see Section 3.3.2) in the cross-stream direction in order to take account of the variation of turbulence length scales (remember that the flow is inhomogeneous in this direction).

As we have already mentioned, a Smagorinsky subgrid model was employed, but the length scale in the direction normal to the walls was modified by the introduction of the *ad hoc* exponential damping factor due to Van Driest (cited by Moin and Kim 1982). Averages were first carried out over planes parallel to the $(x_1 x_3)$ plane, and then improved by means of a rolling average with respect to time. The flow was computed from an arbitrary initial configuration until a steady state was reached.

The numerical results for the mean velocity were found to be in excellent agreement with the accepted experimental results and with the classic 'law of the wall' profile. Computations of turbulence quantities agreed with experiment, at least as well as three representative experiments agreed with each other! However, as well as the mean velocity profile, the other significant achievement was qualitative, in that the simulation revealed coherent structures in the form of alternating low- and high-speed streaks in the region of the wall. This ability to shed light on coherent structures has since been exploited in a number of papers and we shall return to this point in Chapter 11.

10.2.1 *Assessment of subgrid models*

The various LESs which have been carried out are, in effect, tests for subgrid models. But variations from one simulation to another, in terms of choice of filter, filter width, grid resolution, and other numerical factors, can easily obscure differences between subgrid models. In this section we shall discuss investigations which are aimed specifically at testing subgrid models.

Perhaps the most satisfactory aporoach to this whole question is to simulate a flow, with all scales fully resolved, and then to attempt an LES of the same flow using a coarser grid and a subgrid model, but otherwise keeping all the numerical methods the same. This was done by Clark, Ferziger, and Reynolds (1979), who used finite-difference methods to make a 64-cubed simulation of decaying isotropic turbulence. They carried out an LES of the same field on a relatively coarse mesh of 8-cubed points, and, by comparing the two, they were able to predict a value of the Smagorinski constant C (see eqn (3.71)) without having recourse to laboratory experiments. Their overall conclusion was that the agreement between the modelled subgrid stress and the actual subgrid stress was adequate rather than good. However, they were unable to find any model more accurate than the standard Smagorinski form.

A similar—but independent—investigation by McMillan and Ferziger (1979) used an identical filter but employed a pseudo-spectral simulation on the 64-cubed grid. At 32-cubed, their LES simulation was much more finely resolved than that of Clark *et al.*, but broadly speaking their overall conclusions were much the same.

These investigations were both at Taylor–Reynolds numbers in the region of 35–40, and were for the rather unrealistic case of decaying isotropic turbulence. As we have seen, full simulation of channel flow has only been achieved recently, and that at a very low Reynolds number. The problem of extending these validation processes to more realistic situations is clearly not trivial, and has drawn a variety of responses.

For instance, Love (1980) has carried out the same procedure for the Burgers equation (i.e. a one-dimensional analogue of the Navier–Stokes equation, with some properties in common). The fact that the simulation is one-dimensional eases the computational requirements dramatically. Thus Love was able to compare full and large-eddy simulations at (in effect) high values of the Reynolds number. His conclusion was that the Smagorinski model was satisfactory and that the various modifications to it have little net effect on the actual LES.

Kaneda and Leslie (1983), noting the impossibility of carrying out a full simulation of channel flow, proposed that the problem could be tackled by using a low-dimensionality model in the wall region to provide the 'exact' velocity field. From the usual comparison between the modelled and 'exact' subgrid scales, they concluded that the subgrid models which had been used in some of the major simulations were seriously in error. However, on the other hand, they also suggested that their method of generating an exact velocity field may itself be deficient! In other words, their investigation must really be regarded as quite inconclusive. Nevertheless, their paper contains a useful detailed analysis of the various subgrid modelling procedures.

The interaction between the choice of mesh length and the Smagorinski contant was investigated by Mason and Callen (1986). They found that the ratio of the Smagorinski length scale to the channel semi-width is the main parameter determining the Reynolds number of the resolved scales. This led them to an interpretation of the Smagorinski constant as a measure of the numerical resolution of the simulation.

Lastly, we note that Speziale (1985) has drawn attention to the need to chose subgrid models such that the equation of motion for the large eddies is (like the original Navier–Stokes equation) invariant under Galilean transformations. Satisfying this simple requirement may lead to improved subgrid models (Speziale 1985, Germano 1986).

10.3 Application of renormalization methods to the subgrid modelling problem

In this section, we return (as always) to isotropic turbulence and k-space when we want to study the fundamental approaches to the problem. We begin with some of the formalities of setting up an LES in k-space for the case of isotropic turbulence. This material may be regarded as an extension of the opening paragraphs of Section 3.3.2. Then we discuss the application of renormalized perturbation theory to the calculation of the effective subgrid viscosity.

After that, we consider how the concept of LES arises naturally from the renormalization group approach and obtain the subgrid viscosity from the iterative averaging method.

10.3.1 Formulation of spectral LES

The LES equations are usually formulated in **k**-space by dividing up the velocity field at $k = k_c$ into explicit and subgrid scales:

$$u_\alpha(\mathbf{k}, t) = \begin{cases} u_\alpha^<(\mathbf{k}, t) & k \leqslant k_c \\ u_\alpha^>(\mathbf{k}, t) & k > k_c. \end{cases} \tag{10.2}$$

Substitution of this decomposition into the Navier–Stokes equation—in the form of (2.76)—yields an immediate separation into low and high wavenumbers on the l.h.s. However, on the r.h.s. all the modes are coupled, and we have to apply a filter (cut-off at $k = k_c$) to the entire non-linear term. In this way, with an obvious extension of the notation, eqn (2.76) can be resolved into its low-**k** and high-**k** forms:

$$\left(\frac{\partial}{\partial t} + v_0 k^2\right) u_\alpha^<(\mathbf{k}, t) = \sum_{\mathbf{j}} M_{\alpha\beta\gamma}^<(\mathbf{k}) u_\beta^<(\mathbf{j}, t) u_\gamma^<(\mathbf{k} - \mathbf{j}, t) +$$

$$+ \sum_{\mathbf{j}} M_{\alpha\beta\gamma}^<(\mathbf{k}) u_\beta(\mathbf{j}, t) u_\gamma(\mathbf{k} - \mathbf{j}, t)$$

$$(j \text{ and/or } |\mathbf{k} - \mathbf{j}| > k_c) \tag{10.3}$$

$$\left(\frac{\partial}{\partial t} + v_0 k^2\right) u_\alpha^>(\mathbf{k}, t) = \sum_{\mathbf{j}} M_{\alpha\beta\gamma}^>(\mathbf{k}) u_\beta^>(\mathbf{j}, t) u_\gamma^>(\mathbf{k} - \mathbf{j}, t) +$$

$$+ \sum_{\mathbf{j}} M_{\alpha\beta\gamma}^>(\mathbf{k}) u_\beta(\mathbf{j}, t) u_\gamma(\mathbf{k} - \mathbf{j}, t)$$

$$(j \text{ and/or } |\mathbf{k} - \mathbf{j}| > k_c). \tag{10.4}$$

We can use these equations as a basis for the derivation of statistical equations for the explicit and subgrid scales. The procedure is just the same as in the global case. Again, we make an obvious extension of the notation and generalize eqn (2.113a) to yield

$$\left(\frac{\partial}{\partial t} + v_0 k^2\right) Q^<(k; t, t') = P^<(k; t, t') + P^{<>}(k; t, t') \tag{10.5}$$

$$\left(\frac{\partial}{\partial t} + v_0 k^2\right) Q^>(k; t, t') = P^>(k; t, t') + P^{><}(k; t, t') \tag{10.6}$$

where $P^{<>}$ represents the inertial coupling to modes $k < k_c$ from sums over j, with j and/or $|\mathbf{k} - \mathbf{j}| > k_c$, and $P^{><}$ stands for the inertial coupling to modes $k > k_c$ from sums over **j** with j and/or $|\mathbf{k} - \mathbf{j}| < k_c$. Each of $P^<, P^>, P^{<>}$, and $P^{><}$ is defined by eqn (2.113b), with appropriate notational changes.

The standard method of modelling the subgrid-scale energy transfer on the r.h.s. of eqn (10.3) is to assume that it can be represented in terms of an eddy viscosity $v(k|k_c)$. Thus eqn (10.3) can be written in the form

$$\left\{\frac{\partial}{\partial t} + v(k|k_c)k^2\right\} u_\alpha^<(\mathbf{k}, t) = \sum_{\mathbf{j}} M_{\alpha\beta\gamma}^<(\mathbf{k}) u_\beta^<(\mathbf{j}, t) u_\gamma^<(\mathbf{k} - \mathbf{j}, t), \qquad (10.7)$$

where the superscripts indicate the restrictions on the range of the various wavevector variables. Note that this equation differs from (3.58b) in that we have absorbed the molecular viscosity v_0 into the effective subgrid viscosity $v(k|k_c)$.

We now form the energy equation for the explicit scales by multiplying each term in eqn (10.7) by $u_\alpha^<(-\mathbf{k}, t')$ and averaging. Then, setting $\alpha = \alpha'$ and summing over α, we obtain

$$\left\{\frac{\partial}{\partial t} + v(k|k_c)k^2\right\} Q^<(k; t, t') = P^<(k; t, t'), \qquad (10.8)$$

where (as above) $P(k; t, t')$ can be taken to be defined by eqn (2.113b) with appropriate notational changes, i.e. $M \to M^<$ and $u \to u^<$.

Obviously we can obtain a statistical foundation for the subgrid eddy viscosity by imposing the requirement that eqn (10.8) and (10.5) should be identical. This pragmatic approach will be the subject of the next section.

10.3.2   Renormalized perturbation theory

Kraichnan (1976) proposed that the effective eddy viscosity in eqn (10.8) should be defined statistically by the relation

$$T(k|k_c; t) = -2v(k|k_c)k^2 E^<(k, t), \qquad (10.9)$$

where $E^<(k, t)$ is the energy spectrum of the explicit scales, i.e.,

$$E^<(k, t) = 4\pi k^2 Q^<(k, t), \qquad (10.10)$$

and $T(k|k_c; t)$ is defined (see eqn (6.37)) by

$$T(k|k_c; t) = 8\pi k^2 P^{<>}(k; t, t). \qquad (10.11)$$

The prescription of the subgrid eddy viscosity is then completed by the proposal that $T(k|k_c)$ can be obtained from a two-point closure—in Kraichnan's analysis, the test-field model.

Of course, this interpretation of two-point closures in terms of an effective viscosity was not, in itself, new. The fact that a second-order closure for the inertial transfer of energy could be resolved into separate input and output terms was originally recognized by Kraichnan (1959) for the case of DIA, and by Edwards (1964) for EFP; in the latter case, the additional interpretation in terms of an effective turbulent viscosity was also made. A discussion of these aspects will be found in Chapter 6. Later, Kraichnan (1966) showed that the LHDI expression for the energy transfer could be reduced to an eddy viscosity form, provided that the dummy wavevector **j** (see eqn (7.77) of the present work) was much larger than the labelling wavevector **k**.

However, eqn (10.11) differs fundamentally from these earlier interpretations in that it recognizes the need to define an eddy viscosity in terms of the net drain of energy from mode **k**. This was anticipated by McComb (1974), who introduced essentially the same definition as an ansatz for a modified form of EFP theory (see Section 7.3.2).

Equation (10.9) can be used—with an appropriate closure approximation for $T(k|k_c)$—to compute $v(k|k_c)$ directly by numerical means. However, Kraichnan has shown that some physical insights can be gained by treating the limiting cases analytically, and we shall follow this example here, although we shall present a rather different analysis. This is partially because we wish to maintain the consistency of our own notation and partially because a more general treatment allows us to incorporate results obtained by later workers (Leslie and Quarini 1979, Chollet and Lesieur 1981) into the discussion.

We begin by specializing eqn (7.111) for the effective viscosity to the subgrid case. This step involves only the replacement of the general lower bound in the integration by the specific lower bound $j \geqslant k_c$:

$$v(k|k_c) = k^{-2} \int_{j \geqslant k_c, k \leqslant k_c} d^3 j \, L(\mathbf{k}, \mathbf{j}) q(|\mathbf{k} - \mathbf{j}|) \{q(k) - q(j)\} \times$$

$$\times \frac{1}{q(k)\{\omega(k) + \omega(j) + \omega(|k - j|)\}}. \tag{10.12}$$

Notational differences aside, this is exactly the same form as studied by Kraichnan. Also note that, like Kraichnan, we simplify matters by dropping the time dependences.

We are interested in two limiting cases—$k \to 0$ and $k \to k_c$—but we should first say something about the form of the spectrum. We shall take this to be

$$E(k) \sim \begin{cases} k^n & 0 \leqslant k \leqslant k_p \\ k^{-5/3} & k_p \leqslant k \leqslant \infty, \end{cases} \tag{10.13}$$

where k_p marks the boundary between the energy-containing and inertial ranges. This allows us to consider both realistic spectra $(n > 0)$ and the idealized case of $n = -5/3$, as considered by Kraichnan.

The corresponding form for the modal decay rate $\omega(k)$—see eqn (6.121)—can safely be assumed to apply to all wavenumbers, as it is well behaved everywhere. For similar reasons, it will soon be apparent that we do not have to worry about viscous range forms, provided only that the spectrum in this range falls off faster than $1/k$.

From the outset, it is important to recognize that the r.h.s. of (10.12) is made up from 'output' and 'input' terms (i.e. when considered from the point of view of an energy balance for mode **k**), and that the relative magnitude of these two terms is the governing factor in our interpretation of $v(k|k_c)$ as an (apparent) coefficient of viscosity. Accordingly, we emphasize this aspect by rewriting equation (10.12) as

$$v(k|k_c) = k^{-2}A(k|k_c) - \frac{4\pi}{E(k)}B(k|k_c), \tag{10.14}$$

where the coefficients $A(k|k_c)$ and $B(k|k_c)$ are given by

$$A(k|k_c) = \int_{j \geqslant k_c} d^3j\, L(\mathbf{k}, \mathbf{j})q(|\mathbf{k} - \mathbf{j}|)\frac{1}{\omega(k) + \omega(j) + \omega(|\mathbf{k} - \mathbf{j}|)}, \tag{10.15}$$

$$B(k|k_c) = \int_{j \geqslant k_c} d^3j\, L(\mathbf{k}, \mathbf{j})q(|\mathbf{k} - \mathbf{j}|)q(j)\frac{1}{\omega(k) + \omega(j) + \omega(|\mathbf{k} - \mathbf{j}|)}. \tag{10.16}$$

We should note that $A(k|k_c)$ represents the loss of energy from mode **k** to all other modes, whereas $B(k|k_c)$ represents the gain to mode **k** from all other modes.

We are now interested in the situation where $k \to 0$ for fixed k_c. Under these circumstances the dissipating motions and the eddies upon which they act are widely separated in **k**-space. If this separation were to be wide enough, we could expect that an analogy between turbulent dissipation (albeit restricted in this way) and molecular dissipation might hold quite well. In the molecular case, the scales of the dissipating motions (i.e. essentially of the order of the mean free path) are separated by many orders of magnitude from the hydrodynamic scales, where their average effect can be expressed in terms of a constant coefficient of viscosity. For the general turbulent case—without a spectral gap—this analogy (however popular) must necessarily be rather weak.

Let us begin by examining equation (10.14) in a rather qualitative way. It is immediately clear from the nature of the r.h.s. that $v(k|k_c)$ is not positive definite. If the second term should ever be larger than the first, then the effective viscosity would be negative. In fact this would not be the disaster that it would be in the molecular case. There the processes are irreversible and thermodynamics demands that v_0 is always positive. Here we are modelling the conservative—and hence reversible—inertial energy transfer in the Navier–Stokes equation, and a negative coefficient merely implies that the analogy with molecular processes has become a little strained.

We may reach a somewhat more quantitative conclusion if we make the reasonably obvious inference that, as $k \to 0$, the two coefficients $A(k|k_c)$ and $B(k|k_c)$ tend to the same dependence on k. (They will, of course, have different dependences on the fixed cut-off wavenumber k_c). Then the relative magnitudes of the input and output terms will be determined, as $k \to 0$, by the form of $E(k)$ at small k. That is, from eqn (10.13), if $n < 2$ the output term will dominate as $k \to 0$ and $v(0|k_c) > 0$. However, if $n > 2$ the input term will ultimately dominate as $k \to 0$ and $v(0|k_c) < 0$.

This rough analysis tends to suggest that it may be worth our while to discuss various cases according to the value chosen for n (recall that the form of the spectrum is essentially arbitrary in isotropic turbulence). Also, depending on the value of n, we should note that the relative values of k_p and k_c may be significant. Accordingly, we shall discuss three specific cases i.e. $n = -5/3$, $n = 1$, and $n = 4$, as these were the choices made in three different investigations.

Case 1: $n = -5/3$ (or, equivalently, $k_p = 0$). This was the case considered by Kraichnan (1976). For $k \to 0$, he derived the general form

$$v(k|k_c) = \frac{2\pi}{15} \int_{k_c}^{\infty} j^2 \left\{ 7q(j) + \frac{j\, dq(j)}{dj} \right\} \frac{1}{\omega(j) + \omega(j) + \omega(k)} \, dj, \quad (10.17)$$

by (in effect) expanding out $k^{-2} A(k|k_c)$ in powers of $(j - |\mathbf{k} - \mathbf{j}|)$.

The reasons for neglecting the term involving $B(k|k_c)$ are not given in detail. As we have just seen, such a step is only valid once we have made the additional assumption that $n < 2$. In fact, the general case (as $k \to 0$) requires a more general equation than (10.17), with an asymptotic contribution from the input term as well[1].

However, after that Kraichnan assumed that the Kolmogorov power laws

$$E(k) = \alpha \varepsilon^{2/3} k^{-5/3} \tag{2.137}$$

$$\omega(k) = \beta \varepsilon^{1/3} k^{2/3}, \tag{6.121}$$

where α and β are constants, could be applied to all wavenumbers. Then it follows that $B(k|k_c)$ vanishes as $k \to 0$, eqn (10.17) becomes valid, and indeed reduces to the simpler form

$$v(k|k_c) = k^{-2} A(k|k_c)$$

$$= \frac{\alpha}{12\beta} \varepsilon^{1/3} k_c^{-4/3} \qquad k \ll k_c. \tag{10.18}$$

We should note the important feature of this asymptotic subgrid viscosity: it is independent of k.

At the other extreme, we have $k = k_c$. Now, not only can the input term no longer be neglected, but its presence turns out to be vital in order to cancel a singularity in the output. We can see this as follows. Taking the full form for $v(k|k_c)$, as given by eqn (10.12), and substituting (2.137) and (6.121) for the spectra and modal decay rates, we obtain

$$v(k|k_c) = \left(\frac{\alpha\varepsilon^{1/3}}{4\pi\beta}\right)k^{-2}\int_{j\geqslant k_c}d^3j\, L(\mathbf{k},\mathbf{j})|\mathbf{k}-\mathbf{j}|^{-11/3}\times$$

$$\times\frac{k^{-11/3}-j^{-11/3}}{k^{2/3}+j^{2/3}+|\mathbf{k}-\mathbf{j}|^{2/3}}. \tag{10.19}$$

The singularity arises when $k = j = k_c$, such that $|\mathbf{k} - \mathbf{j}| = 0$. However, it can be shown that the factor $k^{-11/3} - j^{-11/3}$ cancels rapidly enough to make the singularity integrable, as $|\mathbf{k} - \mathbf{j}| \to 0$. Accordingly, the integral over $\mathbf{j}$ exists, and can be evaluated by making the change of variables

$$\mathbf{j} = k_c\mathbf{J},$$

so that, for $k = k_c$, eqn (10.19) becomes

$$v(k_c|k_c) = B\left(\frac{\alpha\varepsilon^{1/3}}{\beta}\right)k_c^{-4/3}, \tag{10.20}$$

where B is given by (see also eqns (7.113) and (7.114); note that $B = D''$)

$$B = \frac{1}{4\pi}\int_{J\geqslant 1}d^3j\, L(\mathbf{1},\mathbf{J})|\mathbf{1}-\mathbf{J}|^{-11/3}(1-J^{-11/3})\frac{1}{1+J^{2/3}+|\mathbf{1}-\mathbf{J}|^{2/3}}. \tag{10.21}$$

It follows from eqn (10.18) that we can also write (10.20) in terms of the asymptotic eddy viscosity as

$$v(k_c|k_c) = 12Bv(0|k_c). \tag{10.22}$$

Kraichnan has given this as

$$v(k_c|k_c) = 5.24v(0|k_c), \tag{10.23}$$

which implies the value $B = 0.44$ for the integral.

The constant β can be eliminated by invoking the energy integral—see eqn (6.125)—with the result that (10.18) becomes

$$v(0|k_c) = \left(\frac{0.44}{\alpha}\right)\varepsilon^{1/3}k_c^{-4/3} \qquad k \ll k_c. \tag{10.24}$$

The full numerical calculation of $v(k|k_c)$ is presented in Fig. 10.1 (broken curve), with the value of the Kolmogorov constant chosen to be $\alpha = 1.5$. It can be seen that, as $k \to 0$, the effective viscosity quickly reaches its asymptotic value $v(0|k_c) = 0.29\varepsilon^{1/3}k_c^{-4/3}$, and, for $k \to k_c$, it rises to a cusp where $v(k_c|k_c) = 5.24v(0|k_c)$.

Case 2: $n = 1$, $k_p \leqslant k_c$. Leslie and Quarini (1979) considered the effect of what they called a 'production spectrum', which was supposed to be typical of real shear flows. The form chosen reduced to $E(k) \sim k$ at small values of k,

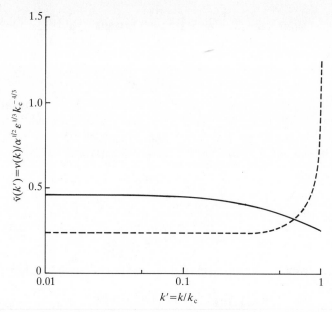

Fig. 10.1. Comparison of subgrid eddy viscosities for isotropic turbulence, based on RPT and RG theory: $----$ RPT value, as calculated from equation (10.12) for $\alpha = 1.5$. $\underline{\hspace{3cm}}$ RG (iterative averaging) value, as calculated from equation (10.26) for $h = 0.8$.

and to the Kolmogorov form for large k. In both cases 'small' and 'large' are relative to k_p. Then eqn (10.12) was used to compute $v(k|k_c)$ for various values of the ratio k_c/k_p.

For large values of this ratio, the computed subgrid eddy viscosity was virtually identical with Kraichnan's result, as indeed one would expect. However, as k_c/k_p was reduced in steps, from 16 down to 2, the cusp at $k = k_c$ softened and reduced while the low-k asymptote was more or less unaffected.

This behaviour confirms Kraichnan's remark that the form of the cusp at $k = k_c$ is not universal. It can also be readily understood in terms of the preceding analysis of Case 1. As k_p tends to zero, the integrand on the r.h.s. of eqn (10.21) can approach the integrable singularity at $|\mathbf{k} - \mathbf{j}| = 0$, with the resulting development of the cusp.

Case 3: $n = 4$, $k_p \leqslant k_c$. Chollet and Lesieur (1981) carried out an analysis, similar to that of Kraichnan, but based on the Eddy-damped quasi-normal Markovian closure. Their expression for $v(k|k_c)$ as $k \to 0$ (their eqn (2.5)) is slightly different from Kraichnan's result, as quoted here in the form of eqn (10.17), but Chollet and Lesieur seem to imply that they are the same.

From our present point of view, the most interesting result from this work is that Chollet and Lesieur found that their eddy viscosities became negative as k_p became small relative to k_c. This behaviour accompanied a choice of initial spectrum which varied as k^4 for small k, and was most marked as $k_p \to k_c$. They also found a reduction of the cusp at $k = k_c$, and their results in this instance were similar to those of Leslie and Quarini.

Recently, Domaradzki, Metcalfe, Rogallo, and Riley (1987) used a similar initial spectrum when they obtained eddy viscosities for isotropic turbulence directly from 64-cubed and 128-cubed numerical simulations. Qualitatively, their results are quite like those of Chollet and Lesieur, with a reduced cusp at $k = k_c$ and negative viscosities for $k < k_c$. However, at a more detailed level, there appears to be very little quantitative agreement between the two investigations.

We should also note that Domaradzki *et al.* cite the form of eddy viscosity (for small k) which is due to Chollet and Lesieur, but erroneously attribute it to Kraichnan (1976). There are, as we have pointed out earlier, small differences in the coefficients between the two forms. Further, Domaradzki *et al.* have incorrectly concluded from this relationship that there will be a cross-over to negative eddy viscosities when the spectrum behaves as k^n, which $n < -5$, for $k < k_c$. However, this is attempting to explain the existence of a cross-over from the properties of the output term alone. As we saw previously, such an explanation must take the input term into account as well, with the result that one predicts a cross-over at $n = 2$.

10.3.3 *Renormalization group*

In Section 9.4, we saw how we could progressively eliminate degrees of freedom by averaging over a series of shells in wavenumber space, and how this led to a reduced form of Navier–Stokes equation defined on the residual interval in **k**-space. The non-linear coupling of the residual modes to the eliminated modes was represented by a renormalized eddy viscosity. This, in turn, was determined by an iteration which reached a fixed point associated with the onset of Kolmogorov scaling behaviour.

While discussing this process, we noted the close resemblance to the ideas associated with the LES technique. In this section we shall consider this point a little more formally, so that the renormalized viscosity will be interpreted as a subgrid viscosity, and this will provide a basis for a comparison with the results from RPT.

In RG, the existence of a fixed point is taken as an indication that universal behaviour has been reached. In the present case, we can interpret this as meaning that we have iterated for a sufficient number of cycles for a universal form of the effective viscosity to have been established. If we denote the number of iteration cycles to the fixed point by $n = N$, for any particular value of h, then, from eqn (9.90), it follows that we have eliminated modes in **k**-space down from k_0 to k_N, where k_N is given by

$$k_N = h^N k_0 \qquad (10.25)$$

and the scaling parameter lies in the range $0 < h < 1$.

Now we can make the connection with the LES formulation quite simply by choosing the cut-off wavenumber to be the wavenumber at the fixed point, i.e.

$$k_c = k_N.$$

Then the equation of motion for the explicit scales is given by (9.93), with all wavevectors satisfying the condition $0 < k, j, |\mathbf{k} - \mathbf{j}| < k$. The renormalized viscosity is given by eqn (9.98), and we can rewrite this as

$$v(k) = \tilde{v}\left(\frac{k}{k_N}\right)\alpha^{1/2}\varepsilon^{1/3}k_N^{-4/3}, \qquad (10.26)$$

where we have dropped the subscripts on eddy viscosities at the fixed point.

Numerical values of $\tilde{v}(k/k_N)$ have been reported (McComb and Shanmugasundaram 1985; McComb 1986), and in Fig. 10.1 we show a typical result for $h = 0.8$, which can be compared with the result from RPT. It can be seen that both forms asymptote to constant values as k' tends to zero, and that

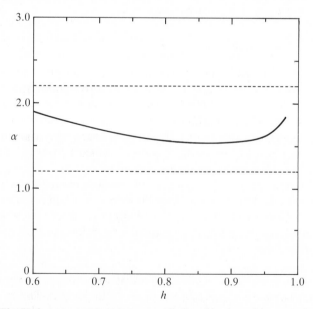

Fig. 10.2. The Kolmogorov constant α, as calculated by iterative averaging from eqn (10.27): variation with the shell thickness parameter h.

their magnitudes are quite comparable. However, at the cut-off $k' = 1$, the RG form demonstrates its local (in wavenumber) character, as it does not give the cusp due to the unphysical behaviour of the Kolmogorov spectrum at low wavenumbers.

Some idea of the quantitative performance of the iterative method can be gained by calculating the Kolmogorov spectral constant α (McComb 1986). This can be done as follows. Substitute from eqn (9.98) for the renormalized viscosity into the dissipation relation for the explicit scales, as given by (9.94). Then, with some rearrangement, we obtain

$$\alpha = \left\{ 2 \int_0^1 \tilde{v}(k')k'^{1/3} \, dk' \right\}^{-2/3}, \tag{10.27}$$

where we again drop the subscript on $\tilde{v}(k')$ at the fixed point.

Equation (10.27) can be evaluated numerically, and the results are plotted in Fig. 10.2 for a range of values of the scaling parameter h. It may be noted that, for these values of h (the range for which the expansion in higher-order moments would converge; see Section 9.4.8), the values of α shown lie comfortably within the most probable experimental range. However, this dependence on the thickness of the eliminated shells is an aspect of the iterative averaging method which still has to be elucidated.

10.4 Miscellaneous simulation methods

From our discussions of the numerical simulation of turbulence, it has become quite clear that there is no fundamental difficulty in solving the Navier–Stokes equation in this way. The techniques are well established and in practice the only limitation is one of computer size. Another aspect of our discussion has, of course, been the ways in which we could reduce the size of a given turbulent computation, so that we can fit it on to an existing computer. In this area we have covered both phenomenological and renormalization methods of modelling subgrid modes, in order to carry out an LES of the Navier–Stokes equation.

In this section, we shall give a brief discussion of some other methods of reducing the computational problem involved in representing turbulence numerically on the computer. These particular methods can be divided roughly into two categories: those which simplify the computation of the Navier–Stokes equation but do not conveniently fit into the mainstream category discussed above, and those which bypass the Navier–Stokes equation altogether. In addition, we shall confine ourselves to methods where some actual computation has taken place and has been reported in the literature. This, for the present, rules out those novel approaches which have not yet got beyond the stage of theoretical speculation.

10.4.1   *Methods based on the Navier–Stokes equation*

We have seen that, for a fixed number of grid points stored on a computer, there has to be a tradeoff between the value of the Reynolds number that we can achieve on the one hand, and the complexity of the flow that we wish to study on the other. This is, of course, just a matter of symmetry. The more symmetric the flow, the higher is the Reynolds number which can be obtained. Thus we rank flows in order of difficulty: isotropic turbulence, followed by flows in slab geometry (themselves ranked according to boundary and initial conditions), and so on.

Now, these symmetries are applicable to the mean behaviour of turbulent flows and hence (despite their geometric nature) are in this context purely statistical concepts. In other words, the problem of representing the raw velocity field, obtained by solving the discretized Navier–Stokes equation on a mesh in a box, is much the same whether the field is isotropic or corresponds to plane Poiseuille flow for example. The differences really only arise when one starts to work out averages (this is an oversimplification, as we are ignoring the practical consequences of different boundary conditions). Thus the question may be asked: can one find even simpler computational problems in which symmetry constraints can be used to reduce the number of points at which one must calculate the unaveraged velocity field?

For quite a long time there has been speculation that the Taylor–Green vortex (Taylor and Green 1937) could provide such a problem. This is a three-dimensional vortex structure with an initial velocity field which is two-dimensional. As time goes on, a three-dimensional velocity field develops, and vortex stretching produces a form of turbulence. The essential computational feature is that symmetries applied to the initial deterministic velocity field are taken to apply for all subsequent times including, ultimately, the resulting turbulent field. In a well-known simulation, Brachet *et al.* (1983) observed a turbulent inertial range with power-law exponent of the order 1.6–2.2 (note that $5/3 = 1.67$) at a Taylor–Reynolds number of $R_\lambda = 110$. The authors state that their TG code with 256-cubed resolution is equivalent to a 64-cubed general spectral code in terms of computational work and storage.

More recently, Kida and Murakami (1987) have identified an even more symmetric field which is essentially just a pragmatic modification of the usual concept of isotropic turbulence in a box. By imposing additional symmetries upon the unaveraged velocity field, these authors state that the value of a single component of the velocity field (as opposed to three such components) at each point of a sub-box, of volume equal to 1/64 of the volume of the original box, is sufficient to represent the whole velocity field. In this way, they were able to make an 85-cubed simulation resolve 3×340-cubed effective modes, thus enabling them to achieve Taylor–Reynolds numbers up to $R_\lambda = 100$ and a decade of Kolmogorov inertial-range spectrum. These results can be compared with those obtained in the 128-cubed conventional simulation of Kerr ($R_\lambda = 82.9$ and a short range of $-5/3$ spectrum) as discussed in Section 10.1.1.

It is clear that the application of deterministic symmetry criteria to a random velocity field begs a good many questions. Clearly, in the language of the subject, these cannot be regarded as 'honest simulations'. This point seems to be recognized by Brachet *et al.*, who really only claim to be studying the small-scale structure of the TG vortex, and who (noting that their values of skewness are much larger than for wind-tunnel turbulence) concluded that their TG simulation may be reproducing certain features of geophysical turbulence at much higher Reynolds numbers. However, the results of this approach are interesting, and clearly it merits further (and more critical) attention.

Another possible way of reducing the number of degrees of freedom is to take a representative sample of the Fourier modes. A standard method is to use random sampling, as in Monte Carlo methods. But, in a spectral simulation of turbulence, this raises technical problems to do with satisfying the various invariance properties. Hosokawa and Yamamoto (1987) have demonstrated that such problems can be overcome, in connection with a numerical simulation of the Hopf characteristic functional. Using a 32-cubed spectral code, they carried out simulations at $R_\lambda = 57.7$, 115, 231, and 577. Their energy spectra demonstrated Kolmogorov scaling and they found the constant of proportionality to be $\alpha = 1.4$. A comparison of their result for the one-dimensional spectrum with the LET theory for $R_\lambda = 533$ (McComb and Shanmugasundaram 1984; see Section 8.2.3) indicated good agreement between the simulation and the analytical theory.

10.4.2 *Allternatives to the Navier–Stokes equation*

In fluid mechanics the study of vortex dynamics has long been regarded as a valid alternative to a direct attack on the equations of motion. The general method is to assume a form of vortex (i.e. to specify the localized velocity distribution) and then to consider an assembly of such vortices. Interactions between the constituent vortices are inductive and can be taken to be described by the Biot–Savart law. In particular cases (e.g. a long vortex filament) there can be a self-inductive effect, with the vortex filament moving in its own induced velocity field.

A classical problem is to represent a vortex sheet by an array of point vortices. In practice, as the sheet rolls up, the individual vortices tend to behave in a chaotic way and destroy the representation of the vortex sheet. This can make it difficult to use ideal flow methods in aerodynamic situations (e.g. see Saffman and Baker 1979), but, in the present context, the possible relevance to the breakdown of shear layers to turbulence is of interest, and additional point is given to that interest by the vortical character of turbulent shear layers.

Simulations of turbulent shear layers have proceeded from a few hundred to many thousand vortices (Kadomtsev and Kostomarov 1972; Ashurst 1979; Aref and Siggia 1980), and have generally managed to reproduce some of the features observed in laboratory experiments on such flows. Other vortex simulations include the turbulent spot (Leonard 1980) and the axisymmetric turbulent jet (Edwards and Morfey 1981). Again, the quantitative aspects are realistic, and—as in the simulations of the shear layer—the results are really more of interest to the study of coherent structures than in the context of quantitative turbulent simulations. Accordingly, we shall not pursue the subject any further here, and the interested reader is referred to the review by Leonard (1984).

From a fundamental point of view, the limiting feature of vortex methods is that they represent continuous distributions of vorticity in a fluid by discontinuous vortices. Nevertheless, they preserve the macroscopic continuum picture of the fluid, and one might enquire whether there is anything to be gained by introducing a discrete representation at some more microscopic level.

On the face of it, the idea of first simulating the Navier–Stokes equation from the molecular level and then simulating turbulence on top of that (at enormously greater length and time scales!) seems a rather clumsy approach to the problem. Certainly simulations of macroscopic behaviour, obtained by using a computer to solve Newton's second law for intermolecular collisions in a gas, were being performed for two decades before the first numerical simulation of turbulence. And in recent years, these methods have been extended to simple non-equilibrium flows (for a review, see Evans and Hoover 1986). But, taking the most advanced simulation of macroscopic fluid flow to date (Rapaport 1987), the prospects for a turbulent simulation do not look good. Rapaport used a molecular dynamics simulation to reproduce the flow of a two-dimensional fluid around a circular obstacle. Known phenomena such as periodic vortex shedding and the formation of a vortex street have been successfully reproduced. However, the author observes that the increase in computer power needed to extend even this simple situation to three dimensions would require the next generation of computers.

It seems, therefore, that one would be justified in totally dismissing the idea of turbulence simulation from microscopic levels. Yet, the recent development of lattice gas models seems to hold out a possibility that such a verdict may be a little premature.

In this approach, the gas is modelled by the movement of particles (all of which have the same constant speed) along the links of a lattice. The rules which decide whether a particle will move along one lattice link rather than another (or remain stationary or collide with another particle) are deterministic and locally determined. Such a set of rules constitutes what is known as a cellular automation (Wolfram 1983). Macroscopic behaviour is then obtained in the usual way by averaging over many 'particles' to obtain the drift velocities.

In two dimensions, it has been shown (Rivet and Frisch 1986) that the lattice gas model reduces to the two-dimensional Navier–Stokes equations, and plane Poiseuille flow has been successfully simulated (d'Humieres and Lallemand 1986). The success of this method depends on the correct choice of lattice type and also the collision rules. In two dimensions, the hexagonal lattice gives the required isotropic macroscopic behaviour, but apparently there is no corresponding lattice in three dimensions. Recently d'Humieres, Lallemand, and Frisch (1986) reported the derivation of the three-dimensional Navier–Stokes equations from two lattice gas models, i.e. a multi-speed model on a three-dimensional cubic lattice and the three-dimensional projection of the four-dimensional Navier–Stokes equations modelled on a face-centred hypercubic lattice.

Much about this technique remains mysterious, but at least there appears to be a possibility of quasi-microscopic simulations with much smaller numbers of particles than in molecular dynamics simulations. Nevertheless, Orszag and Yakhot (1986) have concluded that the computational requirements for the lattice gas models are likely to be much more severe than for the Navier–Stokes equations. Intuitively this is obvious, and is always likely to be the case. Perhaps these models may turn out to be useful in cases where the continuum equations are either unknown or particularly difficult to solve. One thinks of non-Newtonian and two-phase flows, flows with chemical reactions, and flows involving combustion. Certainly this subject is developing so rapidly that any conclusion or firm prediction is likely to be falsified by experience rather quickly.

Note

1. The need for this was pointed out to the present writer by J. K. McKee (personal communication, 1987). In an unpublished analysis, McKee derived the general expression (i.e. including the input term) for the effective viscosity, in the limit $k \to 0$, as

$$v(k|k_{\mathrm{c}}) = \frac{2\pi}{15} \int_{k_{\mathrm{c}}}^{\infty} t(j, j, k) \left\{ 7q(j) + \frac{j \partial q(j)}{\partial j} \right\} j^2 \, \mathrm{d}j -$$

$$- \frac{14\pi}{15q(k)} \int_{k_{\mathrm{c}}}^{\infty} t(j, j, k) q^2(j) j^2 \, \mathrm{d}j,$$

where $t(k, j, l)$ is the correlation time for triple moments and is given by

$$t(k, j, l) = [\omega(k) + \omega(j) + \omega(l)].$$

This expression should be compared with eqn (10.17), where only the output term (i.e. the first term on the r.h.s. above) is included.

References

AREF, H. and SIGGIA, E. D. (1980) *J. fluid Mech.* **100**, 705.
ASHURST, W. T. (1979). In *Turbulent shear flows*, Vol. 1 (eds F. DURST, B. E. LAUNDER, F. W. SCHMIDT, and J. H. WHITELAW), Springer, New York.
BIRINGEN, S. and REYNOLDS, W. C. (1981). *J. fluid Mech.* **103**, 53.
BRACHET, M. E., MEIRON, D. I., ORSZAG, S. A., NICKEL, B. G., MORF, R. H., and FRISCH, U. (1983) J. fluid Mech. **130**, 411.
CHOLLET, J-P. and LESIEUR, M. (1981). *J. atmos. Sci.* **38**, 2747.
CLARK, R. A., FERZIGER, J. H., and REYNOLDS, W. C. (1979). *J. fluid Mech.* **91**, 1.
DOMARADZKI, J. A., METCALFE, R. W., ROGALLO, R. S., and RILEY, J. J. (1987). *Phys. rev. Lett.* **58**, 547.
EDWARDS, S. F. (1964). *J. fluid Mech.* **18**, 239.
EDWARDS, A. V. F. and MORFEY, C. L. (1981). *Comput. Fluids*, **9**, 205.
EIDSON, T. M. (1985). *J. fluid Mech.* **158**, 245.
EVANS, D. J. and HOOVER, W. G. (1986). *Ann. Rev. fluid Mech.* **18**, 243.
GERMANO, M. (1986). *Phys. Fluids* **29**, 2323.
HERRING, J. R. and KERR, R. M. (1982). *J. fluid Mech.* **118**, 205.
HOSOKAWA, I. and YAMAMOTO, K. (1987). *J. Phys. Soc. Jpn* **56**, 521.
d'HUMIERES, D. and LALLEMAND, P. (1986). *C.R. Acad. Sci. Paris, Ser. ii*, **302**, 983.
KADOMTSEV, B. B. and KOSTOMAROV, D. P. (1972). *Phys. Fluids* **15**, 1.
KANEDA, Y. and LESLIE, D. C. (1983). *J. fluid Mech.* **132**, 349.
KERR, R. M. (1985) *J. fluid Mech.* **153**, 31.
KIDA, S. and MURAKAMI, Y. (1987). *Phys. Fluids* **30**, 2030.
KIM, J., MOIN, P. and MOSER, R. (1987). *J. fluid Mech.* **177**, 133.
KRAICHNAN, R. H. (1959). *J. fluid Mech.* **5**, 497.
—— (1966). *Phys. Fluids* **9**, 1728.
—— (1976). *J. atmos. Sci.* **33**, 1521.
LEONARD, A. (1980). *J. comput. Phys.* **37**, 289.
—— (1985). *Ann. Rev. fluid Mech.* **17**, 523.
LESLIE, D. C. and QUARINI, G. L. (1979). *J. fluid Mech.* **91**, 65.
LOVE, M. D. (1980). *J. fluid Mech.* **100**, 87.
McCOMB, W. D. (1974). *J. Phys. A* **7**, 632.
—— (1986) in *Direct and large eddy simulation of turbulence* (eds U. SCHUMANN and R. FRIEDRICH), *Notes on numerical fluid mechanics*, Vol. 15. Vieweg, Braunschweig.
—— and SHANMUGASUNDARAM, V. (1984). *J. fluid Mech.* **143**, 95.
—— and —— (1985). In *Proc. 4th. Conf. on Numerical Methods, Swansea, 1985.* Pineridge Press, Swansea.
McMILLAN, O. J. and FERZIGER, J. H. (1979). *AIAA J.* **17**, 1340.
MASON, P. J. and CALLEN, N. S. (1986). *J. fluid Mech.* **162**, 439.
METCALFE, R. W. and RILEY, J. J. (1981). In *Proc. 7th Int. Conf. on Numerical Methods in Fluid Dynamics* (eds W. C. REYNOLDS and R. W. MacCORMACK). Springer, New York.
MOENG, C-H. (1984). *J. atmos. Sci.* **41**, 2052.
MOIN, P. and KIM, J. (1982). *J. fluid Mech.* **118**, 341.
ORSZAG, S. A. (1971). *Stud. appl. Math.* **50**, 293.
—— and PAO, Y.-H. (1974). *Adv. Geophys.* **18A**, 225.
—— and PATTERSON, G. S. (1972). *Phys. Rev. Lett.* **28**, 76.
—— and YAKHOT, V. (1986). *Phys. Rev. Lett.* **56**, 1691.
RAPAPORT, D. C. (1987). *Phys. Rev. A* **36**, 3288.
RIVET, J.-P. and FRISCH, U. (1986). *C.R. Acad. Sci. Paris, Ser. ii*, **302**, 267.

SAFFMAN, P. G. and BAKER, G. R. (1979). *Ann. Rev. fluid Mech.* **11**, 95.

SCHUMANN, U. and PATTERSON, G. S. (1978a). *J. fluid Mech.* **88**, 685.

—— and —— (1978b). *J. fluid Mech.* **88**, 711.

—— GROTZBACH, G. and KLEISER, L. (1980). In *Prediction methods for turbulent flows* (ed. W. KOLLMAN). Hemisphere, Washington.

SIGGIA, E. D. (1981). *J. fluid Mech.* **107**, 375.

—— and PATTERSON, G. S. (1978) *J. fluid Mech.* **86**, 567.

SPALART, P. R. (1988). *J. fluid Mech.* **187**, 61.

SPEZIALE, C. G. (1985). *J. fluid Mech.* **156**, 55.

TAYLOR, G. I. and GREEN, A. E. (1937). *Proc. R. Soc. Lond. A* **158**, 499.

WOLFRAM, S. (1983). *Rev. mod. Phys.* **55**, 601.

11

COHERENT STRUCTURES

Organizing a logical account of a fashionable and fast-growing subject, like the study of coherent structures (or recognizable 'deterministic' patterns) in turbulent flows, is not without its difficulties. To begin with, quite a variety of such structures has now been identified, and, although different structures can appear to be very different phenomena, clearly they must all be facets of the same basic phenomenon, if only at the level that they are all solutions of the Navier–Stokes equation!

At the present time, the safest way of tackling the problem of taxonomy, it seems to us, is to make a subdivision of the subject into very broad generic classes. For instance, a division into flows without solid boundaries—where coherent structure emerges rather easily as an organized vortex pattern—and flows with solid boundaries—where the coherent structure is not so easily seen, and requires a good deal of careful elucidation—seems likely to prove helpful, if only because such a division has tended to reflect different schools of activity.

We shall also make a further division of wall-bounded flows into those flows where the Reynolds number is intermediate between the values associated with laminar and turbulent flows, and flows where the turbulence is well developed. We should note in passing that, although we shall thereby consider the transition from laminar to turbulent flows, we shall not deal with the subject of hydrodynamic stability, which is an important subject in its own right (see our suggestions for further reading at the end of this chapter).

Then there is the question of historical priority. The current interest in coherent structures stems from certain specific investigations in the late 1960s/ early 1970s. Yet, even from its earliest days, turbulence research has always included—if only as a minority interest—some consideration of deterministic structures such as waves or vortices. Thus, inevitably perhaps, certain of the later results were to some extent rediscoveries, and their value may lie in having made the relevant phenomena more apparent or better known.

However, from our present point of view, our problem is one of deciding whether to discuss the various investigations in chronological order or to adopt (essentially) the flashback method. In the event, we have taken no definite decision on this, but have used whichever method seems best suited to clarity of exposition in a given topic.

Lastly, we should make a few general remarks about vortices and vorticity. Although it is natural to refer to the eddying motions which occur in real fluid flows as 'vortices', it should at least be borne in mind that the term has a precise meaning in classical hydrodynamics. In an ideal fluid there can be a mathematically sharp boundary between rotational and irrotational regions of the flow, so that the concept of a vortex is well defined. In a viscous fluid the concept becomes blurred—literally—as the effect of viscosity is to sustain a velocity gradient and hence to smear out the boundary between rotating and non-rotating fluid. Nevertheless, some investigators feel that the methods of classical hydrodynamics can usefully be extended (with suitable modifications) to real fluids and, of course, this is a particularly attractive approach in the case of coherent structures.

The concept of vorticity is equally available to both real and ideal fluids, being merely the curl of the velocity field. It may be helpful to think of vorticity as a propensity on the part of the fluid to rotate. The direction of the associated spin vector is—by definition—at right angles to the fluid velocity. The reader who is not used to thinking in terms of vorticity should be careful to bear this in mind, and (for instance) recall that the term 'streamwise vorticity' means that component of the spin vector which points in the direction of flow. Thus, with our usual conventions, we would be talking about ω_1, which would (from the definition $\boldsymbol{\omega} = \text{curl } \mathbf{u}$) involve velocity components u_2 and u_3. Some general references on the subject of vortex methods are given at the end of this chapter.

11.1 Coherent structures in free turbulent flows

11.1.1 *Plane mixing layers*

Possibly the most striking of all the coherent structures in turbulent flow is the Brown–Roshko vortex, as found in the plane mixing layer at high Reynolds numbers (Brown and Roshko 1974). The physical situation is illustrated in Fig. 3.9, where the mixing layer (or free shear layer) is formed at the boundary of two flows which are moving at different speeds, and the roll vortices, as observed by means of spark shadowgraphs, are sketched in Fig. 3.10. As we shall see, the remarkably regular 'two-dimensional' quality of these vortices may be uncharacteristic (and even perhaps somewhat fortuitous). Nevertheless, they have played a major part in stimulating interest in the whole subject of coherent structures.

The experiments of Brown and Roshko were carried out with gases of different densities, the most extreme case being nitrogen in one stream and helium in the other, thus giving a density ratio of 7:1 across the mixing layer. The Reynolds number—based (as in boundary layers) on the downstream distance—took values up to 10^6, quite adequate to ensure well-developed turbulence.

If one disregards the small-scale structure due to the turbulence, then the shadowgraphs of the mixing layer look for all the world like a classical vortex street. But the vortex street is, of course, a phenomenon also encountered at low

Reynolds numbers, where the turbulence is not appreciable. It is this fact which made these results so interesting and surprising, and indeed Brown and Roshko pointed out the resemblance between their structures and the vortices found by Freymuth (1966) who studied the laminar instability of the free shear layer.

However, they also noted an important difference between the laminar and turbulent cases. In the former, the spacing of the eddies was constant and related to the wavelength of the perturbation which initiated them, whereas, in the turbulent case, the diameter of eddies and the spacing between them both increased with downstream distance. This behaviour was attributed to the amalgamation of neighbouring eddies as they proceeded downstream.

It was soon established that the roll vortices were not just a property of the mixing layer between gas flows of different densities. It was shown by Winant and Browand (1974) that the effect could be found in water at small Reynolds numbers, and by Dimotakis and Brown (1976) that the roll vortices were present at values of the Reynolds number (based as before on the downstream distance) up to 3×10^6. Both investigations relied on flow visualization, but the latter authors also use a laser anemometer to measure the velocity field. In particular, they measured the autocorrelation of streamwise velocities at a fixed point but with a variable time lag τ.

The basic definition of the autocorrelation can be obtained by specializing eqn (2.36). As usual, we take x_1 to be the direction of flow. Then, putting $\alpha = \beta = 1$, $\mathbf{x} = \mathbf{x}' = x_1$, and $t' = t + \tau$, we have the streamwise autocorrelation defined as

$$Q_{11}(x_1, x_1; t, t + \tau) = R_{11}(\tau)\langle u_1^2 \rangle, \qquad (11.1)$$

where the general correlation tensor $Q_{\alpha\beta}$ is, in turn, defined by eqn (2.24). It should be noted in passing that Dimotakis and Brown actually take the autocorrelation coefficient to be (in effect) Q divided by $(\Delta U/2)^2$, where $\Delta U = U_a - U_b$, and U_a and U_b are the upper and lower stream velocities as shown in Fig. 3.9.

The autocorrelation coefficients obtained were found to reflect the results of the flow visualization in that they showed a clear oscillatory dependence on the lag time. Naturally one would attribute this to a periodic structure being swept past the measuring point in the direction of flow (i.e. x_1) with some mean convention velocity. A plausible choice of convection velocity would be the mean of the upper and lower streams, i.e

$$U_c = \frac{U_a + U_b}{2}, \qquad (11.2)$$

with the convection time scale τ_c following as

$$\tau_c = x_1/U_c. \qquad (11.3)$$

If we denote the time taken for the autocorrelation coefficient to reach its first minimum by $\tau_0/2$, then it turns out that this period scales quite well on the convection time, as given by eqn (11.3). Dimotakis and Brown found that, for a fixed value of the velocity ratio U_a/U_b, and a wide range of values of x_1 and U_a, the scaled periodic time was approximately constant, satisfying the condition

$$0.40 < \tau_0/\tau_c < 0.50.$$

It was pointed out by Bradshaw (1975) that the shadowgraph method of flow visualization (at least, in the configuration used by Brown and Roshko) tends to average out any spanwise variations, and hence will emphasize the two-dimensional structure of the large-scale eddies. Certainly this raises the question of what the flow actually looks like in plan view (the side elevation view is the one shown in Fig. 3.10).

In considering this point, let us begin by noting that 'two-dimensional' in the present context merely means that there is no variation of mean properties in the spanwise direction. If this is the case, the roll vortices sketched in Fig. 3.10 can be interpreted as cross-sections of cylinders of rotating fluid (or, perhaps, vortex tubes), with the 'diameter' of the cylinder being independent of the spanwise coordinate.

In practice, of course, the spanwise extent of the flow must be finite, and so there must be edge effects, with various forms of instability (e.g. kinking) of the vortex tubes being possible. This was found to be the case by Chandrsuda, Mehta, Weir, and Bradshaw (1978) , who made a flow visualization of the plane mixing layer and concluded that the Brown–Roshko vortices were associated with transitional behaiour at low ambient turbulence levels. They also concluded that the Brown–Roshko vortices were rare in practice, with the characteristic structure being fully three-dimensional and hence less obviously ordered. Later investigations of spanwise structure broadly support these conclusions (Browand and Troutt 1980; Jimenez 1983). They also underline just how complicated this simple flow configuration can be, with many factors (including initial conditions) having to be taken into account.

We conclude this section with an interesting investigation by Tavoularis and Corrsin (1987), who sought to eliminate purely systematic (e.g. geometrical) effects by constructing their shear layer from a non-uniform array of turbulent jets and wakes, and surrounding this by the approximately isotropic turbulence generated by a grid. The resulting shear layer was found to be without detectable periodicity, yet weak periodic vortices grew up as the flow moved downstream. The particularly interesting feature of these vortices was that their scale and frequency matched those generated in shear layers which began with strong vortices. This must tend to suggest that the Brown–Roshko vortices may indeed be a universal property of the flow configuration.

11.1.2 *Other free shear flows*

Coherent structures have been found in other free shear flows but are not as striking or as dramatic as the Brown–Roshko roll vortices. In general, their presence has to be established by careful eduction methods, usually based on quantitative measurements rather than on flow visualization alone. Nevertheless, the early work by Grant (1958) was quite specific on the subject. He described the large-eddy structure of turbulent flow as being 'more ordered than has usually been supposed'. Indeed, his photographs of the wake behind a cylinder being towed through still water show evidence of a regular vortex motion which modulates the fine-scale turbulence. Despite the blurring effect of the turbulence, it is quite possible to discern an ordered structure which resembles the vortex street which one would expect at Reynolds numbers too low for turbulence to appear.

The presence of these regular eddies was confirmed by detailed measurement of velocity correlations using hot-wire anemometry (incidentally also confirming the still earlier measurements and speculation of Townsend (1956)).

Further research in this area has mainly concentrated on quantitative methods of elucidating the large vortex structures of the plane wake. Recent accounts (including many references to work carried out in the interim) can be found in Mumford (1983) and Hussain and Hayakawa (1987). Both these investigations used arrays of hot-wire anemometers to map out the coherent structures, but Mumford detected the velocity signal whereas Hussain and Hayakawa made the instantaneous vorticity field the basis of their detection scheme.

The other classical free-shear flows have also yielded their own coherent structures. For instance. Crow and Champagne (1971) found vortex 'puffs' in the transitional region of the round free jet. We shall not go into further detail here, but in the interests of completeness we shall list some representative investigations into coherent structures in various free shear flows as follows: three-dimensional wake behind bluff bodies (Perry and Watmuff 1981, Perry and Steiner 1987, Steiner and Perry 1987); round jets (Bruun 1977, Yule 1977, Sreenivasan, Antonia, and Britz 1979); plane jets (Rajagopalan and Antonia 1981, Mumford 1982, Moum, Kawall, and Keffer 1983).

11.2 Conditional sampling, intermittency and the turbulent–non-turbulent interface

In Section 11.1.1, we noted how the strikingly clear Brown–Roshko vortices could easily be demonstrated by traditional methods: by both flow visualization and correlation of anemometer signals. Not surprisingly, the less clear coherent structures encountered in other free-shear flows require more elaborate techniques, such as pattern-recognition methods, for their elucidation.

A discussion of the technicalities of this subject would take us too far afield, but there are certain aspects which have a general application to the study of turbulence, and hence merit some attention here. To be specific, we shall discuss the topic of conditional sampling.

As an example, let us consider the intermittency which arises at the outer edge of a boundary layer or a jet. As we saw previously in Section 3.2, we can distinguish quite sharply between turbulent and non-turbulent fluid. The fact that the position of the interface between the two regions is a random function of time means that a suitably sited anemometer will experience an intermittent signal. Thus, if we were to take an average of the whole velocity record, we would be including periods when there was no turbulent signal at all.

In one sense, of course, this would be a perfectly reasonable thing to do, and would give an average velocity at the measuring point. However, for other purposes, we might see a more meaningful average as being one in which only the turbulent portions of the signal were included.

Now the idea of conditional sampling is really quite general. It simply means that we sample a data set according to some predetermined criterion. In our present example, an obvious form of conditional sampling would be to accept portions of the velocity record into the averaging if, and only if, they were known to be turbulent.

Alternatively, we could generalize this procedure to situations where there were two anemometers at the same streamwise location, but with one in the intermittent zone and the other deep in the boundary layer. Then we could accept data from inner anemometer, conditional upon there being turbulence at the outer anemometer. (We do not assert that this would be dynamically significant, merely that it would be an example of conditional sampling.) Clearly there are many such possibilities.

In turbulence, the pioneering work in this area is due to Kovasznay, Kibens, and Blackwelder (1970), who introduced an intermittency function $I(\mathbf{x}, t)$ such that

$$I(\mathbf{x}, t) = \begin{cases} 1 & \text{for turbulent flow} \\ 0 & \text{for non-turbulent flow.} \end{cases} \tag{11.4}$$

With this definition, the intermittency function $I(t)$ at a particular value of $\mathbf{x}$ consists of a series of positive-going unit pulses of random interval and duration. It may be related to Townsend's intermittency factor γ (see Section 3.2.1) by time averaging:

$$\gamma(\mathbf{x}, t) = \bar{I}(\mathbf{x}, t) = \lim_{T \to \infty} \frac{1}{T} \int_t^{t+T} I(\mathbf{x}, t') \, \mathrm{d}t'. \tag{11.5}$$

Then, using the terminology of Kovasznay *et al.*, we can introduce new conditional averages, such as 'zone averages', which we shall define as follows. Consider some arbitrarily chosen property of the turbulence, which we shall call $f(t)$. That is, f could be, for example, the pressure or a scalar component of the velocity field at a given point. Evidently it will be a fluctuating quantity, and its time average is given by

$$\bar{f} = \lim_{T \to \infty} \frac{1}{T} \int_t^{t+T} f(t') \, dt'. \tag{11.6}$$

Now, if we perform this time average only during those intervals when $I(t) = 1$, we obtain

$$\bar{\bar{f}} = \left\langle \frac{If}{\gamma} \right\rangle = \lim_{T \to \infty} \frac{1}{T} \int_t^{t+T} f(t') I(t') \, dt'. \tag{11.7}$$

Kovasnay *et al.* called this quantity a 'turbulent zone average'.

Conversely, a 'non-turbulent zone average' can be obtained by averaging only during the intervals when $I = 0$, or

$$\tilde{\tilde{f}} = \frac{\langle (1 - I)f \rangle}{1 - I}$$

$$= \lim_{T \to \infty} \frac{1}{(1 - \gamma)T} \int_t^{t+T} \{1 - I(t')\} f(t') \, dt'. \tag{11.8}$$

It readily follows from (11.6), (11.7), and (11.8) that the conventional time average is just the weighted mean of the two zone averages:

$$\bar{f} = \gamma \bar{\bar{f}} + (1 - \gamma) \tilde{\tilde{f}}. \tag{11.9}$$

Clearly we could extend this approach to define conditional averages in many different ways. For instance. Kovasznay *et al.* go on to introduce 'point averaging', a procedure which relies on detecting the instants at which the turbulent –non-turbulent interface passes the detector probe.

However, we mention this only to illustrate the potential of the method, as we shall not pursue any more specific cases here. Instead, we shall concentrate on the most important general point. That is, how do we specify the intermittency function $I(t)$?

The underlying problem, of course, is how do we answer the question: is there turbulence at a point or not? Indeed, which quantity do we measure in the first place in order to determine $I(t)$: the velocity? its first derivative? the vorticity? Clearly there are many options open.

This problem has been studied by various workers. A good discussion of the problem (along with a summary of other work) will be found in Hedley and Keffer (1974). Other specific approaches (not based upon the velocity field and its derivatives) include use of fluid temperature (in the flow past a heated cylinder) as a criterion upon which to base decisions about intermittency (LaRue and Libby 1976), and the use of simulated pseudo-turbulent signals to study the problems of intermittency detection (Antonia and Atkinson 1974).

Attempts to reduce the arbitrariness of decisions about the turbulent–non-turbulent interface, by working from the equation of motion in order to obtain a conservation equation for the intermittency function (Libby 1976) or conditionally averaged transport equations (Dopazo 1977), are undoubtedly the right way forward. While at present this subject is in its infancy, in the future we can expect to see it develop as an important aspect of the usual turbulence closure problem.

Lastly, we should note that the long tradition of pragmaticism in experimental fluid mechanics is upheld by the frequent practical use of conditional sampling, without apparently worrying unduly about the finer points of the underlying theory. In this connection, an interesting work is that of Fiedler (1975), who compares familiar mean quantities (for example, the radial variation of the three turbulent intensities in a plane jet) with their turbulent-zone-averaged equivalents. As one would expect, there is no difference between the two methods of taking means in the core region of the jet. But, in the intermittent outer region, the conditional average gives (again, as one would expect) higher values.

Evidently, this type of correction could have its implications for turbulent energy balances, as discussed in Section 1.6.2. Such balances involve non-local transfers and may be affected by intermittency if terms which dominate the outer region are not corrected by zone averaging.

11.3 Transitional structures in boundary layers and pipes

In Section 3.2 we noted that the transition from laminar to turbulent flow in a pipe was an intermittent process, involving patches of turbulence alternating with intervals of laminar flow. Indeed, this phenomenon was first observed towards the end of the nineteenth century by Reynolds, who invented the idea of an injected dye line as a means of flow visualization (see Section 1.2 for references) and hence was able to see what he referred to as 'flashes of turbulence'. However, the notion that transitional flows might possess interesting structures is due to the much more recent investigation of Emmons (1951), who observed that the transition to turbulence in a boundary layer over a flat plate was also intermittent, with isolated patches of turbulence forming at random, and growing in number and in size as they moved downstream. These patches of turbulence were referred to by Emmons as 'spots', and the term now seems to be universally employed.

Later on, Elder (1960) obtained the first really clear photograph of a turbulent spot in a laminar boundary layer, revealing a shape which has now been shown, by many subsequent investigations, to be quite characteristic. In Plate II we show a photograph of a typical turbulent spot, which in plan view is shaped rather like a diamond with its major axis in the direction of the flow. This structure can further be resolved into two parts. The front part is very disordered and turbulent, with an envelope which is shaped like an arrowhead whereas the rear part (i.e. the rest of the 'diamond') has a much more ordered 'streaky' structures.

Elder's results also supported the conjecture by Emmons that a turbulent boundary layer is simply the aggregation of a number of turbulent spots. Other views of the turbulent spot, as obtained by flow visualization, also support this conjecture. For instance, Perry, Lim, and Teh (1981) found that a spanwise cross-section (i.e. the end view) of a turbulent spot resembled the corresponding (instantaneous) cross-section of the turbulent boundary layer on a flat plate (as sketched, for example, in Fig. 1.2 of the present work). Longitudinal cross-sections (side views) have been found to have a similar appearance (Gad-el-hak, Blackwelder, and Riley 1981).

It has also been found (Wygnanski, Sokolov, and Friedman 1976) that the locus of points of maximum height of the turbulent spot, as it moves downstream, is approximately equal to the thickness of a hypothetical turbulent boundary layer, which is supposed to have originated at the position where the turbulent spot was generated, and which was initially of the thickness of the laminar boundary layer at that point.

Observations of this kind lend considerable support to the idea that the turbulent spots are, so to speak, the basic building blocks of the turbulent boundary layer. But what about other flows which are bounded by rigid walls? In fact, turbulent spots have been observed, at the transitional range of Reynolds numbers in plane channel flow (Carlson, Widnall, and Peeters 1982), and turn out to look very much like those found in the boundary layer on a flat plate. However, there was one striking difference: in the channel flow, each individual spot not only grew laterally as it moved downstream, but also ultimately split into two separate spots. This latter behaviour has not apparently been seen in boundary layers.

In the case of pipes of round cross-section, one would expect things to be different. Merely as a consequence of the geometry, it is not surprising that the coherent precursors of full turbulence are not isolated structures in the spanwise (i.e. circumferential) direction but appear to occupy the entire cross-section of the pipe. Thus, in practice, one would expect to characterize them by their length in the direction of flow. However, Wygnanski and Champagne (1973), who made a very full investigation of transition in a pipe, found it helpful to distinguish between two different types of coherent structure, which they termed 'turbulent slugs' and 'turbulent puffs'.

These categorizations arose because these investigators (like so many before them, going back to the pioneering work of Reynolds) found that transition depended upon the degree of disturbance at the inlet of the pipe. For the sake of clarity, we shall summarize their initial findings as follows.

(a) *Turbulent slugs.* For the case of a smooth inlet and (at worst) slight disturbances, transition occurred naturally, owing to boundary-layer instabilities at the inlet, for $R > 5 \times 10^4$. The turbulent patches occupied the entire cross-section of the pipe, they had sharply defined leading and trailing edges, and their length was of the same order of magnitude as the length of the pipe. These particular turbulent patches were called 'slugs'.

(b) *Turbulent puffs*. For the case of a large disturbance at the inlet and values of the Reynolds number given by $2000 < R < 2700$, patches of turbulence were convected downstream from the inlet. These patches were less clearly defined than those which occurred naturally, and were called 'puffs'.

Qualitatively, there were some resemblances between the slugs and puffs. For instance, their frequencies of occurrence increased with increasing Reynolds number, passed through a maximum, and then declined. Similarly, their associated intermittency factor γ reached an asymptotic value of unity as the Reynolds number increased. Naturally, in view of the above definitions of the two types of phenomena, the ranges of values of Reynolds number in which these qualitatively similar kinds of behaviour took place were quite different in the two cases.

However, detailed studies using hot-wire anemometry showed that the turbulent structure in the slugs was identical with that in the fully developed pipe flow, whereas later measurements (Wygnanski, Sokolov, and Friedman 1975) indicated that the turbulent activity in puffs is higher in the centre rather than at the walls, as it would be in pipe flow.

11.3.1 *Anatomy of the turbulent spot*

The problems involved in making detailed measurements of the velocity field inside a turbulent spot are very far from trivial. Firstly, the spot is of limited extent and is moving and, secondly, as the spot moves it changes its shape to some extent. The best approach might be to mount one's anemometer on a carriage moving in the streamwise direction at the same speed as the spot, or perhaps even to employ a global method of instantaneously measuring the entire velocity field in the spot, such as laser speckle holography. However, the first technique is not without its difficulties, and the second is very new. So, not surprisingly, at the present time we have only stationary anemometer measurements to call upon.

In practice, this means that an anemometer is placed at a fixed position and the velocity record obtained as a function of the time. Then one obtains velocity records of this kind for a large number of turbulent spots, and obviously this procedure can be repeated for various streamwise and spanwise positions until an adequate statistical picture can be built up for the entire turbulent spot. The main practical problem which can arise is that the fixed position of the anemometer probe, relative to the experimental apparatus say, may not be a fixed position within the turbulent spot. However, this can be compensated for, to a good extent, by judicious conditional sampling.

In this way the ensemble-averaged structure of the turbulent spot has been obtained by anemometry (Coles and Barker 1975; Zilberman, Wygnanski, and Kaplan 1977; Cantwell, Coles, and Dimotakis 1978; Barrow, Barnes, Ross, and Hayes 1984) augmented by measurements of the wall-pressure field associated with the turbulent spot (Mautner and Van Atta 1982) and the use of temperature tagging by heating the boundary layer (Van Atta and Helland 1980). Agreement between all these investigations is fair, with the overall picture of the turbulent spot emerging as follows: (a) its internal structure is similar to that of the turbulent boundary layer; (b) its growth normal to the boundary surface is by entrainment, again like the turbulent boundary layer; (c) its growth in the spanwise direction is not by entrainment, but is believed to be due to local instability.

One interesting result of this work was the proposal by several authors (e.g. Coles and Barker 1975) that the turbulent spot is no more than a large Λ-shaped vortex, of a type well known to form near the wall in turbulent shear flows (these vortices are also known as horseshoe-shaped, hairpin-shaped, and U-shaped but we think that they look more like capital Greek lambda than anything else). To be precise, it is the front (arrowhead) portion of the spot which is supposed to be the large Λ vortex.

Subsequently Perry et al. (1981) argued that such proposals were only true for the ensemble-averaged turbulent spot. On the basis of very detailed flow visualizations, they argued that any individual spot is made up of an array of the usual Λ-shaped vortices. It is fair to say that their arguments are well supported by some very convincing photographs. More recently Itsweire and Van Atta (1984) have changed the way in which anemometer signals are processed in order to obtain the 'statistically most probable' spot. This modified form of ensembled-averaged spot does reveal the presence of several eddies. Evidence of this kind has been forthcoming from various recent investigations (Riley and Gad-el-Hak 1985) and seems to support the view of Perry et al. No doubt progress in this area would be helped by a really satisfactory global method of measuring the fluid velocity.

11.4   Developed structures in boundary layers and pipes

In this section we shall mainly be concerned with the bursting phenomenon, as introduced briefly in Section 3.2.3. This is an important process both in turbulent boundary layers and in well-developed flows through ducts. It is probably the most important of the coherent structures (because it governs the generation of turbulence by shear due to a solid surface) and is certainly the most subtle and difficult to elucidate.

Before turning to a detailed consideration of turbulent bursts, we should note that it is conventional to mention that ideas about large eddies, or other deterministic features, have a long history in the study of the turbulent boundary layer. Indeed, we have already made such an observation as a preliminary to the subject of coherent structures in general. However, among the many investigations and theories which may be said to foreshadow modern work on turbulent bursts (a long, but not exhaustive, list has been given by Corino and Brodkey 1969, p. 25)), there is one particularly intriguing reference. This is a short review article by Corrsin (1957), which does not appear to be at all well known but which has several illustrations which have a distinctly up-to-date look to them. In addition, Corrsin discusses 'streamwise filaments of residual dye', 'axial vorticity near the wall "sweeping" the (dyed) wall fluid into these long narrow strips', and turbulent fluid coming into the wall region in what he calls 'bursts'. The relevance of these comments will emerge more clearly over the next few sections!

11.4.1 Turbulent bursts

As we have already seen, in Section 3.2.3, Kline, Reynolds, Schraub, and Rundstatler (1967) reported a detailed study of the region near the wall in a turbulent boundary layer. Their principal finding was the existence of a streaky structure which indicated the presence of streamwise vorticity. They described a sequence of events in which the streaks (when viewed from the side) would move out from the wall, oscillate, and finally break up. This sequence was called a 'burst'. A similar, nearly regular, 'ejection-sweep' cycle was observed in pipe flow by Corino and Brodkey (1969).

Our purpose in this section is to give a more detailed description of the results of these, and subsequent, investigations. While doing this, we shall find it convenient to express the various length scales involved in units of v/u_τ, where v is the kinematic viscosity of the fluid and u_τ is the friction velocity. For example, we have previously met the scaled distance from the wall x_2^+, as defined by eqn (1.31). This definition can be extended in an obvious way to the streamwise and spanwise directions through the introduction of the analogous quantities x_1^+ and x_3^+.

We should begin by noting that the two characteristic features of the bursting phenomenon are (a) the formation of low-speed streaks near the wall and (b) the near-periodicity of events. It is the latter point which leads to the classification of turbulent bursts as coherent structures.

In itself, this near-periodicity has two aspects. First, the streaks are spaced out in the spanwise direction in a surprisingly regular way. Second, the whole cycle of events, which is known as a burst, repeats itself in fairly regular way. Both these forms of behaviour can be classified as quasi-periodic, and they can be characterized by an appropriate adaptation of the usual parameters of periodic motion. Thus the spanwise distribution of streaks can be described in terms of a 'mean wavelength', while the repetition of bursts has an associated 'mean period'.

The streaky structure is restricted essentially to the viscous sublayer for which $x_2^+ < 5$. At distances from the wall much greater than this, the dyelines are distorted by the presence of turbulence. The streaks are believed to indicate the presence of pairs of counter-rotating vortex filaments, as predicted by Bakewell and Lumley (1967) from their measurements of correlations in the viscous sublayer. The streaks themselves are amalgamations of flow visualization material (e.g. dye or hydrogen bubbles) due to the 'pumping action' of the vortex pairs, alternatively to and from the wall, as one traverses the boundary layer in a spanwise direction.

Kline et al. described the streaks as spending most of their time migrating downstream while, at the same time, moving slowly out from the wall. When a streak reaches a point corresponding to $8 < x_2^+ < 12$, it begins to oscillate. The oscillation then amplifies as the streak moves on outward, ending in an abrupt break-up at a position somewhere in the range $10 < x_2^+ < 30$. Correspondingly, Corino and Brodkey (1969) found that ejections in pipe flow originated in the band $5 < x_2^+ < 15$, with the maximum interaction of these ejections with the outer flow occurring in the range $7 < x_2^+ < 30$. Considering just how subjective these measurements are, this is an impressive show of agreement between these two investigations.

We have previously noted (in Section 3.2.3) that Kline et al. found that the rate of bursting increased when the pressure gradient in the streamwise direction favoured production of turbulence (and vice versa). For the case of pipe flow, Corino and Brodkey found that the number and intensity of bursts increased with increasing Reynolds number. Both these investigations led to the conclusion that the bursting process plays a dominant part in the production of turbulence by the subtraction of energy from the mean motion. Kline et al. also cite various pieces of indirect evidence from other investigations which support this view (ibid. pp. 771–2).

11.4.2 Frequency of turbulent bursts

The mean time between bursts (or, its inverse, the bursting frequency) seems to have become the most controversial quantitative parameter associated with the bursting phenomenon. In particular, there has been serious disagreement on the precise nature of the relationship between the bursting period and the Reynolds number. Early investigations concluded that the bursting period scaled on the outer region variables (U_∞, δ), where U_∞ is the free stream velocity and δ is the boundary-layer thickness. More recent measurements suggest that inner region scaling on the variables (u_τ, v), where u_τ is the friction velocity as defined by eqn (1.29) and v is the kinematic viscosity.

This situation is not altogether surprising, as clearly there is a highly subjective element in judging the beginning and end of a burst. Moreover, it is difficult to exclude all other effects, such as contamination by the influence of a burst other than the one that is being observed. Techniques which have been tried include (a) differentiating and filtering of the signal from a hot-wire anemometer (Rao, Narasimha, and Badri Narayanan 1971), (b) identifying the position of a second maximum in the autocorrelation of the streamwise velocity (Kim, Kline, and Reynolds 1971), (c) conditional sampling such that only the most energetic bursts were included (Lu and Willmarth 1973), (d) measurement of the short-sample-time autocorrelation (Strickland and Simpson, 1975), and (e) application of pattern-recognition techniques to the streamwise fluctuating velocity signal (Wallace, Brodkey, and Eckelmann 1977). Clearly there are many ways of combining these various techniques and this approach is in itself a very active field of research. We shall concentrate here on a few investigations which may be seen as seminal or (in the case of the most recent ones) as especially representative.

We begin with the results of Rao *et al.* (1971), who studied a large amount of data (both from their own and from other investigations) for a wide range of Reynolds numbers. They plotted graphs of the mean bursting period against the Reynolds number, and tried the effect of scaling the bursting periods in various ways. When they used wall variables, the data points from the different investigations clustered together quite well, but there was an obvious systematic dependence upon the Reynolds number. In contrast, when the scaling was done using outer variables, the clustered data points were substantially independent of the Reynolds number. These results can be expressed in the universal form

$$\frac{U_\infty T_B}{\delta} = 5 \pm 2, \tag{11.10}$$

where T_B is the mean interval of time between bursts.

This result might seem to be quite surprising. After all, bursting is an effect of the region very close to the wall. But Kim *et al.* (1971)—in supporting and agreeing with the conclusion of Rao *et al.*—argue that the lifting of low-speed streaks is triggered by large disturbances already in the flow. This would provide a basis for a dependence of bursting rate upon the variables of the outer region of the boundary layer. Certainly the use of the autocorrelation function measured with short sample times (Strickland and Simpson 1975) led to a result much like that given in eqn (11.10).

In recent years the pendulum has swung back, and the evidence seems to be growing in favour of scaling on the inner variables after all (Blackwelder and Haritonidis 1983, Luchik and Tiederman 1987). The first of these two investigations suggests that earlier results (which supported scaling on outer variables) were erroneous because the probes of the hot-wire anemometers used were large enough to cause a degree of spatial averaging. These authors found that only sensors smaller than 20 viscous length-scales were free from inconsistencies due to spatial averaging effects.

The second of the above two investigations tried out various methods of signal processing and concluded that there was good agreement between them, provided that ejections were correctly grouped into bursts. The problem arose because these methods all detected ejections, whereas a burst contains one or more ejections. They also concluded that outer-variable scaling would not work at all, and that inner-variable scaling was more appropriate than mixed scaling. This had previously been suggested by Alfredsson and Johansson (1984), who proposed that the appropriate scaling could be the geometric mean of the inner and outer time-scales. This proposal was less pragmatic than it might appear, as this geometric mean is the same as the Kolmogorov time-scale and hence might have some dynamic significance.[1]

We shall end this section by reiterating the difficulty of avoiding subjective assessments in this area of research. We mentioned the conditional sampling method of Lu and Willmarth (1973). These authors adopted the criterion that they would only count those bursts for which $-u_1 u_2 > 4u_1' u_2'$, where the prime denotes the r.m.s. value, and hence obtained $U_\infty T_B/\delta = 4$, which agrees quite well with the results of other investigations. However, Sabot and Comte-Bellot (1976) have pointed out that the mean time between bursts would decrease if the criterion underlying the conditional averaging was made less severe. That is, if one includes weaker bursts, then one counts more bursts in a given time and, naturally, the time between bursts is reduced. Sabot and Comte-Bellot state that the effect of changing the criterion used by Lu and Willmarth would be to change their results as follows:

$$\frac{U_\infty T_B}{\delta} = \begin{cases} 1.5 & \text{for } -u_1 u_2 > 2u_1' u_2', \\ 0.45 & \text{for } -u_1 u_2 > 0. \end{cases}$$

If one bears in mind that all these criteria are quite arbitrary, then the difficulty of the problem becomes manifest.

11.4.3 Streaky structure and streamwise vortices

From their flow visualizations, Kline et al. (1967) were able to establish that the mean spacing of streaks in the spanwise direction was given by

$$\lambda_3^+ = 100,$$

where λ_3^+ has been made dimensionless by the use of wall variables. As the streaky structure is often regarded as being quasi-periodic in space (just as the bursts are thought of as being quasi-cyclic in the time), λ_3^+ is usually referred to as the (dimensionless) mean wavelength of the streaky structure. This result has been confirmed by several investigators, and (unlike the bursting period) the question of scalings other than on wall variables has scarcely arisen.

Also, according to Blackwelder and Haritonidis (1983), the streaks have an approximate width of $\Delta x_3^+ = 10 - 20$ and a streamwise length in the range $\Delta x_1^+ = 100 - 1000$.

As we remarked earlier, the streaks are believed to be due to the presence of counter-rotating pairs of streamwise vortices. Blackwelder and Haritonidis suggest that these vortices have radii between $20v/u_\tau$ and $50v/u_\tau$. They also state that the streamwise extent is unknown, but earlier Blackwelder and Eckelmann (1979) pointed to the existence of correlations which would suggest $\Delta x_1^+ > 1000$ for these vortices. This value would at least be consistent with the above estimate, for the length of streaks, which would provide a lower bound for the length of the streamwise vortices.

The importance of streamwise vortices was further emphasized by Head and Bandyopahyay (1981), who studied a smoke-filled turbulent boundary layer illuminated by an intense sheet of light. They concluded that the boundary layer consisted of a forest of Λ vortices, each of which was inclined at a fairly constant angle to the boundary surface (about 45°). These vortices were stretched into ever more elongated shapes as they moved downstream. In fact the authors (who themselves refer to 'vortex loops' for the lower Reynolds numbers, and to 'hairpin vortices' for the higher Reynolds numbers) suggest that the variety of terms applied to vortices of this type merely reflects the degree of stretching that a particular vortex loop has suffered.

This proposal is very similar to that of Perry et al. (1981) for the turbulent spot in a laminar boundary layer (see Section 11.3). Another point of resemblance is that the supporting evidence, as obtained by flow visualization, is really quite as impressive as that presented by Perry et al. for the case of the turbulent spot. In view of this, we shall now turn our attention to the subject of streamwise vortices. In particular, we should like to know how these can form at all, in view of the fact that flow over a flat plate only tends to generate spanwise vorticity.

We must begin by emphasizing that we shall present a purely physical description of the probable sequence of events leading to the formation of streamwise vorticity in the turbulent boundary layer. In other words, we shall discuss the physics of transition entirely without reference to the subject of hydrodynamic stability. This means that we shall be ignoring various unanswered questions about mechanisms of instability, but, at the same time, we shall draw attention to their existence. The reader who wishes to pursue this aspect will find a helpful appendix on the subject in the paper by Blackwelder (1983), which should provide a good starting point for further enquiry.

Our discussion of the formation of streamwise vortices will be based very largely on Willmarth and Tu (1967) and Perry et al. (1981). Both papers contain several earlier references on this topic. We begin by referring back to our treatment of the boundary layer on a flat plate, as given in Section 1.4.

We shall use the same notation and terminology here, but now we wish to concentrate on the events leading to the breakdown to turbulence.

Consider, as before, an irrotational flow incident in the direction of x_1 on a flat plate which lies in the $(x_1 x_3)$ plane. Initially a fluid element will possess zero vorticity. But, as it passes over the plate, spanwise vorticity ω_3 will be generated at the boundary surface and will diffuse outwards, in the direction of x_2, into the free stream.

The sense of the vorticity is quite easily visualized if one imagines that a small portion of the fluid has been instantaneously solidified. Then, that part of the resulting solid body which is nearest to the boundary surface will slow down, the solid body will rotate, and its spin vector will point in the direction of x_3.

The question now is: how does streamwise vorticity develop? Referring to Fig. 11.1, a plausible sequence of events may be as follows.

(a) Two-dimensional waves appear, moving downstream. These consist of concentrations of spanwise vorticity. In the figure we show one such 'crest' as stage (a). One can conveniently visualize it as a vortex tube.

(b) Three-dimensional behaviour sets in. A 'wiggle' distorts the vortex tube; this may be the result of a perturbation due to the edge of the plate.

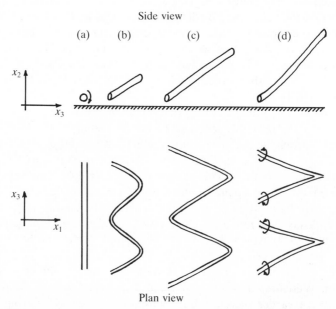

Fig. 11.1. Schematic view of the sequence of events leading to the formation of Λ vortices.

(c) As the vortex tube moves downstream, the wiggle develops into a triangular waveform in the spanwise direction. At the same time, upstream apexes of the triangular wave are convected towards the wall, while downstream apexes are convected away from the wall.

(d) Each upstream apex ultimately becomes anchored to the boundary surface (because of the 'no-slip' boundary condition), while the downstream apex is carried off in a faster-moving region of the fluid. The result is the formation of Λ vortices, each of which is increasingly stretched as it moves downstream. Ultimately, each vortex becomes so extended that it is better described as a hairpin vortex, the legs of which constitute the counter-rotating vortex pair, which in turn organizes the flow visualization material into the low-speed streak (Blackwelder and Eckelmann 1979).

This sequence of events is quite easily deduced from the equations of motion and the boundary conditions, provided that we can postulate step (a); that is, that the initial spanwise vorticity has become concentrated into an actual vortex filament or, in other words, that the tendency to rotate has become an actual rotation. Precisely how, or why, this should happen (although it seems perfectly reasonable that it should!) is a question which lies in the difficult area between linear stability theory and full-blown turbulence theory. Indeed, various writers seem to take the view that stages (a) and (b) might be combined. For instance, Perry and Chong (1982) consider a carpet of spanwise vorticity which is 'wrapped into a vortex' by the process of lifting and stretching of vortex lines.

However, two-dimensional waves—of which stage (a) would be one crest—have been seen in boundary layers. In particular, Gad-el-Hak, Davis, McMurray, and Orszag (1984) observed all the stages shown in Fig. 11.1 in a decelerating laminar boundary layer, but noted that the sequence terminated in a turbulent burst.

The picture of the turbulent boundary layer as an array of Λ vortices is very persuasive, and the flow visualization is undoubtedly convincing. At the same time, apparent conflict between different investigations may be due to an uncertainty about whether the burst is due to the break-up of an actual hairpin vortex (Gad-el-Hak *et al.* 1984) or instead is due to the ejection of low-speed fluid from between the counter-rotating vortices (Grass 1971).

11.4.4 *Relationship between bursts and other types of intermittency*

Turbulent bursting is a form of intermittency, and it is natural to enquire whether there is any connection between it and the other forms of intermittency. As we have seen, there are three of these: transitional intermittency (i.e. spots, slugs and puffs), free-surface intermittency (e.g. mixing layers, free edge of boundary layers, and wakes), and fine-structure intermittency.

In the case of the first of these, the answer seems obvious—at least, at a superficial level. If we consider the boundary layer over a flat plate, then it was postulated from the first that the turbulent spot was the basic building block of the turbulent boundary layer. From our discussions in the preceding sections, it seems likely that a slightly more accurate discription would be that the Λ vortices are the basic building blocks for both the turbulent spot and the fully developed turbulent boundary layer. However, it must be borne in mind that an array of turbulent spots may interact in a strongly non-linear fashion, and it would probably be an unjustified oversimplification to interpret a burst as being a turbulent spot which occurs in a fully turbulent environment.

Continuing with the case of the boundary layer, a relationship to the outer-edge intermittency was postulated at an early stage. For instance, as we saw earlier, Kim *et al.* (1971) argued that the result that the bursting period scaled on the outer variables was reasonable because the lifting of low-speed streaks was triggered by large disturbances already in the flow. At about the same time, this question was examined rather more formally (Kovasznay *et al.* 1970; Blackwelder and Kovasznay 1972) by using conditional sampling to study the relationship between wall-region bursting and the motion of the outer interface between turbulent and non-turbulent fluid. The overall conclusion of this work was that the outer region of the boundary layer was characterized by a large-scale organized motion consisting of 'bulges' of turbulence. It was speculated that these bulges are the end result of the wall-region bursts, and, in support of this view, it was shown that the mean time between bulges is of the same order as the mean time between bursts.

Given our present poor understanding of the dynamics of the motion in a turbulent boundary layer, it is inevitable that there should be a 'chicken and egg' flavour to this kind of conjecture. In effect, it is really only safe to conclude that there must be some coupling between coherent motions in the wall and outer regions.

However, an interesting consideration in this context, is the case of well-developed pipe flow, where there are bursts but no free surface for outer-region intermittency to occur. What happens in this case? Are the bursts any different from those in the boundary layer? Unfortunately there do not seem to be any dramatic differences which might shed light on the effect or otherwise of a free surface. Indeed, Sabot and Comte-Bellot (1976) report the presence of large rotating structures in pipe flow, and note their similarity to the outer-region motions found in boundary layers.

Lastly, we should note that the question of a relationship between large-scale phenomena like bursts and the fine-scale intermittency observed in the energy cascade is unpromising. In addition to the subjective (and other) problems of categorizing bursts, there is the difficulty of establishing relationships between large-scale and small-scale phenomena (see Sections 3.2.1 and 3.2.2). The subject has been studied experimentally (e.g. Ueda and Hinze 1975;

Rajagopalan and Antonia 1984; Sreenivasan 1985), but at the time of writing little has emerged in the way of definite conclusions.

11.5 Theoretical approaches

When we first considered the statistical formulation of turbulence (see Section 1.3), we employed the Reynolds decomposition of the velocity field into mean and fluctuating parts. Hussain (1983) has proposed a triple decomposition which aims to separate out the time-mean value, the coherent part of the velocity field, and the turbulent (purely random) part of the velocity field. (A double decomposition is also discussed, but we shall not pursue that here.)

The triple decomposition can be illustrated for any property of the turbulent system (e.g. pressure or velocity) which we denote in general by $f(\mathbf{x}, t)$:

$$f(\mathbf{x}, t) = F(\mathbf{x}) + f_C(\mathbf{x}, t) + f_T(\mathbf{x}, t), \tag{11.11}$$

where F stands for the time-mean value of f, f_C is the coherent part, and f_T is the turbulent part.

In order to make use of this, we have to introduce the concept of a phase average. This is an ensemble average over structures, taken at the same value of the phase each time. For example, if we considered a regular periodic function, of period T, then the phase average would be

$$\langle f(\mathbf{x}, t) \rangle = \lim_{N \to \infty} \frac{1}{N} \sum_{n=1}^{N} f(\mathbf{x}, t + nT)$$

which is, of course, a rather trivial operation. However, Hussain generalizes this to the quasi-periodic case by writing

$$\langle f(\mathbf{x}, t) \rangle = \lim_{N \to \infty} \frac{1}{N} \sum_{n=1}^{N} f_n(\mathbf{x}, t + T_n), \tag{11.12}$$

where T_n is the randomly varying period of each coherent structure. Then, from eqns (11.11) and (11.12), we have the properties

$$\langle f \rangle = F + f_C \qquad f_T = f - \langle f \rangle,$$

where, for this section only, the angular brackets $\langle \ \rangle$ denote a phase average as defined by eqn (11.12).

Applying the triple decomposition to the Navier–Stokes equation and invoking the above properties, we can proceed in the same way as we derive the Reynolds equations (see Section 1.3.1) and obtain equations of motion for the various mean values (Hussain 1983). Naturally, the resulting equations are much more complicated than the Reynolds equations, and at the present time have not received a great deal of attention. It should also be noted that the triple decomposition is arbitrary (Hussain also considers a double decomposition) and there are many ways of deriving analogues of the Reynolds equations which take coherent structures into account. For instance, a quite different formulation has been given (McComb 1982; see also Section 9.4) using renormalization group methods to average away the smallest scales progressively and replace their mean effect by an effective viscosity.

In all, analytical formulations of this kind pose formidable problems, and it is not surprising that most theoretical progress in coherent structures has been mainly numerical. This will be reflected in the remainder of this section, where we consider, for the most part, computer simulations based on either the full Navier–Stokes equations or on various vortex models.

11.5.1 Numerical simulation of bounded turbulence

Some of the numerical simulations discussed in Chapter 10 have provided a data base which can be exploited to study coherent structures. This material can be used to provide instantaneous information about single realizations, or the raw data can be processed in a number of ways—often involving conditional sampling—in order to provide quantitative measures of the coherent structures. In general, agreement between computer and laboratory experiments has been good enough to reinforce confidence in the former. The consequences of this are likely to be far-reaching when on considers the vast potential of the computational method to generate detailed data and to process it numerically in virtually any way that one would like.

However, we shall concentrate here on the imformation gleaned from simulations on the organized structure near the wall. We begin with the simulations of Moin and Kim: Kim (1985) reported the existence of a pair of counter-rotating streamwise vortices, while Kim and Moin (1986) found that bursting was associated with organized vortical structures inclined at an angle of about $45°$ to the wall. Single-realization pictures of vortex lines indicated Λ vortices which were formed by the sheets of spanwise vorticity rolling up.

More recently, Rogers and Moin (1987) carried out a 128-cubed simulation of the Navier–Stokes equations for homogeneous shear flow, and confirmed the roll-up of spanwise vorticity to produce Λ vortices and (as the Reynolds number was increased) ultimately hairpin vortices. But, if there was no mean shear, these vortices did not appear.

An important result found by these authors was that vorticity existed in coherent filaments, which were stretched and intensified by the mean rate of strain. However, when compressed, these vortex filaments buckled rather than weakened. In the context of the vortex-stretching picture of turbulent energy transfer from large scales to small (see Section 2.7.3), this is a most interesting result. It may well be that one does not need to have recourse to arguments about random walks to suggest that (on average) vortex lines are stretched.

This result (if correct) would appear to suggest that there is no 'vortex-compressing' process.

Further support for the formation of Λ vortices, from initially spanwise vorticity, has come from the study of the transition to turbulence in channel flow based on the numerical simulation of the Navier–Stokes equations (Biringen 1984; Krist and Zang 1987), and from the study of the motion of a vortex filament near a wall (Leonard 1980; see Leonard 1985; Hon and Walker 1988). However, as we shall see in Section 11.5.4, there is also evidence from numerical simulations for a quite different mechanism for generating stream-wise vorticity in a turbulent boundary layer.

Lastly, in view of the controversy about the time-scale for the mean bursting period—as discussed in Section 11.4.2—it is of interest to note the contribution from the computer experiments to the arguments. In fact, Kim and Spalart (1987) have used the data from Spalart's simulation of a turbulent boundary layer (see Chapter 10) to show that the spatial characteristics of the bursting phenomenon (such as mean spacing of bursts) scale convincingly (albeit for very small numbers of points) on the wall variables. From this they infer a similar scaling for the mean time between bursts, in agreement with some of the most recent experimental investigations.

11.5.2 *Numerical simulation of free shear layers*

The pronounced vortical nature of coherent structures in free shear flows (especially the two-dimensional mixing layer) has lead to a resurgence of interest in that great staple of classical hydrodynamics: the vortex. The interested reader will find a clear statement of the terminology of the subject in the review by Saffman and Baker (1979), and a general account of numerical simulation using vortex elements in Leonard (1985). Various free shear flows have been simulated by vortex (and other) methods, but here we shall deal only with the mixing layer.

As representative examples, we shall discuss the simulations by Ashurst (1979) and Aref and Siggia (1980). Each of these investigations is based on the idea that a vortex sheet can be represented for computational purposes by an array of discrete vortices, and both demonstrate observed features of the mixing layer, especially the roll-up of the vortex sheet to form wave-like vortices.

In other respects, the two investigations differ. Ashurst concentrates on the quantitative prediction of various mean quantities (e.g. the Reynolds stresses), and finds good agreement with experiment, while Aref and Siggia are more motivated towards the existence of similarity solutions, scaling behaviour, and, indeed, the statistical mechanics of vortex arrays.

This latter topic perhaps fits in more naturally with the work of the next section (11.5.3), but Ashurst's simulation merits a little more attention here, as it is a particularly good example of how a computer experiment can be more effective than a laboratory experiment. In Section 3.2.3, we mentioned the competing explanations of Brown–Roshko vortices in terms of vortex pairing (Winant and Browand 1974) or vortex tearing (Moore and Saffman 1975). Laboratory experiments—even employing the most elaborate and quantitative methods of analysing flow visualizations (e.g. Hernan and Jiminez 1982)—seem unable to decide between these mechanisms, whereas Ashurst's computer simulation suggests quite unambiguously that all aspects of the vortex-pairing model are borne out in practice.

Numerical simulations of free shear layers have also been based on the Navier–Stokes equation. For example, Riley and Metcalfe (1980), Knight (1981), and Davis and Moore (1985) have all studied the effect of forcing on the thin shear layer and the relevance of subharmonic modes to the observed roll-up of the vortex sheet. In our present context, the paper by Knight is particularly interesting for two reasons. First, he has used a triple decomposition to distinguish the coherent fluctuations from the usual mean and (small-scale) turbulent velocities. This is the same general form as given in eqn (11.11), but does not involves a phase average. Instead, Knight uses the conventional time average, in the same form as eqn (1.6), where the averaging period $(2T)$ is long enough to smooth out the small-scale random motion but is shorter than the coherent time-scales.

The second point of interest is Knight's use of a turbulence model to handle the small-scale turbulence. It must be seen as encouraging that the results of this approach are in detailed agreement with experiment. Attempts—in this vein—to unite the study of coherent structures with the technique of large-eddy simulation (see Section 10.2) could prove to be very fruitful; note the work by Aubry, Holmes, Lumley, and Stone (1988), which is discussed in the next section.

11.5.3 *Deterministic chaos*

In recent years there has been a growth of interest in the behaviour of simple dynamical systems with only a few degrees of freedom. The particular interest is in how a deterministic system behaves in a complicated way (what we would conventionally call 'random' or 'stochastic' behaviour) owing to some element of non-linearity.

As an example, let us consider the motion of a simple pendulum. This is, of course, periodic in both space and time, and can be represented in phase space by an elliptical orbit. Evidently the motion of the pendulum bob is completely predictable (classically) and will follow the same orbit in phase space, irrespective of where and how it starts. That is to say, it is insensitive to the initial conditions.

Now let us suppose that the pendulum bob is made of some magnetic material and is suspended in equilibrium between two identical magnets (which we shall call A and B). Clearly this equilibrium will be unstable, and a small perturbation will result in the bob moving to either A or B, in which case it will then be in a position of stable equilibrium. Consider, however, a large displacement of the bob. In general its motion will no longer be periodic; instead, it will trace out some complicated path and end up at either A or B. Naturally, the corresponding path in phase space will no longer be a closed orbit, and this kind of behaviour is known as 'aperiodic'.

Nevertheless, it should be clear that the path of the pendulum bob— however complicated—is still, in principle, fully deterministic and entirely predictable, provided that the initial conditions can be specified. But, in practice, this turns out to be the snag. When the experiment is actually carried out (experiments of this kind are usually carried out as computer simulations), it transpires that the resulting path depends in a very sensitive way upon the initial conditions. If a particular starting point (from rest) results in the bob ending up at A, then another starting point—only a slight distance away—will result in the bob arriving at B.

Furthermore, if a starting point which leads the bob to A is coloured red and a starting point which leads to B is coloured black, then the resulting red–black pattern is a fractal, exhibiting self-similarity on all scales.

All these features are characteristic of chaotic behaviour by non-linear dynamical systems. In addition, an important route to chaos is by means of a cascade of period-doubling (i.e. subharmonic) bifurcations, with the latter term referring to a two-fold choice of orbits in phase space at a given frequency. This route has universal properties, and is observed in real dynamical systems, most importantly (from our point of view) in certain types of thermally generated turbulence.

An introduction to deterministic chaos can be found in Serra, Andretta, Zanarini, and Compiani (1986), while the reader who wishes to go into further detail will find a collection of seminal papers in Hao Bai-Lin (1984). Our purpose in dealing with the topic here is to do little more than raise the natural question of how far the idea of chaos is relevant to turbulent shear flow?

Certainly turbulence is a chaotic phenomenon, and studies of the motion of small numbers of vortices reinforce the idea that fluid motions can be classified as examples of dynamical chaos (Aref 1983; Spiegel 1985). Yet one is left with the problem that turbulence involves a large number of degrees of freedom, and, on the face of it, it may be difficult to make contact with theories based on a small number of degrees of freedom.

In fact Deissler (1986) has shown that numerical solutions (obtained at low Reynolds numbers) of the Navier–Stokes equation are chaotic, and that the number of active modes (effective degrees of freedom) may be as low as 10. One way of reconciling such an estimate with the enormous number of independently excited Fourier modes (normally interpreted as degrees of freedom) of even low-Reynolds-number turbulence is to suppose that the dynamical processes are dominated by coherent structures.

An idea of this kind appears to be at the root of the very interesting approach by Aubry *et al.* (1988), who use a 'proper orthogonal decomposition' of the turbulence in which only a few modes need be retained as they contain most of the energy and the Reynolds stress. The energy transfer to the rest of the (excluded) modes is represented by a Heisenberg-type effective viscosity. Thus the Navier–Stokes equations can be replaced by a low-dimensional set of ordinary differential equations, which allows the authors to carry out the kind of numerical experiment which demonstrates chaotic behaviour in simple dynamical systems. Indeed, these authors find both intermittent and chaotic behaviour which, they claim, captures the main features of the ejection and bursting events of the boundary layer.

Although this work has an empirical input (the dominant eigenfunctions were determined from experiment), it provides a plausible link between low-dimensional chaos and a realistic turbulent shear flow in the form of the classical boundary layer on a flat plate.

11.5.4 *Wave theories*

Linear stability theory (valid only for small-amplitude disturbances in Navier–Stokes systems) predicts the existence of two-dimensional waves as precursors of the breakdown from laminar to turbulent flow. These are known as Tolmien–Schlichting (TS) waves, and are often observed in association with turbulent spots. This raises the interesting possibility that coherent structures have a wave-like character, which provides an obvious basis for a theoretical approach.

This line of attack has not been completely neglected and, for example, Landahl (1975) has discussed the bursting process in terms of a form of wave breakdown due to secondary instability. Gaster (1975) has tackled a more limited, but rather illuminating, problem in that he has applied linear stability theory to a wave packet and compared the results with those for an experi-mentally generated wave packet (Gaster and Grant 1975). He finds good agreement initially but, as the wave packet moves downstream, irregularities develop and these can be attributed to the growth of non-linear effects.

It might be supposed that such wave packets would be stages on the way to generating a full turbulent spot. Such a view would be supported by the full numerical simulation of the Navier–Stokes equation by Fasel (1984), who concludes that the inclusion of non-linearity leads to the behaviour observed by Gaster and Grant.

However, there can be more than one view of this theory. Quite typically, experimentalists in this field tend to argue that the spots do not result from the breakdown of wave packets but rather that the two phenomena have a common origin (e.g. Chambers and Thomas 1983). Recently there has been a surprising communication from Jimenez, Moin, Moser, and Keefe (1988), who have studied the viscous wall region using a fully resolved numerical simulation of the Navier–Stokes equation. These authors claim that structures in this region are consistent with the classical mechanism involving the TS waves. As this is fundamentally different from the mechanism involving Λ vortices, it is likely to prove a very controversial result.

11.6 Implications for other turbulence concepts

It is obviously far too soon to try to assess the effects that coherent structures will have on our understanding of turbulence, and on our existing concepts and ways of thinking about the phenomenon. Therefore any enquiry in that direction must necessarily be rather tentative. Nevertheless, it is interesting to consider one or two aspects in an appropriately tentative way.

The first thing one might seize upon is that the apparent regularity of coherent structures could simplify the problem of turbulence modelling and allow the use of deterministic equations. This possibility has not been overlooked, and Beljaars, Krishna Prasad, and De Vries (1981) claim that the transport properties of the turbulent boundary layer can be calculated quite well with a deterministic model designed to reflect the large-scale dynamics of the bursting process. However, more fundamental approaches may be possible.

Let us consider the intermittent generation of turbulence in the boundary layer. According to Willmarth and Lu (1972), virtually all the Reynolds shear stress is generated during only 55 per cent of the total time. If one considers the experimental value obtained for the associated correlation coefficient $R_{12} = \langle u_1 u_2 \rangle / u_1' u_2'$ in the light of this intermittency, an interesting conclusion emerges. From Fig. 1.6, it can be seen that typical values of the correlation coefficient are about 0.4–0.55. It follows that while the Reynolds stress is being generated u_1 and u_2 are virtually perfectly correlated (i.e. $R_{12} \doteq 1$), and that the rest of the time they are essentially uncorrelated. That is, a value of about 0.4, say, would appear to be due to a suitably weighted average of unity and zero.

Taking this conclusion (along with the description of events during the bursting process) into account, one is tempted to wonder whether the picture of turbulence upon which Prandtl based the mixing-length theory (see Section 1.5.2) might turn out to be better supported than has usually been thought to be the case. Surprisingly, such a speculation does not appear to have been published until quite recently, when Landahl (1984) analysed the implications of the bursting process for the mixing-length theory. At this stage, some of Prandtl's assumptions are quite well borne out by Landahl's work, while others require further examination. In view of the extreme simplicity of the mixing-length model, taken in conjunction with its remarkable performance, this is clearly a potentially valuable line of research.

In attempting to reconcile the new picture (coherent structures, deterministic motions) with the older picture (the energy cascade, randomness), one should not begin by making one's task unnecessarily difficult by exaggerating the deterministic character of the former. They are often structures with a fair amount of randomness—even on the large scales—and, in quantitative terms, more quasi-cyclic than cyclic.

Even so, at the present time, it is probably best to be pragmatic. As we have seen, both Knight (1981) and Aubry *et al.* (1988) have (in their different ways) isolated the large scales of interest and have then simply represented the effect of the small-scale turbulence as an energy drain, thus perhaps giving a more physical twist to the idea of large-eddy simulation. This seems a reasonable approach, and introduces into turbulence for the first time a convincing ability to separate out dynamically distinct time-scales (as in gas kinetic theory; see Chapter 4). On the experimental side, however, this aspect still has to receive much attention. The only specific investigation of small-scale effects that we know of is that due to Bernal, Breidenthal, Brown, Konrad, and Roshko (1979), who studied the plane mixing layer. They found that the effect of increasing the Reynolds number was to increase the amount of small-scale turbulence in the flow. This had the effect of increasing the mixing within the roll vortices, but did not otherwise seem to affect these structures.

Lastly, we should mention one point which may be obvious, but sometimes can apparently cause confusion. It should be appreciated that concepts like the viscous sublayer, the buffer region, and so on have not in some way been outdated or invalidated by the discovery of coherent structures. These concepts are the result of taking long-time means, and have precisely the same value and validity that they always have had.

11.7   Further reading

In the interests of letting the main ideas stand out, we have presented a rather brief summary of a very large and complex field. It should be understood that many of the references cited are in fact the reports of quite substantial investigations, and the brief citation may give only a very inadequate impression of the many detailed measurements and discussion of mechanisms involved. To take only one example, we have discussed the measurements of the mean time between burst but not the mean duration of bursts. The reader who wishes to go into the subject in more detail should consult the original references, although the review articles by Cantwell (1981), Riley and Gad-el-Hak (1985), and Fiedler (1988) will provide a good starting point.

At times we have impinged slightly on the subject of hydrodynamic stability. Two classical references to this subject are Lin (1955) and Chandrasekhar (1961), and two more modern works are Joseph (1976) and Drazin and Reid (1981).

Note

1. The Kolmogorov length and velocity scales are given by eqns (2.131) and (2.132) respectively. The Kolmogorov time-scale can be defined as the ratio of these two quantities: $\eta/v = (v/\varepsilon)^{1/2}$.

References

ALFREDSSON, P. H. and JOHANSSON, A. V. (1984). *Phys. Fluids* **27**, 1974.

ANTONIA, R. A. and ATKINSON, J. D. (1974). *J. fluid Mech.* **64**, 679.

AREF, H. (1983). *Ann. Rev. fluid Mech.* **15**, 345.

—— and SIGGIA, E. D. (1980). *J. fluid Mech.* **100**, 705.

ASHURST, W. T. (1979). In *Turbulent shear flows*, Vol. 1 (eds F. DURST, B. E. LAUNDER, F. W. SCHMIDT, and J. H. WHITELAW). Springer, New York.

AUBRY, N., HOLMES, P., LUMLEY, J. L., and STONE, E. (1988). *J. fluid Mech.* **192**, 115.

BAKEWELL, H. P. and LUMLEY, J. L. (1967). *Phys. Fluids* **10**, 1880.

BARROW, J., BARNES, F. H., ROSS, M. A. S., and HAYES, S. T. (1984). *J. fluid Mech.* **149**, 319.

BELJAARS, A. C. M., KRISHNA PRASAD, K., and DE VRIES, D. A. (1981). *J. fluid Mech.* **112**, 33.

BERNAL, L. P., BREIDENTHAL, R. E., BROWN, G. L., KONRAD, J. H., and ROSHKO, A. (1979). In *Turbulent shear flows*, Vol. 2 (eds. L. J. S. BRADBURY, F. DURST, B. E. LAUNDER, F. W. SCHMIDT, and J. H. WHITELAW). Springer, New York.

BIRINGEN, S. (1984). *J. fluid Mech.* **148**, 413.

BLACKWELDER, R. F. (1983). *Phys. Fluids* **26**, 2807.

—— and ECKELMANN, H. (1979). *J. fluid Mech.* **94**, 577.

—— and HARITONIDIS, J. H. (1983). *J. fluid Mech.* **132**, 87.

—— and KOVASZNAY, L. S.G. (1972). *Phys. Fluids* **15**, 1545.

BRADSHAW, P. (1975). In *Turbulent mixing in nonreactive and reactive flows* (ed. S. N. B. MURTHY), p. 311. Plenum, New York.

BROWAND, F. K. and TROUTT, T. R. (1980). *J. fluid Mech.* **97**, 771.

BROWN, G. L. and ROSHKO, A. (1974). *J. fluid Mech.* **64**, 775.

BRUUN, H. H. (1977). *J. Fluid Mech.* **83**, 641.

CANTWELL, B. J. (1981). *Ann. Rev. Fluid Mech.* **13**, 457.

—— COLES, D. and DIMOTAKIS, P. (1978). *J. Fluid Mech.* **87**, 641.

CARLSON, D. R., WIDNALL, S. E., and PEETERS, M. F. (1982). *J. fluid Mech.* **121**, 487.

CHAMBERS, F. W. and THOMAS, A. S. W. (1983). *Phys. Fluids* **26**, 1160.

CHANDRASEKHAR, S. (1961). *Hydrodynamic and hydromagnetic stability.* Clarendon Press, Oxford.

CHANDRSUDA, C., MEHTA, R. D., WEIR, A. D. and BRADSHAW, P. (1978). *J. fluid Mech.* **85**, 693.

COLES, D. and BARKER, S. J. (1975). *Turbulent mixing in nonreactive and reactive flows* (ed. S. N. B. MURTHY), p. 285. Plenum, New York.

CORINO, E. R. and BRODKEY, R. S. (1969). *J. fluid Mech.* **37**, 1.

CORRSIN, S. (1957). *Naval Hydrodynamics, Publ. 515*, NAS-NRC.

CROW, S. C. and CHAMPAGNE, F. H. (1971). *J. fluid Mech.* **48**, 547.

DAVIS, R. W. and MOORE, E. F. (1985). *Phys. Fluids* **28**, 1626.

DEISSLER, R. G. (1986). *Phys. Fluids* **29**, 1453.

DIMOTAKIS, P. E. and BROWN, G. L. (1976). *J. fluid Mech.* **78**, 535.

DOPAZO, C. (1977). *J. fluid Mech.* **81**, 433.

DRAZIN, P. G. and REID, W. H. (1981). *Hydrodynamic stability.* Cambridge University Press, Cambridge.

ELDER, J. W. (1960). *J. fluid Mech.* **9**, 235.

EMMONS, H. W. (1951). *J. aero. Sci.* **18**, 490.

FASEL, H. (1984). In *Turbulence and chaotic phenomena in fluids* (ed. T. TATSUMI), p. 31. North-Holland, Amsterdam.

FIEDLER, H. E. (1975). *Turbulent mixing in nonreactive and reactive flows* (ed. S. N. B. MURTHY), p. 381. Plenum New York.

—— (1988). *Prog aerospace Sci.* **25**, 231.

FREYMUTH, P. (1966). *J. fluid Mech.* **25**, 683.

GAD-EL-HAK, M., BLACKWELDER, R. F., and RILEY, J. J. (1981). *J. fluid Mech.* **110**, 73.

——, DAVIS, S. H., McMURRAY, J. T., and Orszag, S. A. (1984). *J. fluid Mech.* **138**, 267.

GASTER, M. (1975). *Proc. R. Soc. Lond. A* **347**, 271.

—— and GRANT, I. (1975). *Proc. R. Soc. Lond.* A **347**, 253.

GRANT, H. L. (1958). *J. fluid Mech.* **4**, 149.

GRASS, A. J. (1971). *J. fluid Mech.* **50**, 233.

HAO BAI-LIN (1984). *Chaos.* World Scientific Singapore.

HEAD, M. R. and BANDYOPADHYAY, P. (1981). *J. fluid Mech.* **107**, 297.

HEDLEY, T. B. and KEFFER, J. F. (1974). *J. fluid Mech.* **64**, 625.

HERNAN, M. A. and JIMENEZ, J. (1982). *J. fluid Mech.* **119**, 323.

HON, T.-L. and WALKER, J. D. A. (1988). *NASA Tech. Memo. 100858.*

HUSSAIN, A. K. M. F. (1983). *Phys. Fluids* **26**, 2816.

—— and HAYAKAWA, M. (1987). *J. fluid Mech.* **180**, 193.

ITSWEIRE, E. C. and VAN ATTA, C. W. (1984). *J. fluid Mech.* **148**, 319.

JIMENEZ, J. (1983). *J. fluid Mech.* **132**, 319.

——, MOIN, P., MOSER, R. and KEEFE, L. (1988). *Phys. Fluids* **31**, 1311.

JOSEPH, D. D. (1976). *Stability of fluid motions.* Springer, New York.

KIM, H. T., KLINE, S. J. and REYNOLDS, W. C. (1971). *J. fluid Mech.* **50**, 133.

KIM, J. (1985). *Phys. Fluids* **28**, 52.

—— and Moin, P. (1986). *J. fluid Mech.* **162**, 339.

—— and SPALART, P. R. (1987). *Phys. Fluids* **33**, 26.

KLINE, S. J., REYNOLDS, W. C., SCHRAUB, F. A., and RUNDSTATLER, P. W. (1967). *J. fluid Mech.* **30**, 741.

KNGHT, D. D. (1981). *Proc. Symp. on Turbulence in Liquids, University of Missouri-Rolla, October 1979.*

KOVASZNAY, L. S. G., KIBENS, V., and BLACKWELDER, R. F. (1970). *J. fluid Mech.* **41**, 283.

KRIST, S. E. and ZANG, T. A. (1987). *NASA Tech. Pap. 2667.*

LANDAHL, M. T. (1975). *SIAM J. appl. Math.* **28**, 735.

—— (1984). *Z. Flugwiss. Weltraumforsch.* **8**, 233.

LARUE, J. C. and LIBBY, P. A. (1976). *Phys. Fluids* **19**, 1864.

LEONARD, A. (1985). *Ann. Rev. fluid Mech.* **17**, 523.

LIBBY, P. A. (1976). *Phys. Fluids* **19**, 494.

LIN, C. C. (1955). *The theory of hydrodynamic stability.* Cambridge University Press, Cambridge.

Lu, S. S. and Willmarth, W. W. (1973). *J. fluid Mech.* **60**, 481.

Luchik, T. S. and Tiederman, W. G. (1987). *J. fluid Mech.* **174**, 529.

McComb, W. D. (1982). *Phys. Rev. A* **26**, 1087.

Mautner, T. S. and Van Atta, C. W. (1982). *J. fluid Mech.* **118**, 59.

Moore, D. W. and Saffman, P. G. (1975). *J. fluid Mech.* **69**, 465.

Moum, J. N., Kawall, J. G. and Keffer, J. F. (1983). *Phys. Fluids* **26**, 2939.

Mumford, J. C. (1982). *J. fluid Mech.* **118**, 241.

—— (1983). *J. fluid Mech.* **137**, 447.

Perry, A. E., Lim, T. T., and Teh, E. W. (1981). *J. fluid Mech.* **104**, 387.

Perry, A. E. and Chong, M. S. (1982). *J. fluid Mech.* **119**, 173.

—— and Steiner, T. R. (1987). *J. fluid Mech.* **174**, 233.

—— and Watmuff, J. H. (1981). *J. fluid Mech.* **103**, 33.

Rajagopalan, S. and Antonia, R. A. (1981). *J. fluid Mech.* **105**, 261.

—— and —— (1984). *Phys. Fluids* **27**, 1966.

Rao, K. N., Narasimha, R., and Badri Narayanan, M. A. (1971). *J. fluid Mech.* **48**, 339.

Riley, J. J. and Gad-el-Hak, M. (1985). In *Frontiers in fluid mechanics* (ed. S. H. Davis and J. L. Lumley), p. 123. Springer-Verlag, New York.

—— and Metcalfe, R. W. (1980). *AIAA Paper 80-0274.*

Rogers, M. M. and Moin, P. (1987). *J. fluid Mech.* **176**, 33.

Sabot, J. and Comte-Bellot, G. (1976). *J. fluid Mech.* **74**, 767.

Saffman, P. G. and Baker, G. R. (1979). *Ann. Rev. fluid Mech.* **11**, 95.

Serra, R., Andretta, M., Zanarini, G., and Compiani, M. (1986). *Physics of complex system.* Pergamon, Oxford.

Spiegel, E. A. (1985). In *Theoretical approaches to turbulence* (eds D. L. Dwoyer, M. Y. Hussaini, and R. G. Voight). *Applied Mathematical Sciences,* Vol. 58. Springer, Berlin.

Sreenivasan, K. R. (1985). *J. fluid Mech.* **151**, 81.

——, Antonia, R. A. and Britz, D. (1979). *J. fluid Mech.* **94**, 745.

Steiner, T. R. and Perry, A. E. (1987). *J. fluid Mech.* **174**, 271.

Strickland, J. H. and Simpson, R. L. (1975). *Phys. Fluids* **18**, 306.

Townsend, A. A. (1956). *The structure of turbulent shear flow* (1st edn). Cambridge University Press, Cambridge.

Tavoularis, S. and Corrsin, S. (1987). *Phys. Fluids* **30**, 3025.

Ueda, H. and Hinze, J. O. (1975). *J. fluid Mech.* **67**, 125.

Van Atta, C. W. and Helland, K. N. (1980). *J. fluid Mech.* **100**, 243.

Wallace, J. M., Brodkey, R. S., and Eckelmann, H. (1977). *J. fluid Mech.* **83**, 673.

Willmarth, W. W. and Lu, S. S. (1972). *J. fluid Mech.* **55**, 65.

—— and Tu, B. J. (1967). *Phys. Fluids Suppl.* **S134**.

Winant, C. D. and Browand, F. K. (1974). *J. fluid Mech.* **63**, 237.

Wygnanski, I. J. and Champagne, F. H. (1973). *J. fluid Mech.* **59**, 281.

——, Sokolov, M. and Friedman, D. (1975). *J. fluid Mech.* **69**, 283.

——, —— and —— (1976). *J. fluid Mech.* **78**, 785.

Yule, A. J. (1977). *Turbulent shear flows,* p. 11, 13. Pennsylvania State University, University Park, PA.

Zilberman, M., Wygnanski, I., and Kaplan, R. E. (1977). *Phys. Fluids* **20**, S258.

12

TURBULENT DIFFUSION:
THE LAGRANGIAN PICTURE

The ability to disperse and mix added substances is one of the characteristic features of fluid turbulence. It is also an aspect which is of enormous environmental and industrial importance.

In this part of the book, our attention will be concentrated mainly on the underlying phenomenon of the diffusive action of turbulence. Indeed, to be more precise, we shall be concerned with the subject of passive scalar convection, where by 'scalar' we mean something like temperature or chemical species concentration and by 'passive' we mean that the added substance does not change the nature of the fluid to the point where the turbulence is appreciably affected.

This subject can be discussed perfectly well in terms of a statistical formulation in a Eulerian (laboratory) coordinate frame, and, in fact, will be treated in just this way in Chapter 13. However, several profound insights have been gained by working in Lagrangian coordinates and, for this reason, we begin by discussing turbulent diffusion from this point of view in the present chapter.

Let us first say what we mean by a Lagrangian frame of reference. Referring to Fig. 12.1, we follow the motion of one fluid particle, which is assumed to be identified in some way. Clearly that is not easily done experimentally, and this is the main snag of the method. However, as we shall shortly see, it has overwhelming advantages for a theoretical approach.

In Fig. 12.1, we denote the position of the marked fluid particle at any time by $\mathbf{X}(t)$, and this is the Lagrangian position coordinate of the particle. Then we can introduce the Lagrangian velocity $\mathbf{V}(t)$ of the particle by the usual rules for differentiating a vector:

$$\mathbf{V}(t) = \lim_{\Delta t \to 0} \frac{\Delta \mathbf{X}}{\Delta t} = \frac{d\mathbf{X}}{dt}. \tag{12.1}$$

That is, the Lagrangian velocity is the instantaneous rate of change of position with respect to time. It should be noted that $\mathbf{V}$ is in the direction of the rate of change (i.e. $\Delta \mathbf{X}$), which is in general not the same direction as that of $\mathbf{X}$ itself.

We should also note that in order to introduce a statistical treatment, we must average over many such particle paths. In principle, at least, this would imply that the Lagrangian coordinates should be tagged in order to distinguish which coordinate applies to which particle. A more general discussion of Lagrangian coordinates can be found in Monin and Yaglom (1971, p. 527).

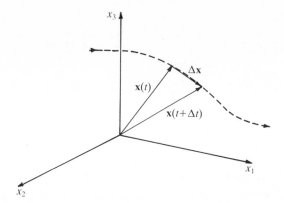

Fig. 12.1. The Lagrangian coordinate system: ---, path taken by the marked particle.

Having introduced Lagrangian coordinates, our next step is to consider Taylor's analysis of diffusion in stationary homogeneous turbulence as this is crucial to an understanding of turbulent diffusion. The rest of the chapter is then taken up with the problems of making practical use of this analysis. These are three: how to express the results in an Eulerian frame of reference, how to take account of finite particle inertia effects, and how to extend the analysis to various shear flows.

12.1 Diffusion by continuous movements

Let us consider an experiment in which we inject a spot of dye into a fluid at rest. As time goes on, the dye will spread out under the influence of the thermal agitation of the fluid molecules. However, if we stir the fluid vigorously, the resulting turbulent eddies will disperse the dye much more rapidly. Then, assuming that the dye moves as part of the fluid, we have a physical realization of the purely hypothetical concept of a marked fluid particle.

For simplicity, we consider the displacement $X(t)$ of a marked fluid particle in one dimension. From eqn (12.1)—appropriately generalized—it follows that the displacement (in one dimension) can be expressed in terms of the velocity field as

$$X(t) = \int_0^t V(t')\,dt'. \tag{12.2}$$

Now, the displacement will be positive as often as it is negative, and so its mean value will be zero. In these circumstances, the lowest-order statistical moment, which does not vanish, is the variance (or mean square) of particle position. We can obtain this by squaring X, as given by (12.2), and averaging to obtain

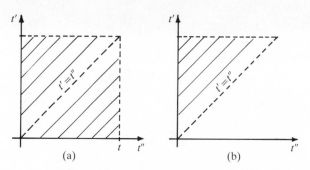

Fig. 12.2. Field of integration for (a) eqn (12.3) and (b) eqn (12.4).

$$\langle X^2(t)\rangle = \left\langle \left\{\int_0^t V(t')\,dt'\right\}\left\{\int_0^t dt''\,V(t'')\right\}\right\rangle$$

$$= \left\langle \int_0^t dt' \int_0^t dt''\,V(t')V(t'')\right\rangle$$

$$= \int_0^t dt' \int_0^t dt''\langle V(t')V(t'')\rangle. \tag{12.3}$$

The averaging procedure is based upon the supposition that we have simultaneously released a large number of particles at $t = 0$, at different points in the fluid, and averaged over all the particle tracks. Naturally, this requires the turbulence field to be spatially homogeneous. We then can interpret the r.h.s. of eqn (12.3) in terms of the Lagrangian correlation function of fluctuating velocities. However, first it is possible to make a useful simplification of (12.3).

The integral on the r.h.s. of eqn (12.3) is symmetric under the interchange of t' and t''. Referring to Fig. 12.2, we see that the integration over the rectangular field of integration specified by the limits $0 \leqslant t'$, $t'' \leqslant t$ can be replaced by twice the integration over the triangular field of integration specified by the limits $0 \leqslant t' \leqslant t$, $0 \leqslant t'' \leqslant t'$. That is, we can then write eqn (12.3) in the form

$$\langle X^2(t)\rangle = 2 \int_0^t dt' \int_0^{t'} dt''\langle V(t')V(t'')\rangle$$

$$= 2 \int_0^t dt' \int_0^{t'} ds\langle V(t')V(t' - s)\rangle \tag{12.4}$$

where we have made the change of variable $t'' = t' - s$.

We now introduce the Lagrangian autocorrelation and integral time-scale. We have previously met the Eulerian form of the autocorrelation function— see eqn (2.36)—and the Lagrangian correlation coefficient R_L can be defined in the same way through the relation

$$\langle V(t)V(t-s)\rangle = \langle V^2(t)\rangle R_L(s). \tag{12.5}$$

The definition of the Lagrangian integral time-scale T_L then follows by obvious analogy with eqn (2.39), for the Eulerian case, and takes the form

$$T_L = \int_0^\infty R_L(s)\,ds. \tag{12.6}$$

If we take the turbulence also to be homogeneous in time in the Lagrangian (as well as the Eulerian) frame, then single-time means are constant in time, and we have

$$\langle V^2(t)\rangle = \langle V^2(0)\rangle = \langle V^2\rangle,$$

from which, along with eqn (12.5), it follows that (12.4) for the mean variance of the displacements can be written as

$$\langle X^2(t)\rangle = 2\langle V^2\rangle \int_0^t dt' \int_0^{t'} ds\, R_L(s). \tag{12.7}$$

A further useful simplification can be obtained if we perform the integration with respect to t' by parts:

$$\int_0^t dt' \int_0^{t'} ds\, R_L(s) = \left[t' \int_0^{t'} R_L(s)\,ds \right]_0^t - \int_0^t t' R_L(t')\,dt'$$

$$= t \int_0^t R_L(s)\,ds - \int_0^t s R_L(s)\,ds. \tag{12.8}$$

It should be noted that, in the first version of the r.h.s., the last term comes about because we differentiated a definite integral with respect to its upper limit, and in the second version we have renamed the dummy variable $t' = s$ in the last term.

Then, substituting from eqn (12.8) into (12.7), we have

$$\langle X^2(t)\rangle = 2\langle V^2\rangle \int_0^t (t-s)R_L(s)\,ds \tag{12.9}$$

as the mean square distance travelled by a diffusing marked particle of the fluid.

This is an exact result for particle diffusion. The problem we now face is that we do not possess a theoretical solution for $R_L(s)$. Also, experimental measurements of Lagrangian quantities are difficult to make, and hence are few and far between. So, all in all, we do not have a great deal of information about $R_L(s)$. Nevertheless, we can obtain two important asymptotic results from the following properties of the correlation coefficient:

$$R_L(0) = 1 \qquad R_L(s) \to 0, \text{ as } s \to \infty.$$

The first of these follows immediately from the definition of $R_L(s)$, in the form of eqn (12.5), and the second is a consequence of the obvious physical fact that events which are widely separated in time (or space) become uncorrelated. In practice, all the available experimental results and theoretical arguments suggest that the latter condition could be put more strongly. That is, $R_L(s)$ falls to zero faster than a power of s as s tends to infinity.

Now let us make use of these properties in eqn (12.9). We shall consider in turn the limiting cases of $t \to 0$ and $t \to \infty$.

(a) *Short diffusion times*: $R_L(s) = 1$. If t is small, then $s < t$ is small, and we can approximate the correlation coefficient by unity. Then the integration on the r.h.s. of eqn (12.9) is easily carried out, and yields

$$\langle X^2(t) \rangle = \langle V^2 \rangle t^2. \tag{12.10}$$

This is, of course, the classical result from Newtonian mechanics: the distance travelled by the representative particle is just equal to its velocity multiplied by the time elapsed. Alternatively, the motion of the particle at short diffusion times is predictable, provided that we know the initial conditions. That is, the motion is deterministic.

(b) *Long diffusion times*. We shall consider times long enough for the correlation coefficient to have fallen to zero. This is of the order of the Lagrangian correlation time, and defines what we mean by a long diffusion time. The important point to note is that, for $t > T_L$, the correlation coefficient falls rapidly to zero and cuts off the integral on the r.h.s. of eqn (12.9), which then becomes

$$\langle X^2(t) \rangle = 2\langle V^2 \rangle (T_L t - B) \qquad \text{for } t \gg T_L, \tag{12.11}$$

where the form of the constant B is readily established by a comparison with eqn (12.9) and the Lagrangian integral time-scale is as defined by eqn (12.6).

Clearly, as t goes on increasing, the second term on the r.h.s. of eqn (12.11) can be neglected in comparison with the first term, and the mean-square particle displacement reduces to

$$\langle X^2(t) \rangle = 2\langle V^2 \rangle T_L t, \tag{12.12}$$

or the r.m.s. particle displacement is proportional to the square root of the time elapsed, which is the same result as one would obtain for the classical random walk of discontinuous movements, where the particle variance is proportional to the square root of the number of steps.

This result can justify our drawing an analogy between the diffusive motions of gas kinetic theory and the diffusive motions of the turbulent eddies at long diffusion times. Thus, taking the square root of both sides of (12.12), we can express the r.m.s. distance travelled by a fluid particle in terms of a turbulent coefficient of diffusion D_T, and thus

$$\langle X^2(t)\rangle^{1/2} = (2D_\mathrm{T} t)^{1/2}, \tag{12.13}$$

where the turbulent diffusion coefficient is given by

$$D_\mathrm{T} = \langle V^2\rangle T_\mathrm{L}. \tag{12.14}$$

If we want to pursue the analogy between turbulent diffusion (at long times) and molecular diffusion, then we should identify $\langle V^2\rangle$ with the mean square 'thermal' velocity of molecules, and T_L is analogous to the mean time between molecular collisions.

This analysis was originally due to Taylor (1922), although the step which leads us from eqn (12.7) to eqn (12.9) is usually attributed to Kampé de Feriet (1939; see Monin and Yaglom (1971) or Hinze (1975)). Its importance is twofold. First, it illustrates very clearly the nature of turbulence as a correlated random walk. Second, it offers some justification for theoretical models (see Section 1.5) based upon an analogy between turbulent and molecular processes, along with some indication of the restricted range of validity of such assumptions.

The analysis of turbulent diffusion by continuous movements can readily be extended to the three-dimensional case. If we denote the general correlation of fluctuating velocities in a Lagrangian frame by

$$K_{\alpha\beta}(t - t') = \langle V_\alpha(t)V_\beta(t')\rangle, \tag{12.15}$$

then the extension of the result given in eqn (12.9) to the case of a three-dimensional diffusion process (Batchelor 1949) is

$$\langle X_\alpha(t)X_\beta(t')\rangle = \int_0^t (t - s)\{K_{\alpha\beta}(s) + K_{\beta\alpha}(s)\}\,\mathrm{d}s. \tag{12.16}$$

It is a simple matter to verify that (12.16) reduces to eqn (12.9) if we put $\alpha = \beta = 1$, say.

12.2 The problem of expressing the Lagrangian analysis in Eulerian coordinates

As we have just seen, the use of Lagrangian coordinates allows us to establish, through Taylor's analysis, some simple but profound results about the nature of turbulent diffusion. Unfortunately, what we really require is information about turbulent diffusion in the everyday laboratory (or Eulerian) coordinate system, and this need has inspired a number of attempts to relate the two coordinate systems. However, it seems likely that—apart from certain asymptotic cases—the possession of such a relationship would imply that one had (in some sense) solved the general turbulence problem. Thus it seems to be arguable that such an aim, although natural, may be somewhat illusory.

Nevertheless, attempts to realize this aim can teach us much about the subject, and accordingly the topic merits a place here.

12.2.1 *Statement of the problem*

In Eulerian coordinates we measure position $\mathbf{x}$ from some reference point, and the velocity $\mathbf{u}(\mathbf{x}, t)$ is measured at $\mathbf{x}$ as a function of time t. Thus the Eulerian velocity is a *field*, and consists of a set of numbers associated with each space–time point $(\mathbf{x}, t)$. Any average of a Eulerian quantity is the result of many fluid particles passing through the measuring point over a period of time, but we do not know anything about the past or future behaviour of any of the individual particles.

However, let us suppose that we are able to follow one particular marked fluid particle $\mathbf{V}(t)$ about the system. Then it is obvious that the (Lagrangian) velocity must be equal to the local Eulerian field value $\mathbf{u}(\mathbf{x}, t)$ when the particle is at $\mathbf{x}$. Thus $\mathbf{V}(t) = \mathbf{u}(\mathbf{x}, t)$ if $\mathbf{X}(t) = \mathbf{x}$, or

$$\mathbf{V}(t) = \mathbf{u}[\mathbf{X}(t), t], \tag{12.17}$$

which is the required general relationship between the Lagrangian and Eulerian velocities for just one particle. It should be borne in mind that not only is $\mathbf{X}(t)$ a random variable, but it also depends upon the velocity field itself. Thus, using eqn (12.2), we can rewrite (12.17) as

$$\mathbf{V}(t) = \mathbf{u}\left[\int_0^t \mathbf{V}(t') \, dt', t\right], \tag{12.18}$$

which rather underlines the difficulty of relating the two coordinate systems.

Let us now consider the problem in a slightly more formal way. Equation (12.16) gives us the r.m.s. dispersion of a marked fluid particle in terms of the Lagrangian velocity correlation. Obviously the next step is to try to relate the latter quantity to its Eulerian equivalent. We can do this by considering the motion of a fluid particle which passes through $\mathbf{X} = 0$ at $t = 0$, and subsequently passes through the point $\mathbf{X}(t) = \mathbf{x}$. Then, from (12.17), we can relate the Lagrangian and Eulerian correlations as follows:

$$\langle \mathbf{V}(0)\mathbf{V}(t) \rangle = \langle \mathbf{u}[0, 0]\mathbf{u}[\mathbf{X}(t), t] \rangle \tag{12.19}$$

where the average is taken over many particle paths.

The average on the l.h.s. is quite straightforward, but that on the r.h.s. is much more difficult. As a result (despite what one sees occasionally in the literature) one cannot re-express the above relationship in terms of correlation functions. For instance, the relationship (e.g. Hinze 1973)

$$K_{\alpha\beta}(t) = Q_{\alpha\beta}[\mathbf{X}(t), t], \tag{12.20}$$

where $Q_{\alpha\beta}$ is the Eulerian correlation tensor as defined by eqn (2.24), is simply not valid.

So much is obvious, for the r.h.s. involves an unaveraged random variable, whereas the l.h.s. is fully averaged. However, the underlying error is that one could only obtain the function $Q_{\alpha\beta}$ by averaging as if $\mathbf{u}[\mathbf{X}(t), t]$ and $\mathbf{X}$ were independent variables. In order to see this, it is helpful to go to a more specific formalism in terms of probability distributions and it is to this that we now turn our attention.

We begin by noting that, in evaluating the average on the r.h.s. of eqn (12.19), we cannot use the distribution $P[\mathbf{u}(\mathbf{x}, t)]$, as defined by (4.6), for that would lead to purely Eulerian quantities. Instead, we must consider the more general distribution

$$P\{\mathbf{u}[\mathbf{X}(t), t]\} \delta u[\mathbf{X}(t), t]$$

$$= p\{\mathbf{u}[\mathbf{X}(t), t] | \mathbf{X}(t)\} \delta u[\mathbf{X}(t), t] F[\mathbf{X}(t)] \delta \mathbf{X}(t), \qquad (12.21)$$

where p is the conditional probability distribution, which is the probability that $\mathbf{u}$ takes a particular value $\mathbf{u}(\mathbf{x}, t)$, given that $\mathbf{x} = \mathbf{X}(t)$, and F is the probability that a particle is on a particular path given by $\mathbf{x} = \mathbf{X}(t)$. Then the formal statement of the relationship between Lagrangian and Eulerian correlation functions is just

$$\langle \mathbf{V}(0)\mathbf{V}(t) \rangle = \int_{\text{all paths}} \left(\int p\{\mathbf{u}[\mathbf{X}(t), t] | \mathbf{X}(t)\} \mathbf{u}[0, 0] \mathbf{u}[\mathbf{X}(t), t] \times \right.$$

$$\times \delta \mathbf{u}[\mathbf{X}(t), t] F[\mathbf{X}(t)] \delta \mathbf{X}(t). \qquad (12.22)$$

The main problem now is that $\mathbf{p}$ and F are not independent of each other. This follows from eqns (12.2) and (12.17), which imply that

$$\mathbf{X}(t) = \int_0^t u[\mathbf{X}(t'), t'] \, dt'. \qquad (12.23)$$

We can illustrate the difficulties involved in treating these two equations by considering two simpler problems. First, suppose that the path followed by any particle is not a random variable, but is instead a prescribed path $\mathbf{X}^*(t)$. Then the integral against $F[\mathbf{X}(t)]$ just gives unity (as all particles move along the (constant) prescribed path) and, in terms of correlation functions, eqn (12.22) becomes

$$K_{\alpha\beta}(t) = Q_{\alpha\beta}[\mathbf{X}^*(t), t], \qquad (12.24)$$

where $Q_{\alpha\beta}$ represents an obvious extension of eqn (2.24) to the case of functional notation.

Next, suppose that $\mathbf{X}(t)$ is a random variable, but its distribution $\tilde{F}[\mathbf{X}(t)]$, say, is independent of p. Then we can evaluate the integral against $p\{\mathbf{u}[\mathbf{X}(t), t] | \mathbf{X}(t)\}$ for each $\mathbf{X}(t)$ to obtain

$$K_{\alpha\beta}(t) = \int Q_{\alpha\beta}[\mathbf{X}(t), t] \tilde{F}[\mathbf{X}(t)] \delta \mathbf{X}(t). \qquad (12.25)$$

This analysis of the problem of relating Eulerian and Lagrangian variables was originally given by Lumley (1962), who also established some useful general relationships connecting Lagrangian and Eulerian statistical quantities. It should be noted that the proofs given by Lumley are only valid for incompressible turbulence, but as that is also the criterion for inclusion in the present volume, it need cause us no problems here.

The first two results are established by expanding the Lagrangian velocity as a power series in time. Assuming that $V(t)$ is analytic, we have

$$V(t) = \sum_{n=0}^{\infty} \frac{t^n}{n!} \frac{\partial^n V}{\partial t^n}\bigg|_{t=0} \tag{12.26}$$

where, because of their localness in time and space, the coefficients are all Eulerian quantities:

$$\frac{\partial^n V}{\partial t^n}\bigg|_{t=0} = \left\{\frac{\partial}{\partial t} + \mathbf{u}(\mathbf{x}, t) \cdot \nabla\right\}^n \mathbf{u}(\mathbf{x}, t)\bigg|_{\substack{t=0 \\ x=0}}. \tag{12.27}$$

It follows therefore that any statistical function of the Lagrangian coordinates can also be represented by a power series in time with purely Eulerian coefficients. Hence we can conclude that

$$\text{Eulerian homogeneity} \rightarrow \text{Lagrangian homogeneity}$$

$$\text{Eulerian isotropy} \rightarrow \text{Lagrangian isotropy.}$$

There are technical difficulties in attempting to reverse these conclusions. The reader who wishes to pursue this point is referred to the original reference (Lumley 1962).

Lastly, it has also been shown that, if we take any function $f[\mathbf{X}(t)]$ of the Lagrangian coordinates and average over the ensemble of particle paths, then, for the special case of homogeneous turbulence, it is possible to prove the theorem

$$\langle f[\mathbf{X}(t), t] \rangle = \langle f(\mathbf{x}, t) \rangle. \tag{12.28}$$

If we then take the specific case

$$f(\mathbf{x}, t) = \exp\{i[\mathbf{Z} \cdot \mathbf{u}]\} \tag{12.29}$$

and, recalling eqn (4.45), note that the average of this particular $f(\mathbf{x}, t)$ is the characteristic functional, then it is readily shown for homogeneous turbulence that the distribution of Lagrangian velocity is identical with that of the Eulerian velocity field.

12.2.2   *Approximations based on the conjecture of Hay and Pasquill*

Although the direct measurement of Lagrangian quantities presents severe problems, a certain amount of information has been available from quite an early stage in the history of the subject. For example, measurements of the distribution of fluid temperature in the region downstream from a heated wire (Taylor 1935) can provide information about the Lagrangian autocorrelation. This is obtained by differentiating both sides of eqn (12.7) twice:

$$\frac{d^2 \langle X^2(t) \rangle}{dt^2} = 2\langle V^2 \rangle R_L(t), \tag{12.30}$$

where $\langle X^2(t) \rangle$ has to be obtained from the measured temperature distributions. Also, in large-scale oceanic (or atmospheric) flows, certain measurements using buoys (or balloons), could—however crudely—involve the tracking of a particular particle and hence have a Lagrangian character.

It was against this kind of background that Hay and Pasquill (1960) made their influential conjecture. They noted that Lagrangian and Eulerian correlation functions had similar shapes but differed in scale. Accordingly, they proposed a similarity hypothesis to relate the two assumptions. This was of the form

$$T_L = \beta T_E \tag{12.31}$$

$$R_L(\beta t) = \beta R_E(t). \tag{12.32}$$

From an analysis of experimental results they concluded that $\beta = 4$, although there was considerable scatter with $1.1 < \beta < 8.5$. Thus, on this basis, Lagrangian integral time scales could be expected to be much longer than their Eulerian equivalents.

Later, Wandel and Kofoed-Hansen (1962) made an analysis of the problem in which the turbulent motions were modelled by an array of rigid (i.e. inextensible) vortex tubes. They found that the Hay–Pasquill conjecture (i.e. eqns (12.31) and (12.32)) appeared as an approximation to their transformation equations, with the relationship

$$\beta = \frac{0.44\bar{U}}{u'}, \tag{12.33}$$

where $\bar{U}$ is the local mean velocity and u' is the r.m.s. velocity fluctuation in the direction of the diffusion process. It was then shown by Thomas (1964) that this relationship agreed well with the experimental results of Angell (cited in Thomas 1964).

A different approach has been taken by Corrsin (1963), who approximated the analytical form of the Lagrangian and Eulerian correlations by assuming that they could each be represented by their appropriate (Kolmogorov) inertial-range power laws. Such an approximation might be quite good in the case of very large Reynolds numbers, and in this way Corrsin found that the integral time-scales were virtually equal. He then argued that this result ran counter to intuition, as one would expect velocities to be more persistent along a particle path than at a fixed measuring point, and hence would expect to find $T_L > T_E$. However, Kraichnan (1964) put forward a contrary view. From a consideration of hypothetical experiments and the application of the direct-interaction approximation (see Section 6.1), he suggested that the Lagrangian velocity may be less persistent than the Eulerian.

12.2.3   Approximations based on Corrsin's independence hypothesis

A more general approach was followed by Corrsin (e.g. Corrsin 1962), who reasoned that at long diffusion times (the 'random walk limit' of Taylor's analysis) the probability distribution of the particle displacements, on the one hand, and that of the Eulerian velocity field, on the other, would become statistically independent of each other. This is the so-called independence hypothesis.

We have previously seen that it is the statistical dependence of $F[\mathbf{X}(t), t]$ upon $P[\mathbf{u}(\mathbf{x}, t)]$—through the relationship given in eqn (12.23)—which is the main difficulty in the way of relating the Lagrangian and Eulerian velocity correlations. Let us rewrite eqn (12.22), which is the general statement of this relationship, in the form

$$K_{\alpha\beta}(t) = \int \langle\langle u_\alpha(0,0)u_\beta(\mathbf{x}, t)\delta[\mathbf{x} - \mathbf{X}(t)] \, d\mathbf{x}\rangle\rangle, \qquad (12.34)$$

where $K_{\alpha\beta}(t)$ is the Lagrangian correlation tensor, the double bracket notation $\langle\langle \ \rangle\rangle$ denotes the joint average over the velocity field and the particle paths, and the introduction of the delta function allows us to restore the ordinary (non-functional) form of the Eulerian velocity field.

Then, for large values of the diffusion time t, the independence hypothesis is equivalent to the assumption that we can factor the joint average $\langle\langle \ \rangle\rangle$ into two separate averages, $\langle \ \rangle\langle \ \rangle$. Therefore, eqn (12.34) for the Lagrangian correlation becomes

$$K_{\alpha\beta}(t) = \int \langle u_\alpha(0,0)u_\beta(\mathbf{x}, t)\rangle \langle\delta[\mathbf{x} - \mathbf{X}(t)]\rangle \, d\mathbf{x}$$

$$= \int Q_{\alpha\beta}[\mathbf{X}(t), t]\tilde{F}[\mathbf{X}(t)]\delta\mathbf{X}(t), \qquad (12.35)$$

where the introduction of $\tilde{F}$ involves the same procedures as in the case of the single-particle distribution function (see eqn (4.1)). Comparison should also be made with the result given in eqn (12.25). In effect, the independence hypothesis is stating that the conditions of the model underlying eqn (12.25) are applicable to the full problem in the limit of long diffusion times.

Saffman (1963) took eqn (12.35) as his starting point in order to derive an approximate relationship between the Lagrangian and Eulerian integral time-scales. The first step was to assume that $\mathbf{X}(t)$ had a normal—or Gaussian—

distribution. This is known to be a good assumption for all diffusion times, and becomes rigorously valid as $t \to \infty$ (Batchelor 1949). Thus, taking the one-dimensional case for analytical simplicity, we can write

$$\tilde{F}[X(t)] = (2\pi \langle X^2(t) \rangle)^{1/2} \exp \left\{ \frac{-X^2}{2 \langle X^2(t) \rangle} \right\}. \tag{12.36}$$

Now, in order to carry out the integration on the r.h.s. of (12.35), it is necessary to have an explicit analytical form for $Q_{\alpha\beta}(\mathbf{x}, t)$. Restricting himself to isotropic turbulence, Saffman made an arbitrary—but physically plausible—choice for the Eulerian correlation at large values of x (i.e. the distance separating the beginning and the end of the random walk). The result was the relationship

$$\beta = \frac{0.8\bar{U}}{u'}, \tag{12.37}$$

where we have adopted the notation of Hay and Pasquill, so that Saffman's result can be compared with that of Wandel and Kofoed-Hansen, as given by eqn (12.33).

An independent approach by Philip (1967) has led to similar results. His main step is to introduce a sub-ensemble Eulerian correlation (we shall call it Q') which is made up from the subset of realizations in which the particle at $(\mathbf{x}, t)$ is the same one which was at $(0, 0)$. Then, a probabilistic argument leads to an equation like (12.35), but involving Q' rather than Q.

The necessary approximation is then obtained by replacing Q' by Q, and arguing that this is a relatively weak approximation under the integral sign. The result for the Lagrangian correlation then becomes the same as that contained in eqns (12.35) and (12.36). This is really not surprising, as Q' is presumably just the same as Q constrained to the particle paths by the delta function, as in eqn (12.35). However, Philip makes the noteworthy point that Saffman's result is two or three times too large (compare eqns (12.37) and (12.33)) and attributes this discrepancy to the arbitrarily chosen Q having transverse length scales which are nearly twice as large as they should be.

More recently, it has been shown (Weinstock 1976, Kraichnan 1977) that Corrsin's independence hypothesis is equivalent to a first-order truncation of certain renormalized perturbation expansions. The result is that corrections to it can be treated systematically, although we shall not pursue that aspect here.

12.2.4 *Experimental measurements of Lagrangian quantities*

Ideally, we would obtain Lagrangian information about the motion of a fluid by seeding it with discrete particles, which would be carefully chosen to behave as if they were part of the fluid. In practice, these requirements are stringent and not easily met, so that most investigations using particles produce results representative of the particles and not of the fluid as such.

For example, in the experiments carried out by Snyder and Lumley (1971), using a grid-generated turbulence with the flow in the vertical direction, of the particles used (hollow glass, glass, corn pollen, and copper), only the hollow glass beads were small and light enough to behave as fluid 'particles'. Even then, the Lagrangian integral time-scale could only be inferred as an asymptotic result extrapolated to the case of zero particle inertia.

We shall return to the results for particles in Section 12.4. Here we note only that Eulerian and Lagrangian correlations were found to have the same shape—thus giving added support to the Hay–Pasquill conjecture—and the ratio of Lagrangian to Eulerian integral time-scales was found to be $\beta = 3$.

Schlien and Corrsin (1974) also studied grid turbulence, although they used the more traditional technique of measuring the dispersion of heat (Taylor 1935) and differentiating the distribution curves in order to obtain the Lagrangian autocorrelation function. Essentially, their main claims for this work were that the scatter in their results was less than that in previous investigations, and that they were able to make detailed comparisons with Eulerian measurements taken previously on the same apparatus (Comte-Bellot and Corrsin 1971). They found that $R_L(t)$ was consistently larger than $R_E(t)$, for all t, and that the ratio of integral scales was $\beta = 1.9$.

These authors also reiterate (with Snyder and Lumley) that there is no reason why Lagrangian and Eulerian correlation curves should have the same shape. However (unlike Snyder and Lumley), they conclude that the two types of correlation are not of the same shape.

Evidently, the development of laser anemometry (see Section 3.1.2), along with modern methods of computer-based data processing, offers new hope of obtaining good Lagrangian statistics for turbulent flows. Recently, Berman (1986) has claimed to have produced such a method by using two laser probe volumes and considering the arrival rates of particles. However, his conclusion that Lagrangian integral scales are about 25 per cent of the corresponding Eulerian scales is very much at variance with the rest of the evidence and must necessarily raise questions about the present status of his method.

We conclude this section with a brief discussion of two numerical simulations (Deardorff and Peskin 1970; Riley and Patterson 1974) which, in the present context, we can regard as numerical experiments. The first of these was based on a large-eddy simulation of plane channel flow, as discussed in Section 3.3.2, whereas the second was an extension of the spectral simulation of 'isotropic turbulence in a box' which was discussed in Section 10.1.1.

In both cases, once the velocity field $\mathbf{u}(\mathbf{x}, t)$ is available from the straightforward turbulence simulation, the particle velocities and trajectories can be found by integrating the equation for particle motion forward in time. From (12.1) and (12.17), we see that the appropriate equation is

$$\frac{d\mathbf{X}}{dt} = \mathbf{u}[\mathbf{X}(t), t] \qquad \text{for } \mathbf{X}(0) = 0. \tag{12.38}$$

The isotropic simulation by Riley and Patterson was not carried on to the point where the 'random walk limit' was established, and in general this simulation—although interesting in itself—is not sufficiently comparable with the foregoing experimental results to allow us to make comparisons. Similarly, although the qualitative features of the work by Deardorff and Peskin (especially the fact that they find $R_L(t) < R_E(t)$ for most values of t) differ from those of Schlien and Corrsin, it is possible that this may be put down to the fact that the two flow situations are not strictly comparable. However, Deardorff and Peskin find the ratio of the integral scales to be $\beta = 4.2$, in good agreement with the experimental investigations.

12.3 Relative diffusion

So far we have dealt with the motion of a single representative marked particle. Now we shall consider the relative motion of a pair of marked particles. Intuitively, we can expect that the random effect of the turbulent eddies will be to move the two marked particles apart. (Alternatively, we can regard this as being a rigorous consequence of the average growth of material lines in a turbulent fluid; see Section 2.7.3.)

With some physical plausibility, we can distinguish three stages in the process of relative diffusion. These are as follows.

(a) The particles are initially close together, and only the smallest eddies can increase their separation.

(b) As the particles move further apart, a greater range of eddy sizes comes into play, with, at all times, the eddies comparable in size to the interparticle separation having the dominant effect.

(c) Ultimately, the distance between the particles becomes greater than the largest turbulent eddy, and the motion of each particle becomes independent of the other. The separation between them is then determined by their own individual random walks. In an infinite system, the interparticle separation can increase without limit.

We should note that stages (a) and (b) are determined by the smallest and intermediate eddy sizes, whereas stage (c) is determined by the largest energy-containing eddies.

We can attempt to attack the present problem by applying the previous analysis for the single marked particle to each member of the particle pair. Let us suppose that particle 1 and particle 2 are released simultaneously at time $t = 0$ and at positions x_1 and x_2 respectively. We shall take the turbulence to be stationary and isotropic, and, for analytical simplicity, we shall also restrict our attention to motion in one dimension.

It may be helpful if we begin by examining the limiting case (i.e. stage (c) above) where the motions of the two particles become uncorrelated. Let the distance between the two particles be $Y(t)$. Then, in a simplified notation, we shall put

$$Y(t) = X_1(t) - X_2(t) \tag{12.39}$$

and the mean square separation is readily seen to be

$$\langle Y^2(t) \rangle = \langle X_1^2(t) \rangle - 2\langle X_1(t)X_2(t) \rangle + \langle X_2^2(t) \rangle. \tag{12.40}$$

Then, for long diffusion times, we have $\langle X_1 X_2 \rangle = 0$ and both $\langle X_1^2(t) \rangle$ and $\langle X_2^2(t) \rangle$ are given by eqn (12.13), and so we obtain the asymmetric mean square separation as

$$\langle Y^2(t) \rangle \rightarrow 2[2D_T]t. \tag{12.41}$$

For the classical random walk, one can obtain the diffusion coefficient as half the time derivative of the mean square distance of travel (e.g. do this with eqn (12.11) or (12.12)). Hence it follows from (12.41) that the relative diffusion coefficient is just twice the single-particle result in the limit of long diffusion times, or

$$\frac{1}{2}\frac{d\langle Y^2(t) \rangle}{dt} = 2D_T \tag{12.42}$$

for $t \rightarrow \infty$.

Now, we apply the Taylor analysis of single-particle motion to the joint motion of two particles in a more formal way. From the initial conditions, and eqn (12.2), we can write the separation of the two particles at any time t as

$$y(t) = (x_1 - x_2) + \int_0^t W(t')\,dt', \tag{12.43}$$

where the relative velocity $W(t)$ is given by

$$W(t) = V(x_1, t) - V(x_2, t) \tag{12.44}$$

and the Lagrangian velocities of the two particles have each been labelled by the position of their respective particles at $t = 0$. At this stage, we can still make use of the techniques of the single-particle analysis (especially the transition from (12.3) to (12.4)) and obtain the mean square separation as

$$\langle Y^2(t) \rangle = \langle (x_1 - x_2)^2 \rangle + 2\int_0^t dt' \int_0^{t'} ds \langle W(t')W(t'-s) \rangle, \tag{12.45}$$

from which the diffusivity immediately follows as

$$\frac{1}{2}\frac{d\langle Y^2(t) \rangle}{dt} = \int_0^t ds \langle W(t)W(t-s) \rangle. \tag{12.46}$$

Unfortunately, this is as far as we can take the rigorous analysis. The problem is that, unlike $V(t)$, the relative velocity $W(t)$ is not a stationary random variable. Accordingly, we can only conclude from (12.46) that the relative diffusivity will depend on the duration of the diffusion process and on the initial separation of the particles. However, there is no elegant way of passing to the limiting case of $t \to \infty$, as previously discussed.

12.3.1 *Richardson's law*

The subject of relative diffusion was first treated by Richardson (1926), who began by studying Fickian diffusion and showed that the relative diffusivity was just equal to twice the single-particle result (compare the asymptotic turbulent case, as given by eqn (12.42) above). He then extended this approach to the turbulent case by postulating the existence of a Fickian-style diffusion equation, but with a diffusivity which depended on the particle separation.

The introduction of such a diffusion equation requires that the underlying process be Markovian—a requirement which is plainly not satisfied until the particles are separated by a distance greater than the integral length scale of the turbulence (Durbin 1980). However, at intermediate times we can still discuss the diffusivity in the sense of equation (12.46).

Having introduced a relative diffusivity, Richardson attempted to find a form for it. By considering the results of various experiments on diffusion in the atmosphere, he obtained the empirical formula

$$\frac{1}{2}\frac{d\langle Y^2(t)\rangle}{dt} = A\langle Y^2(t)\rangle^{2/3} \tag{12.47}$$

which is usually known as 'Richardson's law'.

Alternatively, by integrating once, we can write the mean square separation of the particles as

$$\langle Y^2(t)\rangle \propto t^3. \tag{12.48}$$

It was later shown by Obukhov (1941) that Richardson's law could be derived by dimensional analysis, provided that the particle separation wavenumber (i.e. $2\pi\langle Y^2(t)\rangle^{-1/2}$) was in the inertial range. Further discussion of these ideas will be found in Sections 2.7.2 and 3.2.1. Here, we merely note that dimensional analysis gives

$$\frac{1}{2}\frac{d\langle Y^2(t)\rangle}{dt} = B\varepsilon^{1/3}\langle Y^2(t)\rangle^{2/3}, \tag{12.49}$$

where ε is the rate of energy dissipation per unit mass of fluid and B is a constant. This result was later derived in other ways by Batchelor (1952) and Lin (1960).

12.3.2 *Three-dimensional diffusion*

The formal extension of the above analysis to the three-dimensional case is quite straightforward. For instance, the generalization of (12.46) for the diffusivity is clearly just

$$\frac{1}{2}\frac{d\langle Y_\alpha(t) Y_\beta(t)\rangle}{dt} = \int_0^t ds \langle W_\alpha(t) W_\beta(t-s)\rangle$$

$$= J_{\alpha\beta}(\mathbf{Y}, t), \tag{12.50}$$

where $J_{\alpha\beta}$ is the diffusivity tensor which applies to the relative, or two-particle, case.

One immediate consequence of extending the theory to three dimensions is that we can have two kinds of diffusing motion, i.e. longitudinal, which is along the direction of the separation vector, and transverse, which is at right angles to the separation vector. Thus, just as for the isotropic two-point velocity correlation—see Section 2.4.1—we can write the diffusion tensor in terms of the longitudinal and transverse diffusivities J_L and J_T respectively:

$$J_{\alpha\beta}(\mathbf{Y}, t) = J_L(Y, t) Y_\alpha Y_\beta |\mathbf{Y}|^{-2} + J_T(Y, t) D_{\alpha\beta}(\mathbf{Y}). \tag{12.51}$$

The general three-dimensional problem has been tackled using renormalized perturbation theory (Roberts 1961; Kraichnan 1966), and we shall return to these treatments in the next chapter.

By way of a closing remark, we note that the effect of intermittency corrections (i.e. to inertial-range forms; see Section 3.2.2) on relative diffusion has also been studied (Grossman and Procaccia 1984) and that a discussion of experimental studies of relative diffusion can be found in Monin and Yaglom (1971, pp. 556–67).

12.4   The motion of discrete particles in a turbulent fluid

The general problem of turbulent diffusion can be stated as follows: given the velocity field, what is the average behaviour of a marked fluid particle? If we now add some solid particles to the fluid, we can then state the specific problem of particle diffusion as follows: given the average behaviour of a marked fluid particle, what is the average behaviour of an added solid particle?

In attempting to answer this question, we face two subsidiary problems. First, if the particle is not of the same density as the fluid, it will tend to lag behind fluid motion. Second, even if the particle is of the same density as the fluid, its rigidity (or possibly even its elasticity) will determine its response to fluid eddies on a scale of the same order as its diameter.

In this section we shall base our discussion mainly on the work of Tchen (1947; quoted in Hinze 1975, pp. 460–71), who was apparently the first person to consider this problem in a theoretical way. But first we shall find it helpful to deal with some general aspects of particle motion in a turbulent fluid.

12.4.1 *Some asymptotic results*

We begin by noting that we can apply Taylor's analysis of the marked fluid point to the case of added particles, provided that we work consistently with the Lagrangian coordinates of the added particle. That is (in one dimension, for simplicity), we have $X_p(t)$ and $V_p(t) = dX_p/dt$ as the position and velocity respectively of the added particle. We shall refer to these as 'particle coordinates' and denote them by subscript p.

It is readily shown that the general result for the dispersion of marked fluid points, in the form of eqn (12.9), is equally applicable to particle motion, with only a change of notation being needed. Thus, we have, for the mean square dispersion of added particles

$$\langle X_p^2(t) \rangle = 2 \langle V_p^2 \rangle \int_0^t (t - s) R_p(s) \, ds, \tag{12.52}$$

where the autocorrelation $R_p(s)$ of particle velocities is defined by

$$\langle V_p(t) V_p(t - s) \rangle = \langle V_p^2(t) \rangle R_p(s) \tag{12.53}$$

and the corresponding integral time-scale is just

$$T_p = \int_0^\infty R_p(s) \, ds. \tag{12.54}$$

The results for short—and long— diffusion times are similar to eqns (12.10) and (12.12) for the diffusion of marked fluid points. With our general definition of diffusivity as being half the derivative of the mean square dispersion (e.g. see (12.42), for relative diffusion), we obtain the diffusion coefficient D_p for particles as

$$D_p = \langle V_p^2 \rangle t \qquad t \text{ small} \tag{12.55}$$

$$D_p = \langle V_p^2 \rangle T_p \qquad t \text{ large.} \tag{12.56}$$

It follows from these results, and from eqns (12.10) and (12.12), that the ratio of 'particle' to 'fluid' diffusivity is given by

$$\frac{D_p}{D_T} = \frac{\langle V_p^2 \rangle}{\langle V^2 \rangle} \qquad t \text{ small,} \tag{12.57}$$

$$\frac{D_p}{D_T} = \frac{\langle V_p^2 \rangle}{\langle V^2 \rangle T_L} \qquad t \text{ large.} \tag{12.58}$$

Now, from the first of these results, it is clear that the particle and fluid diffusion coefficients are not equal, as in general $V_p \neq V$ unless the particle and fluid densities happen to be the same. However, eqn (12.58) is rather less obvious. Differences between $\langle V_p^2 \rangle$ and $\langle V^2 \rangle$ can be compensated for by differences between T_p and T_L. In order to obtain a definite asymptotic result, we need to consider spectra rather than correlations.

The time–frequency spectrum was introduced in Section 2.9.1, while the appropriate Fourier transform pair can be obtained from eqns (D8) and (D9) of Appendix D by renaming $x = t$ and $k = \omega$, and other variables as appropriate. Hence we can write the relationship between the Lagrangian (angular) frequency spectrum $E_L(\omega)$ and the Lagrangian correlation function $R_L(t)$, as (compare eqn (D9))

$$E_L(\omega) = \frac{1}{2\pi} \left\langle \frac{V^2}{2} \right\rangle \int_{-\infty}^{\infty} R_L(t) \exp(-\omega t)\, dt \qquad (12.59)$$

with a corresponding result for $E_p(\omega)$ and $R_p(t)$. Then, if we take the zero-frequency case, we obtain a relationship between $E_L(0)$ and the Lagrangian integral time-scale T_L. That is,

$$E_L(0) = \frac{1}{2\pi} \langle V^2 \rangle T_L, \qquad (12.60)$$

and the corresponding relationship between $E_p(0)$ and T_p is simply deduced to be

$$E_p(0) = \frac{1}{2\pi} \langle V_p^2 \rangle T_p, \qquad (12.61)$$

so that the asymptotic ratio given by eqn (12.58) now becomes

$$\frac{D_p}{D_T} = \frac{E_p(0)}{E_L(0)} = 1. \qquad (12.62)$$

The last step follows because, at zero frequency, there can be no difference between fluid and particle motion.

12.4.2   Tchen's analysis

We shall only give the highlights of this work here—a detailed treatment will be found in Hinze (1975).

As well as restricting our attention to isotropic stationary turbulence, we shall consider only particles which satisfy the following criteria.

(a) Their motion relative to the fluid is governed by Stokes' law of resistance.
(b) The particle is spherical and its diameter is small compared with the smallest turbulent eddy.

In addition, Tchen assumed that the fluid surrounding the particle would always be the same fluid. On the face of it, this assumption is quite imponderable. We shall not discuss it at this point, but will return to it after completing the analysis.

Let us consider the motion of a single particle of diameter d and material density ρ_p moving in an infinite fluid of kinematic viscosity v and density ρ. Then, Tchen's starting point was an equation for the particle motion, which we shall write as

$$\frac{dV_p}{dt} = A(V - V_p) + B\frac{dV}{dt} + C\int_{t_0}^{t} G(t - t')\left(\frac{dV}{dt'} - \frac{dV_p}{dt'}\right)dt', \quad (12.63)$$

where

$$G(t - t') = \frac{1}{(t - t')^{1/2}} \quad (12.64)$$

and the constants A, B, and C are given by

$$A = \frac{36v}{(2h + 1)d^2} \qquad B = \frac{3}{2h + 1}$$

$$C = \frac{18(v/\pi)^{1/2}}{2h + 1} \qquad h = \frac{\rho_p}{\rho}. \quad (12.65)$$

In this form, eqn (12.63) may seem a little opaque, but we should note that the three terms on the r.h.s. take account of the Stokes' law of resistance, the effect of pressure gradients in the fluid, and the effect of unsteady fluctuating motions respectively. The last of these terms is known as the 'Basset' term. The application of this equation is hedged about with various restrictions (Corrsin and Lumley 1956; Hinze 1975), but we shall not pursue these technicalities here.

If V_p and V are expressed in terms of Fourier integrals, then eqn (12.63) can be solved for the relationship between the particle and the fluid diffusion coefficients D_p and D_T. For short—and long—diffusion times, the theory correctly recovers the asymptotic relationships as given by eqns (12.57) and (12.62). However, for intermediate times, it is necessary to have an analytic expression for the Lagrangian autocorrelation. For this, Tchen assumed the exponential form

$$R_L(t) = \exp\left(-\frac{t}{T_L}\right). \quad (12.66)$$

He also made the additional assumption that the Basset term could be neglected, and hence was able to obtain the ratio of particle to fluid diffusivities, at intermediate times, as

$$\frac{D_p}{D_T} = 1 + (1 - B^2)(A^2 T - 1)\left\{\exp(-At) - \exp\left(-\frac{t}{T_L}\right)\right\} \times$$

$$\times \frac{1}{1 - \exp(-t/T_L)}. \tag{12.67}$$

In general physical terms, this seems quite a reasonable result. It predicts that the diffusion coefficient for particles is equal to that for marked fluid points, under any of the following circumstances.

(1) $B = 1$: or the density of the particle is equal to that of the fluid.
(2) $t \to \infty$: the limit of long diffusion times.
(3) $A = 1/T_L$: this draws attention to the significance of $1/A$ as the response time of the particle.

However, the sparseness of reliable experimental data makes it difficult to arrive at a quantitative assessment of Tchen's analysis. Snyder and Lumley (1971) (see also Section 12.2.4) found that autocorrelations for particles fell off more rapidly the heavier the particle. This is counter to what one would expect from inertia effects (e.g. Friedlander 1957), and the authors attribute it to a 'crossing trajectories' effect in which particles with appreciable terminal velocities fall from one eddy to another.

Evidently, the existence of this effect must raise questions about Tchen's assumption that the diffusing particle is always surrounded by the same fluid. But the difficulty lies in telling whether or not, on average, this actually matters.

Some indications that it may not matter come from the numerical simulation of Riley and Patterson (1974) (see also see Section 12.2.4), who essentially solve eqn (12.63) with $C = 0$. They found that fluid autocorrelations, taken along particle paths, did not differ very much from the 'fluid point' case. Also, particle autocorrelation coefficients were found to fall off more slowly as the particle inertia increased.

Lastly, we note that detailed investigations of the ranges of validity of various approximations to eqn (12.63) have been given by Hjelmfelt and Mockros (1966) and, more recently, by Maxey and Riley (1983), as well as by Desjonqueres, Gousbet, Berlemont, and Picart (1986).

12.5 Applications of Taylor's analysis to shear flows

The analysis of diffusion by continuous movements, as discussed in Section 12.1, owes at least some of its success to a very considerable degree of idealization. That is, it treats the motion of a marked fluid point in a stationary field which is both isotropic and homogeneous. Nevertheless, it is natural to enquire whether any generalization is possible, in order to employ this successful technique in developing a theory for the numerous important practical problems which involve turbulent dispersion.

So far we have already considered the general problems of relating Lagrangian and Eulerian statistical quantities, and of taking into account particle inertia effects. We shall conclude the work of this chapter by giving a very brief summary of the work carried out on some more specific problems.

We begin by mentioning the formal extension of Taylor's analysis to heat (and mass) transfer in isotropic turbulence by Corrsin (1952), and to particles of finite size by Batchelor, Binnie, and Phillips (1955). The latter work is particularly interesting because the particles used were neutrally buoyant, and hence pure inertia effects were excluded. These authors found that the longitudinal diffusivity of particles in a pipe flow was reduced quite sharply as the particle diameter was increased, although it should be mentioned that the particles used were rather large—in fact the largest had a diameter equal to one-third that of the pipe itself.

Turning now to the problems associated with the actual flow field—rather than with the diffusing contaminant—we are confronted with the inhomogeneity of actual flows in both space and time. However, if we consider the representative case of duct flows, then many important examples will be steady. Also, there is often spatial homogeneity in two coordinate directions. Hence, the irreducible difficulty—even for longitudinal diffusion—is inhomogeneity in the radial direction resulting from the presence of the solid boundary. The need to take into account the presence of the viscous sublayer near the wall has been established by Chatwin (1971), who also gives some useful references to earlier treatments of this problem.

One important point is worth highlighting, and that is the fact that stationarity of the Eulerian velocity field does not necessarily imply that the Lagrangian velocity is a stationary random variable. This latter condition is the really crucial requirement for the validity of the Taylor analysis: the Lagrangian autocorrelation must not depend on absolute time if eqn (12.7) is to be valid.

This would not be the case in a developing flow like an axisymmetric free jet for example. Here, if we start a particle off at time t', streamwise position x', and radial position r', then as it moves downstream, it constantly finds itself in regions with changing statistical properties and hence the correlation $\langle V(t)V(t') \rangle$ will keep changing as time goes on. In other words, the inhomogeneity of the Eulerian field in the streamwise direction implies the nonstationarity of the Lagrangian velocity correlation. This will be true of all developing flows, but it is interesting to note that, in developed flow in a pipe, the Lagrangian velocity correlation—initially non-stationary because of radial inhomogeneity—will become stationary asymptotically, because of the streamwise homogeneity.

The situation for developing flows is not quite as bleak as this discussion might suggest. Batchelor (1957) proposed that, for flows which exhibit similarity (as indicated by Eulerian statistical quantities) in the streamwise direction, it may be possible to find transformations which turn the Lagrangian velocity into a stationary random variable. In other words, Eulerian similarity could imply an underlying Lagrangian similarity. These ideas have been applied with some success to turbulent boundary layers (Cermak 1963).

In view of the above, the application of the theory of diffusion by continuous movements seems most plausible for the case of well-developed flow in ducts. This appears to be borne out in practice, and for general interest we note the engineering application of Taylor's theory to pipe flow by Flint, Kada, and Hanratty (1960), Baldwin and Walsh (1961), and Groenhof (1970). Lastly, the general problem of diffusion from simple sources in complex flows has recently been reviewed by Hunt (1985).

References

BALDWIN, L. V. and WALSH, T. J. (1961). *AIChE J.* **7**, 53.
BATCHELOR, G. K. (1949). *Austr. J. sci. Res.*, **A2**, 437.
—— (1952). *Proc. Camb. phil. Soc.* **48**, 345.
—— (1957). *J. fluid Mech.* **3**, 67.
——, BINNIE, A. M. and PHILLIPS, O. M. (1955). *Proc. phys. Soc. B* **68**, 1095.
BERMAN, N. S. (1986). *AIChE J.* **32**, 782.
CERMAK, J. E. (1963). *J. fluid Mech.* **15**, 49.
CHATWIN, P. C. (1971). *J. fluid Mech.* **48**, 689.
COMTE-BELLOT, G. and CORRSIN, S. (1971). *J. fluid Mech.* **48**, 273.
CORRSIN, S. (1952). *J. appl. Phys.* **23**, 113.
—— (1962). Theories of turbulent dispersion. In *Mecanique de la turbulence. Colloq. Intern. CNRS, Marseille.* Paris, CNRS.
—— (1963). *J. atmos. Sci.* **20**, 115.
—— and LUMLEY, J. L. (1956). *Appl. Sci. Res. A* **6**, 114.
DEARDORFF, J. W. and PESKIN, R. L. (1970). *Phys. Fluids* **13**, 584.
DESJONQUERES, P., GOUSBET, G., BERLEMONT, A., and PICART, A. (1986). *Phys. Fluids* **29**, 2147.
DURBIN, P. A. (1980). *J. fluid Mech.* **100**, 279.
FLINT, D. L., KADA, H., and HANRATTY, T. J. (1960). *AIChE J.* **6**, 325.
FRIEDLANDER, S. K. (1957). *AIChE J.* **3**, 381.
GROENHOF, H. C. (1970). *Chem. eng. Sci.* **25**, 1005.
GROSSMANN, S. and PROCACCIA, I. (1984). *Phys. Rev. A* **29**, 1358.
HAY, J. S. and PASQUILL, F. (1960). *Adv. Geophys.* **6**, 345.
HINZE, J. O. (1973). Some problems in the study of turbulence transport. In *Proc. 15th Congr. Int. Assoc. for Hydrological Research*, Istanbul, 2–7, September.
—— (1975). *Turbulence* (2nd edn). McGraw-Hill, New York.
HJELMFELT, A. T. and MOCKROS, L. F. (1966). *Appl. sci. Res.* **16**, 149.
HUNT, J. C. R. (1985). *Ann. Rev. fluid Mech.* **17**, 447.
KAMPÉ DE FERIET, J. (1939). *Ann. Soc. Sci. Bruxelles* **59**, 145.
KRAICHNAN, R. H. (1964). *Phys. Fluids* **7**, 142.
—— (1966). *Phys. Fluids* **9**, 1937.

—— (1977). *J. fluid Mech.* **81**, 385.

LIN, C. C. (1960). *Proc. Natl Acad. Sci. (U.S.)* **46**, 566.

LUMLEY, J. L. (1962). The mathematical nature of the problem of relating Lagrangian and Eulerian statistical functions in turbulence. In *Mecanique de la turbulence Colloq. Intern.* CNRS, *Marseille. Paris, CNRS.*

MAXEY, M. R. and RILEY, J. J. (1983). *Phys. Fluids* **26**, 883.

MONIN, A. S. and YAGLOM, A. M. (1971). *Statistical fluid mechanics: mechanics of turbulence*, Vol. 1. MIT Press, Cambridge, Ma.

OBUKHOV, A. M. (1941). *Izv. Akad. Nauk, Ser. Geogr. Geofiz.* **5**, 453.

PHILIP, J. R. (1967). *Phys. Fluids Suppl.* **S69**.

RICHARDSON, L. F. (1926). *Proc. Roy. Soc. A* **110**, 709.

RILEY, J. J. and PATTERSON, G. S. (1974). *Phys. Fluids* **17**, 292.

ROBERTS, P. H. (1961). *J. fluid Mech.* **11**, 257.

SAFFMAN, P. G. (1963). *Appl. Sci. Res. A* **11**, 245.

SCHLIEN, D. J. and CORRSIN, S. (1974). *J. fluid Mech.* **62**, 255.

SNYDER, W. H. and LUMLEY, J. L. (1971). *J. fluid Mech.* **48**, 41.

TAYLOR, G. I. (1922). *Proc. Lond. math. Soc.* **20**, 196.

—— (1935). *Proc. Roy. Soc. A* **151**, 465.

THOMAS, D. M. C. (1964). *Q. J. Roy. met. Soc.* **90**, 342.

WANDEL, C. F. and KOFOED-HANSEN, O. (1962). *J. geogr. Res.* **67**, 3089.

WEINSTOCK, J. (1976). *Phys. Fluids* **19**, 11.

13

TURBULENT DIFFUSION:
THE EULERIAN PICTURE

In this chapter we shall treat the diffusive and mixing properties of fluid turbulence in the same way as we have treated the random velocities, that is, by means of Eulerian fields. In this approach the marked fluid point will no longer be quite such a fruitful concept. Instead we shall be interested in the concentration field describing the distribution of the added substance, which is being dispersed from some initial location by the motions of the fluid. For example, we might be interested in the diffusion of heat from, say, a point or line source situated in a flowing gas or liquid, or we might consider the way in which common salt (in aqueous solution) spreads out when injected into a turbulent water flow (or the same problem for helium gas in air). Clearly, there are many such practical examples—indeed, all possible combinations of host liquid, added contaminant, form of source, and geometry of flow!

Broadly speaking, this means that we are now concerned with the transfer of heat and mass from one part of a fluid to another. As before, we shall restrict our attention to situations where this transfer is passive and the presence of the scalar substance does not affect the velocity field of the fluid. Also, although we shall be adopting quite different methods from those of the previous chapter, we shall, from time to time, refer back to the results of the Lagrangian analysis in order to inform—or even guide—our approach in the Eulerian frame of reference.

The value of such cross references may seem intuitively quite clear. However, it is worth establishing formally that the marked particle motions considered in the previous chapter are equivalent to the diffusion problem in Eulerian coordinates, and this we now do.

By analogy with the one-point distribution function in statistical mechanics—see eqn (4.1)—we introduce the probability $f(\mathbf{x}, t)$ that a particle which was at $\mathbf{x} = 0$ at time $t = 0$ subsequently passes through the point $\mathbf{x}$ at time t. This is just the probability that the Lagrangian position vector $\mathbf{X}(t)$ takes the value $\mathbf{x}$ or

$$f(\mathbf{x}, t) = \langle \delta[\mathbf{x} - \mathbf{X}(t)] \rangle, \tag{13.1}$$

where the average is over all realizations of the velocity field. Batchelor (1949, 1952) showed that $f(\mathbf{x}, t)$, defined in this way, was the solution to an equation of the form

$$\frac{\partial f(\mathbf{x}, t)}{\partial t} + \frac{\partial \{f(\mathbf{x}, t) u_\beta(\mathbf{x}, t)\}}{\partial x_\beta} = 0. \tag{13.2}$$

As we shall see very shortly, this is the equation governing the distribution of a passive scalar, provided only that source terms and molecular effects alike have been set equal to zero.

13.1 Heat and mass transfer

The transfer of heat and mass can, along with the transfer of momentum, be considered as transport processes. In the case of the first two entities, transport due to molecular agitation will take place even in fluid at rest. If, for example, we maintain some points in the fluid at a higher temperature T than the rest, so that there is a steady temperature gradient in the x_1 direction, then we must supply heat at a rate q_1 such that

$$q_1 = -\frac{\kappa}{oC_p} \frac{d(\rho C_p T)}{dx_1}$$

$$= -\kappa \frac{d(\rho C_p T)}{dx_1}. \tag{13.3}$$

Here q_1 is the flux of heat in the x_1 direction (units of heat/unit area $\times$ unit time), k is the thermal conductivity, ρ is the density, C_p is the specific heat and κ is the thermal diffusivity.

Similarly, if we introduce some alien substance to the fluid, and characterize it by the number N of alien molecules per unit volume of fluid, then maintaining a gradient of N in the x_1 direction implies that we must supply molecules at a rate

$$j_1 = -D \frac{dN}{dx_1}, \tag{13.4}$$

where j_1 is a current (number of molecules/unit area $\times$ unit time) in the direction of x_1, and D is the molecular diffusivity.

These two equations are Fourier's law of heat transfer and Fick's law of diffusion respectively. They are, of course, well known, but the interested student will find a fuller discussion in Bird, Stewart, and Lightfoot (1960). It should be noted that eqn (13.3) has been rewritten to make it have a closer resemblance to Fick's law. This is also true of Newton's law, which defines the viscosity in the form given in equation (1.3a). We can emphasize this by rearranging it slightly as

$$\sigma_{12} = v \frac{d(\rho U_1)}{dx_2}; \tag{1.3a}$$

that is, the l.h.s. is a momentum flux and is expressed in terms of a diffusivity and a gradient of the momentum.

In the language of transport theory, the kinematic viscosity, the thermal diffusivity, and the molecular diffusivity are all transport coefficients. They have the same units, i.e. L^2T^{-1}, and, for the case of elementary kinetic theory, are all closely related. At a later stage, we shall consider whether this kind of relationship can be preserved when we replace the transport coefficients by their effective turbulent (i.e. renormalized) forms. However, for the moment, we note that is is conventional to express the relative effectiveness of momentum transport to scalar transport in terms of the dimensionless numbers

$$Pr = \nu/\kappa \tag{13.5}$$

$$Sc = \nu/D, \tag{13.6}$$

where Pr is the Prandtl number and Sc is the Schmidt number.

The theory can be extended to unsteady transport processes by considering the rate of change of T, say, in an elementary volume in terms of the difference between the inflow of heat and the outflow of heat through the bounding surface. It is easily shown that Fourier's law implies the result

$$\frac{\partial T}{\partial t} = \kappa \nabla^2 T, \tag{13.7}$$

with a corresponding equation from Fick's law

$$\frac{\partial N}{\partial t} = D \nabla^2 N. \tag{13.8}$$

The further extension to flowing fluids is easily accomplished (Bird *et al.* 1960; Goldstein 1938) if we merely replace the partial derivative with respect to time by the total derivative, which takes into account the effects of convection upon the time dependence. It follows that (13.7) and (13.8) respectively become

$$\frac{\partial T}{\partial t} + \frac{\partial(TU_\beta)}{\partial x_\beta} = \kappa \nabla^2 T \tag{13.9}$$

$$\frac{\partial N}{\partial t} + \frac{\partial(NU_\beta)}{\partial x_\beta} = D \nabla^2 N, \tag{13.10}$$

where we have not included the effect of source terms, nor have we allowed for the effect of viscous dissipation in eqn (13.9). These would merely complicate the issue without adding anything to our forthcoming discussion of turbulent transport processes.

The same consideration applies to a detailed discussion of the boundary conditions. In principle, eqns (13.9) and (13.10)—along with the Navier–Stokes equations—govern the transport of heat, mass, and momentum in the incompressible flow of a Newtonian fluid, providing that we specify the boundary conditions. We note that, in general, these will require both the concentrations and the fluxes to take prescribed forms on certain bounding surfaces.

13.1.1 Statistical formulation

We can treat eqns (13.9) and (13.10) in the same way as we treated the Navier–Stokes equation in Section 1.3, where discussions of the averaging procedures can be found. That is, in addition to the decomposition of the velocity and pressure fields given by eqns (1.7) and (1.8), we write the temperature and scalar concentration fields as the sum of a mean and a fluctuation:

$$T(\mathbf{x}, t) = \bar{T}(\mathbf{x}, t) + \theta(\mathbf{x}, t) \tag{13.11}$$

$$N(\mathbf{x}, t) = \bar{N}(\mathbf{x}, t) + n(\mathbf{x}, t), \tag{13.12}$$

where the fluctuations about the mean satisfy the conditions

$$\langle \theta(\mathbf{x}, t) \rangle = 0 \qquad \langle n(\mathbf{x}, t) \rangle = 0. \tag{13.13}$$

Then, substituting as appropriate in eqns (13.9) and (13.10) and averaging, we obtain the equations for the mean temperature and mean scalar concentration as

$$\frac{\partial \bar{T}}{\partial t} + \frac{\partial (T \bar{U}_\beta)}{\partial x_\beta} = \kappa \nabla^2 \bar{T} - \frac{\partial \langle \theta u_\beta \rangle}{\partial x_\beta} \tag{13.14}$$

$$\frac{\partial \bar{N}}{\partial t} + \frac{\partial (\bar{N} \bar{U}_\beta)}{\partial x_\beta} = D \nabla^2 \bar{N} - \frac{\partial \langle n u_\beta \rangle}{\partial x_\beta}. \tag{13.15}$$

These results can be compared with the Reynolds equation for the mean velocity, as given by eqn (1.12). In particular, we should note the terms $< \theta u_\beta >$ and $\langle n u_\beta \rangle$, which represent the effect of the fluctuating velocity field in transporting heat and mass. Obviously, these terms are analogous to the Reynolds stress $\langle u_\alpha u_\beta \rangle$, which occurs as an augmentation of the molecular stresses in the equation for the mean velocity.

Similarly, we can derive equations for the fluctuations in temperature and scalar concentration which are analogous to eqn (1.4) for the velocity fluctuation. We do this by subtracting eqns (13.14) and (13.15) respectively from eqns (13.9) and (13.10) to obtain

$$\frac{\partial \theta}{\partial t} + \frac{\partial (\bar{T} u_\beta + \theta \bar{U}_\beta + \theta u_\beta - \langle \theta u_\beta \rangle)}{\partial x_\beta} = \kappa \nabla^2 \theta \tag{13.16}$$

$$\frac{\partial n}{\partial t} + \frac{\partial (\bar{N} u_\beta + n \bar{U}_\beta + n u_\beta - \langle n u_\beta \rangle)}{\partial x_\beta} = D \nabla^2 n. \tag{13.17}$$

Equations (13.14)–(13.17) are the basis of our statistical approach to heat and mass transfer. We can use (13.16) and (13.17) to obtain expressions for

$\langle\theta\mathbf{u}\rangle$ and $\langle n\mathbf{u}\rangle$, which occur in the equations for $\bar{T}$ and $\bar{N}$ respectively, but then we encounter the closure problem, just as in the case of the velocity field.

We shall discuss this in detail, in the next two sections, but here we remark only that if we make the hypothesis that transport is dominated by the gradient of mean concentrations, then we can introduce turbulent eddy diffusivities in the same way that we introduced the eddy viscosity through eqn (1.50). Considering the same simple flow as in that case—mean velocity in the x_1 direction with mean properties varying with x_2 in the direction normal to the wall—we introduce the eddy diffusivities κ_T and D_T by

$$-\langle\theta u_2\rangle = \kappa_T \frac{d\bar{T}}{dx_2} \tag{13.18}$$

$$-\langle n u_2\rangle = D_T \frac{d\bar{N}}{dx_2}. \tag{13.19}$$

Even if the gradient hypothesis is not valid, one can still characterize a transport process by the ratio of the turbulent flux to the mean concentration gradient (although see the remarks about the eddy viscosity at the end of Section 1.5.1). Therefore it follows that we can usefully characterize turbulent transport by extending the concepts of Prandtl and Schmidt numbers to the case of eddy diffusivities. Hence, by analogy with eqns (13.5) and (13.6), we can introduce the eddy Prandtl and Schmidt numbers as

$$\text{Pr (turbulent)} = \nu_T/\kappa_T \tag{13.20}$$

$$\text{Sc (turbulent)} = \nu_T/D_T. \tag{13.21}$$

From now on we shall no longer treat both heat and mass transfer in parallel. As the governing equations are identical, it is only necessary to draw a distinction when both processes are involved at the same time. In order to make general points, we shall take mass diffusion as our example, except when we need to make a specific point about heat transfer.

However, before we turn our attention to the problems of closure, we should briefly mention the so-called 'Reynolds's analogy'. It was suggested by Reynolds that, in turbulence, heat and momentum were transferred in the same way. With some restrictions, this led to a relationship between the coefficients of skin friction and of heat transfer. Nowadays the analogy between all three turbulent transport processes is used in engineering applications, and the interested reader will find a detailed discussion in Goldstein (1938, pp. 649–57).

Lastly, this particular topic is not totally moribund at the fundamental level. The extension of the analogy to fluctuations of the temperature and velocity was proposed by Fulachier and Dumas (1976), and later shown to be satisfactory for a variety of boundary-layer flows (Fulachier and Antonia 1984).

13.1.2 *Single-point equations*

We begin by considering the closure problem in the context of the single-point equations, as this is how it is commonly met in engineering applications. Taking the diffusion of mass as our representative scalar process, the equation for the mean concentration $\bar{N}$ is just (13.15). As we have already noted, in order to solve this equation, we must have a prescription for the flux vector $\langle n\mathbf{u} \rangle$. This is, of course, just the closure problem at its lowest level. We shall now examine the next level of the statistical hierarchy by using eqn (13.17) for the concentration fluctuation $n(\mathbf{x}, t)$ to derive equations for the scalar covariance $\langle n^2 \rangle$ and for the scalar flux $\langle n\mathbf{u} \rangle$.

Starting with the covariance, we multiply each term in (13.17) by n and average. As in Section 1.3.2 for the kinetic energy-balance equation, we make repeated use of the chain rule for differentiating a product. We also invoke eqn (13.13) when carrying out the averages. The result is readily found to be

$$\frac{\partial \langle n^2 \rangle}{\partial t} + \bar{U}_\beta \frac{\partial \langle n^2 \rangle}{\partial x_\beta} = -\frac{\partial \{\langle n^2 u_\beta \rangle - D\partial \langle n^2 \rangle/\partial x_\beta\}}{\partial x_\beta} - 2\langle nu_\beta \rangle \frac{\partial \bar{N}}{\partial x_\beta} -$$

$$-2D\left\langle \left(\frac{\partial n}{\partial x_\beta}\right)^2 \right\rangle. \tag{13.22}$$

This equation seems to have first been derived by Corrsin (1952), who compared it with the energy balance for velocity fluctuations—i.e. eqn (1.21) in the present work. We can see that the two equations are very similar, so that we can make use of our discussion of the kinetic energy balance when interpreting the various terms of (13.22). For instance, the l.h.s. is just the total rate of change with time (i.e. local plus convective derivatives) of the scalar covariance.

The r.h.s. of eqn (13.22) then tells us how the various processes govern the rate of change of $\langle n^2 \rangle$. The first two terms are bracketed together to draw attention to the fact that they can both be written as a divergence. Consequently, we can argue, by analogy with the corresponding terms in eqn (1.21), that these represent the diffusion of concentration fluctuations through space by non-linear effects and by effects of molecular agitation. Similarly, the third term, which expresses the interaction of the flux $\langle n\mathbf{u} \rangle$ with the mean concentration gradient $\partial \bar{N}/\partial x_\beta$, can be interpreted as a generation term for concentration fluctuations. Evidently, then the last term must be regarded as representing the irreversible destruction of concentration fluctuations by molecular agitation.

The equation for $\langle n\mathbf{u} \rangle$ is slightly more difficult to obtain. In order to have an evolution equation, we need the time derivative of $\langle n\mathbf{u} \rangle$, which is just

$$\frac{\partial \langle n\mathbf{u} \rangle}{\partial t} = \left\langle \mathbf{u}\frac{\partial n}{\partial t} \right\rangle + \left\langle n\frac{\partial \mathbf{u}}{\partial t} \right\rangle. \tag{13.23}$$

An expression for the first term on the r.h.s. is found by multiplying each term of (13.17) by u_α and averaging:

$$\left\langle u_\alpha \frac{\partial n}{\partial t} \right\rangle + \left\langle u_\alpha \frac{\partial(n\bar{U}_\beta)}{\partial x_\beta} \right\rangle = -\left\langle u_\alpha \frac{\partial(\bar{N}u_\beta)}{\partial x_\beta} \right\rangle - \left\langle u_\alpha \frac{\partial(nu_\beta)}{\partial x_\beta} \right\rangle +$$

$$+ D\langle u_\alpha \nabla^2 n \rangle. \tag{13.24}$$

Similarly, we multiply eqn (1.14) for u_α through by n and average to obtain

$$\left\langle n\frac{\partial u_\alpha}{\partial t} \right\rangle + \left\langle n\frac{\partial(u_\alpha \bar{U}_\beta)}{\partial x_\beta} \right\rangle = -\left\langle n\frac{\partial(\bar{U}_\alpha u_\beta)}{\partial x_\beta} \right\rangle - \left\langle n\frac{\partial(u_\alpha u_\beta)}{\partial x_\beta} \right\rangle -$$

$$-\frac{1}{\rho}\left\langle n\frac{\partial p}{\partial x_\alpha} \right\rangle + \nu\langle n\nabla^2 u_\alpha \rangle. \tag{13.25}$$

Then, substituting eqns (13.24) and (13.25) into (13.23), we follow the same procedure as for (13.22). In some cases, terms can be added in pairs in order to make total differentials. In other cases, we use the chain rule for differentiation of a product. The final result is

$$\frac{\partial \langle nu_\alpha \rangle}{\partial t} + \bar{U}_\beta \frac{\partial \langle nu_\alpha \rangle}{\partial x_\beta}$$

$$= -\left(\frac{\partial}{\partial x_\beta}\right)\left\{ \langle nu_\alpha u_\beta \rangle - (\nu + D)\frac{\partial \langle nu_\alpha \rangle}{\partial x_\beta} \right\} - \langle nu_\beta \rangle \frac{\partial \bar{U}_\alpha}{\partial x_\beta} -$$

$$- \langle u_\alpha u_\beta \rangle \frac{\partial \bar{N}}{\partial x_\beta} - \frac{1}{\rho}\left\langle n\frac{\partial p}{\partial x_\alpha} \right\rangle - 2(\nu + D)\left\langle \left(\frac{\partial n}{\partial x_\beta}\right)\left(\frac{\partial u_\alpha}{\partial x_\beta}\right) \right\rangle -$$

$$- \nu\left\langle u_\alpha \frac{\partial^2 n}{\partial x_\beta^2} \right\rangle - D\left\langle n\frac{\partial^2 u_\alpha}{\partial x_\beta^2} \right\rangle. \tag{13.26}$$

Once again, the l.h.s. is the total rate of change with time—in this case, of the flux vector $\langle n\mathbf{u} \rangle$—and the various terms on the r.h.s. can easily be classified as contributing to the creation, annihilation, or transport of $\langle n\mathbf{u} \rangle$. The discussion proceeds just as for eqn (13.22), and need not be given in detail. The one exception is the term involving the pressure gradient, which has no counterpart in eqn (13.22). But, in Section 2.1, we saw that the pressure could be expressed as a quadratic in the velocities, thus indicating that the correlation $\langle n\partial p/\partial x_\alpha \rangle$ can be interpreted as an inertial transport of concentration fluctuations.

The closure problem now revealed by eqns (13.15), (13.22), and (13.26) is quite intriguing. We note that there is no dependence on higher orders of $n(\mathbf{x}, t)$

such as n^2, n^3, and so on. Instead, this set of equations requires the mixed third-order moments $\langle n^2 u_\beta \rangle$ and $\langle n u_\alpha u_\beta \rangle$ for closure. Thus the closure problem for the case of scalar diffusion is really just the closure problem for the velocity field—a result which might have been anticipated from the linearity of the basic scalar equation of motion (e.g. eqn (13.10)).

As we saw in Sections 1.5 and 3.3.3, single-point closures for the velocity field—however elaborate they may be—really amount to 'gradient transport' models. Given the non-local continuum nature of fluid turbulence, such approaches may be pragmatic but, from a fundamental point of view, seem to run counter to intuition.

However, the application of such concepts to passive scalar diffusion may seem somewhat less counter-intuitive. The Lagrangian analysis tells us that the turbulent diffusion process becomes Fickian at long diffusion times. Hence a hypothesis like (13.19)—essentially just Fick's law as applied to the turbulent case—can appear very plausible. Indeed, Batchelor (1949) has argued that the experimental evidence indicates that the probability distribution function, for fluid particle displacements, is Gaussian for both short and long diffusion times. Hence the diffusion process can be represented by an equation of the form of (13.8), but with a diffusion coefficient which initially depends on time, in accordance with Taylor's theory, as given in Section 12.1.

Various attempts have also been made to extend mixing-length ideas to scalar diffusion (e.g. Csanady 1966; Berkowicz and Prahm 1980), but with generally only indeterminate results. Indeed, one gains the impression that this field is rather undeveloped at a fundamental level, but, as Corrsin (1974) has pointed out, attempts to model turbulent diffusion by analogies with gas kinetic theory fail to meet many of the underlying restrictions imposed upon the latter. For instance, the important restriction that mean concentrations should vary by only a negligible amount over the equivalent of one mean free path (i.e. the Lagrangian integral length scale) would be difficult to satisfy in most turbulent shear flows. As a result, attempts of this kind are rather fraught, and any apparent success may be fortuitous.

In engineering applications there has been more activity, but the outlook seems just as gloomy. Recently, Antonia (1985) measured heat flux budgets in a plane jet and compared his results with the predictions of various theoretical models. His conclusion is that gradient-type models offer only rough qualitative predictions: their quantitative predictions were found to be rather poor.

13.1.3 *Two-point equations*

A two-point statistical formulation of turbulent heat transfer has been given by Deissler (1963) for the case of homogeneous turbulence, as a preliminary to a 'weak turbulence' approximation. However, we shall base our treatment here on the later more general formulation due to Kraichnan (1964).

Although a two-point formulation is superficially more complicated, it is actually rather more straightforward than the single-point form. The essential difference is that differential operators with respect to $\mathbf{x}$ do not act on functions of $\mathbf{x}'$. This allows us to set up the statistical equations in an uncomplicated way without having to worry about perfect differentials or the chain rule. This further allows us to follow the labour-saving device of treating the scalar field as the 'zero' component of the velocity field (Kraichnan 1964), that is,

$$N(\mathbf{x}, t) \equiv U_0(\mathbf{x}, t). \tag{13.27}$$

With this step, we can extend the existing formulation for velocity fields to include the scalar case (Kraichnan also includes Boussinesq convection, but we shall continue to make the restriction to passive scalar convection). Accordingly, this section should be read in conjunction with Sections 2.1, 2.2, and 8.5.1.

It follows from (13.27) that the various moments of the scalar field can be written as

$$\bar{N}(\mathbf{x}, t) = \bar{U}_0(\mathbf{x}, t), \tag{13.28}$$

$$Q_{00}(\mathbf{x}, x'; t, t') = \langle n(\mathbf{x}, t) n(\mathbf{x}', t') \rangle, \tag{13.29}$$

$$Q_{\alpha 0}(\mathbf{x}, \mathbf{x}'; t, t') = \langle u_\alpha(\mathbf{x}, t) n(\mathbf{x}', t') \rangle$$
$$= Q_{0\alpha}(\mathbf{x}', \mathbf{x}; t', t), \tag{13.30}$$

and so on. Note that the single-point form of (13.29) is just $\langle n^2 \rangle$ and that of (13.30) is $\langle n\mathbf{u} \rangle$.

We now introduce a tensor diffusivity $K_{\alpha\beta}$, such that

$$K_{11} = K_{22} = K_{33} = \nu$$

$$K_{00} \equiv D \qquad K_{\alpha\beta} = 0 \quad (\alpha \neq \beta), \tag{13.31}$$

and a generalised forcing term $F_\alpha(\mathbf{x}, t)$, such that

$$F_\alpha(\mathbf{x}, t) = -\frac{\partial P_{\text{ext}}}{\partial x_\alpha} \qquad F_0(\mathbf{x}, t) = 0. \tag{13.32}$$

In addition, the inertial transfer operators now have the following properties:

if $(\alpha, \beta, \gamma) \neq 0$, $M_{\alpha\beta\gamma}(\mathbf{V})$ is given by (2.13)

if any of $(\alpha, \beta, \gamma) = 0$,   then $M_{\alpha\beta\gamma}(\mathbf{V}) = 0$,

except $M_{00\alpha} = M_{0\alpha0} = \partial/\partial x_\alpha$ \hfill (13.33)

and

if $(\alpha, \beta) \neq 0$, $L_{\alpha\beta}(\mathbf{V})$ is given by (2.12)

if either of $(\alpha, \beta) = 0$,   then $L_{\alpha\beta}(\mathbf{V}) = 0$. \hfill (13.34)

Hence, eqns (8.20) and (8.21)—the general two-point two-time equations for the mean and covariance of the velocity field—are readily generalized to the scalar case as

$$\left(\frac{\partial}{\partial t} - K_{\alpha\beta}\nabla^2\right)\bar{U}_\beta(\mathbf{x}, t) = F_\alpha(\mathbf{x}, t) - L_{\alpha\beta}(\mathbf{V})\bar{U}_\beta(\mathbf{x}, t) +$$

$$+ M_{\alpha\beta\gamma}(\mathbf{V})\{\bar{U}_\beta(\mathbf{x}, t)\bar{U}_\gamma(\mathbf{x}, t) + Q_{\beta\gamma}(\mathbf{x}, \mathbf{x}; t, t)\} \quad (13.35)$$

$$\left(\frac{\partial}{\partial t} - K_{\alpha\beta}\nabla^2\right)Q_{\beta\sigma}(\mathbf{x}, \mathbf{x}'; t, t') = -L_{\alpha\beta}(\mathbf{V})Q_{\beta\sigma}(\mathbf{x}, \mathbf{x}'; t, t') +$$

$$+ 2M_{\alpha\beta\gamma}(\mathbf{V})\{\bar{U}_\beta(\mathbf{x}, t)Q_{\gamma\sigma}(\mathbf{x}, \mathbf{x}'; t, t')\} +$$

$$+ M_{\alpha\beta\gamma}(\mathbf{V})Q_{\beta\gamma\sigma}(\mathbf{x}, \mathbf{x}, \mathbf{x}'; t, t, t') \quad (13.36)$$

for $\alpha, \beta, \gamma, \sigma = 0$.

We now consider two aspects of this formulation. First, how do we recover the statistical equations for scalar diffusion in recognizable form? This is quite simple, and we shall demonstrate the procedure by performing it explicitly for the mean concentration $\bar{N}(\mathbf{x}, t)$. In eqn (13.35), put $\alpha = \beta = 0$. This gives us

$$\left(\frac{\partial}{\partial t} - K_{00}\nabla^2\right)\bar{U}_0(\mathbf{x}, t) = F_0(\mathbf{x}, t) - L_{00}(\mathbf{V})\bar{U}_0(\mathbf{x}, t) +$$

$$+ M_{00\gamma}(\mathbf{V})\{\bar{U}_0(\mathbf{x}, t)\bar{U}_\gamma(\mathbf{x}, t) +$$

$$+ Q_{0r}(\mathbf{x}, \mathbf{x}; t, t)\}. \quad (13.37)$$

Then, invoking (13.28) and (13.30)–(13.33), we can further reduce this to

$$\left(\frac{\partial}{\partial t} - D\nabla^2\right)\bar{N}(\mathbf{x}, t) = \left(\frac{\partial}{\partial x_\gamma}\right)\{\bar{N}(\mathbf{x}, t)\bar{U}_\gamma(\mathbf{x}, t) + \langle n(\mathbf{x}, t)u_\gamma(\mathbf{x}, t)\rangle\}, \quad (13.38)$$

which, with a little rearrangement, is the same as (13.15). Of course, this is what we would expect, as the equation for the mean concentration is inherently a single-point equation in either formulation.

Equations for the general two-point correlations $\langle n(\mathbf{x}, t)n(\mathbf{x}', t')\rangle$ and $\langle n(\mathbf{x}, t)u_\alpha(\mathbf{x}', t')\rangle$ are obtained by putting $\beta = \sigma = 0$ and $\beta = 0, \sigma = 0$ respectively in each term of eqn (13.36).

The second aspect, which we wish to consider here, is the application of the Eulerian DIA closure to the scalar problem. All the equations needed for this purpose will be found in Section 8.5.1. The essential step is to generalize the response tensor (as defined by eqn (8.22)) to include the scalar case. This is easily accomplished by letting both tensor indices take on the value zero, but we must be careful about one aspect. The tensor $G_{0\beta}(\mathbf{x}, \mathbf{x}'; t, t')$ gives the response of the scalar field at $(\mathbf{x}, t)$ to an infinitesimal mechanical force at the point $(\mathbf{x}', t)$. However, $G_{\alpha 0}(\mathbf{x}, \mathbf{x}'; t, t')$ gives the response of the velocity field at

$(\mathbf{x}, t)$ to an infinitesimal heat source at $(\mathbf{x}', t')$. This case is excluded from our formulation, and accordingly we put

$$G_{\alpha 0}(\mathbf{x}, \mathbf{x}'; t, t') = 0 \tag{13.39}$$

in order to ensure that we consider only passive scalar convection.

Then the governing equation for $G_{\alpha\beta}(\mathbf{x}, \mathbf{x}'; t, t')$ is extended to the scalar case by treating eqn (8.23) in the same way as we treated eqn (8.21) in order to derive (13.36). The DIA closure is then completed by invoking eqns (8.25) and (8.26) for the triple moments.

13.1.4   *Some experimental measurements in pipes and jets*

There are, of course, very many reported investigations of heat and mass transfer in a variety of practical situations. Here we shall be interested only in establishing a few general points about turbulent diffusion in simple classical flows with a high degree of symmetry.

We have previously mentioned the conclusion that Batchelor (1949) drew from an examination of the results of various experiments, i.e. that the probability density function (p.d.f.) of particle displacements is Gaussian for all homogeneous flows. In fact, it seems that this is true for simple shear flows as well, provided (possibly) that the diffusion is from a spatially restricted (point or line) source. For example, mean temperature profiles downstream of a heated wire were found to take a Gaussian form (Flint, Kada, and Hanratty 1960; Baldwin and Walsh 1961). Similar results have been found for mass transfer from a point source in pipe flow using sodium chloride in water (Groenhof 1970; McComb and Rabie 1982) and nitrous oxide in air (Quarmby and Anand 1969).

In Table 13.1 we give some representative values for the eddy diffusivities divided by the friction velocity multiplied by the pipe diameter—a scaling which is universal with regard to Reynolds number for values of $R > 20\,000$. It can be seen that there is reasonable agreement between the values for heat and mass (and both agree with the specimen value given for the transfer of momentum, i.e. the eddy viscosity).

Turning now to the case of free jets, we find a broadly similar situation. In a classic investigation of the round free jet, Hinze and Van Der Hegge Zijnen (1949) measured the radial and axial distributions of mean velocity, temperature, and trace gas concentration in an air jet issuing into still air. They found that both kinds of scalar spread out at the same rate (i.e. the eddy diffusivities for heat and mass had the same value), and that this rate of spread was greater than that of the mean velocity. The radial distributions of mean temperature and concentration were found to be (like the mean velocity) self-similar with respect to downstream distance. One particularly interesting observation is

TABLE 13.1 *Some experimental values of turbulent transport coefficients in pipe flow*

$R \times 10^{-3}$	Fluid	Property transferred	$D_T^+ \times 10^2$	Reference
500	Air	Momentum	3.5	Laufer (1954)
280–640	Air	Heat	3.0	Baldwin & Walsh (1961)
18–71	Air	Heat	4.0	Johnk & Hanratty (1962)
25–75	Water	NaCl	4.0	Groenhof (1970)
20–130	Air	NO	3.4	Quarmby & Anand (1969)
45	Water	NaCl	4.0	McComb & Rabie (1982)

Scaled turbulent diffusivity $D_T^+ = D_T/u_\tau d$, where D_T stands for any turbulent diffusivity (mass, momentum, or heat), u_τ is the friction velocity, and d is the diameter of the pipe.

that the eddy diffusivities were found to be about a thousand times larger than the molecular transport coefficients.

Similar results were obtained for an air jet by McComb and Salih (1977), but the novel feature of this work was the use of small solid particles (of titanium dioxide) as a tracer, combined with a laser anemometer to measure the mean velocity, concentration, and size of the particles. The measured concentration profiles agreed quite well with those of Hinze and Van Hegge Zijnen.

13.2 Scalar transport in homogeneous turbulence

In the fundamental approach to turbulence, we restrict our attention to homogeneous turbulence fields. This means that statistical quantities do not depend upon their absolute position in space. In the case of the velocity field, this implies that the mean velocity is a constant. Accordingly, we normally work in a system of coordinates in which the constant mean velocity is zero.

When we consider the scalar field, the same simplification is not available to us. Therefore, realistically, we must consider fluctuations about a constant mean level of concentration, which we shall call N_0. Thus, in this section, the general decomposition of eqn (13.12) is specialized to

$$N(\mathbf{x}, t) = N_0 + n(\mathbf{x}, t). \tag{13.40}$$

Then, eqn (13.17), for the scalar fluctuation, becomes (with some rearrangement)

$$\left(\frac{\partial}{\partial t} - D\nabla^2\right) n(\mathbf{x}, t) = -\frac{\partial(nu_\beta)}{\partial x_\beta}, \tag{13.41}$$

where we have $\bar{\mathbf{U}} = 0$, $\partial N_0/\partial x_\beta = 0$, and $\partial \langle nu_\beta\rangle/\partial x_\beta = 0$ for homogeneous fields.

We now introduce the Fourier representation of the scalar field. This can be done by closely following our earlier treatment of the velocity field as an example. Thus, substituting from (2.71) for the Fourier expansion of the velocity field and introducing an analogous form for the concentration fluctuations

$$n(\mathbf{x}, t) = \sum_{\mathbf{k}} n(\mathbf{k}, t) \exp(i\mathbf{k} \cdot \mathbf{x}), \tag{13.42}$$

we can write eqn (13.41) as

$$\left(\frac{\mathrm{d}}{\mathrm{d}t} + Dk^2\right) n(\mathbf{k}, t) = -ik_\beta \sum_{\mathbf{j}} n(\mathbf{k} - \mathbf{j}, t) u_\beta(\mathbf{j}, t). \tag{13.43}$$

This can be compared with eqn (D.32), the equivalent result for the Navier–Stokes equation.

The statistical treatment in wavenumber space also goes through in the same way as for the velocity field. To begin with, we introduce the spectral covariance for turbulence in a cubical box of side L:

$$\langle n(\mathbf{k}, t)n(-\mathbf{k}, t')\rangle = \left(\frac{2\pi}{L}\right)^3 Z(k; t, t'). \tag{13.44}$$

We shall also make the restriction to isotropic fields; note that the function Z therefore only depends upon the scalar magnitude of wavevector $\mathbf{k}$.

We can give a physical meaning to $Z(k; t, t')$ by considering its relationship to the correlation of scalar fluctuations. This also allows us to introduce the change to continuous variables corresponding to an infinitely large box. That is, the continuous form of $Z(k; t, t')$ is given by the Fourier transform of the correlation:

$$Z(k; t, t') = \left(\frac{1}{2\pi}\right)^3 \int \mathrm{d}^3r\, Z(r; t, t') \exp(-i\mathbf{k} \cdot \mathbf{r}), \tag{13.45}$$

where $Z(r; t, t')$ is defined by

$$Z(r; t', t') = \langle n(\mathbf{x}, t)n(\mathbf{x} + \mathbf{r}, t')\rangle \tag{13.46}$$

and the inverse Fourier transform is

$$Z(r; t, t') = \int \mathrm{d}^3k\, Z(k; t, t') \exp(i\mathbf{k} \cdot \mathbf{r}). \tag{13.47}$$

Let us now consider the special case where $\mathbf{x} = \mathbf{x}'$ (i.e. $|\mathbf{r}| = 0$), $t = t'$, and for simplicity we take the concentration field to be stationary in time. Then, from eqns (13.46) and (13.47), we obtain the scalar variance $\langle n^2 \rangle$ as

$$\langle n^2 \rangle = \int d^3k\, Z(k) = \int_0^\infty 4\pi k^2 Z(k)\, dk = \int_0^\infty F(k)\, dk, \qquad (13.48)$$

where the last step defines $F(k)$ as the spectrum of scalar variance. If the concentration field is not stationary, (13.48) will hold at any time.

We can, in fact, derive an evolution equation for the non-stationary spectrum of scalar variance $F(k, t)$. Multiply both sides of (13.43) by $n(-\mathbf{k}, t)$, average, and take the limit of infinite system size. The result can be written as

$$\left(\frac{d}{dt} + 2Dk^2 \right) F(k, t) = T_s(k, t), \qquad (13.49)$$

where the scalar transfer spectrum $T_s(k, t)$ is given by

$$T_s(k, t) = \lim_{L \to \infty} \left[-8\pi k^2 \left(\frac{L}{2\pi} \right)^3 \sum_{\mathbf{j}} (ik_\beta) \langle n(\mathbf{k} - \mathbf{j}, t) u_\beta(\mathbf{j}, t) n(-\mathbf{k}, t) \rangle \right]. \quad (13.50)$$

This is the required balance equation for the spectral distribution of concentration fluctuations. It can be interpreted in the same way as the energy-balance equation—see Section 2.7.1—and hence it can be shown that the inertial transfer term possesses the conservation property (see eqn (2.126)),

$$\int_0^\infty T_s(k, t)\, dk = 0. \qquad (13.51)$$

Hence, if we integrate both sides of (13.49), with respect to k, and invoke eqn (13.51), we obtain an expression for the rate χ at which scalar variance is destroyed by molecular diffusivity:

$$\frac{d\langle n^2 \rangle}{dt} = \chi = -2D \int_0^\infty k^2 F(k, t)\, dk. \qquad (13.52)$$

Clearly we can draw an analogy between this quantity and the viscous dissipation rate ε, as defined by eqn (2.121).

Our next step is to pursue this and the other analogies between the scalar variance and the energy spectrum in order to obtain a form for $F(k)$ which is equivalent to the Kolmogorov energy spectrum, as given by eqn (2.137). This seems quite a natural thing to do, for the scalar field is essentially just a 'labelling' of the velocity field organized into a distribution with respect to wavenumber only because the velocity field is itself organized in this way.

At the same time, we should note that the scalar case presents us with new complications. Indeed, it should be clear from the outset that the ratio of the viscosity to the diffusivity (i.e. the Prandtl or Schmidt number) will be of key importance. For instance, if D is very much larger than v, then the scalar spectrum will be destroyed by molecular diffusion at wavenumbers low enough for the energy spectrum not to be subject to viscous effects. If, however, v is much larger than D, at large wavenumbers the scalar variance spectrum will fall off less rapidly than the energy spectrum, because the latter is being damped out by viscous effects.

In the remaining parts of this section, we shall consider only the case where the Reynolds number is very large, such that there is a wide range of wavenumbers beyond the energy-containing range.

13.2.1   *The inertial-convective range of wavenumbers*

The idea that the Kolmogorov hypotheses could be applied to the spectrum of scalar variance was due to Obukhov (1949) and, independently, to Corrsin (1951). In the same way that the viscosity produces a cut-off in the energy spectrum, these authors proposed a diffusion cut-off. This took the form

$$k_c = \left(\frac{\varepsilon}{D^3}\right)^{1/4},$$
(13.53)

and, without going into their detailed arguments, we can note that this expression is an obvious extension of the form for the Kolmogorov dissipation wavenumber as given by eqn (2.133).

Now we consider wavenumbers where the effect of both the molecular diffusivity and the kinematic viscosity can be neglected, that is to say, wavenumbers that are much smaller than either k_d or k_c. As in the case of the Kolmogorov spectrum, we use simple dimensional analysis on the assumption that the spectrum is a power law. The result is that we can write the scalar variance spectrum as

$$F(k) = \beta\chi\varepsilon^{-1/3}k^{-5/3},$$
(13.54)

where β is known as the Obukhov–Corrsin constant. Any reader who wishes to verify this will find it helpful to note that the dimensions of $F(k)$ and χ are $\langle n^2 \rangle L$ and $\langle n^2 \rangle T^{-1}$ respectively, as can be deduced from eqns (13.48) and (13.52).

We note finally that, in the case of scalar transport, the analogues of the inertial and dissipation ranges of wavenumbers are the convective and diffusive ranges. Thus, for eqn (13.54) to hold, k must be in the inertial and convective ranges. It is usual to refer to wavenumbers satisfying both conditions as being in the 'inertial–convective' range.

13.2.2 *Universal forms of the scalar spectrum*

The spectrum given in eqn (13.54) is the only result which can be obtained by the use of Kolmogorov-type arguments and simple dimensional analysis. Nevertheless, the picture presented by the Obukhov–Corrsin theory is something of an oversimplification. It was pointed out by Batchelor (1959) that the proposed diffusion cut-off k_c was only valid for $v \ll D$. From his previous work on the stretching of material lines and surfaces, he concluded that for $v \gg D$ the cut-off would be

$$k_B = \left(\frac{\varepsilon}{vD^2}\right)^{1/4},\tag{13.55}$$

where k_B is usually referred to as the Batchelor wavenumber. By distinguishing these two cases and various combinations of wavenumber ranges, we are forced to consider quite a complicated picture. In order to present it as simply as possible, we shall not always follow the chronological order in which papers were published.

Case I. $v \ll D$; (Pr, Sc $\ll 1$); $k_c \ll k_d$. Let us first divide up the high wave-numbers into two ranges: the inertial–convective range, in which $k \ll k_c \ll k_d$, and the inertial–diffusive range, in which $k_c \ll k \ll k_d$. The solution for the former is given by eqn (13.54) and we now consider the procedure due to Batchelor, Howells, and Townsend (1959) for the inertial–diffusive range.

We begin by rewriting eqn (13.43) in the equivalent, but more convenient, form

$$\left(\frac{d}{dt} + Dk^2\right)n(\mathbf{k}, t) = -i\sum_{\mathbf{j}} j_\beta u_\beta(\mathbf{k} - \mathbf{j})n(\mathbf{j}).\tag{13.56}$$

Batchelor *et al.* argued that the convolution sum on the r.h.s. will be dominated by $j < k_c$, as the scalar spectrum will fall off rapidly for wavenumbers higher than this. Therefore it follows that for $k \gg k_c$, we must have $|\mathbf{k} - \mathbf{j}|$ approximately equal to k. All the other approximations to be used stem from this one.

The first of these is a Markovian approximation. It is argued that, on the time-scales of the $n(\mathbf{j}, t)$, the $u_\beta(\mathbf{k} - \mathbf{j}, t)$ are approximately constant, so that (13.56) looks like a stationary balance for mode k with the steady input from the inertial term's being balanced by the molecular diffusion. This is justified because the diffusive relaxation time $1/Dk^2$ is much smaller than the local eddy-turnover time $1/\varepsilon^{1/3}k^{2/3}$. Accordingly, for this particular set of wave-numbers, the time derivative can be dropped, as can the time label on the Fourier components.

It follows that eqn (13.56) can readily be solved for $n(\mathbf{k})$, as can an equivalent equation for $n(-\mathbf{k})$. Hence an expression for the scalar covariance is obtained as

$$D^2 k^4 \langle n(\mathbf{k})n(-\mathbf{k}) \rangle = -\sum_{\mathbf{j}} \sum_{\mathbf{l}} j_\beta l_\gamma \langle u_\beta(\mathbf{k} - \mathbf{j}) u_\gamma(-\mathbf{k} + \mathbf{l}) n(\mathbf{j}) n(-\mathbf{l}) \rangle. \qquad (13.57)$$

Now, we again use the argument that the r.h.s. of (13.57) is dominated by $j \ll |\mathbf{k} - \mathbf{j}| \simeq k$ in order to justify the treatment of $\mathbf{u}(\mathbf{k} - \mathbf{j})$ and $n(\mathbf{j})$ as statistically independent (with the same condition for $\mathbf{u}(-\mathbf{k} + \mathbf{l})$ and $n(-\mathbf{l})$), from which it follows that the quadruple moment can be factored out as

$$\langle u_\beta(\mathbf{k} - \mathbf{j}) u_\gamma(-\mathbf{k} + \mathbf{l}) n(\mathbf{j}) n(-\mathbf{l}) \rangle = \langle u_\beta(\mathbf{k} - \mathbf{j}) u_\gamma(-\mathbf{k} + \mathbf{l}) \rangle \langle n(\mathbf{j}) n(-\mathbf{l}) \rangle. \qquad (13.58)$$

Kraichnan (1968) has referred to this as a 'quasi-normality approximation'.

Once we make this approximation on the r.h.s. of (13.57), we evaluate the velocity covariance by eqn (2.83), thus introducing a Kronecker delta which vanishes unless $j = l$. Eliminating the summation over $\mathbf{l}$ and the Kronecker delta together, we can write (13.57) as

$$D^2 k^4 \langle n(\mathbf{k})n(-\mathbf{k}) \rangle = -D_{\beta\gamma}(\mathbf{k}) Q(k) \int d^3 j\, j_\beta j_\gamma \langle n(\mathbf{j}) n(-\mathbf{j}) \rangle. \qquad (13.59)$$

Then, invoking (13.44) for the scalar covariance and integrating with respect to angles, (13.59) becomes

$$D^2 k^4 Z(k) = -\frac{2}{3} Q(k) \int j^2 F(j)\, dj, \qquad (13.60)$$

where we have used the isotropy of the scalar field and the fact that tr $D_{\alpha\beta}(\mathbf{k}) = 2$.

Now multiply both sides by $4\pi k^2$ in order to turn Z and Q into their respective spectra and also substitute from (13.52) for the integral to obtain

$$D^2 k^4 F(k) = -\frac{2}{3} E(k) \left(-\frac{\chi}{2D} \right). \qquad (13.61)$$

Lastly we substitute from (2.137) for the Kolmogorov energy spectrum, with the result

$$F(k) = \frac{\alpha}{3} \chi \varepsilon^{2/3} D^{-3} k^{-17/3} \qquad k_{\mathrm{c}} \ll k < k_{\mathrm{d}} \qquad (13.62)$$

as the inertial–diffusive spectrum of scalar variance.

Case II. $v \gg D$; (Pr, Sc $\gg$ 1); $k_{\mathrm{d}} \ll k_{\mathrm{B}}$. For this kind of fluid, we have three ranges of wavenumber to consider. In addition to the inertial–convective range—for which eqn (13.54) gives the solution—we have the viscous–convective range, in which $k_{\mathrm{d}} \leqslant k \ll k_{\mathrm{B}}$, and the viscous–diffusive range, in which $k_{\mathrm{d}} \ll k \leqslant k_{\mathrm{B}}$.

We shall now discuss a model proposed by Batchelor (1959) in which the latter two ranges are treated in a unified fashion. For the moment it will be simpler to ignore the distinction between the two viscous ranges, and merely note that we are concerned with scalar transport at wavenumbers larger than the Kolomogorov dissipation wavenumber.

The basis of the model is the proposal that scalar fluctuations in the wavenumbers of interest are concentrated into blobs of fluid. Each blob is assumed to have linear dimensions of the order of the Kolmogorov microscale (i.e. eqn (2.131)), and, on this scale, variations in the velocity gradients can be taken as negligible. Characteristic times for the change in magnitude or for rotation of the principal axes of the local strain relative to the fluid are assumed to be long compared with the lifetime of a typical scalar fluctuation on this scale.

Accordingly, Batchelor argued that the blobs were subject to local rate-of-strain fields which were constant in space and time. Randomness was taken into account by a variation in the direction of the principal axes of strain from one blob to another. Under these circumstances, the gradient of the scalar in a blob becomes aligned with the axis of the least principle rate of strain.

We shall not go into any great detail here, but we note that the key step is to approximate the convection term in eqn (13.41) by

$$u_\beta \frac{\partial n}{\partial x_\beta} \to Sz \frac{dn}{dz}, \tag{13.63}$$

where S is the least principal rate of strain which is taken to have its axis in the z-direction. With this approximation, eqn (13.41) can be used to obtain a solution for the scalar covariance spectrum with the result

$$F(k) = -\frac{\chi}{Sk} \exp\left(\frac{Dk^2}{S}\right) \qquad k \geqslant k_d. \tag{13.64}$$

From an analysis of some experimental observations, Batchelor estimated the effective least principal rate of strain to be given by

$$S = -0.5 \left(\frac{\varepsilon}{\nu}\right)^{1/2}. \tag{13.65}$$

With the substitution of this expression for S, eqn (13.64) for the spectrum of scalar covariance, for $k > k_d$, becomes

$$F(k) = 2 \left(\frac{\nu}{\varepsilon}\right)^{1/2} \frac{\chi}{k} \exp\left\{-2\left(\frac{k}{k_B}\right)^2\right\}, \tag{13.66}$$

where k_B is given by eqn (13.55).

If we then consider the case where $k \ll k_B$, the exponential factor in the r.h.s. of (13.66) tends to unity, and $F(k)$ reduces to the particularly simple form

TABLE 13.2   *High-wavenumber forms of the scalar covariance spectrum*

Range of wavenumbers	Form of spectrum
Case I: $\nu \ll D$; (Pr, Sc $\ll$ 1); $k_c \ll k_d$	
Inertial-convective $(k \ll k_c \ll k_d)$	$F(k) = \beta \chi \varepsilon^{-1/3} k^{-5/3}$
Inertial-diffusive $(k_c \ll k \ll k_d)$	$F(k) = \dfrac{\alpha}{3} \chi \varepsilon^{2/3} D^{-3} k^{-17/3}$
Case II: $\nu \gg D$; (Pr, Sc $\gg$ 1); $k_d \ll k_B$	
Inertial-convective $(k \ll k_d \ll k_B)$	$F(k) = \beta \chi \varepsilon^{-1/3} k^{-5/3}$
Viscous-convective $(k_d \lesssim k \ll k_B)$	$F(k) = \left(\dfrac{\nu}{\varepsilon}\right)^{1/2} \dfrac{\chi}{2k} \exp\left\{-2\left(\dfrac{k}{k_B}\right)^2\right\} \rightarrow$ $\rightarrow \left(\dfrac{\nu}{\varepsilon}\right)^{1/2}\left(\dfrac{\chi}{2k}\right)$
Viscous-diffusive $(k_d \ll k \lesssim k_B)$	$F(k) = \left(\dfrac{\nu}{\varepsilon}\right)^{1/2}\left(\dfrac{\chi}{2k}\right) \exp\left\{-2\left(\dfrac{k}{k_B}\right)^2\right\}$

$$F(k) = \left(\frac{\nu}{\varepsilon}\right)^{1/2}\left(\frac{\chi}{2k}\right) \qquad k_d \lesssim k \ll k_B. \qquad (13.67)$$

Thus our overall picture for this kind of fluid is of a scalar covariance spectrum which falls off as $-5/3$ in the inertial–convective range, then changes to a -1 law in the viscous–convective range and finally rolls off exponentially in the viscous–diffusive range of wavenumbers.

The assumption that S should be a constant in space and time is, of course, very restrictive. However, Kraichnan (1968) has investigated the effect on the scalar spectrum of letting the magnitude of S fluctuate. He concludes that the $1/k$ law is unchanged but that, in the viscous–diffusive range, the spectrum falls off less rapidly than the exponential decay predicted by Batchelor.

The experimental situation has been surveyed by Hill (1978), who concludes that the $k^{-5/3}$ power law in the inertial–convective range is well supported, with a value of the Obukhov–Corrsin constant in the range $0.68 < \beta < 0.83$. The k^{-1} viscous–convective law also seems to be experimentally established but the diffusive ranges present experimental difficulties which have not been adequately resolved. The various theoretical expressions for $F(k)$ are summarized in Table 13.2.

13.2.3   *Renormalized perturbation theory*

There has been relatively little use of renormalization methods in the area of passive scalar convection. What work there has been is mainly restricted to the direct-interaction family of theories (as discussed in Chapters 6 and 7).

In an early application, Roberts (1961) used the original Eulerian form of DIA to derive closed equations for the probability of marked fluid elements. The classical asymptotic results were recovered satisfactorily by the theory. But, for relative diffusion, where the diffusion coefficient depends upon the inertial-range energy spectrum at intermediate times, the theory was limited by the incorrect DIA prediction of the inertial-range power law as $k^{-3/2}$.

Herring and Kerr (1982) have carried out a direct numerical simulation of scalar convection in an isotropic turbulence field for a Prandtl number Pr = 0.5, and have used the resulting data to assess their numerical predictions from DIA and the TFM. They found that both theories gave good predictions of the scalar covariance spectrum at low wavenumbers but that the DIA was markedly superior to the TFM at high wavenumbers. From this, and other results, they concluded that the DIA gave better overall agreement. As the Taylor–Reynolds number of their simulation was limited by $R_\lambda < 30$, this is not altogether surprising (see Section 8.2 of the present work).

As we would expect, the improved behaviour of the LHDI in predicting the inertial-range energy spectrum at high Reynolds numbers also shows up in the scalar-convection problem. Kraichnan (1968) found that, with suitable assumptions, the LHDI theory led to the $k^{-5/3}$ inertial–convective law, the k^{-1} viscous–convective range, and the $k^{-17/3}$ inertial–diffusive range. Unfortunately the quantitative behaviour was not as good, with predictions of the spectrum levels being too small by a factor of 2–3 (Kraichnan 1968).

13.3   The motion of discrete particles

We have seen previously that the motion of marked fluid points is most naturally described in terms of Lagrangian position and velocity coordinates. Also, we saw (Section 12.4) that the motion of discrete particles could be described in a similar way using 'particle' position and velocity coordinates. Now we wish to describe the motion of added particles in a Eulerian framework, and we immediately run into both conceptual and practical difficulties. These arise because the coupling between the motion of the individual added particle and the fluid particles surrounding it cannot be ignored in a Eulerian framework.

Let us consider the conceptual difficulty first. This is solely due to the fact that we wish to treat the cloud of added particles as a continuum. However, if we require the convection to be passive, so that the motion of the host fluid is unaffected by the added particles, then the density of particles must be small enough to rule out interactions between them. This also goes for interactions at a distance, which are mediated by the host fluid. If we admitted such interactions, then we would have to consider the possibility that the combined system of fluid plus particles would act as a non-Newtonian fluid.

However, there is a way in which the fluid may mediate continuum behaviour for the cloud of particles. This is purely statistical and follows from the existence of the Eulerian field $U(x, t)$. As we know, such a field exists in the statistical sense that observed behaviour at the point (x, t) is the aggregate behaviour of many fluid 'particles'. Evidently, in this sense, we can also associate a Eulerian velocity field with the cloud of particles carried along by the host fluid. We shall refer to this particle velocity field as $V(x, t)$, and we shall characterize the particle cloud by this and other continuum concepts, such as a continuously distributed number density $N(x, t)$.

13.3.1   *Governing equations for particles*

The host fluid is taken to be incompressible, with uniform density ρ, and Eulerian velocity and pressure fields $U(x, t)$ and $P(x, t)$ which jointly satisfy eqns (1.1) and (1.2). To this fluid, we add a cloud of particles, as specified in Section 12.4.2.

We now wish to obtain equations which express conservation of mass and momentum for the added particles. We shall begin with the former, as it is the more straightforward.

We already know the equation for conservation of mass of marked fluid particles; it is given by eqn (13.10). From a similar consideration of the detailed balance between inflow, outflow, and rate of change of the number of particles in an elementary volume, we can generalize this equation to the case of elementary particles:

$$\frac{\partial N}{\partial t} + \frac{\partial (N V_\beta)}{\partial x_\beta} = D V^2 N, \tag{13.68}$$

where D is the diffusion coefficient for added particles due to molecular motions of the host fluid (i.e. it represents the Brownian motion of the added particles). It should be noted that the above equation reduces back to (13.10), in the limit where the added particles have the same material density as the host fluid and the particle diameter is much smaller than any relevant scale of the fluid motion, as under these circumstances $V(x, t) = U(x, t)$.

For the case of turbulent flows, the appropriate generalization of (13.68) can be obtained by making the Reynolds-type decomposition (as previously applied to the other variables) of the velocity field for particles, that is,

$$V_\alpha(x, t) = \bar{V}_\alpha(x, t) + v_\alpha(x, t) \tag{13.69}$$

$$\langle v_\alpha(x, t) \rangle = 0, \tag{13.70}$$

whence the equivalent of eqn (13.15) for marked particles can by derived from (13.68) as

$$\frac{\partial \bar{N}}{\partial t} + \frac{\partial (\bar{N} \bar{V}_\beta)}{\partial x_\beta} = D \nabla^2 \bar{N} - \frac{\partial \langle n v_\beta \rangle}{\partial x_\beta}. \tag{13.71}$$

If we now wish to consider (as we did before, in the case of marked fluid particles) possible applications to two-dimensional mean flows, then the corresponding generalization of eqn (13.19) is readily seen to be

$$-\langle nv_2 \rangle = D_p \frac{dN}{dx_2},$$ (13.72)

where D_p is a coefficient representing tthe diffusive effects of turbulent motions and the subscript indicates that this is for discrete particles rather than marked fluid particles.

The problem of obtaining the conservation law of momentum for the particle cloud is rather more difficult. Our starting point is just the same as for the fluid continuum, in that we seek to apply Newton's second law of motion. However, this time, we have to apply it to a single particle (which does not interact directly with the other added particles) rather than to an elementary volume of the fluid continuum.

Paradoxically, it turns out to be easier to begin directly with the Lagrangian-type frame of reference of the moving particle, obtain an equation of motion, and then transform the results to the laboratory frame. We shall base our discussion of this on the treatment due to Hinze (1972).

We start by noting that it is deceptively easy to write down Newton's second law for a single particle:

$$\ddot{\mathbf{X}}_p(t) = \mathbf{f},$$ (13.73)

where $\mathbf{X}_p(t)$ is the position of the particular particle at any time t, the dots denote time differentiation in the frame of reference of the moving particle, and $\mathbf{f}$ represents the resultant of all the forces acting on the particle. Naturally, these are mainly due to the interaction of the particle with the host fluid and include forces due to relative motion (both rectilinear and rotational), along with forces due to pressure gradients and viscous shear effects. However, we can also include the effects of external fields such as gravity. In principle, we can complete our derivation of a Eulerian equation of motion for the particle cloud by transforming to the fixed coordinate system in which the velocity field of the particles is $\mathbf{V}(\mathbf{x}, t)$. If we denote the operation of time differentiation in the moving frame of reference of the particle by $(d/dt)_p$, then the appropriate transformation to the fixed frame is given by

$$\left(\frac{d}{dt}\right)_p = \frac{\partial}{\partial t} + V_p(\mathbf{x}, t)\frac{\partial}{\partial x_\beta}.$$ (13.74)

In practice, the difficulty arises when we try to treat the interaction terms on the r.h.s. of (13.73). Unfortunately, even in the Lagrangian frame of reference, there is no agreed general form.

However, with our present restriction to small particles, with low velocities relative to the host fluid, we can simply take over Tchen's result (see eqns (12.63)–(12.65)) and use the transformation contained in eqn (13.74) to obtain

$$\left(\frac{\partial}{\partial t} + V_\beta \frac{\partial}{\partial x_\beta}\right) V_\alpha(\mathbf{x}, t) = A\{U_\alpha(\mathbf{x}, t) - V_\alpha(\mathbf{x}, t)\} +$$

$$+ B\left\{\frac{\partial}{\partial t} + V_\beta \frac{\partial}{\partial x_\beta}\right\} U_\alpha(\mathbf{x}, t) +$$

$$+ C \int_{t_0}^{t} G(t - t')\left\{\frac{\partial}{\partial t'} + V_\beta(\mathbf{x}, t')\frac{\partial}{\partial x_\beta}\right\} \times$$

$$\times \{U_\alpha(\mathbf{x}, t') - V_\alpha(\mathbf{x}, t')\}\, dt', \tag{13.75}$$

where $G(t - t')$ and A, B, C are as given by eqns (12.64) and (12.65).

13.3.2   *Interpretation in terms of a diffusion coefficient*

The introduction of a diffusion equation—along with a time-dependent diffusivity—for marked fluid particles was discussed in Section 13.1.2, where it was justified in terms of the observed Gaussian p.d.f.s for tagged particles at all diffusion times. However, as Batchelor (1949) pointed out, such a diffusion equation cannot be expected to hold for inhomogeneous turbulent fields. We shall now discuss this limitation for the case of discrete added particles, where, as we shall see, it takes a more severe form than for marked fluid particles.

Let us consider the turbulent dispersion of particles in a steady flow field. We shall also make the simplifying assumption—usually valid—that we can neglect the effects of molecular diffusion in comparison with those due to turbulence. Then eqn (13.71) can be written as

$$\frac{\partial \bar{N}}{\partial t} + \bar{V}_\beta \frac{\partial \bar{N}}{\partial x_\beta} + \bar{N}\frac{\partial \bar{V}_\beta}{\partial x_\beta} = -\frac{\partial \langle nv_\beta \rangle}{\partial x_\beta} \tag{13.76}$$

where we have expanded out the differential of a product on the l.h.s.

The first point to be made about this equation is the difference between the forms it takes for marked fluid particles and for discrete added particles. In the case of marked fluid particles, the third term on the l.h.s.—involving, as it does, the divergence of the velocity field—will be zero because of the incompressibility condition. This is, of course, an appreciable simplification, and one which is in general not available for added particles. The divergence of the *particle* velocity field is not automatically zero, and in this respect the cloud of particles can be thought of as resembling a compressible fluid.

Additional simplifications are possible for the dispersion of marked fluid points, if we consider homogeneous turbulence. These arise because the mean velocity must then be constant. Hence (13.76) can be transformed to a coordinate system moving with velocity $\bar{V}_\beta$ (which is just $\bar{U}_\beta$, as we are now discussing fluid particles). Then, with a gradient hypothesis (i.e. like that of eqn (13.72)) for the flux vector, eqn (13.76) can be shown to take the form of a standard diffusion equation for a random walk, although, as Batchelor pointed out, with a diffusion equation which depends on the elapsed time.

No such general simplification exists for the diffusion of discrete particles, because the velocity field $\bar{V}_\beta$ is not usually constant in space. This can readily be understood in terms of a concrete example.

Suppose that we consider the dispersion of particles released from a point source in a unidirectional homogeneous flow. In such circumstances the particles could be expected to spread out as they move downstream, thus forming a 'particle jet' within the host fluid. It follows that in this jet neither the mean density $\bar{N}$ nor the mean velocity $\bar{V}$ of particles will be constant in space. In other words, the particle jet constitutes a form of inhomogeneous turbulent field within the homogeneous turbulent field of the continuous phase.

It further follows from all this that the downstream mean concentration profiles in any particular experiment will depend not only on the nature of the source and the value of the diffusion coefficient (as introduced through (13.72)) but also on the mean motion effects due to the inertial terms on the l.h.s. of eqn (13.76).

Having established that the diffusion process is not entirely governed by it, what can we say about the particle diffusion coefficient itself? In Section 12.4.1, we deduced from the application of Taylor's analysis to discrete particles that the particle diffusion coefficient D_p would equal the turbulent diffusion coefficient of the fluid for long diffusion times. This result also emerges form Tchen's analysis.

Hinze (1972) mentions an analysis due to Peskin (1962; cited by Hinze), which suggests that $D_p/D_T < 1$. But, with a restriction to small separations between discrete and fluid particles (not to mention the assumption of a Gaussian p.d.f. for the velocities!), there seems little reason to regard this result as being in conflict with eqn (12.62). A more interesting possibility is that of $D_p/D_T > 1$, which may occur if the discrete particles are of appreciable size on the scale of turbulent eddies. The occurrence of particle diffusivities greater than fluid diffusivities has been attributed, with some plausibility, to a filtering effect (Hinze 1972) in which large particles are assumed not to respond to turbulent eddies on a scale smaller than the particle diameter.

There is an unfortunate lack of fundamental work in this area—although, in the interests of completeness, we should mention the application of Eulerian DIA to the diffusion of particles (Reeks 1980)—and so we are left with mostly only phenomenological approaches. We shall discuss two rather pragmatic methods, in which the diffusion of particles is assumed from the first to be a random walk. Then we shall conclude this section with a brief examination of the experimental situation.

13.3.3 *Random walk models of particle diffusion*

A novel approach to turbulent diffusion was taken by Hutchinson, Hewitt, and Dukler (1971), where it was seen as one aspect of the problem of predicting the deposition of liquid droplets or solid particles from turbulent gas flows in pipes. These authors argued that the diffusing particles travelled downstream at (on average) the mean velocity of the fluid, while simultaneously executing a random walk on a disc at right angles to the tube axis and moving with the flow. As each particle would take many steps to reach the wall (where it is supposed to be absorbed; this is the boundary condition for the problem), this was felt to justify the application of Chandrasekhar's (1943) form of the diffusion equation

$$\frac{\partial W}{\partial t} = D_c \nabla^2 W + S, \tag{13.77}$$

where $W(r, t)2\pi r\,dr$ is the probability of finding a particle in the annulus between r and $r + dr$ at any time t, D_c is the associated diffusivity for a macroscopic random walk, and S is a source term.

Chandrasekhar has given the diffusivity D_c (in one dimension) as

$$D_c = \frac{n\langle l_p^2 \rangle}{2} \tag{13.78}$$

where n is the number of displacements per unit time and $\langle l_p^2 \rangle$ is the mean square value of these displacements.

Hutchinson *et al.* point out that the validity of (13.77) is achieved in the limit of many finite displacements. That is, individual steps are not required to be differentially small, nor need the p.d.f. for the magnitude of any given displacement be Gaussian.

The diffusion coefficient given by eqn (13.78) can then be related to measurable turbulence quantities by writing the frequency n as

$$n = U_e/l_e, \tag{13.79}$$

where U_e and l_e are characteristic velocity and length scales for large eddies. These are taken to be the friction velocity u_τ (as defined in this book by eqn (1.29)) and Townsends' length scale for large eddies respectively. In fact, we shall just continue to represent the latter quantity by l_e.

The particle displacement is taken to be related to l_e by

$$l_p = hl_e \tag{13.80}$$

where h is to be determined. Then, eqn (13.78) can be written as

$$D_c = \frac{u_\tau l_e \langle h^2 \rangle}{2}. \tag{13.81}$$

The friction velocity and the Townsend length scale can readily be expressed in terms of the global parameters of the pipe flow, i.e. the radius, the friction factor, the kinetic energy of fluctuations, and the dissipation rate. Therefore the problem reduces to one of finding $\langle h^2 \rangle$.

At this point we encounter one of the more interesting and ingenious features of the work. Hutchinson *et al.* simply bypass the analytical difficulties inherent in modelling particle motion and assume that, during any particular displacement, the drag force exerted by the fluid on the particle is given by the phenomenological law (see eqn (1.43))

$$\frac{dU_p}{dt} = C_f \frac{A_p \rho}{2m_p} (U_e - U_p)^2, \tag{13.82}$$

where U_p is the velocity of the particle during one particular displacement, and A_p and m_p are the cross-sectional area and the mass of the particle. Then, $\langle h^2 \rangle$ is obtained by a numerical simulation based on eqn (13.82) and the use of a curve fitted to the experimental results for C_f (as a function of Reynolds number) as an input to the calculation.

It should be emphasized that this procedure—like the argument leading to eqn (13.81)—has its arbitrary features. However, all the steps are physically plausible, and the results turn out to be good. In Fig. 13.1 we show a compari-

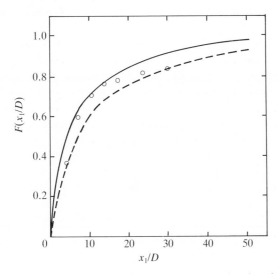

Fig. 13.1. Theoretical prediction of the fraction F of water droplets deposited from a turbulent airstream as a function of downstream distance x_1, as measured from the point of entrainment (Hutchinson *et al.* 1971). Theory: ———, particle radius $a_p = 10$ μm; ————, particle radius $a_p = 15$ μm. Experiment: o, air flow in a pipe with diameter $d = 0.047$ m, $R = 1.28 \times 10^5$, and water droplets with radii in the range 10 μm $< a_p < 15$ μm.

son of theory and experiment which can be taken as representative of the much fuller comparisons carried out by Hutchinson *et al.* It can be seen to illustrate a major problem in this work, which is the uncertainty over particle size. Here, the water droplets had radii in the range 10–15 μm, and the calculations have been carried out for these two extremes. Just to complicate matters further, there may also be some tendency for the size distribution to change as the droplets move downstream. Nevertheless, despite these uncertainties, agreement between theory and experiment is good.

The possibility that the finite-step diffusion coefficient D_c could be extended to turbulent diffusion in a more fundamental (i.e. less arbitrary) way was subsequently studied by applying it to the limiting case where the discrete particle becomes a 'fluid point' (McComb 1974). It was shown that, at long diffusion times, D_c—as given by eqn (13.78)—was equivalent to the result from Taylor's theory (see eqn (12.14)). Thus, under certain circumstances, Chandrasehkar's and Taylor's theories give the same result for turbulent diffusion. This is hardly suprising, as Taylor's theory becomes equivalent to a random walk of many finite steps at long diffusion times.

However, from a practical point a view, the advantage of Chandrasekhar's theory is that it is formulated as a random walk in a fixed coordinate system. Therefore it can be applied to turbulence without reference to the difficult task of transforming from Lagrangian to Eulerian coordinates. The problem then becomes one of expressing (13.78) in terms of measurable turbulence quantities.

There is, at present, no accepted rigorous method of doing this. Nevertheless, given the simple universal form of (13.78), it seems likely that its form, when expressed in terms of the statistical parameters of fluid turbulence, should also be universal. That is, if we could work out an expression for D_c in one particular kind of turbulent flow, then that expression might also be easily generalized to other turbulent flows. Accordingly, it was suggested by McComb (1974) that we should attempt to obtain D_c for the special case of homogeneous turbulence, where we know the exact result in Lagrangian coordinates and have a good idea of what it should be in Eulerian coordinates.

That is, we take the exact result for long diffusion times and use the Hay–Pasquill conjecture (see Section 13.2.2), along with the replacement $\langle V^2 \rangle = \langle u^2 \rangle$ (valid for homogeneous turbulence), in order to transform the expression for D_T into the Eulerian frame. Thus, from (12.14) and (12.31), we obtain

$$D_T = \beta \langle u^2 \rangle^{1/2} L_E, \tag{13.83}$$

where we have also made the replacement $\langle u^2 \rangle^{1/2} T_E = L_E$ and L_E is an appropriate Eulerian integral length scale associated with correlation in the direction of the diffusion process.

Now, when we apply (13.78) for the diffusion coefficient in an Eulerian framework, it is plausible to suggest that the fluid particle has an average displacement cL_E, where c is a constant, and the frequency of such displacements is just $n = T_E^{-1}$. Hence, making the replacement $L_E = \langle u^2 \rangle^{1/2} T_E$, we obtain (13.78) as

$$D_c = \frac{c}{2} \langle u^2 \rangle^{1/2} L_E, \tag{13.84}$$

which agrees with (13.83) for homogeneous turbulence provided that $c = 2\beta \approx 2$, taking β to be of order unity in homogeneous turbulence.

The weakness in then asserting that (13.84) should apply to any flow lies in the fact that β is not a universal constant (see, for instance, eqn (12.33)). Nevertheless, Davidson and McComb (1975) have found that the diffusion coefficient given by (13.84) can be used to make a successful predication of tracer distribution in a jet. Typical results are shown in Fig. 13.2, where they are compared both with experimental data and with an earlier calculation (Hinze and Van Der Hegge Zijnen 1949) based on a constant velocity. It should be noted that the results are presented in self-similar form, with the radial variation in x_2 scaled by the downstream distance x_1.

Davidson and McComb also extended their calculations to particles with

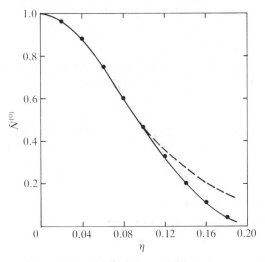

Fig. 13.2. Theoretical distribution of scalar contaminant in a round free jet, as predicted on the basis of eqn (13.84) for the diffusion coefficient (Davidson and McComb 1975): ———, D_c given by eqn (13.84); ————, constant diffusivity (Hinze and Van Der Hegge Zijnen 1949); ●, experimental data of Hinze and Van Der Hegge Zijnen (1949).

finite inertia by means of perturbation theory applied to the governing equations for $\bar{N}$ and $\bar{V}$. Full details are given in the original reference, and so we shall just sketch out the general outline here.

For a round free jet in the x_1 direction, with x_2 as the radial coordinate, we can expect similarity solutions of the form

$$-U_\alpha(\mathbf{x}) = \frac{GU_0 d}{x_1}\, \tilde{U}_\alpha\!\left(\frac{x_2}{x_1}\right),$$ (13.85)

where G is a constant, U_0 is the nozzle exit velocity, and d is the nozzle diameter. Forms like this are usually valid for all statistical quantities at downstream distances such that $x_1/d > 8$.

If particles emerge from the jet nozzle with number density N_0 and velocity U_0, then we can write similarity solutions for $\bar{V}_\alpha$ and $\bar{N}$ as follows:

$$V_\alpha(\mathbf{x}) = \frac{GU_0 d}{x_1}\, \tilde{V}_\alpha\!\left(\frac{x_2}{x_1}\right)$$ (13.86)

$$N(\mathbf{x}) = \frac{HN_0 d}{x_1}\, \tilde{N}(x_2/x_1)$$ (13.87)

where H is another constant. Like G, it can be fixed from a consideration of global momentum and particle flux balances.

Davidson and McComb solve the governing equations ((13.75) and (13.76) in this book) using perturbation expansions for mean velocity, mean number density, and the diffusion coefficient for particles, as given by[1]

$$\tilde{V}_\alpha = \tilde{U}_\alpha + \epsilon \tilde{V}_\alpha^{(1)} + O(\epsilon^2)$$ (13.88)

$$\tilde{N} = \tilde{N}^{(0)} + \epsilon \tilde{N}^{(1)} + O(\epsilon^2)$$ (13.89)

$$D_p = D_T + \epsilon D^{(1)} + O(\epsilon^2)$$ (13.90)

with the expansion parameter

$$\epsilon = G\!\left(\frac{d}{x_1}\right)^2 S.$$ (13.91)

Here S is the Stokes number; thus

$$S = \frac{U_0 \tau}{d},$$ (13.92)

where τ is the momentum relaxation time and is given by

$$\tau = \frac{2\rho_p A_p}{9\pi\mu}.$$ (13.93)

Note that this is the same as the inverse of the constant A, which appears in

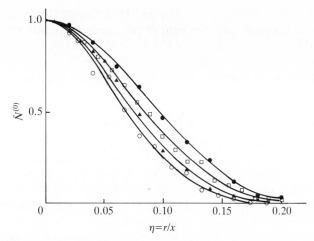

Fig. 13.3. Dispersion of solid particles in a turbulent air jet (McComb and Salih 1978): ————, theoretical predictions (Davidson and McComb 1975): Experimental measurements: $\bullet$, $\epsilon = 0$; $\square$, $\epsilon = 0.10$; $\blacktriangle$, $\epsilon = 0.20$; $\circ$, $\epsilon = 0.30$.

eqn (12.63), for the case where the material density of the particle is much greater than that of the fluid.

The results which can be obtained by this method are limited by the need to restrict values of the expansion parameter to the case where the Stokes drag law is valid. Yet a surprisingly wide range of particle inertia effects can be included. In Fig. 13.3 we show the results of these calculations for the radial distribution of the mean number density for $\epsilon \leqslant 0.30$. The predicted profiles can be seen to compare favourably with the corresponding values measured by laser anemometry (McComb and Salih 1978).

13.3.4 *Measurements of particle motion in turbulent flows*

The difficulties of making an experimental study of the motion of discrete particles in turbulent flow are well known and many expedients have been tried. The mechanical sampling techniques previously used with some success in measuring the dispersion of tracers in turbulent flow have also been extended to particles (e.g. Laats 1966, Bragg and Bednarik 1975), but inevitably there must be appreciable flow perturbation with these methods. While it has been proposed that this disturbance can be reduced by counting impacts on a conventional anemometer probe (Goldschmidt and Eskinazi 1966), this method can only be used with liquid drops or gas bubbles and is generally limited by the well-known fragility of hot-wire probes.

Optical methods have also been tried, but generally such approaches have tended to be somewhat qualitative. As examples of some more quantitative investigations, we note the work of Becker, Hottel, and Williams (1965), who related the intensity of scattered light to the number of scatterers, the work of Jones, Chao, and Shirazi (1967), who used luminous particles and an array of photomultipliers to follow particle movements, and a study of particle trajectories using electronic flash illumination and a battery of cameras (Snyder and Lumley 1971).

In more recent years there has been a rapid development in the technique of laser-Doppler anemometry (LDA) (see Section 3.1.2 for a brief general account), which is a way of measuring the velocity which depends on the scattering of light from particles in the fluid. It was suggested by Farmer as long ago as 1972 that, as well as measuring particle velocity, a suitably adjusted LDA could also be used to measure the size and number density of the scattering particles. This would rely on the dependence of the modulation depth (or visibility) of the LDA output signal on either the number density or (depending on the way in which the fringe spacing has been adjusted) the size of the particles. Other similar proposals have been made (e.g. She and Lucero 1973, Durst and Umhauer 1975).

This basic approach has been developed in practice by Fristrom, Jones, Schwar, and Weinberg (1973) and by Hong and Jones (1976) to measure particle sizes, and by McComb and Salih (1977) to measure velocity, size and number density of tracer particles. The use of LDA to measure the velocity distribution of large particles in a flow is an obvious application (Arundel, Hobson, Lalor, and Weston 1974; Popper, Abuaf and Hetsroni 1974; McComb and Salih 1978; Tsuji and Morikawa 1982), although the investigation by McComb and Salih was mainly concerned with the measurement of concentration profiles of particles (see Fig. 13.3).

Apart from its non-perturbative character, the LDA has another great advantage. It is possible to discriminate between the scattered light signals from particles of appreciably different size. This means that one can simultaneously obtain the velocity of small particles (which act as tracers for the continuous phase) and the velocity of the large particles which may be the main object of a particular study. This would allow one to investigate the relative importance of mean motion and diffusive effects, as discussed in Section 13.3.2.

The application of LDA to particle tracking can be expected to grow rapidly as this equipment is now becoming widely available. A survey of some recent work in this area has been given by Larsen, Engelund, Sumer, and Lading (1980).

13.4   Turbulent mixing

In practical situations, the most important aspect of turbulent diffusion is very often seen to be its mixing ability. For instance, if we consider two different reacting chemical species placed at different locations in a turbulent flow, then the rate at which they react will be affected by the rate at which the diffusive action of the turbulence mixes them together. Indeed, if the chemical reaction involved is very much faster than the turbulent mixing, then we have the interesting situation that the reaction is 'diffusion limited' (Lin and O'Brien 1974).

The subject of turbulent mixing can be treated by an extension of some of the fundamental ideas concerning turbulent diffusion, provided that there are suitable restrictions on the nature of the chemical reactions involved. For instance, the spectral theories of Section 13.2.1 have been extended to isotropic turbulence with a first-order chemical reaction (Corrsin 1958, 1961). However, the chemical species must be so dilute that the heat of reaction has no effect on the reaction rate. Certainly we must have a general requirement that the turbulence is unaffected by the presence of the reaction.

In practice, conditions of this kind are often unlikely to be satisfied. Turbulent mixing is of importance not only in chemical processes, but in a whole host of situations ranging through combustion to oceanography and meteorology. As a result, there must inevitably be a great need for practical solutions in problems which are remote from the idealized situations which we have been discussing in this (and other) chapters. As a further result, turbulent mixing is virtually a specialized subject in its own right.

Accordingly, the present note is included only for completeness, and we shall not go into the subject in any greater detail. The reader who wishes to delve further will obtain a good start from the following review articles: Hill (1976), Libby and Williams (1981), O'Brien (1981), and Jones and Whitelaw (1982). Also, direct numerical simulation has recently been extended to include chemical reactions and mixing (Leonard and Hill 1988).

Note

1. It should be noted that this is an unusual kind of perturbation theory. The 'zero-order' terms are the exact solutions for the mean turbulent moton of the fluid. Naturally, the corresponding functions are only accessible by experiment. The experimental values can be incorporated in the calculation by obtaining them in the form of coefficients in polynomials of suitably high order. Then the 'perturbation' is the small difference between the motion of the added particles and that of the fluid points.

References

ANTONIA, R. A. (1985). *Int. J. heat mass Transfer* **28**, 1805.
ARUNDEL, P. A., HOBSON, C. A., LALOR, M. J. and WESTON, W. (1974). *J. Phys. D* **7**, 2288.
BALDWIN, L. V. and WALSH, T. J. (1961). *AIChE J.* **7**, 53.
BATCHELOR, G. K. (1949). *Aust. J. sci. Res.*, **A2**, 437.
—— (1952). *Proc. R. Soc. A* **213**, 349.

—— (1959). *J. fluid Mech.* **5**, 113.
——, HOWELLS, I. D., and TOWNSEND, A. A. (1959). *J. fluid Mech.* **5**, 134.
BECKER, A. H., HOTTEL, H. C., and WILLIAMS, G. C. (1965). *J. fluid Mech.* **30**, 259.
BERKOWICZ, R. and PRAHM, L. P. (1980). *J. fluid Mech.* **100**, 433.
BIRD, R. B., STEWART, W. E. and LIGHTFOOT, E. N. (1960). *Transport phenomena*. Wiley, New York.
BRAGG, G. M. and BEDNARIK, H. V. (1975). *Int. J. heat mass Transfer* **18**, 443.
CHANDRASEKHAR, S. (1943). *Rev. mod. Phys.* **15**, 1.
CORRSIN, S. (1951). *J. appl. Phys.* **22**, 469.
—— (1952). *J. appl. Phys.* **23**, 113.
—— (1958). *Phys. Fluids* **1**, 42.
—— (1961). *J. fluid Mech.* **11**, 407.
—— (1974). *Adv. Geophys.* **18A**, 25.
CSANADY, G. T. (1966). *J. atmos. Sci.* **23**, 667.
DAVIDSON, G. A. and McComb, W. D. (1975). *J. aerosol Sci.* **6**, 227.
DEISSLER, R. G. (1963). *Int. J. heat mass Transfer* **6**, 257.
DURST, F. and UMHAUER, H. (1975). *Proceedings of the LDA Symp. Copenhagen*, University of Denmark.
FARMER, W. M. (1972). *Appl. Optics* **11**, 2603.
FLINT, D. L., KADA, H., and HANRATTY, T. J. (1960). *AIChE J.* **6**, 325.
FRISTROM, R. M., JONES, A. R., SCHWAR, M. J. R., and WEINBERG, F. J. (1973). *Faraday Symp. Chem. Soc.*, no. 7, 183.
FULACHIER, L. and ANTONIA, R. A. (1984). *Int. J. heat mass Transfer* **27**, 987.
—— and DUMAS, R. (1976). *J. fluid Mech.* **77**, 257.
GOLDSCHMIDT, V. and ESKINAZI, S. (1966). *J. appl. Mech., Trans. ASME* 735.
GOLDSTEIN, S. (1938). *Modern developments in fluid dynamics*. Clarendon Press, Oxford. (Reprinted by Dover Publications, New York, 1965.)
GROENHOF, H. C. (1970). *Chem. eng. Sci.* **25**, 1005.
HERRING, J. R. and KERR, R. M. (1982). *J. fluid Mech.* **118**, 205.
HILL, J. C. (1976). *Ann. Rev. fluid Mech.* **8**, 135.
HILL, R. J. (1978). *J. fluid Mech.* **88**, 541.
HINZE, J. O. (1972). *Prog. heat mass Transfer* **6**, 433.
—— and VAN DER HEGGE ZIJNEN (1949). *Appl. sci. Res.* **A1**, 435.
HONG, N. S. and JONES, A. R. (1976). *J. Phys. D* **9**, 1839.
HUTCHINSON, P., HEWITT, G. F., and DUKLER, A. E. (1971). *Chem. eng. Sci.* **26**, 419.
JOHNK, R. E. and HANRATTY, T. J. (1962). *Chem. eng. Sci.* **17**, 867.
JONES, B. G., CHAO, B. T. and SHIRAZI, M. A. (1967). *Developments in mechanics, 10th Mid-western Conf., Colorado*, Vol. 4, p. 1249.
JONES, W. P. and WHITELAW, J. H. (1982). Combust. Flame, **48**, 1.
KRAICHNAN, R. H. (1964). *Phys. Fluids* **7**, 1048.
—— (1968). *Phys. Fluids* **11**, 945.
LAATS, M. K. (1966). *Inz.-Fiz. Zh.* **10**, 11.
LARSEN, P. S., ENGELUND, F., SUMER, B. M. and LADING, L. (1980). *J. fluid Mech.* **99**, 641.
LAUFER, J. (1954). *NACA Tech. Rep. 1174.*
LEONARD, A. D. and HILL, J. C. (1988). *J. Sci. Comput.* **3**, 25.
LIBBY, P. A. and WILLIAMS, F. A. (1981). *AIAA J.* **19**, 261.
LIN, C-H. and O'BRIEN, E. E. (1974). *J. fluid Mech.* **64**, 195.
McComb, W. D. (1974). *J. Phys. A* **7**, L164.
—— and RABIE, L. H. (1982). *Chem. eng. Sci.* **37**, 1759.
—— and SALIH, S. M. (1977). *J. aerosol Sci.* **8**, 171.
—— and SALIH, S. M. (1978). *J. aerosol Sci.* **9**, 299.

O'BRIEN, E. E. (1981). *AIAA J.* **19**, 366.

OBUKHOV, A. M. (1949). *Izv. Akad. Nauk SSSR, Geogr. Geofiz.*, **13**, 58.

POPPER, J., ABUAF, N., and HETSRONI, G. (1974). *Int. J. multiphase Flow*, **1**, 715.

QUARMBY, A. and ANAND, R. K. (1969). *J. fluid Mech.* **38**, 433.

REEKS, M. W. (1980). *J. fluid Mech.* **97**, 569.

ROBERTS, P. H. (1961). *J. fluid Mech.* **11**, 257.

SHE, C. Y. and LUCERO, J. A. (1973). *Opt. Commun.* **9**, 300.

SNYDER, W. H. and LUMLEY, J. L. (1971). *J. fluid Mech.* **18**, 41.

TSUJI, Y. and MORIKAWA, Y. (1982). *J. fluid Mech.* **120**, 385.

14

NON-NEWTONIAN FLUID
TURBULENCE

There may appear to be some element of ambiguity about the title of this chapter. Does the adjective 'non-Newtonian' qualify the fluid? Or, should we read 'fluid turbulence' as a single entity, and take it to be non-Newtonian? In fact, this ambiguity is really rather more general in character. A purely Newtonian fluid flowing in the presence of solid boundaries which are flexible may behave quite differently from the case where the boundaries are rigid. It might therefore be appropriate to regard the combination of Newtonian fluid and flexible boundaries as an example of a fluid flow which should be described as non-Newtonian.

Another example (and arguably the most important) is the phenomenon of drag reduction by additives (see Section 3.4). As we shall see later, the turbulence which occurs in the presence of drag-reducing additives is different from the turbulence which occurs in the solvent alone. Indeed, in some cases of very dilute polymer solutions, the anomalous (i.e. less dissipative) turbulence is probably the only detectable non-Newtonian effect.

It is really only the existence of this truly remarkable phenomenon which justifies the inclusion of a discussion of non-Newtonian effects here. The development of constitutive equations for non-Newtonian fluids is a scientific problem which is fully comparable in difficulty with the basic problems in turbulence. Accordingly, the combined problem posed by turbulence in a non-Newtonian fluid—while perhaps a possible candidate for the title of 'most difficult problem in physics'—does not, as yet, lead to very much in the way of good science. Nevertheless, in order to provide a context for our discussions of turbulent structure in drag-reducing structures, we begin with a brief and rather summary treatment of the non-Newtonian aspects of the subject.

14.1 Non-Newtonian fluid flow

We have defined a Newtonian fluid as one satisfying the relationship between stress and rate of strain, given by equation (1.3), where the kinematic viscosity v is a constant. That is, while the viscosity may depend on the temperature (for instance), it does not depend on the rate of strain. It follows that any fluid which does not satisfy that condition is, by definition, a non-Newtonian fluid.

This definition opens up a rather wide field. There are various classes of non-Newtonian behaviour, with perhaps very many kinds of fluid in each.

Also, there may be difficulties in assigning any particular fluid to one class rather than another. The result is that we are confronted with a large and complicated subject. Our response to this situation will be to narrow down our approach to discuss the non-Newtonian properties of only those kinds of polymer solutions and fibre suspensions which may be expected to be drag reducing under the appropriate circumstances.

14.1.1 *Rheological aspects*

The general equation of motion for the (incompressible) fluid continuum is given by eqn (1.2). For a particular fluid, the stress tensor $s_{\alpha\beta}$ must be specified in terms of the relative fluid motions (from which it arises in the first place). In the case of a Newtonian fluid, the stress tensor is expressed in terms of the velocity gradients (formally, the quantity within brackets is known as the rate-of-strain or deformation tensor; see eqn (14.4) below) and takes the reduced form given in eqn (1.3a) for a simple shear flow.

The general problem of the subject of rheology is to express the stress tensor in terms of the rate-of-strain tensor for any given fluid. There are different ways of approaching this, but a popular practical expedient is the idea of a 'power-law fluid', in which (for simple shearing flow) the stress tensor can be written as

$$s_{12} = K\left(\frac{dU_1}{dx_2}\right)^n, \tag{14.1}$$

where K is a constant. Then, if we introduce an apparent viscosity μ_A through

$$s_{12} = \mu_A\left(\frac{dU_1}{dx_2}\right), \tag{14.2}$$

eqn (14.1) yields

$$\mu_A = K\left(\frac{dU_1}{dx_2}\right)^{n-1}. \tag{14.3}$$

If the exponent n is less than unity, the fluid is said to be 'shear thinning'. This means that it flows faster, the more rapidly it is sheared. That is, the apparent viscosity decreases as the shear rate increases. In practice, this is the most common case (especially for polymeric materials) and the opposite case, a 'shear-thickening' fluid (i.e. one for which $n > 1$), is relatively uncommon.

Non-Newtonian fluids can also show viscoelastic behaviour, which essentially means that, under certain circumstances, they can exhibit an elastic response (e.g. in the form of a recoil). Such fluids can be described by models which are forms of eqn (1.3) for the Newtonian case modified by the addition of time-dependent terms with associated relaxation times. Popular models include the Maxwell model (characterized by the Newtonian viscosity and a stress relaxation time) and the three-parameter Oldroyd model (characterized by the Newtonian viscosity and separate times for the relaxation of both stress and strain).

Both shear-thinning and viscoelastic behaviour have been found in aqueous solutions of polyacrylamide and poly(ethylene oxide) (e.g. Bruce and Schwarz 1969, Darby 1970). These are, of course, the two most common groups of synthetic polymers used for turbulent friction reduction. Two general points are worth making about this kind of measurement. First, viscoelastic properties may be measured under circumstances of transient response, where the rates of strain are not particularly large. This means that the fluid which is being studied can be represented by, for example, a Maxwell fluid model which is linear with respect to the rate-of-strain tensor. However, experiments which show a reduction of apparent (shear) viscosity with increasing strain rate are demonstrating a non-linear relationship between stress and rate of strain. The need to include both cases means that the constitutive equation is unlikely to be simple.

Our second general point concerns the difficulty of relating continuum non-Newtonian behaviour on the one hand and turbulent drag reduction on the other. Although rheologists talk about dilute polymer solutions, by this they mean, typically, polymer concentrations of the order of thousands of parts per million. In contrast, drag reduction is a large effect, with polymer concentrations of only a few tens of parts per million. At that sort of level, non-Newtonian effects can be difficult to detect.

However, an important exception to this comment arises when we consider the resistance of these solutions to stretching. Let us introduce the rate-of-strain tensor $e_{\alpha\beta}$ through the relationship

$$2e_{\alpha\beta} = \frac{\partial U_\alpha}{\partial x_\beta} + \frac{\partial U_\beta}{\partial x_\alpha}. \tag{14.4}$$

Now take a circular cylinder of fluid, with its axis in the direction of flow x_1 and subject to a stretching velocity gradient in the direction of flow, $e_{11} = dU_1/dx_1$. Then it follows from the continuity equation (1.1) and the imposition of axial symmetry, that the transverse normal stresses e_{22} and e_{33} must satisfy the relationship

$$e_{22} = e_{33} = -e_{11}/2. \tag{14.5}$$

We assume that the off-diagonal elements of the rate-of-strain tensor are zero, and that correspondingly we need consider only the diagonal elements of the stress tensor $s_{\alpha\beta}$. In fact, only the differences between the normal stresses (i.e. $s_{11} - s_{22}$ and $s_{11} - s_{33}$) have rheological significance, as the absolute levels are determined by the hydrostatic pressure and can be fixed arbitrarily. Accordingly, it is usual to introduce the coefficient of resistance to extensional motion λ (or the extensional viscosity, for conciseness) through the relationship

$$\lambda = \frac{s_{11} - (s_{22} + s_{33})/2}{e_{11}}. \tag{14.6}$$

For Newtonian fluids, the extensional viscosity can be expressed in terms of the shear viscosity μ through the familiar relationship $\lambda = 3\mu$. However, for polymer solutions it was originally reported by Metzner and Metzner (1970), and later confirmed by other investigators (e.g. Bragg and Oliver 1973), that extensional viscosities could be as large as 10 000 times the shear viscosity. This result was found in both polyacrylamide and polyethyleneoxide solutions, and, even at drag-reducing concentrations, ratios as high as a 1000 were found. Thus, even although there are some problems in producing viscometric flows which satisfy eqn (14.5), and indeed in practice some simplifying assumptions have to be made in order to interpret the measurements, the ratios found are so enormous that it seems clear that here is a non-Newtonian effect which may well be relevant to drag reduction in turbulent flow.

It was later shown by Mewis and Metzner (1974) that the extensional flow of suspensions of macroscopic fibres led to similar anomalous results. They found that, for fibres with aspect ratios (i.e. length-to-diameter ratios) in the range 280–1260, the measured stress levels were one to two orders of magnitude larger than in shear flow. Shear thinning has also been reported in suspensions of macroscopic fibres, but (in contrast with the polymer case) this was at drag-reducing concentrations (Lee, Vaseleski, and Metzner 1976; Jeffrey and Acrivos 1976). Lastly, we note, for completeness, that this is still an active area of research in rheology (Goto, Nagazono, and Kato 1986).

14.1.2 Composite systems: Newtonian fluid with modified boundary conditions

Many, if not most, non-Newtonian fluids, ranging from slurries through suspensions to polymer solutions, would merit the description of being a Newtonian fluid with modified boundary conditions. This is because the non-Newtonian behaviour arises as a result of the no-slip boundary condition of the fluid on each of the suspended particles or molecules, leading to collective behaviour which is mediated by the host fluid (Edwards and Freed 1974). This is an instance of the composite fluid properties being altered by the imposition of microscopic or internal boundary conditions.

However, in this section we are concerned with the macroscopic or external boundary conditions and with modified forms of these, in which the composite system—Newtonian fluid plus modified boundary conditions—may behave in a non-Newtonian way.

The case which has been most studied is that of a moving fluid confined or bounded by flexible walls, an idea which was inspired by the supposition that this was the property underlying the dolphin's ability to reduce its hydro-dynamic drag (see Section 3.4). Early theoretical studies indicated that elastic boundary walls could delay transition from laminar to turbulent flow and hence reduce the overall drag on the boundary surface (Benjamin 1960; Landahl 1962). Although the effect was expected to be small, the potential practical importance in aerodynamics is great, so that many practical studies have been made. A thorough review has been given by Bushnell, Hefner, and Ash (1977), and some more recent references are to be found in Bushnell (1980).

Again, reflecting the practical significance of the possibility of reducing turbulent frictions, various static methods of modifying turbulence have been tried. These are usually modifications of the boundary surface, which are intended to change the turbulence to some less dissipative form. The devices which have been tried include ribs, striations, and fences (Bushnell 1978). Unfortunately, while these devices may reduce the local turbulent friction, at the same time their very presence may generate additional parasitic drag which can cancel out the supposed saving in friction drag. Recent research in this area draws some pessimistic conclusions about the possible use of such devices (Sahlin, Johansson, and Alfredsson 1988).

Lastly, another way of modifying the boundary conditions is to heat the boundary surface in water flows. This avoids the problem of parasitic drag, at least, and appears to afford quite reasonable reductions in overall drag (Reshotko 1977).

14.1.3   Flow in pipes

In Chapter 1 we discussed some of the empirical methods of correlating experimental data for flows in boundary layers and through pipes. Here, our intention is to give a very brief account of the ways in which these methods can be extended to non-Newtonian fluids.

The earliest attempts to do this (Metzner and Reed 1955; Shaver and Merrill 1959; Dodge and Metzner 1959) were based on the assumption that the non-Newtonian fluid involved was shear thinning, and this its apparent viscosity could be described by a power-law form as given in eqn (14.3). Their various approaches differ in terms of details (such as which empirical relationship is to be generalized to the non-Newtonian case), but the basic method is the same in all cases, and where we require an example we shall refer to the work of Shaver and Merrill.

Our first step is to generalise the Reynolds number—as defined by eqn (1.5)—to

$$R_A = \frac{U\,d\rho}{\mu_A}, \tag{14.7}$$

where U is the bulk mean velocity, d is the diameter of the pipe, ρ is the density of the fluid, and μ_A is the apparent viscosity.

Our second step is to express the apparent viscosity, as given by eqn (14.3), in terms of the bulk flow variables as

$$\mu_A \simeq K\left(\frac{U}{d}\right)^{n-1}. \tag{14.8}$$

Then, we substitute (14.8) into (14.7) and achieve the generalization which we are seeking, albeit with a sign of proportionality, rather than equality, in eqn (14.8).

The basic strategy is now quite simple. We use (14.7) and (14.8) to obtain the same form of resistance law for laminar flow in non-Newtonian fluid as in Newtonian fluid, that is, it is well known that $f = 16/R$ for Poiseuille flow. Accordingly, we choose R_A such that the resistance law of a non-Newtonian fluid in laminar flow is given by $f = 16/R_A$. Hence, Shaver and Merrill find that the generalized Reynolds number is given by

$$R_A = \left(\frac{d^n U^{2-n}\rho}{K}\right)8\left\{2\left(3 + \frac{1}{n}\right)\right\}^{-n}. \tag{14.9}$$

It may be noted that the first term in parentheses on the r.h.s. is obtained by substituting from (14.8) into (14.7).

Shaver and Merrill then show that experimental data for a very wide range of values of both R_A and n can be correlated by an *ad hoc* modification of the Blasius resistance law for turbulent flow through pipes, which takes the form

$$f = \frac{0.079}{n^5 R_A^m}, \tag{14.10}$$

where the exponent m is given by

$$m = \frac{2.63}{(10.5)^n}. \tag{14.11}$$

For the case $n = 1$, the apparent viscosity reduces to the Newtonian form, eqn (14.9) reduces to eqn (1.5), and the resistance law given by (14.10) becomes the Blasius resistance law, i.e. $f = 0.079/R^{1/4}$.

Other approaches have been made to the problems of correlating turbulence data in non-Newtonian fluids (e.g. Van Driest 1970; Huang 1974), and indeed this remains an active field of research (Shenoy and Talathi 1985). However, we shall not go any further into the technical details here, but rather will turn our attention to the qualitative features of turbulent structure in non-Newtonian fluids.

Interestingly, Shaver and Merrill (1959) anticipated many of the conclusions which were later to be drawn about drag-reducing flows. To begin with, they found that the friction factors were lower in shear-thinning flows, and that this was associated with blunter mean velocity profiles. They also noted that there was negligible slip at the wall, that there was reduced turbulent intensity and reduced turbulent mixing, and, most prescient of all, that the rate of formation of horsehoe (or lambda) vortices at the wall was much reduced in turbulent flow of shear-thinning fluids.

There are, of course, difficulties in the way of using pitot tubes and hot-wire (strictly, hot-film in liquids) anemometers in non-Newtonian fluids. These problems have been circumvented in various ways. Shaver and Merrill used photographs, in conjunction with dye injection, and later Seyer and Metzner (1969) and Rollin and Seyer (1972) carried out similar procedures using small air bubbles as the tracer. Both investigations confirmed the presence of flatter mean velocity profiles in non-Newtonian flows. In addition, Seyer and Metzner measured the turbulence intensities and found that the streamwise intensities were increased above the values for water, whereas the radial intensities were effectively unchanged.

More recently, Allan, Greated, and McComb (1984) used laser anemometry to confirm earlier results for both mean velocity and turbulent intensity profiles. In addition, these authors measured one-dimensional turbulent spectra and found an interesting effect. At small polymer concentrations (of the kind associated with drag reduction, but not with non-Newtonian behaviour) suitably scaled spectra were much the same as in water. However, at higher (i.e. non-Newtonian) concentrations, spectra were found to be attenuated at the very highest (dissipation range) wavenumbers, with a corresponding enhancement at lower (energy-containing) wavenumbers.

14.1.4 *Structural turbulence*

Structural turbulence is a term which has long been established in rheology, and refers to an unsteady motion which appears at Reynolds numbers that are too low for true turbulence. It is thought to depend on the structure of the fluid, and probably reflects some kind of viscoelastic instability in the neighbourhood of the solid boundaries of the system. (Such an instability can be visualized by the simple experiment of sliding a rubber eraser across a desk top. With some care, the eraser can be made to progress in a series of bounds, in a sort of 'slip–stick' motion. During the slip phase the motion is 'viscous', whereas during the stick phase the motion is elastic.)

The phenomenon is also known as 'early turbulence' and shows some of the characteristics of true turbulence, with the apparent viscosity becoming larger with increasing rate of shear while the flow is still well below the critical Reynolds number.

In recent years there has been some study of the effect in drag-reducing polymer solutions. One of the earliest of these studies (Carver and Nadolink

1965) observed a sinuous movement of tracer particles in a laminar boundary layer when drag-reducing polymers were present. This particular observation may well have shown up the basic viscoelastic instability, but other investigations have been concerned with more global aspects based upon measurements of pressure drops and flow rates through pipes, and indicate that structural turbulence occurs in polymer solutions at low concentrations (Little and Wiegard 1970; Forame, Hansen, and Little 1972). Later measurements, using laser-Doppler anemometry (Zakin, Ni, Hansen and Reischman 1977; Abernathy, Bertschy, Chin, and Keyes 1980) have confirmed this behaviour, and also indicate that the onset of structural turbulence is accompanied by a flattening of the mean velocity profile (as in true turbulence). Generally, investigators in this field seem to feel that structural turbulence is a viscoelastic effect with a character which is intermediate between laminar flow and normal turbulence.

Lastly, as well as having an intrinsic interest, structural turbulence is also important because its viscoelastic fluctuations may tend to indue true turbulence and hence cause transition at lower Reynolds numbers. This possibility is of particular interest when we recall that an early explanation of drag reduction was that the polymers delayed the transition from laminar to turbulent flow. However, there seems to be an adequate body of evidence to suggest that dissolved polymers can induce a transition to turbulence at subcritical Reynolds numbers (Paterson and Abernathy 1972; Zakin *et al.* 1977; Hoyt 1977).

14.1.5 *Isotropic turbulence*

Naturally enough, isotropic turbulence provides us with the obvious test problem for fundamental approaches to well-developed turbulence in non-Newtonian fluids. Even here, however, we are hampered by two major problems. First, if we wish to make a theoretical attack on the problem, there is the question of finding an equation of motion with which to replace the Navier–Stokes equation. Second, there is also the lack of a rational approximation to the Navier–Stokes statistical hierarchy. Without this, it is difficult to argue that we even have a suitable starting point for a non-Newtonian theory.

Nevertheless, some primitive attempts have been made to predict changes in the energy spectrum as a result of non-Newtonian effects, and we shall briefly discuss these here in the context of the small number of relevant experimental results.

The first attempt to assess the effect of a non-Newtonian fluid on the turbulent cascade was possibly that of Lumley (1964), who studied the dissipation of energy in a Reiner–Rivlin fluid, in which the viscosity is a linear function of the rate of strain but there are no elastic effects. Energy and dissipation spectra were evaluated on the assumption that the strain-rate tensors have a Gaussian joint-probability distribution. He concluded that the model fluid dissipates its energy in the same way as a Newtonian fluid, but has an effective viscosity which depends on the mean square shear. It was concluded that inertial transfer was unaffected, and hence there were no anomalous effects in a Reiner–Rivlin fluid. Anomalies such as drag reduction were attributed to viscoelastic effects.

However, the use of Gaussian probability distributions is incompatible with inertial transfer (which depends on triple non-linearities), and so we are not convinced that this conclusion follows in an obvious way. We shall return to this point shortly.

Chow and Saibel (1967) analysed the decay of isotropic turbulence in an Oldroyd fluid, which is characterized by five material constants and which incorporates both elastic and non-linear effects. They based their approach on the Karman–Howarth equation (see eqn (2.118) for the spectral form of this equation) and, assuming that the elasticity of the fluid is (in some sense) smaller than the viscosity, used perturbation theory to predict changes in th turbulence due to non-Newtonian effects.

Their overall conclusion was that anomalous effects could occur in the decay of turbulence in an Oldroyd fluid, but that these would depend on the precise nature of the fluid (i.e. the relative values of the various material parameters). From our present point of view, the most interesting effect would be a reduction in the rate of decay, as this is the nearest analogue to drag reduction that we can hope for in the free decay of isotropic turbulence. Chow and Saibel found that this would happen if there is a positive normal stress effect (or Weissenberg effect), which is due to the quadratic non-linearity in the constitutive equation. This is, in fact, the non-linear term considered by Lumley (1964).

Similar conclusions have been reached by other workers in the field. Kuo and Tanner (1972) studied the one-dimensional Burgers equation, which was modified by a viscosity depending linearly on the rate of strain (i.e. the stress depended quadratically on the rate of strain), thus giving a one-dimensional analogue of the Reiner–Rivlin fluid or the quadratic non-linearity found in the Oldroyd fluid. From numerical calculations of the equation for the energy spectrum, they found that (a) the rate of decay was actually greater in the non-Newtonian case and (b) the spectrum fell off more steeply, with less energy in the high wavenumbers, compared with the Newtonian case. However, one must be cautious about the possible extrapolation of these results to three-dimensional turbulence.

The full three-dimensional case was treated again by McComb (1974, 1976), who used a limited form of renormalized perturbation theory to study free decay of isotropic turbulence in first a Maxwell fluid (McComb 1974) and then a non-linear fluid (McComb 1976) (it should be noted that this is essentially the Reiner–Rivlin fluid model or the quadratic non-linearity of the Oldroyd equation). It was found that in both models the inertial transfer was affected in the same way, with predictions of higher spectral energy levels, reduced rate of decay, and a steepening of the energy spectrum at high wavenumbers.

The first two of these predictions are in agreement with those of Chow and Saibel. However, the idea that the spectrum should show a deficiency at high wavenumbers is only supported by the one-dimensional analysis of Kuo and Tanner, and this agreement is weakened by their reverse conclusion about the decay rate.

Unfortunately, the experimental results for the free decay of grid turbulence in non-Newtonian fluids tend to suffer from a variety of problems, ranging from atypical results for Newtonian fluids in the first place, through the use of polymer solutions which were too dilute to be appreciably non-Newtonian, to the uncertainties of using hot-film anemometry in polymer solutions. A discussion of most of these problems has been given by McComb, Allan, and Greated (1977), who used laser-Doppler anemometry in both water and polymer solutions of relatively high polymer concentration.

These investigators found that the polymer additive reduced both energy levels and rate of decay downstream from the grid. This result is in broad agreement with those of other experiments, and evidently supports the second of the above predictions but not the first. In the case of energy spectra taken in polymer solutions at typical drag-reducing concentrations, no difference was found from the results for water alone. However, at higher concentrations, spectra showed noticeable attenuation in the dissipation range of wavenumbers. This appeared to be a threshold effect, with onset at a polymer concentration of between 100 and 250 wt ppm. Flow visualization also showed this effect, and photographs of dye traces in water and polymer solutions at various concentrations are shown in Fig. 14.1. Here, the onset concentration for the suppression of small eddies was somewhere between 50 and 100 wt ppm in fair agreement with the onset concentration for observable non-Newtonian behaviour in these polymer solutions.

The trouble with this subject at the present time is that it is simply under-developed. To the present writer, the time seems ripe for a much more determined attack on both theoretical and experimental aspects. The three-parameter version of the Oldroyd model due to Lumley (1971) would make a good starting point for a fluid model. Certainly, this was used successfully to predict the effects of anomalous extensional viscosity in polymer solutions on the damping and pulsation of small gas bubbles in viscoelastic liquids (McComb and Ayyash 1980). A model of this kind, when allied with the developments in renormalization methods and computational methods de-

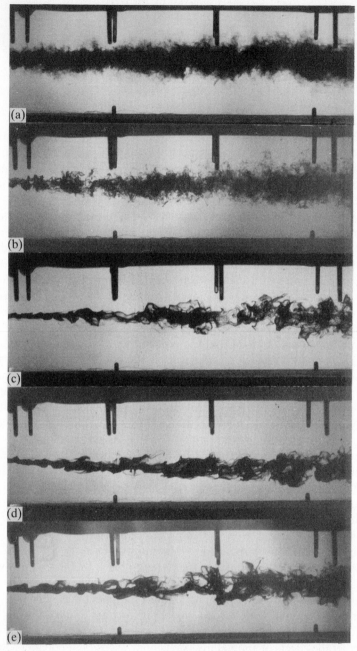

Fig. 14.1. The effect of dissolved Polyox WSR301 on the injected dye trace in grid-generated turbulence (flow direction is from left to right): (a) water; (b) 50 wt ppm Polyox WSR301; (c) 100 wt ppm; (d) 250 wt ppm; (e) wt ppm. (After McComb *et al.* 1977.)

scribed in earlier chapters, should be capable of quite extensive predictions. Given the wide availability of LDA nowadays, there should be no difficulty about making the appropriate checks.

14.2 Turbulent structure in drag-reducing polymer solutions

In this section (and in Section 14.3) we are concerned with drag reduction by additives. Therefore we should emphasize that this is pre-eminently a phenomenon involving the reduction of friction drag at a solid surface. Accordingly, although we shall not totally ignore effects on free turbulence, our main interest will be in turbulence bounded by one or more solid surfaces. Moreover, it appears that the effect is mainly associated with two-dimensional (or near two-dimensional) flows, and so this is a further restriction which we shall impose.

We should also make a general observation to the effect that traditional methods of measurement (such as pitot tubes and hot-film anemometers) rely on non-linear operating relationships, and hence could be affected in unpredictable ways by non-Newtonian fluid behaviour, if used in liquids containing polymers or macroscopic fibres. It follows that those investigations which are based upon such methods cannot—however carefully they are carried out—be regarded as giving us reliable information about the turbulent structure. In fact only optical methods can be regarded as potentially reliable, and, if we require quantitative measurements, this usually boils down to laser-Doppler anemometry.

Taking into account the above remarks, we shall now precede our discussion of the measurements of turbulent structure by a brief general appreciation of the main investigations in this field. We shall begin by listing together several references, which will only be treated here and will not be part of our later detailed discussion: Rudd (1972), Logan (1972), Chung and Graebel (1972), and Mizushina and Usui (1977).

Of these, the first two report LDA measurements taken in ducts with square cross-section. Rudd found that the mean velocity profiles in polymer solutions were blunter than in water, while the axial turbulent intensity u_1'/U was larger, at least near the wall. Logan confirmed Rudd's results for the axial turbulent intensity. In addition, he found that the 'radial' intensity u_2'/U and the normalized Reynolds stress $\langle u_1 u_2 \rangle/U^2$ were both reduced compared with their values in water alone. However, the presence of secondary flows in such configurations makes their results of doubtful relevance to two-dimensional flows (e.g. see the discussion of this point by Reischman and Tiederman (1975)).

In contrast, Chung and Graebel (1972) and Mizushina and Usui (1977) both studied developed turbulent flows in pipes of round cross-section; thus in both cases their results should at least be representative of two-dimensional flows. They also both confirm the general result that velocity profiles are blunter in the polymer solutions, in accordance with what one expects with pipe flow of non-Newtonian fluids (see Section 14.1.3). Unfortunately, unlike the investigations by Rudd and Logan, they both report lower axial intensities in the polymer solutions, in qualitative disagreement with accepted results for non-Newtonian turbulence established by flow visualization. In the case of Chung and Graebel, this appears to be due to refraction errors which limited the reliability of the measurements, except in the core region of the pipe. The measurements by Mizushina and Usui are also limited by the fact that they obtain axial intensities in water which are lower than the accepted values. This is probably due to their LDA having a relatively large probe volume, and clearly the explanation of the low intensities in polymer solutions may be the same.

In this section, our main discussion will be based on the investigations by Reischman and Tiederman (1975), McComb and Rabie (1982a, b), Allan, Greated, and McComb (1984), Willmarth, Wei, and Lee (1987), and Usui, Maeguchi and Sano (1988). The last named differ from the others, insofar as they use video measurement of tracer particle motion rather than LDA. Nevertheless, they manage to obtain believable quantitative results.

14.2.1   Mean velocity distributions

The various measurements of the mean velocity, in pipe or two-dimensional channel flow, all conclude that the profile (i.e. the distribution in the direction normal to the wall) is blunter—or flatter—in polymer solutions than it is in water alone. Here, we wish to supplement this rather qualitative conclusion with some of the quantitative methods introduced in Chapter 1.

We begin by recalling the nature of the mean velocity distribution in duct flow. Taking a simplified view of the discussion presented in Sections 1.4.2 and 1.4.3, we can divide the mean velocity profile into two regions. First, there is the thin layer near the wall, where the effect of the viscosity predominates and the mean velocity distribution is given by eqn (1.37), which for convenience we repeat here:

$$U_1^+ = x_2^+.$$

Second, we have the logarithmic layer, which occupies the rest of the cross-section of the duct and in which the mean velocity is given by (eqn (1.34))

$$U_1^+ = A \ln x_2^+ + B$$

where A and B are taken to be universal constants.

There is also a transitional (or buffer) layer, where the viscous and turbulent stresses are of comparable importance, but in practice this is often glossed over. However, it has been proposed that the mean velocity profile in polymer solutions can be described by a three-layer model (Van Driest 1970; Virk, Mickley, and Smith 1970) in which the region between the viscous sublayer and the logarithmic layer becomes the 'elastic' or 'interactive' layer, and has an extent which depends upon the amount of drag reduction. In this picture, the viscous sublayer is unaffected by the polymer additives, as is the constant A in eqn (1.34). However, the constant B in (1.34) must vary with the amount of drag reduction, in order to allow for the increased mean velocity at constant shear stress.

The mean velocity profile in the interactive layer is also assumed to take the logarithmic form, and can be written as

$$U_1^+ = C \ln x_2^+ + D, \tag{14.12}$$

where the constants C and D remain to be determined. Then, when there is no drag reduction, eqn (14.12) reduces to the Newtonian form given by (1.34).

This provides a lower bound on eqn (14.12). An upper bound can be found from a consideration of the ultimate drag reduction profile (Virk *et al.* 1970) (see Section 3.4.4). The existence of a resistance law of the form of eqn (3.87), for the case of the maximum possible drag reduction, implies an ultimate mean velocity profile given by

$$U_1^+ = 11.7 \ln x_2^+ - 17. \tag{14.13}$$

The existence of the three-layer profile of mean axial velocity was confirmed by Reischman and Tiederman (1975) who used premixed polymer solutions,

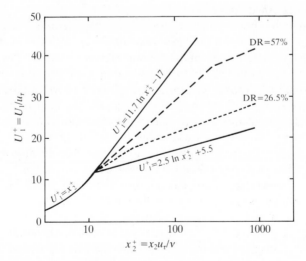

Fig. 14.2. Mean velocity profiles in a pipe during the injection of Polyox WSR301 ($R = 3.5 \times 10^4$) (data taken from McComb and Rabie 1982).

and by McComb and Rabie (1982a, b) who studied pipe flows in which the polymer solution was injected at either the wall or the centre line. An indication of the general picture is given in Fig. 14.2, which shows two typical profiles corresponding to drag reductions of 26.5 and 57 per cent.

Both these investigations agree that the viscous sublayer is unchanged in the polymer solutions, but there is some apparent disagreement about the logarithmic region. Reischman and Tiederman find that the constant A is unchanged and that the constant B in eqn (1.34) should be increased with increasing drag reduction. In contrast, McComb and Rabie find that both A and B must be increased as the drag reduction increases. However, this is probably due to the fact that the former authors only considered drag reductions DR in the range 24–41 per cent, whereas McComb and Rabie worked with drag reductions spanning the range $DR = 26.5$–67 per cent. At $DR = 41$ per cent, they found that the constant A had not changed much.

14.2.2   Turbulent intensities

Reischman and Tiederman found that the ratio of the r.m.s. axial turbulent fluctuation to the friction velocity (i.e. u_1'/u_τ) was increased in the neighbourhood of the wall in the presence of dissolved polymers. Also, the intensity peak—which is rather narrow in Newtonian fluids—tended to be spread out towards the centre of the flow. Similar results were obtained by McComb and Rabie, who noted that the amount of the increase in both the velocity ratio and the peak broadening increased with increasing drag reduction. Such results are consistent with the three-layer model of the mean velocity profile, which was described above.

There are two other aspects of these results which deserve a mention. First, we note that as the ratio of the r.m.s. velocity to the friction velocity does not remain constant, irrespective of the amount of drag reduction, this underlines the fact that the nature of the turbulence is in some way different in these polymer solutions. Second, we cannot immediately assert that the turbulence has not been suppressed by the additive, because the increase in u_1'/u_τ with increasing drag reduction may be due to the corresponding reduction in the friction velocity. However, if we consider the results of McComb and Rabie for $DR = 67$ per cent and use the data given to make the appropriate conversion, we find that the peak value of the true turbulence intensity u_1'/U is about 12 per cent larger in polymer solution than in water. Thus, as U is the same for both cases, we have to conclude that the axial component of the fluctuating velocity has increased relative to the Newtonian case. This behaviour has also been confirmed by Allan et al. (1984), who studied premixed polymer solutions in pipe flow, and who plot the true turbulence intensity as a function of the distance from the wall and the polymer concentration.

Similar increases in axial turbulent fluctuations were later found by Willmarth et al. (1987) and Usui et al. (1988). Also, both these investigations reported substantial reductions in fluctuations u_2'/u_τ normal to the wall and in the Reynolds shear stress $\langle u_1 u_2 \rangle/u_\tau^2$. The latter result is, of course, to be expected in view of the reduced drag.

14.2.3 *Spectra and correlations*

When one probes more deeply into the effect of polymers on turbulent structure, by considering two-point quantities, the results can be rather disappointing. If energy spectra are scaled using Kolmogorov variables, then there is very little difference between results for water and those taken in drag-reducing polymer solutions, although, in the latter case, there does tend to be some enhancement of the energy-containing range of wavenumbers (McComb and Rabie 1982a, b). While qualititative changes in the spectrum have been found in polymer solutions, this was only at concentrations large enough for there to be continuum non-Newtonian behaviour (Allan *et al.* 1984).

Autocorrelations do tend to be more persistent, and integral scales are generally very much larger in polymer solutions (McComb and Rabie 1982a, b), in reasonable agreement with earlier measurements made using mass-transfer probes (Butson and Glass 1974). This behaviour is, of course, consistent with the observation that the energy-containing range of the spectrum is enhanced in polymer solutions; see eqn (2.108) and the discussion in Section 2.6.5 for the relationship between the integral scale and the one-dimensional spectrum.

One aspect of turbulent structure which does seem to be sensitive to the presence of drag-reducing polymers is the formation of bursts and streaks near the wall. We have previously noted that Shaver and Merrill (1959) observed that the rate of formation of horseshoe vortices near the wall was much reduced in flows of non-Newtonian fluids. Also, one of the more plausible suggestions for a theory of drag reduction was that of Gadd (1965), to the effect that the resistance of the polymers to extensional flows was sufficient to stabilize the boundary layer such that the frequency of turbulent bursts was reduced. It has been shown by the use of hologram interferometry (Achia and Thompson 1977) that drag-reducing polymer additives tended to suppress the formation of low-speed streaks and the eruption of bursts. The available results from LDA also indicate a very much increased mean time between bursts during drag reduction (Mizushina and Usui 1977; McComb and Rabie 1982a, b).

14.2.4 *The importance of the region near the wall*

It has been traditional in the study of turbulence to attach a lot of importance to the region near the wall. As we have seen in Chapter 1, the main generation and dissipation processes in turbulent shear flow are believed to occur in the buffer layer, which is a thin layer near the wall. Naturally, therefore, the essential mechanism of drag reduction has been thought to reside in or near the buffer layer, and the three-layer mean velocity profile has been taken as evidence that this is so.

However, in view of the intrinsic non-localness of turbulence, this is not conclusive evidence and we really need some more convincing demonstration that the mechanism by which polymers reduce drag is located in or near the buffer layer.

The obvious approach to this problem is to inject polymer into the flow so that the effect of drag-reducing polymer in one location can be compared with the effect of that in another. In fact, injection techniques have often been employed as a practical expedient (for example, in the case of flows around solid bodies), but the first direct indication that the presence of the polymer in the wall region is crucial was found by Wells and Spangler (1967). They found that, when polymer was introduced into the flow near the wall, the wall shear stress was reduced immediately downstream of the injection point, whereas when the polymer was injected at the centre of the flow, no effect was observed until it had diffused to the neighbourhood of the wall. In view of the rather incomplete nature of this investigation, it could be regarded as a direct, but only qualitative, demonstration of this point.

Later, McComb and Rabie (1978, 1979, 1982a, b) confirmed this general conclusion when they made a much more detailed and quantitative investigation in which polymer solutions were injected at either the centre line or the wall of a well-developed pipe flow. Numerous closely spaced pressure tappings were used to monitor the variation of wall shear stress (measured as a local drag reduction) in the streamwise direction, while a trace of salt in the injected solution enabled a sampling system to measure the radial distribution of the polymer at various downstream locations. In this way the rate at which the polymer spread out radially could be correlated with the way in which the local drag reduction built up downstream from the injection point. From an extensive set of measurements, they deduced that the polymer was effective in an annulus bounded by $15 < x_2^+ < 100$, which corresponds roughly to the buffer layer.

The lower of these bounds was later modified by Tiederman, Luchik, and Bogard (1985), who studied the injection of polymer solutions through slots in the wall into flow through a plane channel. They concluded that the effective zone was $10 < x_2^+ < 100$, although they were not able to verify the upper limit. They also noted that their results for bursting rates were similar to those of McComb and Rabie (1982a, b) who, among other observations, had noted that the bursting process apparently modulated the rate at which the injected polymer solutions were carried from the wall to the buffer zone (McComb and Rabie 1978).

Lastly, we shall briefly consider whether the drag reduction produced by polymer injection is in any sense different from, or unrepresentative of, the conventional drag reduction with premixed polymer solutions, which we are trying to understand. This question might arise anyway, but we are forced to consider it because of the work of Vleggaar and Tels (1973), who studied the injection of concentrated polymer solutions into the core region of a turbulent pipe flow and concluded that they had found a new form of drag reduction. They called it 'heterogeneous drag reduction' in order to distinguish it from the usual kind of drag reduction in premixed dilute polymer solutions, which they referred to as 'homogeneous drag reduction'.

Their main reason for thinking that they had produced a new kind of drag reduction was based on flow visualization showing the injected polymer forming a long thread, which resisted dispersion and extended downstream for a distance of more than 200 tube diameters. Their results indicated that substantial drag reduction occurred before the polymer reached the wall region.

Vleggaar and Tels also noted that the amount of drag reduction was significantly greater than in premixed solutions, when conditions were otherwise identical, and that there did not appear to be an onset effect (i.e. a threshold value of shear stress; see Section 3.4.3). These quantitative features of drag reduction due to injected polymers have also been found by McComb and Rabie (1978, 1979, 1982a, b), Bewersdorff (1982) and Usui *et al.* (1988). Indeed, these last-named authors have proposed that their results, taken along with the earlier results of McComb and Rabie (1982a, b) for the minimum friction factor in injected polymer solutions, should be taken as a maximum drag reduction asymptote, analogous to—but not equal to—the drag-reduction asymptote (for premixed flows) due to Virk *et al.* (1970) (see Section 3.4.4).

However, McComb and Rabie also found that time-averaged measurements of polymer concentration showed a Gaussian distribution of injected polymer over the cross-section of the pipe. This would not appear to support the idea of discrete polymer threads. As in alternative explanation, McComb and Rabie (1982a, b) have suggested that the injection of polymer solutions may lead to a high proportion of supermolecular aggregates, of a kind which are known to occur in premixed polymer solutions. This is undoubtedly a topic which will repay further study.

14.2.5 *Free turbulence*

We conclude this section with a very brief discussion of the effect of drag-reducing polymers on turbulent structure in free turbulence, where the effect of solid boundaries is, at most, an aspect of the initial conditions.

It is well known that polymer additives tend to suppress small eddies in all types of turbulent field. However, as we saw in Section 14.1.5, in order to find a quantitative effect on the turbulent spectrum of grid turbulence, it is necessary to use higher concentrations of the polymer at a level where it is known that the resulting fluid is non-Newtonian (McComb *et al.* 1977).

As well as this work, we know of LDA measurements of turbulent structure in dilute polymer solutions in the round free jet (Barker 1973) and in the plane mixing layer (Hibberd, Kwade, and Scharf 1982). In the first of these investigations, an LDA registered the negative result that there was no change in mean or r.m.s. velocity, or in their radial distributions as result of the presence of Polyox WSR301 in concentrations of 50 and 100 wt ppm. In the second case, the LDA results are again rather negative, but this is overshadowed by results from flow visualization. Here, the action of Separan AP30 (at a concentration of 50 wt ppm) in suppressing small eddies has the happy result of enhancing the roll vortices (i.e. coherent structures) of the mixing layer.

14.3 Turbulent structure in drag-reducing fibre suspensions

In dealing with suspensions of macroscopic fibres, we face all the problems that confronted us in the case of solutions of microscopic polymers. However, in fibre suspensions, there are two additional difficulties, which mean that laser-Doppler anemometry is no longer the panacea that it was for polymer solutions. First, many fibre suspensions are so opaque that any optical method is ruled out. Second, even if the suspension is adequately transparent, the macroscopic fibres can scatter light and hence contribute to the output signal of the LDA. This must inevitably raise questions about interpretation of the LDA signal.

As a result, we are forced to cast our net more widely, and consider that other methods (such as impact tubes and hot-film anemometers) may be no more unsatisfactory than optical methods in this particular context.

However, before turning to specific measurements, we should note that we shall have frequent occasion to refer to the very effective drag-reducing asbestos fibres produced by the Turner Brothers Asbestos Company Limited (see Section 3.4.5). For conciseness, we shall refer to these as 'TBA asbestos fibres' or simply 'TBA fibres', throughout.

14.3.1 *Mean velocity distributions*

Mean velocities can be measured quite satisfactorily by means of an impact tube (or total-head pitot tube), even in non-Newtonian fluids. The practical problem which arises in fibre suspensions is that the fine tube, necessary in order to minimize perturbation of the flow, may become blocked by lumps of fibrous material. It was shown by Mih and Parker (1967) that this problem could be circumvented by using a 'purged' impact tube (for details see the original reference), and this device has since been used in other investigations.

Mih and Parker measured mean velocity distributions in suspensions of rayon fibres or wood pulp in flow through a pipe. They analysed their results by plotting them in semi-logarithmic form in 'law of the wall' coordinates (i.e. as was done for polymer solutions in Fig. 14.2). As the results lay on straight lines, they could then compare them with the universal logarithmic profile as given by eqn (1.34). In general, they found that the mean velocity profiles found in suspensions had a steeper slope (i.e. a larger value of the constant A) than in water alone, with the steepness of the slope increasing with increasing additive concentration.

Later investigations using the same technique of the purged impact tube have led to conflicting results. Kale and Metzner (1976) studied the flow of asbestos fibre suspensions both with and without the addition of drag-reducing polymer. In all cases they found that the slope of the velocity profile was the same in the suspensions as in the Newtonian case, but that increased velocities could be taken account of by an incremental change in the constant B. However, Sharma, Seshadri, and Malhotra (1979), although using the same techniques and additives—the only significant difference was their use of the TBA asbestos fibres, which could be expected to be much more effective drag reducers (by a factor of 2 or 3) than the asbestos fibres used by Kale and Metzner—found that the slope (or the value of A) varied with fibre concentration and the bulk velocity of the flow. Thus their results were in qualitative agreement with those of Mih and Parker.

Measurements of mean velocity profiles by LDA have been carried out by Ek, Moller, and Norman (1978) in suspensions of paper pulp, and by McComb and Chan (1979, 1985) who studied TBA asbestos fibre suspensions (with, in some cases, the addition of drag-reducing polymers). Ek *et al.* have presented only very limited results for one flow. They plot their mean velocities on linear axes, and show that the mean velocity profile in the suspension is flatter in the core region than that for water alone, whereas near the wall the mean velocity in the suspension varies more rapidly than in water. However, Ek *et al.* present no comparison with accepted results for Newtonian flow, and make no attempt to validate their experimental procedures. Accordingly, in the absence of such validation, it is difficult to have confidence in their results and we shall not discuss this investigation any further.

McComb and Chan present results for Reynolds numbers in the range $9 \times 10^3 - 5.3 \times 10^4$, and for a range of percentage drag reductions. The variation in DR was obtained by recycling a given suspension through the apparatus a number of times in order to make a virtue of the fact that these suspensions are quite susceptible to degradation, with the amount of drag reduction being smaller on each pass through the pipe. Results for the mean velocity are given as semi-logarithmic plots, and although the constants A and B are not given explicitly, it is clear that the slope of the curves (and hence A) increases with increasing drag reduction. Thus these results are in qualitative agreement with those of Mih and Parker and those of Sharma *et al*, but not with those of Kale and Metzner.

One interesting feature of these results is that some velocity profiles lie beyond the ultimate profile given by Virk *et al*. This only occurs at the two lower Reynolds numbers (9×10^3 and 1.4×10^4), and for the first two passes of the fibre suspension through the pipe. It does not occur at $R = 3.2 \times 10^4$, even though the drag reduction in pass 1 is, at DR = 71 per cent, higher than that for $R = 9 \times 10^3$.

14.3.2   *Intensities, correlations, and spectra*

The picture which emerges from our consideration of mean velocity profiles is that the slope of the profile (in semi-logarithmic coordinates) in drag-reducing fibre suspensions is different from the value for Newtonian fluids, with the amount of the difference depending on the amount of drag reduction. In the 'mixing-length' model, the slope of the profile is determined by momentum transfer in the direction normal to the wall. Thus a change in this slope implies a change in the rate at which momentum is transported in the radial direction by turbulent fluctuations. This seems only to be expected, in view of the fact that drag reduction implies a reduction in turbulent momentum transport in the radial direction, but now we look to more structural measurements to see whether the turbulence has changed its nature in any sense.

We begin with the work of Bobkowicz and Gauvin (1967), which leads us straight away to an intriguing paradox. These authors based their measurements on the relationship between turbulent diffusion and the velocity field. By using a thermistor probe to detect temperature changes, they were able to map the radial dispersion of hot water injected from a point source. When they applied this technique to the turbulent flow of a drag-reducing suspension of nylon fibres in water, they found that the radial diffusivity, and hence the radial turbulent intensity u_2'/U, was increased.

The apparent contradiction is that radial transport is actually increased, whereas we would expect that momentum transport in the radial direction should be reduced (in order to be consistent with the observed reduction in drag). However, Bobkowicz and Gauvin point out that another investigation (using an impact tube coupled to a capacitative pressure transducer) report decreased axial turbulent intensities u_1'/U. Thus the increase in radial intensity is not incompatible with a reduction in the Reynolds stress $\langle u_1 u_2 \rangle$, and hence with a reduction in drag.

Reduced axial intensities in drag-reducing suspensions have also been reported by Pirih and Swanson (1972) who used a hot-film anemometer with a suspension of rigid rod-like crystals, which apparently did not cause the calibration drift previously found with hot-film instruments in non-Newtonian fluids, and by Kerekes and Garner (1982) who used LDA in a wood pulp suspension. The former authors also found that one-dimensional energy spectra showed a greater concentration of energy at low wavenumbers, and that the integral scales were increased (compared, of course, with results for Newtonian fluids).

A more complicated picture emerges from the LDA results of McComb and Chan (1979, 1985). We have previously seen that mean velocity profiles for the first one or two passes of the suspension through the pipe at Reynolds numbers of 9×10^3 and 1.4×10^4 lay beyond the ultimate velocity profile. In these cases, the axial turbulent intensity u_1'/U was reduced below that for water (in agreement with the results from other investigations discussed above), whereas, for $R = 1.4 \times 10^4$, the transverse turbulent intensity u_3'/U was increased above that for water. For simplicity, we shall call this behaviour 'fibre like', in order to distinguish it from another form of behaviour in fibre suspensions, which, as we shall see, would seem to merit the term 'polymer like'.

This polymer-like behaviour was found by McComb and Chan at the higher Reynolds numbers studied, i.e. 3.2×10^4 and 5.3×10^5, and in solutions which had suffered some degradation (i.e. after two or three passes through the pipe) at the two lower Reynolds numbers. In this regime, the mean velocity profiles lay within the ultimate profile for drag reduction, and axial intensities (relative to the friction velocity) were higher than in water at the same Reynolds number. Thus, qualitatively at least, this is the sort of behaviour found in polymer solutions. It is, of course, the exact reverse of what is found in the fibre-like regime, as discussed in the preceding paragraph.

McComb and Chan further point out that mean velocity profiles in the fibre-like region depend on the Reynolds number, whereas profiles in the polymer-like region are similar at the same value of drag reduction, and hence are independent of the Reynolds number. Moreover, in this region, the comparison with polymer solutions indicates that the quantitative agreement is also quite good. The mean velocity profile for $R = 5.3 \times 10^4$, with DR = 40 per cent, is found to agree quite closely with that measured by Reischmann and Tiederman (1975) in a polymer solution at $R = 5.25 \times 10^4$ with DR = 35.3 per cent.

Other indicators of turbulent structure measured in this investigation showed the same qualitative features as in other studies of drag-reducing flows in both fibre suspensions and polymer solutions. That is, when compared with the pure solvent, correlations were more persistent (indicating increased inte-

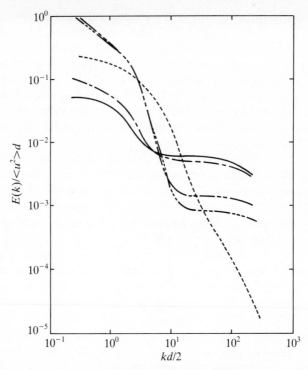

Fig. 14.3. One-dimensional turbulent energy spectra in an aqueous suspension of asbestos fibres (300 wt ppm) at various levels of drag reduction for $R = 1.4 \times 10^4$. All spectra were measured at the centreline of the pipe: ———— first pass, DR = 69 per cent; ———, second pass, DR = 76 per cent; ———— third pass, DR = 56 per cent; —————— fourth pass, DR = 55 per cent; ————————, water. (After McComb and Chan 1981.)

gral length and time scales), bursting rates were much reduced, and spectra were more energetic at low wavenumbers. We shall return to the topic of bursting rates in Section 14.5, when we compare fibres and polymers as drag-reducing additives. Here, we conclude the present discussion by considering a unique spectral feature as reported by McComb and Chan (1981, 1985).

Referring to Fig. 14.3, which shows one-dimensional energy spectra taken at $R = 1.4 \times 10^4$ in a TBA fibre suspension, we see that the spectra have a curious kinked appearance. In particular, they are reduced below the value obtained in water over an intermediate range of wavenumbers. McComb and Chan considered the possibility that these spectra might be due to the scattering of light by the fibres, but concluded that the most probable explanation was that the energy spectrum was reduced below the Newtonian level by resonant absorption of turbulent energy by the fibres. This view was supported by the fact that it turned out to be possible to correlate the amount of drag reduction with an interaction length scale deduced from the spectra by using the following procedure.

The region of spectrum showing the reduced level was characterized by an interaction wavenumber k_m, which was calculated by noting the two wave-mumbers marking the beginning and end of this region, and taking the arithmetic mean. Hence the mean interaction length scale was obtained as $l_m = k_m^{-1}$. Values of l_m for the various experiments were obtained. These ranged from 0.84 to 2.4 mm, and were reasonably close to the mean value of 1.4 mm which was obtained by photographic studies of undegraded suspensions (McComb and Chan 1981, 1985). By making a semi-log plot of drag reduction against the interaction length scale (divided by the mean fibre diameter $d = 40$ nm), it was found that the data for all four values of Reynolds number could be correlated by the empirical relationship

$$DR = \left\{ 0.70 \log_{10}\left(\frac{l_m}{d}\right) - 2.60 \right\} \times 100 \text{ per cent} \tag{14.14}$$

An odd feature of these spectra was that there appeared to be no significant difference between spectra measured in the fibre-like regime and those taken in the polymer-like regime, nor was there any statistically significant dependence on Reynolds number.

14.3.3 *Mixed fibre-polymer suspensions*

In Section 3.4.5, we mentioned the discovery by Lee *et al.* (1974) that the addition of drag-reducing polymer to an already drag-reducing fibre suspension resulted in a combined effect greater than the sum of the two independent effects. In this way, very large reductions in drag (greater than 95 per cent) can be obtained.

The only measurements of turbulent structure in such mixed suspensions seem to be those of McComb and Chan (1981, 1985), who studied an aqueous suspension of TBA fibres (at a concentration of 300 wt ppm) to which had been added Separan AP30 at a concentration of 150 wt ppm. Presumably, this combination was a good deal less than optimum, for the maximum drag reduction obtained was of the order of 60 per cent. However, this particular mixture showed amazing resistance to degradation, with the eighth pass through the system giving DR = 59 per cent, and only after the ninth pass did the drag reduction fall off gently with increasing number of passes.

The mean velocity profiles for the first eight passes of the suspension all lay close to the ultimate drag-reduction asymptote, but did not actually cross it. Thus, on the face of it, the behaviour of the mixture was polymer like. However, the measured axial intensities tell a different story, with all results up to (and including) the eighth pass clustering together about a line showing a marked reduction below the value for water, thus indicating fibre-like drag reduction. Transition to the polymer-like regime accompanied the onset of degradation. Evidently, the main effect of the polymer was to stabilize the fibre suspension against degradation due to shear.

14.4   The effect of drag-reducing additives on turbulent transport

In flow through pipes, the friction drag (or equivalently the axial pressure drop) is governed by the rate at which turbulent fluctuations transport axial momentum in the radial direction. This transport is in the direction from fluid to wall, corresponding to a net loss of streamwise momentum from the fluid. As we have previously noted, a reduction in drag—due to additives—is necessarily associated with a reduction in the radial transport of axial momentum. In this section, we shall be interested in the implications for heat and mass transfer, to the extent that they are also controlled by turbulent fluctuations.

For the most part, we shall consider only the flow of heat, or amount of scalar contaminant, from wall to fluid (or, less commonly, in the opposite direction). Also, we shall divide the topic into two parts. First, we shall briefly consider the type of measurements which are customary in engineering applications and which involve global flux balances, and then we shall go on to the more fundamental study of diffusion from restricted sources, in which a detailed map of the concentration profiles is made as a function of both distance and time.

14.4.1   *Heat and mass transfer*

The treatment of the effect on scalar transfer of drag-reducing additives is scanty compared with that of the actual friction reduction. Nevertheless, it has been clearly established that, for polymer solutions at least, radial transport of scalar contaminant is drastically reduced when compared with the Newtonian case. Some recent references for the reduction in heat transfer are Monti (1972) and Mizushina and Usui (1977), and for the reduction in mass transfer are Sidahmed and Griskey (1972), Virk and Suraiya (1977), and Vassiliadou, McConaghy, and Hanratty (1986).

This effect is easily understood in terms of our previous discussions of drag reduction in polymer solutions. In Section 14.2.2 we saw that axial turbulent intensities were increased compared with the Newtonian case, whereas the radial intensities were reduced. It is, of course, the radial fluctuations in velocity which control the radial scalar transport. Thus, in polymer solutions, it is inevitably the case that scalar transport from wall to fluid (or in the reverse direction) must also be reduced.

It should be noted that this argument does not necessarily work for fibre suspensions. As we saw in Section 14.3.2, the effect of macroscopic fibres on turbulence appears to be the reverse of that due to polymers. In fibre suspensions, it seems that axial turbulent intensities are reduced, and radial and circumferential components are increased. Therefore, one might expect increased scalar transport, in line with the results of Bobkowicz and Gauvin (1967) for radial diffusion. Unfortunately, the investigation by Moyls and Sabersky (1978) found that radial heat transfer was much reduced in suspensions of TBA fibres (at a concentration of 300 wt ppm). However, as these results seem to have been taken at a minimum value of the Reynolds number not less than 2×10^4, a possible explanation is that Moyls and Sabersky were working in the polymer-like regime, as identified by McComb and Chan.

We conclude this discussion by noting that it is usual to treat the transfer of heat and mass by using phenomenological methods based on analogies between transfer of scalars and transfer of momentum. Hence the heat and mass transfer coefficients are related in some way to the friction factor, which is the transfer coefficient for streamwise momentum. It turns out that the effect of drag-reducing additives on momentum and scalar transport is sufficiently different in each case that established predictive methods have to be modified in drag-reducing fluids. To go further into this would take us into a very specialized field and beyond the scope of this book. The reader who wishes to pursue the topic further will find a good starting point in the review by Dimant and Poreh (1976).

14.4.2 Turbulent diffusion

The study of tracer dispersion in drag-reducing flows is an attractive field for research, but, as usual, the paucity of fundamental investigations means that we are unable to resolve all the conflicts of evidence.

In cases where the tracer material was injected with a concentrated polymer solution into a flow of water, the emphasis has usually been on the practical applications, and as a result the injection method is often such that the results depend on the nature of the injector as much as anything else. We note that such solutions have been injected through porous plates (Walters and Wells 1972) and through slots in the wall (Wu 1972, Ramu and Tullis 1976, Fruman and Tulin 1976), and although the generality of the results may be limited, there seems to be a consensus that there was always a large suppression of turbulent diffusion when the polymer was injected into the boundary layer.

Turning to more fundamental studies, we find an apparent conflict. Taylor and Middleman (1974) and McComb and Rabie (1982a, b) both studied the radial dispersion of a tracer from a point source in pipe flow. The only substantial difference between the two approaches was that the former authors injected their tracer directly into the turbulent flow of a premixed aqueous polymer solution, whereas McComb and Rabie injected a tracer mixed with a concentrated polymer solution into water flowing through a pipe. However, Taylor and Middleman concluded that the eddy diffusivity was increased in drag-reducing flows (surprisingly, in view of the above results for heat and mass transfer), while McComb and Rabie came to the opposite conclusion—that the diffusivities were reduced.

In fact, this disagreement may be more apparent than real. Taylor and Middleman concluded that the radial intensity of the turbulence was reduced in the polymer solution, in agreement with the usual picture, and that the dye plumes obtained in the polymer solution were narrower than in water. The usual interpretation of this result would be that the variance (and hence the eddy diffusivity) was reduced. Accordingly, the anomaly may simply be a matter of the diffusion coefficients being incorrectly evaluated from the diffusion data.

Lastly, we note that Bryson, Arunachalam, and Fulford (1971) studied axial dispersion in polymer solutions by injecting pulses of tracer material into pipe flow. They concluded that axial dispersion was enhanced in polymer solutions, and of course this is what one would expect in view of the increased axial intensity of the turbulence.

14.5   Comparison of polymers and macroscopic fibres as drag-reducing additives

In Section 3.4.5 we stated the desirable physical properties of macroscopic fibres in order that they should be effective at reducing turbulent friction. We also noted the superficial resemblance between these characteristics and the desirable physical properties of drag-reducing polymers, as outlined in Section 3.4.2. We now return to the question of whether or not these two types of additive reduce turbulent drag by the same mechanism.

Naturally, we hope that the measured velocity distributions will shed some light on this question. However, a comparison of mean velocity profiles in the two cases is rather less than illuminating. In both cases, the profiles are quite similar, although in the fibre suspensions there is less evidence of the three-layer structure found in the polymer solutions.

However, the picture obtained from the turbulent intensities is rather more indicative of a different mechanism. We can summarize the two cases (relative to water) as follows.

(a) In polymer solutions the axial intensity is increased (especially near the wall), and the radial intensity is reduced.

(b) In fibre suspensions the axial intensity is reduced, and the radial and circumferential intensities are increased.

However, we should bear in mind that, according to McComb and Chan, (b) only holds for a fibre-like regime, and in fibre suspensions there may be a polymeric regime (depending on the Reynolds number, the fibre characteristics, and the amount of degradation).

Nevertheless, the changes in intensity indicate quite clearly that the nature of the turbulence in polymer solutions differs from that in fibre suspensions. In both cases, the underying turbulent structure is different from that in water alone. These facts, taken in conjunction with the synergistic combined effect in polymer–fibre mixtures, are surely convincing evidence for the existence of two separate and distinct mechanisms for the reduction of turbulent drag.

Our final point concerns the mean time between bursts, which is widely regarded as an important parameter in turbulent drag reduction. As we saw in Section 14.2.3, the reduced drag in polymer solutions is associated with substantial increases in the time between bursts. McComb and Chan have found that the bursting period is also increased in TBA fibre suspensions. Thus behaviour of the two types of additive is, in this respect, qualitatively similar. However, a quantitative analysis reveals a subtle difference. If the mean time between bursts is scaled on outer variables (i.e. $T_B U/a$, where a is the radius of the pipe) and plotted against the amount of drag reduction, the results for polymers and for fibres in the polymeric regime both lie on a smooth curve which increases from the value for water (i.e. about 6.8 at DR $= 0$) to a value of $T_B U/a = 110$ at DR ≈ 70 per cent. In contrast, the results obtained in the fibre-like regime (i.e. the first and second passes at $R = 1.4 \times 10^4$) lie far below those in the polymeric regime (e.g. $T_B U/a = 68$ at DR $= 76$ per cent), only rejoining it asymptotically as the amount of drag reduction falls off with increasing number of passes through the system. Again, this is very persuasive evidence for separate and distinct drag-reduction mechanisms in polymer solutions and fibre suspensions.

14.6 Further reading

Background material on rheology can be found in many standard works. However, the treatment of the subject by Tanner (1985) is modern and readable, and has a section on turbulent drag reduction. The material presented in this chapter is very much restricted to fundamental questions of turbulent structure and is only a tiny fraction of a large body of published work. A broader appreciation of the entire topic of drag reduction by additives can be obtained from the reviews by Hoyt (1972) Lumley (1973) and Radin et al. (1975). In addition, there are various conference proceedings which are referred to in the present chapter and, lastly, a very useful publication is the comprehensive bibliography compiled by White and Hemmings (1976).

References

ABERNATHY, F. H., BERTSCHY, J. R., CHIN, R. W., and KEYES, D. E. (1980). *J. Rheol.* **24**, 647.

ACHIA, B. U. and THOMPSON, D. W. (1977). *J. fluid Mech.* **81**, 439.

ALLAN, J. J., GREATED, C. A., and McCOMB, W. D. (1984). *J. Phys. D* **17**, 533.

BARKER, S. J. (1973). *J. fluid Mech.* **60**, 721.

BENJAMIN, T. B. (1960). *J. fluid Mech.* **9**, 513.

BEWERSDORFF, H. W. (1982). *Rheol. Acta* **21**, 587.

BOBKOWICZ, A. J. and GAUVIN, W. H. (1967). *Chem. eng. Sci.* **22**, 229.

BRAGG, R. and OLIVER, D. R. (1973). *Nature phys. Sci.* **241**, 131.

BRUCE, C. and SCHWARZ, W. H. (1969). *J. polymer Sci. Pt A-2*, **7**, 909.

BRYSON, A. W., ARUNACHALAM, V. R., and FULFORD, G. D. (1971). *J. fluid Mech.* **47**, 209.

BUSHNELL, D. M. (1978). *NASA Tech. Memo.* 78688.

—— (1980). In *Viscous flow drag reduction* (ed. G. R. HOUGH), pp. 387–90. AIAA, New York.

——, HEFNER, J. N., and ASH, R. L. (1977). *Phys. Fluids* **20**, S31.

BUTSON, J. and GLASS, D. H. (1974). *Proc. 1st Int. Conf. on Drag Reduction, Cambridge.* Paper A3, BHRA Fluid Engineering, Bedford.

CARVER, C. E. and NADOLINK, R. H. (1965). Measurement of velocity profiles with non-Newtonian additives using photomicroscopy. *Tech. Rep.* 1. Fluid Mechanics Laboratory, University of Massachusetts.

CHOW, P. L. and SAIBEL, E. (1967). *Int. J. eng. Sci.* **5**, 723.

CHUNG, J. S. and GRAEBEL, W. P. (1972). *Phys. Fluids* **15**, 546.

DARBY, R. (1970). *Trans. Soc. Rheol.* **14**, 185.

DIMANT, Y. and POREH, M. (1976). *Adv. heat Transfer*, **12**, 77.

DODGE, D. W. and METZNER, A. B. (1959). *AIChE J.* **5**, 189.

EDWARDS, S. F. and FREED, K. F. (1974). *J. chem. Phys.* **61**, 1189.

EK, R., MOLLER, K. and NORMAN, B. (1978). *Dynamic Measurements in Unsteady Flow, Proc. Dynamic Flow Conf.*, p. 745. Marseille and Baltimore (1979).

FORAME, P. C., HANSEN, R. J. and LITTLE, R. C. (1972). *AIChE J.* **18**, 213.

FRUMAN, D. H. and TULIN, M. P. (1976). *J. ship Res.* **20**, 171.

GADD, G. E. (1965). *Nature (Lond.)* **206**, 463.

GOTO, S., NAGAZONO, H., and KATO, H. (1986). *Rheol. Acta* **25**, 119.

HIBBERD, M., KWADE, M., and SCHARF, R. (1982). *Rheol. Acta* **21**, 582.

HOYT, J. W. (1977). *Nature (Lond.)* **270**, 508.

HUANG, T. T. (1974). *Phys. Fluids* **17**, 298.

JEFFREY, D. J. and ACRIVOS, A. (1976). *AIChE J.* **22**, 417.

KALE, D. D. and METZNER, A. B. (1976). *AIChE J.* **22**, 669.

KEREKES, R. J. and GARNER, R. G. (1982). *Trans. Canad. Pulp Paper Assoc.* **TR 53**.

KUO, Y. and TANNER, R. I. (1972). *Trans. ASME Ser. E*, **39**, 661.

LANDAHL, M. T. (1962). *J. fluid Mech.* **13**, 609.

LEE, W. K., VASELESKI, R. C., and METZNER, A. B. (1974). *AIChE J.* **20**, 128.

LITTLE, R. C. and WIEGARD, M. (1970). *J. appl. polym. Sci.* **14**, 409.

LOGAN, S. E. (1972). *AIAA J.* **10**, 962.

LUMLEY, J. L. (1964). *Phys. Fluids* **7**, 335.

—— (1971). *Phys. Fluids* **14**, 2282.

McCOMB, W. D. (1974). *Proc. R. Soc. Edin. A* **72**, 226.

—— (1976). *Int. J. eng. Sci.* **14**, 239.

—— and AYYASH, S. Y. (1980). *J. Phys. D* **13**, 773.

—— and CHAN, K. T. J. (1979). *Nature (Lond.)* **280**, 45.

—— and —— (1981). *Nature (Lond.)* **292**, 520.

—— and —— (1985). *J. fluid Mech.* **152**, 455.

—— and RABIE, L. H. (1978). *Nature (Lond.)* **273**, 653.

—— and —— (1979). *Phys. Fluids* **22**, 183.

—— and —— (1982a). *AIChE J.* **28**, 547.

—— and —— (1982b). *Chem. eng. Sci.* **37**, 1759.

——, ALLAN, J. J., and GREATED, C. A. (1977). *Phys. Fluids* **20**, 873.

METZNER, A. B. and METZNER, A. P. (1970). *Rheol. Acta* **9**, 174.

—— and REED, J. C. (1955). *AIChE J.* **1**, 434.

MEWIS, J. and METZNER, A. B. (1974). *J. fluid Mech.* **62**, 593.

MIH, W. and PARKER, J. (1967). *Tappi* **50**, 237.

MIZUSHINA, T. and USUI, H. (1977). *Phys. Fluids* **20**, S100.

MONTI, R. (1972). In *Progress in Heat Transfer*, Vol. 5, p. 239. Pergamon, Oxford.

MOYLS, A. L. and SABERSKY, R. H. (1978). *Int. J. heat mass Transfer*, **21**, 7.

PATERSON, R. W. and ABERNATHY, F. H. (1972). *J. fluid Mech.* **51**, 177.

PIRIH, R. J. and SWANSON, W. M. (1972). *Can. J. chem. Eng.* **50**, 221.

RAMU, K. L. V. and TULLIS, J. P. (1976). *J. Hydronaut.* **10**, 55.

REISCHMAN, M. M. and TIEDERMAN, W. G. (1975). *J. fluid Mech.* **70**, 369.

RESHOTKO, E. (1977). *Proc. 2nd Int. Conf. on Drag Reduction, Cambridge, August 1977*, Paper E2. BHRA Fluid Engineering, Bedford.

ROLLIN, A. and SEYER, F. A. (1972). *Can. J. chem. Eng.* **50**, 714.

RUDD, M. J. (1972). *J. fluid Mech.* **51**, 673.

SAHLIN, A., JOHANSSON, A. V. and ALFREDSSON, P. H. (1988). *Phys. Fluids* **31**, 2814.

SEYER, F. A. and METZNER, A. B. (1969). *AIChE J.* **15**, 426.

SHARMA, R. S., SESHADRI, V., and MALHOTRA, R. C. (1979). *Chem. eng. Sci.* **34**, 703.

SHAVER, R. G. and MERRILL, E. W. (1959). *AIChE J.* **5**, 181.

SHENOY, A. V. and TALATHI, M. M. (1985). *AIChE J.* **31**, 520.

SIDAHMED, G. H. and GRISKEY, R. G. (1972). *AIChE J.* **18**, 138.

TANNER, R. I. (1985). *Engineering rheology*. Clarendon Press, Oxford.

TAYLOR, A. R. and MIDDLEMAN, S. (1974). *AIChE J.* **20**, 454.

TIEDERMAN, W. G., LUCHIK, T. S., and BOGARD, D. G. (1985). *J. fluid Mech.* **156**, 449.

USUI, H., MAEGUCHI, K. and SANO, Y. (1988). *Phys. Fluids* **31**, 2518.

Van DRIEST, E. R. (1970). *J. Hydronaut.* **4**, 120.

VASSILIADOU, E., MCCONAGHY, G. A., and HANRATTY, T. J. (1986). *AIChE J.* **32**, 381.

VIRK, P. S. and SURAIYA, T. (1977). *Proc. 2nd Int. Conf. on Drag Reduction, Cambridge, August 1977*, Paper G3. BHRA Fluid Engineering, Bedford.

——, MICKLEY, H. S., and SMITH, K. A. (1970). *Trans. ASME, J. Appl. Mech.* 488.

VLEGGAAR, J. and TELS, M. (1973). *Chem. eng. Sci.* **28**, 965.

WALTERS, R. R. and WELLS, C. S. (1972). *J. Hydronaut.* **6**, 69.

WELLS, C. S. and SPANGLAR, J. G. (1967). *Phys. Fluids* **10**, 1890.

WHITE, A. and HEMMINGS, J. A. G. (1976). *Drag reduction by additives: a review and bibliography*. BHRA Fluid Engineering, Bedford.

WILLMARTH, W. W., WEI, T., and LEE, C. O. (1987). *Phys. Fluids* **30**, 933.

WU, J. (1972). *J. Hydronaut.* **6**, 46.

ZAKIN, J. L., NI, C. C., HANSEN, R. J. and REISCHMAN, M. M. (1977). *Phys. Fluids* **20**, S85.

APPENDIX A

Creation and dissipation of kinetic energy in a viscous fluid

We are interested in the work done by external forces and in the dissipation of the resultant kinetic energy by viscous effects in the incompressible flow of a Newtonian fluid. To this end, we shall consider the fluid to occupy a volume V with surface area S. The velocity field is taken to be $U_\alpha(\mathbf{x}, t)$, where the index α takes the values 1, 2, or 3.

The total kinetic energy of the flowing fluid is given by

$$E_T = \frac{1}{2} \sum_\alpha \int_V \rho U_\alpha^2 \, dV, \tag{A.1}$$

where E_t is the total kinetic energy of the fluid and ρ is the density.

We wish to know how the kinetic energy varies with time. Therefore the first step is to differentiate both sides of eqn (A.1) with respect to time. Thus

$$\frac{dE_T}{dt} = \rho \int_V U_\alpha \frac{\partial U_\alpha}{\partial t} \, dV, \tag{A.2}$$

where we have used the fact that ρ is constant in both space and time.

We can obtain an expression for the r.h.s. of (A.2) by invoking the Navier–Stokes equation, which expresses conservation of momentum per unit volume of fluid. That is, rearranging equation (1.2) and adding an external force per unit volume F_α, we have

$$\rho \frac{\partial U_\alpha}{\partial t} = -\rho U_\beta \frac{\partial U_\alpha}{\partial x_\beta} - \frac{\partial p}{\partial x_\alpha} + \frac{\partial s_{\alpha\beta}}{\partial x_\beta} + F_\alpha, \tag{A.3}$$

along with the continuity eqn (1.1) and the expression for the deviatoric stress tensor $s_{\alpha\beta}$, which is given by eqn (1.3). For convenience, we repeat these equations here as

$$\frac{\partial U_\beta}{\partial x_\beta} = 0 \tag{A.4}$$

and

$$s_{\alpha\beta} = \rho v \left(\frac{\partial U_\alpha}{\partial x_\beta} + \frac{\partial U_\beta}{\partial x_\alpha} \right) \tag{A.5}$$

respectively, where v is the kinematic viscosity of the fluid.

Now multiply both sides of eqn (A.3) by $U_\alpha(\mathbf{x}, t)$ and integrate each term over the volume V. Then, using (A.2) to rewrite the l.h.s., we obtain

$$\frac{dE_T}{dt} = -\int_V \left(\rho U_\alpha U_\beta \frac{\partial U_\alpha}{\partial x_\beta} + U_\alpha \frac{\partial p}{\partial x_\alpha} \right) dV + \int_V U_\alpha \frac{\partial s_{\alpha\beta}}{\partial x_\beta} + \int_V U_\alpha F_\alpha \, dV. \tag{A.6}$$

$$\overset{\text{(a)}}{} \qquad \overset{\text{(b)}}{} \qquad \overset{\text{(c)}}{}$$

We can simplify this result by considering each of the terms labelled (a), (b), and (c) in turn as follows:

$$(a) = \rho U_\alpha U_\beta \frac{\partial U_\alpha}{\partial x_\beta} = \frac{\rho}{2} U_\beta \frac{\partial U_\alpha^2}{\partial x_\beta}$$

$$= \frac{\partial\{(\rho/2)U_\beta U_\alpha^2\}}{\partial x_\beta}, \tag{A.7}$$

where we have used eqn (A.4) between the first and second stages on the r.h.s. The continuity equation can also be used to simplify (b):

$$(b) = U_\alpha \frac{\partial p}{\partial x_\alpha} = \frac{\partial(U_\alpha p)}{\partial x_\alpha}$$

$$= \frac{\partial(U_\beta p)}{\partial x_\beta}, \tag{A.8}$$

as we can rename the dummy suffix $\alpha = \beta$. Lastly, (c) takes the form

$$(c) = U_\alpha \frac{\partial s_{\alpha\beta}}{\partial x_\beta} = \frac{\partial(U_\alpha s_{\alpha\beta})}{\partial x_\beta} - \frac{\partial U_\alpha}{\partial x_\beta} s_{\alpha\beta} \tag{A.9}$$

by the rule for differentiating a product.

Then, substituting the results (A.7)–(A.9) into eqn (A.6), we collect the three terms which take the general form $\partial g_\beta/\partial x_\beta$ (i.e. which take the form of a divergence) into the parentheses:

$$\frac{dE_T}{dt} = - \int_V \frac{\partial}{\partial x_\beta}\left(\frac{\rho}{2} U_\beta U_\alpha U_\alpha + U_\beta p - U_\alpha s_{\alpha\beta}\right)dV -$$

$$- \int_V \frac{\partial U_\alpha}{\partial x_\beta} s_{\alpha\beta}\,dV + \int_V U_\alpha F_\alpha\,dV. \tag{A.10}$$

At this stage, we invoke the divergence theorem in order to make a major simplification. That is, we can transform the volume integral involving the terms in parentheses into a surface integral as follows:

$$\int_V \frac{\partial}{\partial x_\beta}\left(\frac{\rho U_\beta U_\alpha U_\alpha}{2} + U_\beta p - U_\alpha s_{\alpha\beta}\right)dV = \int_S \left(\frac{\rho U_\beta U_\alpha U_\alpha}{2} + U_\beta p - U_\alpha s_{\alpha\beta}\right)dS = 0. \tag{A.11}$$

The last result is due to the boundary conditions, which generally require the velocity to vanish at infinity or on any solid surfaces.

Hence the rate of change of kinetic energy is given by

$$\frac{dE_T}{dt} = \int_V U_\alpha F_\alpha\,dV - \int_V \frac{\partial U_\alpha}{\partial x_\beta} s_{\alpha\beta}\,dV, \tag{A.12}$$

or in words

Rate of change in kinetic energy = rate of doing work by external forces—rate at which energy is dissipated by viscous effects.

Substituting from (A.5) for $s_{\alpha\beta}$, we obtain the explicit form

$$\frac{dE_T}{dt} = \int_V U_\alpha F_\alpha\,dV - \rho v \int_V \frac{\partial U_\alpha}{\partial x_\beta}\left(\frac{\partial U_\alpha}{\partial x_\beta} + \frac{\partial U_\beta}{\partial x_\alpha}\right)dV, \tag{A.13}$$

and, with some rearrangement,

$$\frac{dE_T}{dt} = \int_V U_\alpha F_\alpha \, dV - \frac{\rho v}{2} \int_V \left(\frac{\partial U_\alpha}{\partial x_\beta} + \frac{\partial U_\beta}{\partial x_\alpha} \right)^2 dV. \tag{A.14}$$

Finally, we introduce the external force f_α per unit mass of fluid as

$$f_\alpha = F_\alpha / \rho, \tag{A.15}$$

and the rate ε of energy dissipation per unit mass of fluid per unit time as

$$\varepsilon = \frac{v}{2} \sum_{\alpha, \beta} \left(\frac{\partial U_\alpha}{\partial x_\beta} + \frac{\partial U_\beta}{\partial x_\alpha} \right)^2, \tag{A.16}$$

and substitute into (A.14) to obtain the compact form

$$\frac{dE_T}{dt} = \int_V \rho U_\alpha f_\alpha \, dV - \int_V \rho \varepsilon \, dV. \tag{A.17}$$

Three points about this relationship are worthy of special note.

(a) A steady state exists when $U_\alpha f_\alpha = \varepsilon$.
(b) When the external forces are zero, the kinetic energy of the fluid flow decays away at a rate given by ε.
(c) From (A.11) it can be seen that the non-linear and pressure terms (i.e. the inertial terms) do no net work on the system. In other words, these forces are—as one would expect—conservative in their effect on the fluid.

We conclude by considering the formal extension of eqn (A.16) for the dissipation rate to the case of turbulence. We can introduce the mean dissipation rate by averaging both sides of (A.16), and introducing the decomposition of the velocity field into mean and fluctuating variables. From equation (1.7), and recalling that $\langle \bar{U}u \rangle = 0$, we readily find the general result for turbulent flows:

$$\langle \varepsilon \rangle = \frac{v}{2} \sum_{\alpha, \beta} \left(\frac{\partial \bar{U}_\alpha}{\partial x_\beta} + \frac{\partial \bar{U}_\beta}{\partial x_\alpha} \right)^2 + \frac{v}{2} \sum_{\alpha, \beta} \left\langle \left(\frac{\partial u_\alpha}{\partial x_\beta} + \frac{\partial u_\beta}{\partial x_\beta} \right)^2 \right\rangle. \tag{A.18}$$

For the particular case of homogeneous turbulence, we can set the mean velocity gradients to zero and consider only the dissipation of the kinetic energy of the fluctuating motions. That is, eqn (A.18) becomes

$$\langle \varepsilon \rangle = \frac{v}{2} \sum_{\alpha, \beta} \left\langle \left(\frac{\partial u_\alpha}{\partial x_\beta} + \frac{\partial u_\beta}{\partial x_\alpha} \right)^2 \right\rangle$$

$$= v \sum_{\alpha, \beta} \left\langle \left(\frac{\partial u_\alpha}{\partial x_\beta} \right)^2 \right\rangle, \tag{A.19}$$

where it can be shown that the last step follows from the particular properties of homogeneous turbulence (Hinze 1975, p. 75).

Reference

HINZE, J. O. (1975). *Turbulence* (2nd edn). McGraw-Hill, New York.

APPENDIX B

Probability and statistics

We summarize here a few definitions and theorems which will be helpful in the statistical study of turbulence. We begin with relatively simple situations where discrete events occur randomly and all possible outcomes are equally likely.

Discrete stochastic events

We begin by defining probability. Suppose that an experiment has N possible outcomes, all mutually exclusive and each equally likely to happen. Further, suppose that each one of n of these outcomes would be regarded as a success. Then, for large enough values of N and n, the probability of success in the given experiment is just n/N.

If we denote the probability of the occurrence of n successes by $p(n)$, the above statement can be written as the equation

$$p(n) = n/N. \tag{B.1}$$

Clearly, the probability is a fraction, which in general satisfies $0 \leqslant p \leqslant 1$. If success is certain, then $n = N$ and $p = 1$. Likewise, if success is impossible, then $n = 0$ and $p = 0$.

Now consider a more complicated experiment involving N trials, in which any outcome can contain an event A, or an event B, or both A and B, or neither A nor B. If we represent the number of outcomes in which A only occurs by $n(A, 0)$, then, with an obvious extension of the notation, we have

$$n(A, 0) + n(0, B) + n(A, B) + n(0, 0) = N, \tag{B.2}$$

and, by eqn (B.1), the probability of A occurring is given by

$$p(A) = \frac{n(A, 0) + n(A, B)}{N}, \tag{B.3}$$

with a similar result for $p(B)$.

With this example, we can introduce some compound probabilities.

Joint probability $p(AB)$: this is the probability of both A and B occurring. It is just

$$p(AB) = \frac{n(A, B)}{N}. \tag{B.4}$$

Conditional probability $p(A|B)$: this is the probability of A occurring, given that B has already occurred:

$$p(A|B) = \frac{n(A, B)}{n(0, B) + n(A, B)}. \tag{B.5}$$

We can now use our simple example to illustrate two general theorems of probability.

Addition theorem: let the probability of either A or B occurring be represented by $p(A + B)$. Then we can readily deduce the general result

$$p(A + B) = p(A) + p(B) - p(AB).$$
(B.6)

If A and B are mutually exclusive, it follows that $p(AB) = 0$, and hence (B.6) becomes

$$p(A + B) = p(A) + p(B),$$
(B.7)

which is the addition theorem.

Multiplication theorem: the general result for the joint probability of A and B occurring is expressible in terms of the relevant conditional probabilities as

$$p(AB) = p(A|B)p(B) = p(B|A)p(A).$$
(B.8)

If A and B are statistically independent, it follows that the probability of A occurring does not depend on whether or not B has already occurred. In this case, the conditional probability reduces to a simple probability:

$$p(A|B) = p(A).$$
(B.9)

Hence, from eqns (B.8) and (B.9), we have

$$p(AB) = p(A)p(B)$$
(B.10)

for statistically independent A and B.

Distributions

These results can be generalized to any number of events A, B, C, ... making up one outcome, and to any number N of outcomes. However, the classic extension is to processes with two mutually exclusive and statistically independent outcomes.

Consider, for example, the experiment of tossing a coin, with the result either a head (h) or a tail (t), a large number N of times. We can write down any one realization of this experiment as a sequence of N letters, say

thhtththt...h,

with the probability 1/2 that any particular sequence occurs. Suppose that we want the probability $p(r)$ that exactly r heads occurred. Then, of the $N!/(N - r)!$ possible permutations of t and h, the number of arrangements—irrespective of order—with exactly r heads is given by the binomial coefficient $_NC_r = N!/r!(N - r)!$ Thus the probability that exactly r heads will occur is

$$p(r) = {_NC_r}(1/2)^r.$$
(B.11)

This result can be further generalized to the case where the probability of a success is w, say, and the probability of failure is $1 - w$. What then is the probability of exactly r successes and $N - r$ failures? We can answer this question for the previous case, where $w = 1/2$, and then make the appropriate generalization. That is, we write eqn (B.11) as

$$p(r) = {_NC_r}(1/2)^r(1 - 1/2)^{N-r} = {_NC_r}w^r(1 - w)^{N-r}.$$
(B.12)

In its more general form this is the binomial distribution (sometimes referred to as the Bernoulli distribution). A more detailed treatment will be found in Reichl (1980, p. 146 *et seq*).

Continuous random variables

The above result for discrete stochastic variables can be extended to include the important general case of continuous random variables. In the limit of large N and large pN, it can be shown that the r.h.s. of eqn (B.12) goes over into the normal or Gaussian distribution:

$$p(x) = \frac{1}{\sigma\sqrt{2\pi}}\exp\left(\frac{-x^2}{2\sigma^2}\right), \tag{B.13}$$

where $x = r - N/w$ and

$$\sigma = \{Nw(1 - w)\}^{1/2}$$

is a measure of the width of the distribution and is usually known as the standard deviation (from the mean).

As the distribution goes over into the continuous form, we have to generalize our definition of probability to include the case of a continuous random variable X. Then we make the definition

$p(X = x) =$ probability that X takes on a value in the range $x < X < x + dx$,
$$\tag{B.14}$$

although usually one often just writes $p(X = x) = p(x)$.

Moments and cumulants of a distribution

Probability distributions found in practice are not necessarily Gaussian in form, and so it is helpful to have a method of characterizing distributions. This can be done in terms of the moments. In particular, this offers a particularly simple method of identifying asymmetry, or other departures from Gaussian behaviour. We begin by noting that the mean value of x is evaluated from the distribution by the operation

$$\langle x \rangle = \int p(x)x \, dx, \tag{B.15}$$

with the immediate generalization to any well-behaved function of x

$$\langle f(x) \rangle = \int p(x)f(x) \, dx \tag{B.16}$$

and, in particular,

$$\langle x^n \rangle = \int p(x)x^n \, dx, \tag{B.17}$$

where $\langle x^n \rangle$ is the nth-order moment of the distribution $p(x)$.

We can express any distribution in terms of its moments by introducing the characteristic function $m(k)$, which is just the Fourier transform of $p(x)$, that is,

$$m(k) = \int p(x)\exp(ikx)\,dx$$

$$= \langle\exp(ikx)\rangle. \tag{B.18}$$

Then, expanding out the exponential in powers of k, we obtain

$$m(k) = 1 + ik\langle x\rangle - \frac{k^2\langle x^2\rangle}{2!} + \cdots\frac{(ik)^n\langle x^n\rangle}{n!} + \cdots. \tag{B.19}$$

Thus, from a knowledge of the moments $\langle x^n\rangle$, we can reconstruct $m(k)$ from (B.19) and hence $p(x)$ from (B.18).

It is also usual to introduce a variant of the moment which is known as a cumulant of the distribution. The cumulants C_n of a distribution can be defined by expanding the logarithm of the characteristic function:

$$ln\{m(k)\} = ikC_1 - \frac{k^2 C_2}{2!} + \cdots\frac{(ik)^n C_n}{n!} + \cdots. \tag{B.20}$$

Comparison of (B.19) and (B.20) then leads to the following relationships between cumulants and moments:

$$C_1 = \langle x\rangle = \bar{x}$$
$$C_2 = \langle x^2\rangle - \langle x\rangle^2 = \sigma^2$$
$$C_3 = \langle(x - \langle x\rangle^3\rangle$$
$$C_4 = \langle(x - \langle x\rangle)^4\rangle$$
$$\vdots$$

(B.21)

Flatness and skewness of a distribution

In turbulence the deviation from a Gaussian distribution is sometimes quite small and can be characterized in terms of the third- and fourth-order moments. For simplicity (and also because of its relevance to the fundamental problem of isotropic turbulence) let us restrict our attention to a random variable with zero mean and with a Gaussian distribution centred on the origin $x = 0$. It is therefore an obvious consequence of symmetry that odd-order moments—as defined by eqn (B.17)—all vanish. Further, the even-order moments can be expressed to all orders in terms of $\langle x^2\rangle$:

$$\langle x^n\rangle = (n - 1)\langle x^2\rangle^{n/2} \qquad n \text{ even}. \tag{B.22}$$

It is this result which is often stated as 'a Gaussian can be specified in terms of its second moment'.

Now, if an actual distribution is symmetric, but differs from the Gaussian form, then this means that it is either 'peakier' or 'flatter' than a Gaussian. Such behaviour can be associated with a change in the relationship between the higher-order moments and the variance $\langle x^2\rangle$. The lowest order at which we can check for a departure from

Gaussian behaviour in this way is clearly the fourth. Thus it is usual to define the flatness factor F for a distribution as

$$F = \frac{\langle x^4 \rangle}{\langle x^2 \rangle^2}. \tag{B.23}$$

For a Gaussian distribution, eqn (B.22) tells us that this takes the value $F = 3$.

If $p(x)$ is not symmetric, then it follows from eqn (B.17) that the asymmetric part does not affect the even-order moments. One detects the existence of the asymmetry by the non-vanishing of the odd-order moments, and particularly the third-order moment. In practice one defines a skewness factor S as

$$S = \frac{\langle x^3 \rangle}{\langle x^2 \rangle^{3/2}}. \tag{B.24}$$

Thus, in all, the general criterion for a Gaussian distribution is that we should have $F = 3$ and $S = 0$.

Central limit theorem

The existence of the odd-order moments in the statistical hierarchy based on the Navier–Stokes equation guarantees that turbulence processes are inherently non-Gaussian. Yet, despite this, Gaussian distributions can arise in the following way. Suppose that we consider a random variable x, with distribution $p(x)$, characterized by a mean $\langle x \rangle$ and variance $\langle x^2 \rangle$. Now, make N measurements of x and form an average (not the mean $\langle x \rangle$), which we denote by y_N; thus

$$y_N = \frac{1}{N} \sum_{n=1}^{N} x_n. \tag{B.25}$$

It can be shown (Reichl 1980, p. 152) that, irrespective of the form of $p(x)$, the probability distribution $q(y_N - \langle x \rangle)$ approaches a Gaussian form as N becomes large:

$$q(y_N - \langle x \rangle) = (\sqrt{N}/\sigma\sqrt{2\pi}) \exp\{-N(y_N - \langle x \rangle)^2/2\sigma^2\}. \tag{B.26}$$

This result is known as the central limit theorem, and its only requirements are that the distinct x_n are statistically independent and that moments of $p(x)$ are finite.

The theorem can be extended in a restricted way to turbulence, especially if the velocity field has been operated on by either an integral or a filter. The ways in which these ideas can be applied to the turbulent velocity field are discussed by Tennekes and Lumley (1972, pp. 216–21).

References

REICHL, L. E. (1980). *A modern course in statistical physics*. Edward Arnold, London.
TENNEKES, H. and LUMLEY, J. L. (1972). *A first course in turbulence*. MIT Press, Cambridge, Ma.

APPENDIX C

Symmetry and invariance

Symmetry under rotation of coordinate axes

In isotropic turbulence there is no preferred direction, which means that statistical quantities must be invariant under arbitrary rotations of coordinate axes. Consider the implications for the single-point covariance $\langle u_\alpha(\mathbf{x}, t)u_\beta(\mathbf{x}, t)\rangle$.

Let the coordinate axes be rotated clockwise through an angle of $\pi/2$ about the x_3 axis and denote the operation by R_3. In terms of the position coordinates, R_3 produces the mapping

$$x_1 \rightarrow -x_2, x_2 \rightarrow x_1, \text{ and } x_3 \rightarrow x_3,$$

and hence the operator R_3 is given by

$$R_3 = \begin{pmatrix} 0 & -1 & 0 \\ 1 & 0 & 0 \\ 0 & 0 & 1 \end{pmatrix}. \tag{C.1}$$

Now apply R_3 to the velocity vector $\mathbf{u}$. We have

$$R_3\mathbf{u} = \begin{pmatrix} 0 & -1 & 0 \\ 1 & 0 & 0 \\ 0 & 0 & 1 \end{pmatrix}\begin{pmatrix} u_1 \\ u_2 \\ u_3 \end{pmatrix} = \begin{pmatrix} -u_2 \\ u_1 \\ u_3 \end{pmatrix}. \tag{C.2}$$

Hence, operating with R_3 upon u_1 yields for the diagonal elements of the covariance

$$\langle R_3 u_1 R_3 u_1 \rangle \rightarrow \langle u_2 u_2 \rangle$$

$$= \langle u_1 u_1 \rangle, \tag{C.3}$$

where the last step follows from the requirement of invariance under rotations. Therefore we have

$$\langle u_1^2 \rangle = \langle u_2^2 \rangle, \tag{C.4}$$

and a similar rotation about x_1 then establishes the corresponding result

$$\langle u_2^2 \rangle = \langle u_3^2 \rangle. \tag{C.5}$$

Symmetry under reflection in planes

For isotropic turbulence this is again a property of statistical quantities, and is taken with respect to all possible planes. Let us consider the specific case of the $(x_1 x_3)$ plane, as shown in Fig. C.1. The components of the velocity vector to be reflected are denoted by u_1 and u_2. The third component u_3 is not shown. The corresponding reflections in the $(x_1 x_3)$ plane are u_1' and u_2', and are shown as broken lines. It is clear from the figure

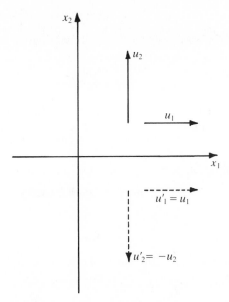

Fig. C.1. Reflection of velocity vectors in the $(x_1 x_3)$ plane.

that the result of reflection in the $(x_1 x_3)$ plane is

$$u_1' = u_1 \qquad\qquad\qquad (C.6)$$

$$u_2' = -u_2. \qquad\qquad\qquad (C.7)$$

Now, invariance under reflection implies that we must have

$$\langle u_1 u_2 \rangle = \langle u_1' u_2' \rangle, \qquad\qquad\qquad (C.8)$$

whereas, from (C.6) and (C.7), we actually have

$$\langle u_1 u_2 \rangle = -\langle u_1' u_2' \rangle \qquad\qquad\qquad (C.9)$$

and hence

$$\langle u_1 u_2 \rangle = 0. \qquad\qquad\qquad (C.10)$$

It readily follows that $\langle u_2 u_3 \rangle = 0$ also, and reflection in either the $(x_1 x_2)$ or $(x_2 x_3)$ planes will further establish that $\langle u_1 u_3 \rangle = 0$.

This particular symmetry is a weaker requirement than rotational symmetry and can also be applied to inhomogeneous problems. For instance, in the channel flow discussed in Sections 1.4.5 and 2.3, the $(x_1 x_3)$ and $(x_1 x_2)$ planes are still planes of symmetry, even though the field is inhomogeneous in the x_2 direction. The above arguments go through again to establish that

$$\langle u_1 u_3 \rangle = 0 \qquad \langle u_2 u_3 \rangle = 0, \qquad\qquad\qquad (C.11)$$

but reflection of $\langle u_1 u_2 \rangle$ in the $(x_1 x_3)$ plane now gives

$$\langle u_1 u_2 \rangle = -\langle u_1' u_2' \rangle \neq 0 \tag{C.12}$$

which implies that $\langle u_1 u_2 \rangle$ is an odd function of the variable x_2.

Galilean invariance of the Navier–Stokes equation

It is a cardinal principle that laws of physics are unaltered by transformations from one inertial frame to another. For velocities which are small compared with the speed of light, we use the Galilean transformations connecting the laboratory frame S to a frame $\tilde{S}$ moving with velocity c relative to S.

From Fig. C.2, it is readily seen that the Galilean transformations are given by

frame S	frame $\tilde{S}$	transformations
x	$\tilde{x}$	$x = \tilde{x} + \mathbf{c}t$
t	$\tilde{t}$	$t = \tilde{t}$
$\mathbf{u}(\mathbf{x}, t)$	$\tilde{\mathbf{u}}(\tilde{\mathbf{x}}, \tilde{t})$	$\mathbf{u} = \tilde{\mathbf{u}} + \mathbf{c}.$

$$\tag{C.13}$$

From these we can obtain the transformation laws of the derivatives:

$$\frac{\partial \tilde{u}_\alpha}{\partial x_\beta} = \frac{\partial \tilde{u}_\alpha}{\partial \tilde{x}_\beta} \frac{\partial x_\beta}{\partial x_\beta} + \frac{\partial \tilde{u}_\alpha}{\partial \tilde{t}} \frac{\partial \tilde{t}}{\partial x_\beta}$$

$$= \frac{\partial \tilde{u}_\alpha}{\partial \tilde{x}_\beta}, \tag{C.14}$$

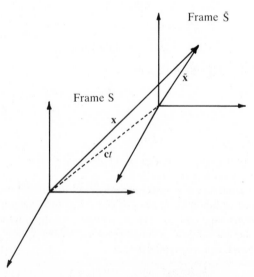

Fig. C.2. Relative configuration of moving and laboratory frames for the Galilean transformation.

where we have used the fact that neither t nor $\tilde{t}$ depends on $\mathbf{x}$. Similarly, we have

$$\frac{\partial \tilde{u}_\alpha}{\partial t} = \frac{\partial \tilde{u}_\alpha}{\partial \tilde{x}_\beta} \frac{\partial \tilde{x}_\beta}{\partial t} + \frac{\partial \tilde{u}_\alpha}{\partial \tilde{t}} \frac{\partial \tilde{t}}{\partial t}$$

$$= -c_\beta \frac{\partial \tilde{u}_\alpha}{\partial \tilde{x}_\beta} + \frac{\partial \tilde{u}_\alpha}{\partial t}. \tag{C.15}$$

Now the Navier–Stokes equation in frame S is just

$$\left(\frac{\partial}{\partial t} + u_\beta \frac{\partial}{\partial x_\beta} \right) u_\alpha = -\frac{\partial p}{\partial x_\alpha} + v\nabla^2 u_\alpha, \tag{C.16}$$

and, with the substitution of (C.13) for u_α, it becomes

$$\left(\frac{\partial}{\partial t} + (\tilde{u}_\beta + c_\beta) \frac{\partial}{\partial x_\beta} \right) (\tilde{u}_\alpha + c_\alpha) = -\frac{\partial \tilde{p}}{\partial \tilde{x}_\alpha} + v\nabla^2 (\tilde{u}_\alpha + c_\alpha). \tag{C.17}$$

Note that we transform the pressure gradient as an invariant. This is justified because the pressure itself can differ between the two frames by, at most, an additive constant.

Then, using (C.14) and (C.15) to transform the derivatives acting on the velocity field, we obtain

$$\frac{\partial \tilde{u}_\alpha}{\partial \tilde{t}} - c_\beta \frac{\partial \tilde{u}_\beta}{\partial x_\beta} + \tilde{u}_\beta \frac{\partial \tilde{u}_\alpha}{\partial \tilde{x}_\beta} + c_\beta \frac{\partial \tilde{u}_\alpha}{\partial \tilde{x}_\beta} = -\frac{\partial \tilde{p}}{\partial \tilde{x}_\alpha} + v\tilde{\nabla}^2 \tilde{u}_\alpha,$$

and, upon cancelling the terms on the l.h.s. containing $\mathbf{c}$, the invariant form of the Navier–Stokes equation in the moving frame:

$$\left(\frac{\partial}{\partial \tilde{t}} + \tilde{u}_\beta \frac{\partial}{\partial \tilde{x}_\beta} \right) \tilde{u}_\alpha = -\frac{\partial \tilde{p}}{\partial \tilde{x}_\alpha} + v\tilde{\nabla}^2 \tilde{u}_\alpha. \tag{C.18}$$

Similar demonstrations of form invariance under Galilean transformation can be given for the various equations of the moment hierarchy, as discussed in Chapters 1 and 2.

APPENDIX D

Application of Fourier methods and Green's functions to the Navier–Stokes equation

We start with a brief summary of Fourier methods. There is no pretence to either rigour or completeness. The interested reader can find these qualities in one of the standard references, such as Whittaker and Watson (1965) or Lighthill (1964). Also, for a quite general treatment along with numerous examples and applications, the monograph by Stuart (1961) is worth consulting.

As an introduction, consider standing waves in a stretched string of length L. The wavelengths of the various harmonics are given by L divided by an integer:

$$\lambda = L/n, \tag{D.1}$$

where λ is the wavelength of any particular harmonic. Then, if we define the wave-number (or spatial frequency) by

$$k = \frac{2\pi}{\lambda}, \tag{D.2}$$

we have

$$k = \frac{2\pi n}{L} \qquad n = 1, 2, 3, \dots . \tag{D.3}$$

We can represent harmonics by $a_k \cos(kx)$ or $b_k \sin(kx)$ or by the complex form

$$c_k \exp(ikx) = a_k \cos(kx) + ib_k \sin(kx), \tag{D.4}$$

and use this for the basis of an expansion. We rely on the most important theorem in the subject, which is to the effect that any reasonably well-behaved function, defined on the interval $-L/2 < x < L/2$, can be written as

$$f(x) = \sum_{k=-\infty}^{\infty} F_k \exp(ikx), \tag{D.5}$$

where the Fourier coefficients F_k are given by

$$F_k = \frac{1}{L} \int_{-L/2}^{L/2} f(x) \exp(-ikx)\,dx. \tag{D.6}$$

We can plot F_k against the spatial frequency k as a sequence of discrete spectral lines occurring at a constant separation of $k = 2\pi/L$. Clearly, if we let L become infinite, then the spacing between the spectral lines becomes zero, and the spectrum becomes a continuous function. We shall take this limit in the following way. The index n takes on all integer values between $-\infty$ and $+\infty$. We let $L \to \infty$ such that $k \to \infty$ as $n \to \infty$.

The continuous spectrum function $F(k)$ can be introduced through the definition

$$F(k) = \frac{L}{2\pi} F_k. \tag{D.7}$$

Then it follows that (D.5) can be written as

$$f(x) = \lim_{L \to \infty} \sum_{k=-\infty}^{\infty} \frac{2\pi}{L} F(k) \exp(ikx)$$

$$= \int_{-\infty}^{\infty} F(k) \exp(ikx) \, dk. \tag{D.8}$$

Also, multiplying both sides of eqn (D.6) by $L/2\pi$ and taking the limit of $L \to \infty$, we obtain

$$F(k) = \frac{1}{2\pi} \int_{-\infty}^{\infty} f(x) \exp(-ikx) \, dx. \tag{D.9}$$

The functions $f(x)$ and $F(k)$ make up a Fourier transform pair (i.e. each is the transform of the other, although note the change of sign in the exponent from one transformation to the other). This leads to an important relationship. Substitute (D.9) for $F(k)$ in (D.8) to obtain

$$f(x) = \left(\frac{1}{2\pi}\right) \int_{-\infty}^{\infty} \exp(ikx) \left\{ \int_{-\infty}^{\infty} f(x') \exp(-ikx') \, dx' \right\} dk. \tag{D.10}$$

We note that this relationship will hold for all x if

$$\frac{1}{2\pi} \int_{-\infty}^{\infty} \exp\{ik(x - x')\} \, dk = \delta(x - x'), \tag{D.11}$$

where $\delta(x)$ is the Dirac delta function and has the properties

$$\int_{-\infty}^{\infty} \delta(x) \, dx = 1 \tag{D.12}$$

and

$$\int_{-\infty}^{\infty} \delta(x) h(x) \, dx = h(0) \tag{D.13}$$

where $h(x)$ is any reasonably behaved function.

Two general properties of Fourier transforms will be of use to us. First, the differential coefficient (of any order) can readily be expressed in terms of the Fourier transform; thus from (D.8) we can easily show that

$$\frac{d^m f}{dx^m} = \int_{-\infty}^{\infty} (ik)^m F(k) \exp(ikx) \, dx. \tag{D.14}$$

Second, the Fourier transform of a product can be evaluated as the convolution of the individual Fourier transforms—a result known as the convolution theorem. Consider a function $p(x)$ which is defined by the product

$$p(x) = f(x)g(x), \tag{D.15}$$

where $F(k)$ and $G(k)$ are the Fourier transforms of f and g respectively. What then is $P(k)$, the Fourier transform of $p(x)$? It can be shown by using (D.9) to transform p, and by direct substitution of the Fourier integral representations for f and g, that

$$P(k) = \int_{-\infty}^{\infty} F(j)G(k-j)\,dj, \tag{D.16}$$

where j is a dummy wavenumber and this type of integral is known as the convolution of the functions F and G.

Now let us consider an entirely fictitious non-linear differential equation, as an illustration of the way in which Fourier methods can be applied to the Navier–Stokes equation:

$$\frac{d\{f^2(x)\}}{dx} + \frac{d^2f}{dx^2} = f(x). \tag{D.17}$$

Substitute for each term its Fourier representation

$$\int_{-\infty}^{\infty} (ik)P(k)\exp(ikx)\,dx - \int_{-\infty}^{\infty} k^2 F(k)\exp(ikx)\,dx = \int_{-\infty}^{\infty} F(k)\exp(ikx)\,dx, \tag{D.18}$$

where $P(k)$ stands for the transform of the product $f \times f$. Clearly the relationship must hold for arbitrary values of k, and so we can equate the integrands and cancel the common factor $\exp(ikx)$. The result is

$$ikP(k) - k^2 F(k) = F(k), \tag{D.19}$$

and using the convolution theorem (i.e. eqns (D.15) and (D.16), with $g = f$ and $G = F$) we obtain the final form

$$\int_{-\infty}^{\infty} ikF(j)F(k-j)\,dj - k^2 F(k) = F(k). \tag{D.20}$$

We should reiterate that there is no special significance in either eqn (D.17) or its Fourier transformed state. It is merely intended as a simple illustration of how Fourier methods can be applied to a more complicated equation such as the Navier–Stokes equation.

We can complete this brief revision of Fourier methods by listing the three-dimensional version of (D.8) and (D.9):

$$f(\mathbf{x}) = \int d^3k\, F(\mathbf{k})\exp(i\mathbf{k}\cdot\mathbf{x}) \tag{D.21}$$

$$F(\mathbf{k}) = \left(\frac{1}{2\pi}\right)^3 \int d^3x\, f(\mathbf{x})\exp(-i\mathbf{k}\cdot\mathbf{k}). \tag{D.22}$$

It may also be helpful to note the generalization of the rule for differentiation in three dimensions. That is, eqn (D.14) is replaced by

$$\frac{\partial^m f}{\partial x_\alpha^m} = \int (ik_\alpha)^m \cdot d^3k\, F(\mathbf{k})\exp(i\mathbf{k}\cdot\mathbf{x}), \tag{D.23}$$

where we note that $\mathbf{k}\cdot\mathbf{x} = k_\alpha x_\alpha$, with the summation convention for repeated indices.

In order to make contact with the work in Section 2.6.1, we shall briefly discuss the use of the Green function $G(\mathbf{x}, \mathbf{x}')$ to obtain the general solution of inhomogeneous differential equations. We take the subject of electrostatics as an example, partly because of its clear physical interpretation and partly because of the relevance of Laplace's equation to the procedure by which the pressure is eliminated from the Navier–Stokes equation.

Therefore we begin with Laplace's equation for the potential V_L in a region:

$$\nabla^2 V_L(\mathbf{x}) = 0. \tag{D.24}$$

For a particular region and set of boundary conditions, the solution to this equation is unique. However, when we consider the inhomogeneous Poisson equation

$$\nabla^2 V_P(\mathbf{x}) = \rho(\mathbf{x}), \tag{D.25}$$

there are as many solutions for $V_P(\mathbf{x})$ as there are arbitrary prescriptions for the charge distribution $\rho(\mathbf{x})$.

If, instead of solving (D.24), we solved the inhomogeneous equation for a point charge, we could hope to synthesize the solution to (D.25), with $\rho(\mathbf{x})$ as a distribution of such charges. We call the solution for the case of the point charge, $G(\mathbf{x}, \mathbf{x}')$ and obtain it as the solution of

$$\nabla^2 G(\mathbf{x}, \mathbf{x}') = \delta(\mathbf{x} - \mathbf{x}'), \tag{D.26}$$

where we employ a set of units in which permittivities and factors of 4π alike are unity.

Then it is a general result that, for the same region of space and the same boundary conditions, the solution to (D.25) can be written as an integral equation in the form:

$$V_P(\mathbf{x}) = \int d^3\mathbf{x}' \, G(\mathbf{x}, \mathbf{x}')\rho(\mathbf{x}'). \tag{D.27}$$

This is readily verified by substituting (D.27) into the r.h.s. of (D.25) and using (D.26).

In order to make comparisons with the work of Chapter 2, we restrict our attention to the case of spherical symmetry in an infinite medium. Then the use of the Fourier transforms as discussed above allows us to establish the solution of (D.25) directly as

$$V_P(k) = -\frac{\rho(k)}{k^2}. \tag{D.28}$$

This is of course the same result as is obtained by using Fourier methods to solve (D.26), with the Green function in k-space as

$$G(k) = -1/k^2 \tag{D.29}$$

and the convolution theorem applied to (D.27) yielding

$$V_P(k) = G(k)\rho(k). \tag{D.30}$$

Now we are in a position to tackle the Navier–Stokes equations. From (2.2) we can write

$$\left(\frac{\partial}{\partial t} - \nu\nabla^2\right)u_\alpha = -\frac{\partial p}{\partial x_\alpha} - \frac{\partial(u_\alpha u_\beta)}{\partial x_\beta}, \tag{D.31}$$

where we use u and p to denote that we are considering fluctuations only in a field with zero mean, and the units are chosen such that the density is unity. Using the Fourier decomposition as given by (2.71), with an analogous expression for the pressure, we can follow the procedures used in conjuction with our simple illustration discussed above and obtain (D.31) in the form

$$\left(\frac{\partial}{\partial t} + vk^2\right)u_\alpha(\mathbf{k}) = -ik_\alpha p(\mathbf{k}) - ik_\beta \sum_{\mathbf{j}} u_\alpha(\mathbf{k} - \mathbf{j})u_\beta(\mathbf{j}). \tag{D.32}$$

As in Section 2.1, we can use the continuity relation to eliminate the pressure. Multiply each term of (D.32) by k_α, sum over α, and invoke continuity in the form of (2.74). The l.h.s. vanishes, and this leaves us with

$$k^2 p(\mathbf{k}) = -k_\alpha k_\beta \sum_{\mathbf{j}} u_\alpha(\mathbf{k} - \mathbf{j})u_\beta(\mathbf{j}). \tag{D.33}$$

It is not difficult to see that this equation is just the Fourier transform of (2.3) for the case of the infinite system.

We complete our programme by substituting for $p(\mathbf{k})$ from (D.33) into (D.32). First, however, we note that we are using α as a dummy suffix in (D.33) and to avoid confusion we rename it γ. Then the Fourier-transformed Navier–Stokes equation becomes

$$\left(\frac{\partial}{\partial t} + vk^2\right)u_\alpha(\mathbf{k}) = ik_\alpha k_\beta k_\gamma |k|^{-2} \sum_{\mathbf{j}} u_\beta(\mathbf{j})u_\gamma(\mathbf{k} - \mathbf{j}) - ik_\beta \delta_{\alpha\gamma} \sum_{\mathbf{j}} u_\beta(\mathbf{j})u_\gamma(\mathbf{k} - \mathbf{j}), \tag{D.34}$$

where we have used the Kronecker delta as a substitutional symbol to write

$$u_\alpha(\mathbf{k} - \mathbf{j}) = \delta_{\alpha\gamma} u_\gamma(\mathbf{k} - \mathbf{j})$$

in the second term on the r.h.s. of (D.34). A little further manipulation shows that (D.34) is the same as equation (2.76) in the main text.

References

LIGHTHILL, M. J. (1964). *Fourier analysis and generalised functions.* Cambridge University Press, Cambridge.

STUART, R. D. (1961). *An introduction to Fourier analysis.* Methuen, London.

WHITTAKER, E. T. and WATSON, G. N. (1965). *A course of modern analysis.* Cambridge University Press, Cambridge.

APPENDIX E

Evaluation of the coefficients $L(\mathbf{k}, \mathbf{j})$ and $L(\mathbf{k}, \mathbf{k} - \mathbf{j})$

We begin by noting the two alternative forms of the inertial transfer coefficient. $L(\mathbf{k}, \mathbf{j})$ first occurs as eqn (2.162) and is given by

$$L(\mathbf{k}, \mathbf{j}) = \{\mu(k^2 + j^2) - kj(1 + 2\mu^2)\}(1 - \mu^2)kj\frac{1}{k^2 + j^2 - 2kj\mu}. \tag{E.1}$$

$L(\mathbf{k}, \mathbf{k} - \mathbf{j})$ first occurs as eqn (6.46) and is given by

$$L(\mathbf{k}, \mathbf{k} - \mathbf{j}) = \frac{(k^4 - 2k^3 j\mu + kj^3\mu)(1 - \mu^2)}{k^2 + j^2 - 2kj\mu}. \tag{E.2}$$

In both cases, $\mu = \mathbf{k} \cdot \mathbf{j}/kj$.

We now have three aims. These are as follows:

(a) to show how the alternative forms of the inertial transfer coefficient arise as a consequence of our choice of the dummy wavevector to be eliminated;
(b) to give details of the derivation of the coefficients;
(c) to make the connection between the present coefficients and those of the formulation due to Kraichnan.

Essentially, our overall objective is to show that eqn (2.160) becomes (2.161). To do this, we write (2.160) in terms of two explicitly dummy variables $\mathbf{j}$ and $\mathbf{l}$, rather than $\mathbf{j}$ and $\mathbf{k} - \mathbf{j}$, by using the identity

$$\int d^3j \int d^3l\, f(k, j, l)\delta(\mathbf{k} - \mathbf{j} - \mathbf{l}) = \int d^3j\, f(k, j, |\mathbf{k} - \mathbf{j}|) \tag{E.3}$$

and (2.97) for the isotropic spectral tensor to obtain

$$T(k, t) = 4\pi k^2 \int d^3j \int d^3l\, \delta(\mathbf{k} - \mathbf{j} - \mathbf{l}) \times$$

$$\times \int_{-\infty}^{t} ds\, \exp\{-v(k^2 + j^2 + l^2)(t - s)\} \times$$

$$\times \{A(\mathbf{k}, \mathbf{j}, \mathbf{l})Q(j, s)Q(l, s) - B(\mathbf{k}, \mathbf{j}, \mathbf{l})Q(k, s)Q(l, s) - B(\mathbf{l}, \mathbf{k}, \mathbf{j})Q(k, s)Q(j, s)\}, \tag{E.4}$$

where the coefficients are given by

$$A(\mathbf{k}, \mathbf{j}, \mathbf{l}) = -2M_{\alpha\beta\gamma}(\mathbf{k})M_{\alpha\rho\delta}(\mathbf{k})D_{\rho\beta}(\mathbf{j})D_{\delta\gamma}(\mathbf{l}) \tag{E.5}$$

$$B(\mathbf{j}, \mathbf{k}, \mathbf{l}) = -2M_{\alpha\beta\gamma}(\mathbf{k})M_{\beta\rho\delta}(\mathbf{j})D_{\rho\alpha}(\mathbf{k})D_{\delta\gamma}(\mathbf{l})$$

$$= -2M_{\rho\beta\gamma}(\mathbf{k})M_{\beta\rho\delta}(\mathbf{j})D_{\delta\gamma}(\mathbf{l}) \tag{E.6}$$

$$B(\mathbf{l}, \mathbf{k}, \mathbf{j}) = -2M_{\alpha\beta\gamma}(\mathbf{k})M_{\beta\rho\delta}(\mathbf{l})D_{\rho\alpha}(\mathbf{k})D_{\delta\beta}(\mathbf{j})$$

$$= -2M_{\rho\beta\gamma}(\mathbf{k})M_{\gamma\rho\delta}(\mathbf{l})D_{\delta\beta}(\mathbf{j}). \tag{E.7}$$

It should be noted that the second form of the r.h.s. in both (E.6) and (E.7) requires the contraction property

$$D_{\alpha\beta}(\mathbf{k})D_{\beta\gamma}(\mathbf{k}) = D_{\alpha\gamma}(\mathbf{k}). \tag{E.8}$$

In Section 2.7.1 we proved the general result that $T(k, t)$ must vanish when integrated over all $\mathbf{k}$ in order to conserve energy. Here we establish the necessary relationship between the coefficients A and B for this property to hold.

The proof that the non-linear term conserves energy relies on the relationship

$$[M_{\alpha\beta\gamma}(\mathbf{k}) + M_{\beta\alpha\gamma}(\mathbf{j}) + M_{\gamma\alpha\beta}(\mathbf{l})]\langle u_\alpha(\mathbf{k})u_\beta(\mathbf{j})u_\gamma(\mathbf{l})\rangle$$

$$= 0 \quad \text{iff } \mathbf{k} - \mathbf{j} - \mathbf{l} = 0. \tag{E.9}$$

This is easily proved, using the continuity equation, by expanding out the Ms and considering vectors with the same subscripts in pairs. An obvious corollary of this relationship is given by

$$\{M_{\alpha\beta\gamma}(\mathbf{k}) + M_{\beta\alpha\gamma}(\mathbf{j}) + M_{\gamma\alpha\beta}(\mathbf{l})\}\{D_{\alpha\rho}(\mathbf{k})D_{\beta\sigma}(\mathbf{j})D_{\gamma\delta}(\mathbf{l})\}$$

$$= 0 \quad \text{iff } \mathbf{k} - \mathbf{j} - \mathbf{l} = 0 \text{ for arbitrary } \rho, \sigma, \delta. \tag{E.10}$$

Using this relationship it is not hard to show that the coefficients A and B must satisfy the condition

$$- A(\mathbf{k}, \mathbf{j}, \mathbf{l}) = B(\mathbf{j}, \mathbf{k}, \mathbf{l}) + B(\mathbf{l}, \mathbf{k}, \mathbf{j}). \tag{E.11}$$

In order to facilitate numerical calculations, these coefficients can be expressed in terms of trigonometric functions and vector magnitudes. We do this by first noting that the vectors $\mathbf{k}$, $\mathbf{j}$, and $\mathbf{l}$ form the sides of a triangle. If we denote the cosines of the angles opposite to $\mathbf{k}$, $\mathbf{j}$, and $\mathbf{l}$ by x, y, and z respectively, then these cosines can be written as

$$x = -\frac{\mathbf{l}\cdot\mathbf{j}}{lj} \qquad y = \frac{\mathbf{k}\cdot\mathbf{l}}{kl} \qquad z = \frac{\mathbf{k}\cdot\mathbf{j}}{kj}. \tag{E.12}$$

Then we invoke (2.77) for the Ms and (2.78) for the Ds, and expand out each of (E.5)–(E.7) in terms of the inner products of the various vectors. Next we substitute for the inner products from (E.12) and, after a great deal of algebra, we obtain the forms

$$A(\mathbf{k}, \mathbf{j}, \mathbf{l}) = k^2(1 - xyz + 2y^2z^2) \tag{E.13}$$

$$B(\mathbf{j}, \mathbf{k}, \mathbf{l}) = kj(z^3 + xy)$$

$$= \frac{k^2\{(z + xy)(z^3 + xy)\}}{1 - x^2} \tag{E.14}$$

$$B(\mathbf{l}, \mathbf{k}, \mathbf{j}) = kl(y^3 + xy)$$

$$= \frac{k^2\{(y + xz)(y^3 + xz)\}}{1 - z^2}. \tag{E.15}$$

These result allow us make a connection with the coefficients $a(k, j, l)$ and $b(k, j, l)$ in Kraichnan's formulation (Leslie 1973, pp. 356–62). From (E.13) and the corresponding equation for $a(k, j, l)$ (Leslie 1973, (A.30)) we have

$$A(\mathbf{k,j,l}) = -2k^2 a(k, j, l),\tag{E.16}$$

and from (E.14), (E.15), and the corresponding equation for $b(k, j, l)$ (Leslie 1973, (A.24)) we also have

$$B(\mathbf{j,k,l}) = k^2 b(k, j, l)\tag{E.17}$$

$$B(\mathbf{l,k,j}) = k^2 b(k, l, j)\tag{E.18}$$

Now, with the aid of (E.11) we can write eqn (E.4) for $T(k, t)$ as

$$T(k, t) = 4\pi k^2 \int d^3j \int d^3l\, \delta(\mathbf{k - j - l}) \times$$
$$\times \int_{-\infty}^{t} ds\, \exp\{-v(k^2 + j^2 + l^2)(t - s)\} \times$$
$$\times [B(\mathbf{j,k,l})Q(l,s)\{Q(j,s) - Q(k,s)\} +$$
$$+ B(\mathbf{l,k,j})Q(j,s)\{Q(l,s) - Q(k,s)\}].\tag{E.19}$$

Further, interchanging dummy variables $\mathbf{j}$ and $\mathbf{l}$ on the r.h.s. gives us

$$T(k, t) = 4\pi k^2 \int d^3j \int d^3l\, \delta(\mathbf{k - j - l}) \times$$
$$\times \int_{-\infty}^{t} ds\, \exp\{-v(k^2 + j^2 + l^2)(t - s)\} \times$$
$$\times 2B(\mathbf{j,k,l})Q(l,s)\{Q(j,s) - Q(k,s)\}.\tag{E.20}$$

At this stage we have a choice. When we eliminate the delta function we can also eliminate either $\mathbf{j}$ or $\mathbf{l}$. Evidently the final result will depend on which of these we do. If we eliminate $\mathbf{j}$, then $T(k, t)$ becomes

$$T(k, t) = 4\pi k^2 \int d^3j \int_{-\infty}^{t} ds\, \exp\{-v(k^2 + j^2 + |\mathbf{k - j}|^2)(t - s)\} \times$$
$$\times 2L(\mathbf{k,j})Q(|\mathbf{k - j}|, s)\{Q(j,s) - Q(k,s)\}\tag{E.21}$$

where $L(\mathbf{k,j})$ is given by

$$L(\mathbf{k,j}) = B(\mathbf{j,k,k-j}),\tag{E.22}$$

whereas, if we eliminate l, then $T(k, t)$ becomes

$$T(k, t) = 4\pi k^2 \int d^3l \int_{-\infty}^{t} ds\, \exp\{-v(k^2 + |\mathbf{k - l}|^2 + l^2)(t - s)\} \times$$
$$\times 2L(\mathbf{k,k-l})Q(l,s)\{Q(|\mathbf{k - l}|, s) - Q(k,s)\}\tag{E.23}$$

where $L(\mathbf{k,k-l})$ is given by

$$L(\mathbf{k,k-l}) = B(\mathbf{k-l,k,l}).\tag{E.24}$$

Then, substituting (E.12) for the cosines in (E.14), and using $\mathbf{l = k - j}$, we obtain (E.22) as

$$L(\mathbf{k}, \mathbf{j}) = \frac{\{z(k^2 + j^2) - kj(1 + 2z^2)\}kj(1 - z^2)}{|\mathbf{k} - \mathbf{j}|^2}. \qquad \text{(E.25)}$$

Similarly, from (E.12), (E.15), and $\mathbf{j} = \mathbf{k} - \mathbf{l}$, we obtain (E.24) as

$$L(\mathbf{k}, \mathbf{k} - \mathbf{l}) = \frac{(k^4 - 2k^3 ly + kl^3 y)(1 - y^2)}{|\mathbf{k} - \mathbf{l}|^2}. \qquad \text{(E.26)}$$

In order to obtain eqn (2.162) from (E.25), we simply rename z as μ; similarly, renaming $\mathbf{l}$ as $\mathbf{j}$ and y as μ in (E.26) gives us eqn (6.46).

Reference

LESLIE, D. C. (1973). *Developments in the theory of turbulence.* Clarendon Press, Oxford.

APPENDIX F

Optical background to laser-Doppler anemometry

The subject of laser-Doppler anemometry (LDA) was discussed in Section 3.1.2. Here we fill in some of the background details for those readers who are unfamiliar with the subject of physical optics. We begin by noting that light is a transverse electromagnetic wave, with its electric and magnetic field vectors at right angles to the direction of propagation. The phrase 'transverse electromagnetic' is often abbreviated to TEM.

Coherence

Consider two wave trains

$$f_1 = A_1 \exp\{i(kx - \omega t + \delta_1)\},$$

where f is the field at position x and time t, A_1 is the amplitude, k is the wavenumber, ω is the angular frequency, and δ_1 is the value of the phase at $x = t = 0$, and

$$f_2 = A_2 \exp\{i(kx - \omega t + \delta_2)\},$$

which arrive together at a particular point. Without loss of generality, we consider the point $x = 0$, $t = 0$ and use the principal of superposition

$$f = f_1 + f_2 \tag{F.1}$$

to obtain the total field at the point. Then we obtain the intensity I of light at the point in the usual way as

$$I = ff^* = (f_1 + f_2)(f_1^* + f_2^*)$$
$$= A_1^2 + A_2^2 + 2A_1 A_2 \cos(\delta_1 - \delta_2). \tag{F.2}$$

Clearly the resultant intensity will depend on the phase relation between the two wave trains. If the waves are from the same source, then fluctuations in δ_1 will match fluctuations in δ_2. If the two waves arrive in phase, then they will remain in phase with $\delta_1 = \delta_2$, and the resultant intensity will be given by

$$I = (A_1 + A_2)^2. \tag{F.3}$$

Conversely, if the waves arrive out of phase, then they remain out of phase with $\delta_1 = \delta_2 + \pi$, and the resultant intensity will be

$$I = (A_1 - A_2)^2. \tag{F.4}$$

However, if the waves are from different sources, the phase difference will fluctuate randomly. Accordingly, we work with the average intensity and find that

$$\langle I \rangle = A_1^2 + A_2^2 + 2A_1 A_2 \langle \cos(\delta_1 - \delta_2) \rangle$$
$$= A_1^2 + A_2^2 \tag{F.5}$$

as $\langle \cos(\delta_1 - \delta_2) \rangle = 0$. Such wave trains are said to be incoherent, and—as we can see—in the case of incoherent waves the resultant intensity is obtained by simply adding the separate intensities.

In practice, we may have an intermediate situation (even when the waves are derived from the same source) in which the phases will be correlated over some time or length. This is known as partial coherence.

Lasers

Optical anemometry is really only possible because of the unique coherence properties of the continuous-wave gas laser. This device can readily be understood as a superior form of gas discharge lamp. In the conventional gas discharge lamp, light is emitted spontaneously when electrons in the gas atoms drop from excited states to lower energy levels. The initial excitation is due to the application of a strong electric field, which is the source of energy for the lamp.

However, emission can be stimulated by shining light at the correct frequency through the discharge. The important characteristic of such stimulated emission is that the light so emitted is in phase with the stimulating radiation. Hence, reflecting light back and forth through the lasing medium produces highly coherent light. In practice this is achieved by placing partial reflectors to form a resonant cavity from which some of the light can leak out to give the laser beam. If the distance between the mirrors is L, then the light inside the laser forms a standing wave whose wavelength λ satisfies the relation

$$L = \frac{q\lambda}{2}$$

where q is an integer.

Of course this is a standing wave in three dimensions, and for a cylindrical geometry we have the general result

$$E(r, \theta) = E_0 \left(\frac{r}{a}\right)^l L_p^{(l)}\left(\frac{2r^2}{a^2}\right) \exp\left(\frac{-r^2}{a^2}\right) \cos(l\theta), \tag{F.6}$$

where r and θ are cylindrical coordinates, a is the beam radius, E_0 is the intensity on the optical axis, and $L_p^{(l)}$ is a generalized Laguerre polynomial.

A solution of Maxwell's equations of the form of eqn (F.6) is known as a TEM_{plq} mode. If we consider the lowest-order case, this is the TEM_{00q} mode, which is often abbreviated to TEM_{00} mode. From (F.6) we see that the TEM_{00} mode is given by

$$E(r) = E_0 \exp\left(\frac{-r^2}{a^2}\right) \tag{F.7}$$

as the zero-order Laguerre polynomial is normalized to unity. In other words, the lowest-order mode has a Gaussian profile. This is the usual case.

Interference

Consider light from two sources S_1 and S_2 reaching a screen. We assume that the sources are perfectly in phase with each other at $x = 0$, and we examine the pattern on the screen at time $t = 0$.

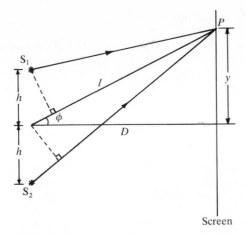

Fig. F.1. Interference due to two coherent sources S_1 and S_2.

Referring to Fig. F.1, we can see that interference at the point P will be due to the path difference $S_1 P - S_2 P$. It follows from the geometry of the figure that the distances from each source to the screen are given by $S_2 P = l + h \sin \phi$ and $S_1 P = l - h \sin \phi$ respectively. Then, the two wavetrains arriving at P at time t are given by

$$f_1 = A_1 \exp\{-ik(l - h \sin \phi)\} \qquad (F.8)$$

$$f_2 = A_2 \exp\{-ik(l + h \sin \phi)\}, \qquad (F.9)$$

and the resulting intensity is given by

$$I = A_1^2 + A_2^2 + 2A_1 A_2 \cos(2kh \sin \phi). \qquad (F.10)$$

As before, the fringes are defined by the maxima and minima of I, which in turn are controlled by the values of the argument of the cosine term in eqn (F.10). If we make the approximation $\sin \phi = y/D$, then it is easily shown that the distance of the nth bright fringe from the optical axis is given by

$$y_n = \frac{n\lambda D}{2h}. \qquad (F.11)$$

In principle, therefore, the intensity distribution on the screen is a cosine variation between the constant levels $(A_1 + A_2)^2$ and $(A_1 - A_2)^2$. However, as we have seen in eqn (F.7), the distribution of light intensity across the beam is actually Gaussian, and this variation also modulates the intensity distribution of the fringes. A typical result can be seen in Fig. 3.5. Also, in practice, the beams are focused by a lens to form a small measuring volume, as shown in Fig. 3.3. Thus, if we make the further approximation of putting $h/D = \sin \theta$, where the angle between the beams is 2θ, eqn (F.11) leads to the usual expression for the distance d between fringes:

$$d = y_n - y_{n-1} = \frac{\lambda}{2 \sin \theta}. \qquad (F.12)$$

Analysis in terms of the Doppler effect

The original analysis of the operating principle of the LDA was based on the Doppler effect. Light scattered by a moving particle is shifted in frequency. If $\mathbf{k}_0$ is a unit vector in the direction of the incident light as $\mathbf{k}_s$ is a unit vector in the direction of the scattered light, and the particle velocity satisfies the condition $|\mathbf{u}| \ll c$, where c is the speed of light, we have

$$f_D = \frac{\mathbf{u} \cdot (\mathbf{k}_s - \mathbf{k}_0)}{\lambda_0}, \tag{F.13}$$

where f_D is the Doppler frequency shift and λ_0 is the wavelength of the incident light.

If two incident beams $\mathbf{k}_{01}$ and $\mathbf{k}_{02}$ intersect, then a moving particle passing through the point of intersection will cause a net difference in Doppler shift, which from eqn (F.13) is given by

$$f_D = f_{D2} - f_{D1} = \frac{\mathbf{u} \cdot (\mathbf{k}_{01} - \mathbf{k}_{02})}{\lambda_0}, \tag{F.14}$$

which is independent of the direction of the scattered light $\mathbf{k}_s$. This is very useful in practice, as it allows scattered light to be collected over a large solid angle. A particularly useful version of this is the symmetrical case (as shown in Fig. 3.3), in which it can be shown that the net Doppler shift, as given by eqn (F.14), takes the form

$$f_D = f_{D2} - f_{D1} = \frac{2u_1 \sin \theta}{\lambda_0}, \tag{F.15}$$

which is (not surprisingly) the same result as for the real fringe case.

Measurement of the Doppler frequency: heterodyning

It is difficult to measure optical frequencies directly as they are very large, and so one normally resorts to non-linear mixing at the photodetector. Let us suppose that the photodetector has a square-law response so that

$$i(t) = aE^2, \tag{F.16}$$

where $i(t)$ is the output current, a is the efficiency of the photodetector, and E is the amplitude of the light wave at the photocathode.

Now, heterodyning is the non-linear mixing of two waves with different frequencies. That is, if E is determined by the superposition of two waves

$$E = E_1 \cos(\omega_1 t) + E_2 \cos(\omega_2 t), \tag{F.17}$$

then the output current is determined by E^2, as given by

$$E^2 = E_1^2 \cos^2(\omega_1 t) + E_2^2 \cos^2(\omega_2 t) + E_1 E_2 \cos\{(\omega_1 + \omega_2)t\} +$$
$$+ E_1 E_2 \cos\{(\omega_1 - \omega_2)t\}. \tag{F.18}$$

However, although this should be the output signal in principle, it will not be observed in practice, as the detector is unable to respond to the optical frequencies and in effect rectifies the high-frequency part of the signal. But it will usually respond to the difference (Doppler) frequency, and this behaviour can be represented by averaging the output signal over a time T such that

$$\frac{2\pi}{\omega_1} \ll T \ll \frac{2\pi}{\omega_1 - \omega_2}. \tag{F.19}$$

Averaging over one complete cycle results in

$$i(t) = a\langle E^2 \rangle = a\left(\frac{E_1^2}{2} + \frac{E_2^2}{2}\right) + aE_1 E_2 \cos\{(\omega_1 - \omega_2)t\}$$

$$= i_{DC} + i_D, \tag{F.20}$$

where we have used $\langle \cos^2 \theta \rangle = 1/2$ and $\langle \cos \theta \rangle = 0$. The first term on the r.h.s. of eqn (F.20) is usually referred to as the pedestal level and i_D is the Doppler current. The corresponding frequency spectrum will have a peak at the origin (the DC peak) and a peak centered on the Doppler frequency (the AC peak).

Signal processing

In a steady laminar flow, the velocity at a point is constant and hence the output signal from an LDA would be at a constant value of the Doppler frequency. Therefore we could measure the velocity by using a spectrum analyser to determine the centre frequency of the AC peak.

For unsteady flows (including turbulence) the velocity at a point will vary with time. This means that we need some way of measuring the Doppler frequency very quickly (i.e. quicker than the fastest fluctuation in the velocity).

There are many ways to doing this, but the technique has been dominated by only two methods. If there are many scattering centres in the fluid, then the LDA output signal will be more or less continuous in nature, and we use a frequency tracker. Alternatively, if there are not many scattering centres, then we should use a counter processor.

The initial stages of signal processing are common to both methods. First the LDA output signal is filtered to remove the DC pedestal and any high-frequency noise. Then the signal is fed to a device called a Schmidt trigger, which essentially turns the cosine (see equation (F.20)) into a series of positive pulses. The essential feature—the Doppler frequency—is of course unchanged. Then, if the signal is essentially continuous it can be processed by a frequency tracker (which consists basically of a voltage-controlled oscillator and a feedback loop) in order to give an output which is a voltage linearly proportional to the Doppler frequency. However, if the LDA signal consists mainly of individual bursts, then the signal can be handled by a counter processor. This device uses a crystal-controlled high-frequency oscillator to count and time a number of fringes in the individual burst signal. In this way, very rapid measurements can be made of the speed of individual particles in the fluid.

A list of further reading will be found at the end of Section 3.1.2.

APPENDIX G

Second-order term in the perturbation series as an example of the diagram calculus

We begin with eqn (5.73):

$$D_{\alpha\beta}(\mathbf{k})Q(k;\omega,\omega') = D_{\alpha\beta}(\mathbf{k})Q_0(k;\omega,\omega') +$$

$$+ \left(\frac{2\pi}{L}\right)^3 [\lambda^2 \{\langle u_\alpha^{(0)}(\mathbf{k},\omega)u_\beta^{(2)}(-\mathbf{k},\omega')\rangle +$$

$$+ \langle u_\alpha^{(1)}(\mathbf{k},\omega)u_\beta^{(1)}(-\mathbf{k},\omega')\rangle + \langle u_\alpha^{(2)}(\mathbf{k},\omega)u_\beta^{(0)}(-\mathbf{k},\omega')\rangle \} +$$

$$+ O(\lambda^4)]. \tag{G.1}$$

We shall evaluate the second-order terms one by one using (5.76a) and (5.78) for $u^{(1)}$ and $u^{(2)}$:

$$\langle u_\alpha^{(0)}(\mathbf{k},\omega)u_\beta^{(2)}(-\mathbf{k},\omega')\rangle$$

$$= 2G_0(-\mathbf{k},\omega')M_{\beta\delta\sigma}(-\mathbf{k})\sum_{\mathbf{j}}\sum_{\omega''} G_0(\mathbf{j},\omega'')M_{\delta\rho\gamma}(\mathbf{j}) \times$$

$$\times \sum_{\mathbf{l}}\sum_{\omega'''} \langle u_\alpha^0(k,\omega)u_\rho^{(0)}(\mathbf{l},\omega''')u_\gamma^{(0)}(\mathbf{j}-\mathbf{l},\omega''-\omega''')u_\sigma^{(0)}(-\mathbf{k}-\mathbf{j},\omega'-\omega'')\rangle$$

$$= 2G_0(k,\omega')M_{\beta\delta\sigma}(-\mathbf{k})\sum_{\mathbf{j}}\sum_{\omega''} G_0(j,\omega'')M_{\delta\rho\gamma}(\mathbf{j}) \times$$

$$\times \sum_{\mathbf{l}}\sum_{\omega'''} \{\langle u_\alpha^{(0)}(\mathbf{k},\omega)u_\rho^{(0)}(\mathbf{l},\omega''')\rangle\langle u_\gamma^{(0)}(\mathbf{j}-\mathbf{l},\omega''-\omega''')u_\sigma^{(0)}(-\mathbf{k}-\mathbf{j},\omega'-\omega'')\rangle +$$

$$+ \langle u_\alpha^{(0)}(\mathbf{k},\omega)u_\gamma^{(0)}(\mathbf{j}-\mathbf{l},\omega''-\omega''')\rangle\langle u_\rho^{(0)}(\mathbf{l},\omega''')u_\sigma^{(0)}(-\mathbf{k}-\mathbf{j},\omega'-\omega'')\rangle +$$

$$+ \langle u_\alpha^{(0)}(\mathbf{k},\omega)u_\sigma^{(0)}(-\mathbf{k}-\mathbf{j},\omega'-\omega'')\rangle\langle u_\rho^{(0)}(\mathbf{l},\omega''')u_\gamma^{(0)}(\mathbf{j}-\mathbf{l},\omega''-\omega''')\rangle \}$$

where we have noted that $G_0(-\mathbf{k},\omega')$ only depends on $|-\mathbf{k}| = k$ and $u^{(0)}$ has Gaussian statistics thus giving the factorization into second moments which we saw in Chapter 2. The last pair of second moments does not contribute as

$$\left(\frac{2\pi}{L}\right)^3 \langle u_\rho^{(0)}(\mathbf{l},\omega''')u_\gamma^{(0)}(\mathbf{j}-\mathbf{l},\omega''-\omega''')\rangle = D_{\rho\gamma}(\mathbf{l})Q(l;\omega''',\omega''-\omega''')\delta(\mathbf{j})$$

and thus when we sum over $\mathbf{j}$ we obtain $M_{\delta\rho\gamma}(0)$ which vanishes. Hence,

$$\langle u_\alpha^{(0)}(\mathbf{k},\omega)u_\beta^{(2)}(-\mathbf{k},\omega')\rangle$$

$$= 2G_0(k,\omega')M_{\beta\delta\sigma}(-\mathbf{k})\sum_{\mathbf{j}}\sum_{\omega''} G_0(j,\omega'')M_{\delta\rho\gamma}(\mathbf{j})\left(\frac{L}{2\pi}\right)^6 \times$$

$$\times \sum_{\mathbf{l}}\sum_{\omega'''} \{D_{\gamma\sigma}(-\mathbf{k}-\mathbf{j})Q_0(|-\mathbf{k}-\mathbf{j}|;\omega''-\omega''',\omega'-\omega'') \times$$

$$\times \delta(-\mathbf{k}-\mathbf{l})D_{\alpha\rho}(\mathbf{k})Q_0(k;\omega,\omega''')\delta(\mathbf{k}+\mathbf{l}) +$$

$$+ D_{\alpha\gamma}(\mathbf{k})Q_0(k;\omega,\omega''-\omega''')\delta(\mathbf{k}+\mathbf{j}-\mathbf{l})D_{\rho\sigma}(-\mathbf{k}-\mathbf{j}) \times$$

$$\times Q_0(|\mathbf{k}+\mathbf{j}|;\omega''',\omega'-\omega'')\delta(-\mathbf{k}-\mathbf{j}+\mathbf{l})\} \tag{G.2}$$

Similarily,

$$\langle u_\alpha^{(2)}(\mathbf{k}, \omega) u_\beta^{(0)}(-\mathbf{k}, \omega') \rangle$$

$$= 2G_0(k, \omega) M_{\alpha\delta\gamma}(\mathbf{k}) \sum_{\mathbf{j}} \sum_{\omega''} G_0(j, \omega'') M_{\delta\rho\sigma}(\mathbf{j}) \times$$

$$\times \sum_{\mathbf{l}} \sum_{\omega'''} \langle u_\rho^{(0)}(\mathbf{l}, \omega''') u_\sigma^{(0)}(\mathbf{j} - \mathbf{l}, \omega'' - \omega''') u_\gamma^{(0)}(\mathbf{k} - \mathbf{j}, \omega - \omega'') u_\beta^{(0)}(-\mathbf{k}, \omega') \rangle$$

$$= 2G_0(k, \omega) M_{\alpha\delta\gamma}(\mathbf{k}) \sum_{\mathbf{j}} \sum_{\omega''} G_0(j, \omega'') M_{\delta\rho\sigma}(\mathbf{j}) \times$$

$$\times \sum_{\mathbf{l}} \sum_{\omega'''} \{ \langle u_\rho^{(0)}(\mathbf{l}, \omega''') u_\sigma^{(0)}(\mathbf{j} - \mathbf{l}, \omega'' - \omega''') \rangle \langle u_\gamma^{(0)}(\mathbf{k} - \mathbf{j}, \omega - \omega'') u_\beta^{(0)}(-\mathbf{k}, \omega') \rangle +$$

$$+ \langle u_\rho^{(0)}(\mathbf{l}, \omega''') u_\gamma^{(0)}(\mathbf{k} - \mathbf{j}, \omega - \omega'') \rangle \langle u_\sigma^{(0)}(\mathbf{j} - \mathbf{l}, \omega'' - \omega''') u_\beta^{(0)}(-\mathbf{k}, \omega') \rangle +$$

$$+ \langle u_\rho^{(0)}(\mathbf{l}, \omega''') u_\beta^{(0)}(-\mathbf{k}, \omega') \rangle \langle u_\sigma^{(0)}(\mathbf{j} - \mathbf{l}, \omega'' - \omega''') u_\gamma^{(0)}(\mathbf{k} - \mathbf{j}, \omega - \omega'') \rangle \}$$

$$= 2G_0(k, \omega) M_{\alpha\delta\gamma}(\mathbf{k}) \sum_{\mathbf{j}} \sum_{\omega''} G_0(j, \omega'') M_{\delta\rho\sigma}(\mathbf{j}) \left(\frac{L}{2\pi} \right)^6 \times$$

$$\times \sum_{\mathbf{l}} \sum_{\omega'''} \{ D_{\rho\gamma}(\mathbf{k} - \mathbf{j}) Q_0(|\mathbf{k} - \mathbf{j}|; \omega - \omega'', \omega''') \delta(\mathbf{k} - \mathbf{j} + \mathbf{l}) \times$$

$$\times D_{\sigma\beta}(-\mathbf{k}) Q_0(k; \omega', \omega'' - \omega''') \delta(-\mathbf{k} + \mathbf{j} - \mathbf{l}) + D_{\rho\beta}(-\mathbf{k}) Q_0(k; \omega', \omega''') \delta(\mathbf{l} - \mathbf{k}) \times$$

$$\times D_{\sigma\gamma}(\mathbf{k} - \mathbf{j}) Q_0(|\mathbf{k} - \mathbf{j}|; \omega - \omega'', \omega'' - \omega''') \delta(\mathbf{k} - \mathbf{l}) \}, \tag{G.3}$$

and

$$\langle u_\alpha^{(1)}(\mathbf{k}, \omega) u_\beta^{(1)}(-\mathbf{k}, \omega') \rangle$$

$$= G_0(k, \omega) M_{\alpha\delta\gamma}(\mathbf{k}) G_0(k, \omega') M_{\beta\rho\sigma}(-\mathbf{k}) \times$$

$$\times \sum_{\mathbf{j}} \sum_{\omega''} \sum_{\mathbf{l}} \sum_{\omega'''} \langle u_\delta^{(0)}(\mathbf{j}, \omega'') u_\gamma^{(0)}(\mathbf{k} - \mathbf{j}, \omega - \omega'') u_\rho^{(0)}(\mathbf{l}, \omega''') u_\sigma^{(0)}(-\mathbf{k} - \mathbf{l}, \omega' - \omega''') \rangle$$

$$= 1 G_0(k, \omega) M_{\alpha\delta\gamma}(\mathbf{k}) G_0(k, \omega') M_{\beta\rho\sigma}(-\mathbf{k}) \left(\frac{L}{2\pi} \right)^6 \times$$

$$\times \sum_{\mathbf{j}} \sum_{\omega''} \sum_{\mathbf{l}} \sum_{\omega'''} \{ D_{\delta\rho}(\mathbf{j}) Q_0(j; \omega'', \omega''') \delta(\mathbf{j} + \mathbf{l}) D_{\gamma\sigma}(\mathbf{k} - \mathbf{j}) \times$$

$$\times Q_0(|\mathbf{k} - \mathbf{j}|; \omega - \omega'', \omega' - \omega''') \delta(-\mathbf{j} - \mathbf{l}) +$$

$$+ D_{\delta\sigma}(\mathbf{j}) Q_0(j; \omega'', \omega' - \omega''') \delta(\mathbf{j} - \mathbf{k} - \mathbf{l}) D_{\gamma\rho}(\mathbf{k} - \mathbf{j}) \times$$

$$\times Q_0(|\mathbf{k} - \mathbf{j}|; \omega - \omega'', \omega''') \delta(\mathbf{k} - \mathbf{j} + \mathbf{l}) \} \tag{G.4}$$

factorizing as before and using the fact that the term contributing at $\mathbf{k} = 0$ vanishes. We now use the symmetry of $M_{\alpha\beta\gamma}(\mathbf{k})$ under $\beta \leftrightarrow \gamma$ and the fact that $\omega' - \omega''$ and ω'' are interchangeable, provided that the summation is over ω'', to simplify each of the right-hand sides above into a single term. Now putting (G.2), (G.3), and (G.4) into (5.73) gives

$$D_{\alpha\beta}(\mathbf{k})Q(k;\omega,\omega')$$

$$= D_{\alpha\beta}(\mathbf{k})Q_0(k;\omega,\omega') + \lambda^2\left(\frac{L}{2\pi}\right)^3 \times$$

$$\times \Bigg\{ 4G_0(k,\omega')M_{\beta\delta\sigma}(-\mathbf{k})\sum_{\mathbf{j}}\sum_{\omega''}\sum_{\omega'''} G_0(j,\omega'')M_{\delta\rho\gamma}(\mathbf{j}) \times$$

$$\times D_{\alpha\rho}(\mathbf{k})D_{\gamma\sigma}(-\mathbf{k}-\mathbf{j})Q_0(|-\mathbf{k}-\mathbf{j}|;\omega''-\omega''',\omega'-\omega'')Q_0(k;\omega,\omega''') +$$

$$+ 4G_0(k;\omega')M_{\alpha\delta\gamma}(\mathbf{k})\sum_{\mathbf{j}}\sum_{\omega''}\sum_{\omega'''} G_0(j,\omega'')M_{\delta\rho\sigma}(\mathbf{j}) \times$$

$$\times D_{\beta\rho}(-\mathbf{k})D_{\sigma\gamma}(\mathbf{k}-\mathbf{j})Q_0(k;\omega',\omega''')Q_0(|\mathbf{k}-\mathbf{j}|;\omega-\omega'',\omega''-\omega''') +$$

$$+ 2G_0(k,\omega)M_{\alpha\delta\gamma}(\mathbf{k})\sum_{\mathbf{j}}\sum_{\omega''}\sum_{\omega'''} G_0(k,\omega')M_{\beta\rho\sigma}(-\mathbf{k}) \times$$

$$\times D_{\delta\sigma}(\mathbf{j})D_{\gamma\rho}(\mathbf{k}-\mathbf{j})Q_0(j;\omega'',\omega'-\omega''')Q_0(|\mathbf{k}-\mathbf{j}|;\omega-\omega'',\omega''')\Bigg\} +$$

$$+ O(\lambda^4). \tag{G.5}$$

If we use $(L/2\pi)^3 \sum_{\mathbf{j}} \to \int d^3j$ then this is just eqn (5.79).

We shall now perform the preceding calculation diagrammatically. Starting from the diagram expansion of $u_\alpha(\mathbf{k},t)$ in Fig. 5.8,

$$Q_{\alpha\beta}(k;\omega,\omega')=\langle u_\alpha(\mathbf{k},\omega)u_\beta(-\mathbf{k},\omega')\rangle$$

truncating at second-order and setting $\lambda = 1$.

To illustrate how to interpret a graph algebraically, consider the last term on the r.h.s. of equation (G.5) above:

Labelling this

$$2 \xleftarrow{\quad \mathbf{k},\omega \quad} \underset{\substack{\omega'' \\ +(\mathbf{j})-}}{\overset{\substack{+(\mathbf{k}-\mathbf{j})- \\ \omega-\omega''\;\times\;\omega'''}}{\Big\langle\Big\rangle}} \xrightarrow{\quad -\mathbf{k},\omega' \quad}_{\omega'-\omega''}$$

gives

$$2G_0(k,\omega)M_{\alpha\delta\gamma}(\mathbf{k})G_0(k,\omega')M_{\beta\rho\sigma}(-\mathbf{k}) \times$$

$$\times \int d^3j \sum_{\omega''}\sum_{\omega'''} \{D_{\delta\sigma}(\mathbf{j})D_{\gamma\rho}(\mathbf{k}-\mathbf{j})Q_0(j;\omega'',\omega'-\omega''') \times$$

$$\times Q_0(|\mathbf{k}-\mathbf{j}|;\omega-\omega'',\omega''')\}. \tag{G.6}$$

APPENDIX H

The Novikov functional formalism

The idea of random stirring forces with delta function correlations in time was introduced into turbulence theory by Edwards (see Section 6.2) and later applied by Novikov (1965) to the functional formalisms, as discussed in Section 4.2 of this book.

Novikov generally worked in configuration space and considered stirring forces $f_\alpha(\mathbf{x}, t)$, which have zero mean and Gaussian statistics, satisfying the condition

$$\langle f_\alpha(\mathbf{x}, t) f_\beta(\mathbf{x}', t') \rangle = F_{\alpha\beta}(\mathbf{x} - \mathbf{x}') \delta(t - t'). \tag{H.1}$$

The turbulence is, of course, taken to be homogeneous and isotropic. Then, for forces satisfying (H.1), he proved the general theorem

$$\langle f_\alpha(\mathbf{x}, t) R[\mathbf{f}] \rangle = \int F_{\alpha\beta}(\mathbf{x} - \mathbf{x}') \left\langle \frac{\delta R[\mathbf{f}]}{\delta f_\beta(\mathbf{x}', t)} \right\rangle \mathrm{d}^3 x', \tag{H.2}$$

where R is any functional of $\mathbf{f}$.

The proof relies on R's being analytic in $\mathbf{f}$. We expand R about $\mathbf{f} = 0$, and evaluate correlations of $\mathbf{f}$ to all orders using the assumption that $\mathbf{f}$ has a Gaussian distribution. Further details will be found in the original reference. Here we are concerned with the particular case which arises when we multiply the Navier–Stokes equation through by $u_\alpha(\mathbf{x}, t)$. This situation is discussed at various points in the main part of the present book; see, for example, eqns (4.88) and (6.32).

Accordingly, we wish to evaluate the cross-correlation $\langle f_\alpha(\mathbf{x}, t) u_\alpha(\mathbf{x}, t) \rangle$. To do this, we substitute for $R[\mathbf{f}]$ in (H.2) to obtain

$$\langle f_\alpha(\mathbf{x}, t) u_\alpha(\mathbf{x}, t) \rangle = \int F_{\alpha\beta}(\mathbf{x} - \mathbf{x}') \left\langle \frac{\delta u_\alpha(\mathbf{x}, t)}{\delta f_\beta(\mathbf{x}', t)} \right\rangle \mathrm{d}^3 x'. \tag{H.3}$$

Then the variational derivative of the velocity with respect to the stirring force is readily found by writing the solution of the Navier–Stokes equation as

$$u_\alpha(\mathbf{x}, t) = u_\alpha(\mathbf{x}, 0) + \int_0^t A_\alpha[\mathbf{u}(\mathbf{x}, s), \mathbf{x}] \, \mathrm{d}s + \int_0^t f_\alpha(\mathbf{x}, s) \, \mathrm{d}s \tag{H.4}$$

where A_α represents the effect of the non-linear terms of the Navier–Stokes equation. Its form can be deduced from eqn (2.15), which also requires the forcing term $f_\alpha(\mathbf{x}, t)$ to be added to its r.h.s. It then follows that the general variational derivative of $\mathbf{u}$ with respect to $\mathbf{f}$ is obtained as

$$\frac{\delta u_\alpha(\mathbf{x}, t)}{\delta f_\beta(\mathbf{x}', t')} = \int_{t'}^t \mathrm{d}s \frac{A_\alpha[\mathbf{u}(\mathbf{x}, s), \mathbf{x}]}{\delta f_\beta(\mathbf{x}', t')} + \Theta(t - t') \delta_{\alpha\beta} \delta(\mathbf{x} - \mathbf{x}'), \tag{H.5}$$

where Θ is the unit step-function which has the properties

$$\Theta(t - t') = \begin{cases} 1 & t' < t \\ 1/2 & t' = t \\ 0 & t' > t. \end{cases} \tag{H.6}$$

The required derivative in (H.3) is evaluated at $t = t'$. In this case, the first term on the r.h.s. of eqn (H.5) vanishes, and from (H.6) we can write

$$\frac{\delta u_\alpha(\mathbf{x}, t)}{\delta f_\beta(\mathbf{x}', t)} = \frac{1}{2}\delta_{\alpha\beta}\delta(\mathbf{x} - \mathbf{x}').$$ (H.7)

Then, substituting this result on the r.h.s. of (H.3) yields

$$\langle f_\alpha(\mathbf{x}, t)u_\alpha(\mathbf{x}, t)\rangle = \frac{1}{2}\delta_{\alpha\beta}F_{\alpha\beta}(0)$$

$$= \frac{1}{2}F_{\alpha\alpha}(0).$$ (H.8)

For the turbulence to be stationary, the rate of doing work by the stirring forces must be just equal to the dissipation rate (see Appendix A). Hence a corollary of equation (H.8) is that

$$F_{\alpha\alpha}(0) = 2\varepsilon$$ (H.9)

for stationary turbulence.

Novikov goes on to generalize the derivation of the Hopf equation for the characteristic functional to include the effect of the additon of random stirring forces $\mathbf{f}(\mathbf{x}, t)$ to the r.h.s. of the Navier–Stokes equation. The result can be given in the notation of Chapter 4 in the following way. Write the Hopf equation for $M[\mathbf{Z}(\mathbf{x}), t]$ as

$$\frac{\partial M[\mathbf{Z}(\mathbf{x}), t]}{\partial t} = (L_2 + \nu L_1)M[\mathbf{Z}(\mathbf{x}), t],$$ (H.10)

where the form of the operators L_2 and L_1 can be determined by comparison with eqn (4.79). Novikov shows that the addition of random stirring forces is taken into account by replacing (H.10) by

$$\frac{\partial M[\mathbf{Z}(\mathbf{x}), t]}{\partial t} = (L_2 + \nu L_1 + L_0)M[\mathbf{Z}(\mathbf{x}), t],$$ (H.11)

where L_0 is given by

$$L_0 = -\frac{1}{2}\int\int F_{\alpha\beta}(\mathbf{x} - \mathbf{x}')Z_\alpha(\mathbf{x})Z_\beta(\mathbf{x}')\,d^3x\,d^3x'.$$ (H.12)

This result is equivalent to the corresponding term in the Edwards–Fokker–Planck equation for the probability distribution (see eqn (6.66)).

Reference

NOVIKOV, E. A. (1965). *Sov. Phys.–JETP* **20**, 1290.

AUTHOR INDEX

SUBJECT INDEX